D0845245

JUVENILE JUSTICE

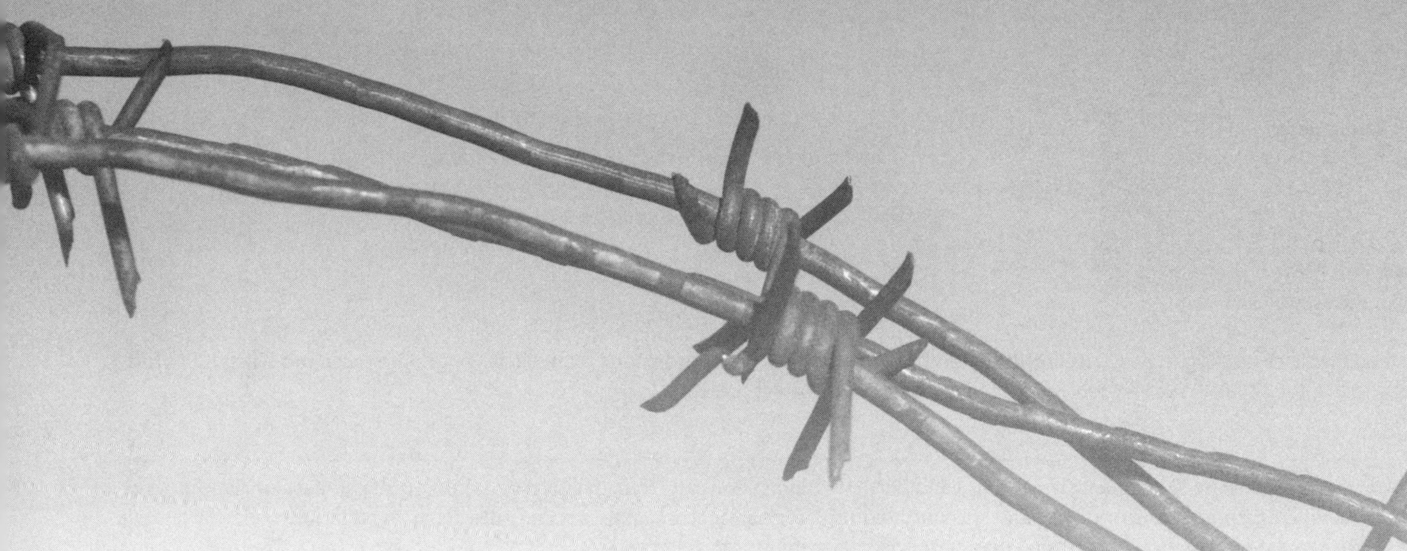

JUVENILE JUSTICE

A Social, Historical, and Legal Perspective

THIRD EDITION

Preston Elrod, PhD
Professor, Department of Criminal Justice
Eastern Kentucky University

R. Scott Ryder, JD
Tribal Court Administrator and ICWA Attorney
Nottawaseppi Huron Band of the Potawatomi

JONES AND BARTLETT PUBLISHERS
Sudbury, Massachusetts
BOSTON TORONTO LONDON SINGAPORE

World Headquarters
Jones and Bartlett Publishers
40 Tall Pine Drive
Sudbury, MA 01776
978-443-5000
info@jbpub.com
www.jbpub.com

Jones and Bartlett Publishers Canada
6339 Ormindale Way
Mississauga, Ontario L5V 1J2
Canada

Jones and Bartlett Publishers International
Barb House, Barb Mews
London W6 7PA
United Kingdom

Jones and Bartlett's books and products are available through most bookstores and online booksellers. To contact Jones and Bartlett Publishers directly, call 800-832-0034, fax 978-443-8000, or visit our website, www.jbpub.com.

Substantial discounts on bulk quantities of Jones and Bartlett's publications are available to corporations, professional associations, and other qualified organizations. For details and specific discount information, contact the special sales department at Jones and Bartlett via the above contact information or send an email to specialsales@jbpub.com.

Copyright © 2011 by Jones and Bartlett Publishers, LLC

All rights reserved. No part of the material protected by this copyright may be reproduced or utilized in any form, electronic or mechanical, including photocopying, recording, or by any information storage and retrieval system, without written permission from the copyright owner.

This publication is designed to provide accurate and authoritative information in regard to the Subject Matter covered. It is sold with the understanding that the publisher is not engaged in rendering legal, accounting, or other professional service. If legal advice or other expert assistance is required, the service of a competent professional person should be sought.

Production Credits
Publisher, Higher Education: Cathleen Sether
Acquisitions Editor: Sean Connelly
Associate Editor: Megan R. Turner
Senior Production Editor: Renée Sekerak
Production Assistant: Jill Morton
Associate Marketing Manager: Jessica Cormier
Manufacturing and Inventory Control Supervisor: Amy Bacus
Composition: Glyph International
Cover and Title Page Design: Kristin E. Parker
Photo Research and Permissions Manager: Kimberly Potvin
Cover Images: Courtroom, © Spencer Grant/PhotoEdit, Inc.; Barbed wire, © Dmitriy Shironosov/Dreamstime.com
Printing and Binding: Malloy Incorporated
Cover Printing: Malloy Incorporated

Library of Congress Cataloging-in-Publication Data
Elrod, Preston.
Juvenile justice : a social, historical, and legal perspective / by Preston Elrod and R. Scott Ryder. — 3rd ed.
p. cm.
Includes bibliographical references and index.
ISBN-13: 978-0-7637-6251-3
ISBN-10: 0-7637-6251-2
1. Juvenile justice, Administration of—United States. I. Ryder, R. Scott. II. Title.
KF9779.E44 2010
345.73'08—dc22

2009035575

6048

Printed in the United States of America
13 12 11 10 09 10 9 8 7 6 5 4 3 2 1

To my parents, Herman and Catherine, and to my children, Colin and Ryan; you make it all worthwhile.
P.E.

To my parents, Polly and Dick, my wife Denise, and my children, Joshua, Adam, Tim, and Kirby; thank you for your love and support.
R.S.R.

Brief Contents

Contents

Preface

This book is intended to serve as a comprehensive introduction to the field of juvenile justice. It is designed to help readers understand the complexities of present juvenile justice practice by presenting a thorough examination of the social, historical, and legal context within which delinquency and juvenile justice practice occurs. We have included a broad range of information so that those without experience in the field will be able to understand the multiple decision-making stages that comprise juvenile justice, the interconnections between agencies involved in juvenile justice, and the factors that influence case processing. We examine the relevant literature on the effectiveness of various juvenile justice interventions, as well as a variety of practical, political, economic, philosophical, ethical, and legal issues that invariably arise when juvenile justice agencies become involved in the lives of youths and their families. We also examine a variety of common myths about juvenile offenders and juvenile justice practice.

Our mission in this edition, as it has been in previous editions, is to make youths who are involved in the juvenile justice process, the institutions that make up juvenile justice, and juvenile justice decision makers accessible to readers. We have, of course, relied heavily on the scientific literature in developing the contents of this book. However, we have also relied on our own experiences as juvenile justice practitioners (our combined years as practitioners total 40), as well as our experiences as teachers and instructors of juvenile justice practitioners and both graduate and undergraduate university students (our combined years of experience total 47).

Despite the many hours of work that have gone into this project, we recognize that the final product is not perfect. Indeed, there are several aspects of juvenile justice that create difficulties for anyone attempting to develop a comprehensive introductory text in this area. One significant challenge has been to present the diversity of current approaches to juvenile justice. Juvenile justice is a complex undertaking and there is considerable variation in juvenile justice practice across jurisdictions. Nevertheless, we are confident that we have produced a text that will help the reader understand the basic operation of juvenile justice regardless of where they live, and will help them understand many of the nuances that define contemporary juvenile justice practice. Moreover, we have worked diligently to present a balanced view of juvenile justice. We want readers to recognize that juvenile justice continues to face a number of problems, and we encourage readers to critically examine past and present juvenile justice operations. We believe that a critical examination of juvenile justice theory and practice is important for the future development of juvenile justice practice. However, we have also attempted to acknowledge the many positive things that are done for children and families in juvenile courts and other juvenile justice agencies each day. We hope that we have succeeded.

Acknowledgments

Many people have contributed in significant ways to the completion of this book. We would like to acknowledge Carol Elrod and Denise Ryder; and our children Colin Elrod, Ryan Elrod, Joshua Ryder, Adam Ryder, Tim Ryder, and Kirby Jarzebiski, who supported our efforts in countless ways. Special thanks go to Carol Elrod for her assistance with editing the early versions of the manuscript. Her efforts have been a significant contribution to this project. Also, we thank the court staff and judges of the Family Division of the Ninth Circuit Court, Kalamazoo County, Michigan, and the many capable and caring individuals we have come across who are dedicated to serving the needs of children and families. Their work has been an inspiration to both of us. We particularly acknowledge the support of Carolyn Williams, former presiding judge, Family Division of the Ninth Circuit Court, Kalamazoo County, Michigan, and Doug Slade, former court administrator, for their support of this project. We also thank the many individuals who were willing to spend time with us and share their experiences in juvenile justice. Without their help, this book would not have been possible. We give special thanks to Brenda Foley, Stephanie Sims, Janet Snow, Jennifer Fielder, Janice Marcum, Steve Parson, Jeff Cantrell, Amber Wells, Mavis Poe, and Brandon Griffith who provided various types of support for this project. Finally, we would like to thank a number of individuals at Jones and Bartlett Publishers who have assisted us with this project, particularly Chambers Moore and Carolyn Rogers, who provided invaluable assistance with the previous editions, and Megan Turner, Associate Editor for Criminal Justice, who, along with Jill Morton and Maria Townsley, have provided invaluable assistance in the preparation of this edition. Each of these persons has been a pleasure to work with. Thank you all! Special thanks also go to several reviewers who provided helpful critiques of our work:

Theodore Darden
College of DuPage

Traqina Q. Emeka
University of Houston–Downtown

Amy P. Harrell
Nash Community College

Suman Kakar
Fellow of the Honors College
Florida International University

Jo Ann M. Short
Northern Virginia Community College

About the Authors

Preston Elrod received his BA in History from Presbyterian College and his MA and PhD in Sociology from Western Michigan University. He has taught at Texas Christian University and UNC–Charlotte. He currently serves as a professor in the Department of Criminal Justice at Eastern Kentucky University where he teaches courses on juvenile justice, crime prevention, and school safety. Among his published works are studies on citizens' attitudes toward the death penalty, juvenile justice policy development, public attitudes toward electronic monitoring, the effectiveness of interventions for juvenile probationers, and the experiences of adolescent jail inmates. He is the former codirector of a model school-based delinquency reduction program and he has worked in juvenile justice as a court intake officer and as the supervisor of a juvenile probation department. He served as the first chairperson of the Madison County Delinquency Prevention Council and he continues to be actively involved in a variety of community activities designed to assist at-risk youths and their families. He also serves as a prison visitor, the codirector of a community delinquency prevention project, and the Madison County Public Schools Safe Schools Evaluator. His present research focuses on school crime and victimization. When not at work, he likes spending time with his family, and he enjoys a variety of outdoor activities.

R. Scott Ryder graduated magna cum laude from Wittenberg University, Springfield, Ohio, in 1971 with a BA in History. He attended Indiana University School of Law in Bloomington, Indiana, where he received his Juris Doctor degree in 1974. He was admitted to practice law in Michigan on January 17, 1975 and continues in the practice of law to the present day. He began his involvement in juvenile justice in 1975 while working as an assistant prosecuting attorney in Shiawassee County, Michigan with primary responsibility for all proceedings in the juvenile court. His involvement in the juvenile justice system continued after leaving Shiawassee County. He served as the chief hearing referee and then research referee for 25 years at the Kalamazoo County, Michigan, juvenile court and later at the family court. After retiring from his referee position in May of 2004, he became the juvenile court director for St. Joseph County, Michigan, a position he held until 2007 when he went to work for the Nottawaseppi Huron Band of the Potawatomi as their tribal court administrator. Presently, he serves the Tribe as tribal court administrator and represents the Tribe in state courts as their Indian child welfare attorney. In addition to working in the courts, he has extensive training and teaching experience.

He was an instructor/trainer for the Michigan Judicial Institute, the Michigan Department of Human Services, and the Michigan Supreme Court Administrator's Office. He taught at the college level as an adjunct assistant professor at Western Michigan University and as an instructor at Glenn Oaks Community College. He also published in the area of juvenile justice and has a limited private legal practice. When he is not working, teaching, or writing, his hobbies include officiating soccer, playing golf, and reading. He is married to Denise and has three adult sons and an adult step-daughter.

The Context of Juvenile Justice: Defining Basic Concepts and Examining Public Perceptions of Juvenile Crime

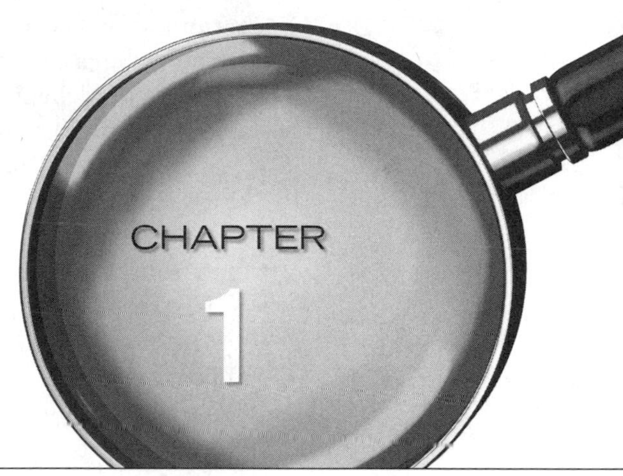

CHAPTER OBJECTIVES

After studying this chapter, you should be able to

- Define basic concepts necessary for understanding the juvenile justice process
- Present a legal definition of delinquency
- Describe the shortcomings of legal definitions of delinquency
- Explain why it is important to understand diversity within the population of juvenile offenders
- Explain why age is the primary criterion for defining a juvenile from a legal perspective
- Describe the problems associated with using age to define the clients of the juvenile justice process
- Describe those factors that contribute to variability and conflict in juvenile justice practice
- Explain why "juvenile justice system" may be a misleading term
- Describe the role that public perception plays in responding to the "delinquency problem"

CHAPTER OUTLINE

- Introduction
- Defining Delinquency
- Defining Juveniles: The Legal Perspective
- The Juvenile Justice System
- Public Perceptions of Delinquency and the Politics of Juvenile Justice
- Chapter Summary
- Key Concepts
- Review Questions
- Additional Readings
- Notes

Introduction

The creation of separate institutions for the treatment of youths who engage in illegal or immoral behavior is a recent historical development. From the beginning of the colonial period to the early 1800s, youths were subject to the same criminal justice process as adults. Consequently, children who were apprehended for crimes were tried in the same courts and, when found guilty, were often given the same punishments as adults.[1]

Today, most Americans are aware that a separate justice process exists for juveniles. However, many people lack an understanding of how this process works or how effective juvenile justice programs are, and others hold views that are based more on myth than reality. Moreover, most lack an understanding of the social, political, and economic contexts that shape juvenile crime and juvenile justice practice.

Although many citizens lack an understanding of the operation and effectiveness of juvenile justice programs, they often express strong opinions about the causes of delinquent behavior and how such behaviors should be handled. Such opinions are found in letters to local newspapers, in radio talk show programs, in television commentaries, and in the everyday discussions that people have about current events. Some members of the public call for increased efforts to prevent delinquency and rehabilitate young offenders. Others call for harsher punishments for young "thugs" and demand that youths who commit "adult" crimes be treated like adults. Similar sentiments are found among political leaders as well. Although rhetoric designed to capture the public's attention may be seen by some as "good politics," it is problematic for several reasons. First, it contributes little to the public's understanding of juvenile crime and the development of effective methods of responding to juvenile offenders. Second, **labeling** youths as "thugs" dehumanizes them and makes it easier to respond to them in ways that protect neither their interests nor the interests of the community. Third, inflammatory rhetoric about juvenile crime often results in poorly conceived and ineffective policies that squander precious resources and lead to increased public cynicism about our ability to deal with juvenile crime.

labeling Defining and responding to a youth as a delinquent. Labeling can lead to stigmatization, lost opportunities, and additional delinquent behavior.

Young people do commit serious offenses, although most juvenile crime does not result in serious injury or property loss. The question is, what is the best response to the illegal behavior of youths? Which responses are more likely to help youths learn from their mistakes and make more socially productive decisions in the future? Which responses are more likely to represent a sound investment of public funds and protect community safety? These are not easy questions to answer, but any viable response to juvenile crime must be predicated on sound knowledge of the causes of delinquency as well as a clear understanding of past and present responses to the juvenile crime problem, including the effectiveness of those responses. A primary purpose of this text is to help readers gain such an understanding.

FYI

For Your Information

Throughout this text, important points of information that help you understand juvenile justice practice are highlighted. In addition, myths about the operation of juvenile justice are explored in sidebars separated from the main text.

FYI

Juvenile Justice Practice
Ultimately, juvenile justice involves many people making decisions, sometimes very difficult decisions that are often made under arduous conditions. Moreover, it is important to remember that these decisions can have a profound influence on the lives of others.

Although this text is intended to help students understand the social, historical, and legal context of juvenile crime and juvenile justice, and to understand present juvenile justice responses to juvenile crime, bear in mind that juvenile justice is ultimately a human endeavor in which a variety of individuals, from police officers to correctional personnel, have responsibility for making decisions about how to respond to youths' behavior. Indeed, what makes juvenile justice an interesting, challenging, and potentially very rewarding career is that many youths and their families are helped by the many committed, highly trained, and caring individuals who work in juvenile justice. However, other youths are not helped, despite the best efforts of juvenile justice professionals. In other instances, youths and their families are harmed by those who are ostensibly charged with protecting their well-being. Unfortunately, in some instances, those who work in juvenile justice lack the knowledge, training, support, and commitment necessary for effective practice. Our goal in this text is to point out the many positive things that occur in juvenile justice, but also to balance our presentation by critically examining many of the problems that have historically plagued juvenile justice operations in the United States. We also provide descriptions of juvenile justice practices in other countries in sections that focus on **comparative juvenile justice** throughout the text in order to give readers insights on juvenile justice practices in other countries. Globalization can already be seen in juvenile justice practice, and we can expect the sharing of ideas about juvenile justice to expand in the future. While people in other countries have learned from our experiences in juvenile justice, both good and bad, we can learn important lessons from others.

comparative juvenile justice The examination of juvenile justice history and practice in one country or culture by comparing it to the history and practice of juvenile justice in another country or culture.

COMPARATIVE JUVENILE JUSTICE FOCUS

According to Dammer and Fairchild in their book, *Comparative Criminal Justice Systems*, there are three practical reasons for studying criminal justice in other countries or cultures: "(1) to benefit from the experience of others, (2) to broaden our understanding of different cultures and approaches to problems, and (3) to help us deal with the many transnational crime problems that plague our world today."[2]

This chapter is intended to introduce the basic concepts necessary for understanding present juvenile justice practice. It begins by defining delinquency. Next, it provides a profile of juvenile offenders, examines public perceptions of delinquency (which form an important part of the contemporary context of juvenile justice), and explores the concept of a juvenile justice system.

Defining Delinquency

delinquency Behaviors that fall within the jurisdiction of the juvenile court and result in processing by official juvenile justice agents.

From a legal standpoint, **delinquency** consists of those behaviors that are prohibited by the family or juvenile code of the state and that subject minors (i.e., persons not legally adults) to the jurisdiction of the juvenile court. Behaviors prohibited by juvenile codes can be grouped into two general categories: (1) behaviors that would be defined as criminal offenses if committed by adults (malicious destruction of property, larceny, robbery, motor vehicle theft, etc.), and (2) behaviors that are prohibited only for minors, which are called **status offenses** (school truancy, running away from home, incorrigibility, etc.).

status offenses Acts that are considered illegal when committed by a child but not when committed by an adult (e.g., running away, school truancy, and failure to obey parents' directions).

Although the preceding definition of delinquency is technically accurate, like all legal definitions, it fails to completely capture the complex human dimension of delinquency and juvenile justice responses to youths' illegal behaviors. For example, police respond to only some of the actions that are legally defined as delinquent. Police often ignore some (typically minor) illegal behaviors that are prohibited by legal codes. Also, as noted in Chapter 6, how police respond to youths alleged to be engaged in illegal behavior can vary considerably from one community to another. Police in one community may arrest youths who do not go to school, whereas, in another community, failure to attend school receives little attention by the authorities. Moreover, the delinquent activities of some youths tend to be more visible than the activities of others, thus increasing the likelihood that certain youths will come to the attention of the police and be labeled delinquents. Defining delinquency as behavior that violates the legal code ignores these nuances in juvenile justice practice. In recognition of these realities, some have suggested that a more useful definition of delinquency would focus on those acts that official agents select for enforcement rather than on all legally prohibited behaviors.[3] From this perspective, delinquency actually represents a sample of those behaviors prohibited by state law and delinquents are, for all practical purposes, youths who are "caught" and subject to formal processing by the authorities.

Another problem with using a legal definition is that such definitions cover an extremely broad range of behaviors, from incorrigibility (i.e., not obeying one's parents) to serious criminal actions (e.g., homicide). From a legal standpoint, almost all minors could be considered delinquents because most youths engage in at least one illegal behavior at some time during their juvenile years. For example, research conducted by the U.S. Centers for Disease Control and Prevention revealed that approximately 82% of high school seniors reported using alcohol during their lifetime and approximately 48% reported using marijuana.[4] The percentage of youths who fail to obey their parents, also illegal in many **jurisdictions**, is likely to be even higher. According to a strict legal definition, most youths would be considered delinquents, even though most people would not consider many of these youths to be delinquents, nor would they consider their actions to be illegal.

jurisdiction A geographic area subject to a particular law or court.

There are additional problems with legal definitions. Legal definitions do not make a distinction between those who are caught and those who engage in delinquent behavior. Yet, this distinction may be important because those who are caught may be subject to the loss or the restriction of their freedom. Moreover, we should not assume that our reactions to juvenile offenders will necessarily lead to a cessation of their illegal behavior. As Harold Garfinkel notes, the process of labeling a youth a delinquent may be seen as a status degradation ceremony through which the youth's identity is (possibly) transformed into a lesser form of humanity.[5] Not only may labeling cause the individual to see him- or herself differently (e.g., as a troublemaker, thief, or delinquent), but it also may cause others to respond differently or avoid the person, leading to rejection and the restriction of law-abiding opportunities. This, in turn, can increase the likelihood of further delinquency.[6]

FYI

The Saints and the Roughnecks

In a classic study of delinquency, "The Saints and the Roughnecks," William Chambliss pointed out that lower-class youths, who tend to be visible to the community, nonmobile, and not very adept at meeting the social expectations of authorities, are more likely to be selected for punishment than affluent youths, who are generally less visible and more mobile. By observing the Saints, eight white males from solid middle-class families, and the Roughnecks, six lower-class white males, Chambliss discovered that the Saints actually engaged in more frequent delinquent behaviors than the Roughnecks, who engaged in somewhat more serious behaviors, such as fighting and property offenses. The Roughnecks, however, were more likely to be seen as delinquents destined for lives of trouble than the Saints, who were seen as upstanding students with bright futures.

In explaining the difference in the reaction to these two groups, Chambliss stated that the Roughnecks' activities took place within the purview of the community because the Roughnecks did not have access to cars. In contrast, the Saints, who had access to cars, were able to travel to the periphery of town or to other towns. Thus, their delinquent behaviors were less visible. Furthermore, Chambliss noted that, during contact with the authorities, the demeanor of the Saints allowed them to avoid difficulty. The Saints generally responded in ways that were felt to be more acceptable by authority figures. As a result, the authorities possessed a perceptual bias that led them to see lower-class youths as more problematic—a bias reinforced by the visibility of the Roughnecks and their lack of social skills.[7]

Source: Courtesy of William J. Chambliss

A number of studies have uncovered the problems faced by persons on whom certain negative labels have been placed, such as mental patients, youths identified as delinquents, and those convicted of crimes. For example, Charles Frazier documented the problems experienced by "Ken," a young man who lived in a small town and was labeled a "criminal" at a public trial. Subsequent to his official labeling, people began to see previous events in Ken's life as indications of deviance. In addition, former friends and associates began to reject Ken, and their rejection led Ken to see himself as a criminal.[8] Research by Christine Bodwitch that examined school disciplinary practices revealed that students who were seen as delinquents by school administrators were more likely to receive more severe disciplinary responses such as suspension, transfer to another school, or even expulsion.[9] Moreover, studies in the United States and in the Netherlands have highlighted the problems faced by "offenders" in their efforts to seek employment.[10] In these studies, prospective employers were given job applications of fictitious persons that contained varying amounts of information regarding criminal involvement. Not surprisingly, both studies found that employers were more likely to consider a prospective candidate with no criminal history than a candidate with a criminal history.

The labeling perspective predicts that system involvement may lead to negative outcomes because it can limit youths' education, social, and employment opportunities. Support for this argument was found in research conducted in Rochester, New York, that examined a representative sample of males over a nine-year period from the time that the subjects were approximately age 13 years until they were approximately age 22 years. The researchers found that involvement in the juvenile justice process decreases the odds that youths will graduate from high school and educational success is related to future employment prospects. In addition, involvement in the juvenile justice process was found to increase the odds that those youths will engage in crime in their young adult years.

MYTH VS REALITY

Punishment is Not Always an Effective Response to Youths' Behavior

Myth—Youths should always be punished when they violate the law.

Reality—Sometimes our responses to youths can increase the likelihood of further deviant behavior, which sociologist Edwin Lemert called "secondary deviance." Lemert claimed that persons may engage in initial acts of deviance, such as delinquency, for a variety of reasons. Lemert termed such deviance "primary deviance." However, through repeated interaction between someone identified as deviant and authority figures, a process that may involve labeling and stigmatization, the individual may reorganize his or her identity around a more deviant role, thus increasing the likelihood of further acts of deviance. This secondary deviance is not a product of the original factors that produced the initial acts of deviance, but an adjustment to or a means of defense against societal reactions to the primary deviance.[12]

Moreover, the negative effects of system involvement were particularly strong for economically disadvantaged and African American youths.[11]

Although the preceding studies do not address possible positive effects of labeling, such as the avoidance of negative behaviors out of fear of receiving a negative label or the potential deterrent effects of shame and embarrassment associated with criminal justice involvement, they do challenge the notion that a formal response to a juvenile offender is always beneficial. While some youths are helped by their involvement in juvenile justice, there are other instances in which youths are placed in correctional programs where they are exposed to more hardened offenders and are victimized by other youths and/or staff. Moreover, juvenile justice responses may set some youths on paths that increases their odds of future criminality.

Another difficulty with legal definitions of delinquency is that they obscure potentially important differences between youths involved in illegal behavior. Some youths become involved in the juvenile justice process because of their involvement in status offenses; others become involved because of criminal offending. However, making distinctions between status and criminal offenders still does not take into account the many ways in which youths differ from one another. Indeed, youths who engage in illegal behavior constitute a diverse population. They vary considerably in terms of their psychological and biological characteristics and the social context within which they live. Moreover, these variations in psychological and biological characteristics, as well as social context, need to be considered when making decisions about their treatment.

Finally, legal definitions of delinquency present it as an either/or phenomenon and ignore both the frequency and the seriousness of delinquent conduct. In reality, most youths engage in infrequent and minor types of delinquent behavior; only a small

PINS Persons in need of supervision; children who engage in status offenses.

CHINS Children in need of supervision; see PINS.

FYI

Distinguishing Between Status Offenders and Delinquents

Most states make a distinction between youths who engage in criminal behaviors by designating them "delinquent offenders" or "public offenders" and those who engage in status offenses, who are called **PINS** (persons in need of supervision), **CHINS** (children in need of supervision), or "undisciplined children."

percentage of the juvenile population engages in repetitive and serious delinquent actions. Consequently, it would arguably be more accurate to view delinquency as a form of behavior that falls on a behavioral continuum ranging from extreme conformity to extreme nonconformity.[13]

In fact, juvenile justice practitioners spend a considerable amount of time attempting to determine how they should respond to youths who engage in minor to very serious forms of delinquency. How many resources should we devote to minor offenders? If we fail to devote resources to minor offenders, will they become serious offenders? How many resources should we devote to serious offenders? Which juvenile offenders represent a threat to the public and which are "good kids" who have made a mistake? These are questions that juvenile justice practitioners wrestle with daily. Moreover, how juvenile justice practitioners answer these questions can have profound effects, not only on youths who come to the attention of the authorities, but also on their families and the wider community. Consider the actual case of Marcus (see Box 1-1). How has the court responded to this case? From Marcus's point of view, what have been the pros and cons of court involvement? Do you think the court responded appropriately?

BOX 1-1 Interview: James, a Delinquent Offender

Q: How old were you when you first became involved in the juvenile court system, and how old are you now?

A: I was 15 when I was first arrested, and I am 16 years old now.

Q: What was your presenting delinquent offense? Were you alone or with someone?

A: A friend and I robbed a convenience store, and we had a pistol; the charge was armed robbery.

Q: What was your sentence or disposition?

A: I was sentenced in July of 1997 to probation and put into the day treatment program. They monitor me 24 hours a day, including before and after school; but I get to live at home.

Q: Since being sentenced, have you committed any new delinquent crimes or violated your probation?

A: I violated my probation by trying to buy some marijuana after a few months on probation. I got 24 hours in detention and in-home detention for Valentine's weekend, so I couldn't spend Valentine's Day with my girlfriend. I learned my lesson and have not violated since!

Q: What are the benefits that you believe you are receiving from your involvement with the juvenile court?

A: My involvement with the court has really helped turn my life around. The referee who sentenced me made me attend soccer workouts all summer as part of my probation. I ended up playing on my high school's junior varsity soccer team and was elected captain and selected as most valuable player. These good things that have happened to me have helped my confidence. I feel good about them. My day treatment worker, teachers, and soccer coaches have been very positive persons in my life. The structure and the pressure of having rules that I must follow helps keep me out of trouble. In addition to soccer, the day treatment program gives good rewards for obeying the rules, like traveling to amusement parks. We also do stuff in the community.

Q: What don't you like about your involvement with the court?

A: It's not too bad so long as you obey the rules, so long as you don't mess up. I don't like the tracking and monitoring part of day treatment because it really takes away my freedom; but I know if I violate the rules, I'll only get deeper into trouble and deeper into the system.

Q: How do you feel about committing a crime that hurt someone?

A: I feel bad about it. I know it was wrong; I know I shouldn't have done it. I got the idea from a bunch of guys.

(continues)

Q: How are you doing in school?
A: My grades are improving because I know I have to be eligible to play soccer next fall.
Q: When do you think you will be off court supervision?
A: I am not sure of a date, but I must complete my community service hours. I have about 3 1/2 hours left. Probably not until the school year is finished, maybe longer.
Q: What goals for your future do you have now?
A: I want to graduate from high school with a 3.0 grade point average. I want to play soccer in college at Indiana University or some other Division I university. I want to major in athletic administration and be a coach.
Q: What have you learned from your court involvement?
A: Even though you make a real big mistake, there's always someone out there to give you another chance. You need to take advantage of that chance or you'll get locked back up.
Q: Has your experience with the court been enough to keep you from committing another delinquent offense?
A: Yes!

Source: James (personal communication, April 28, 1998)

Which definition of delinquency is best? Should we use a legal definition and include all youths who violate juvenile laws, or should we employ a definition based on the system's identification and response to particular youths?

As noted earlier, strictly legal definitions of delinquency ignore much of its complexity. Moreover, delinquency can be defined in more than one way. Because this text focuses on the processing of youths by official agencies that make up the juvenile justice process, a legal and justice agency response definition will be used. According to this definition, delinquency consists of behaviors that fall within the jurisdiction of the juvenile court and result in processing by official agents of juvenile justice. Keep in mind that this definition does not take into account the differences that exist among youths who engage in delinquency, nor does it take into account the differences that exist among the types of delinquency that youths commit. These differences are important, however, because they help determine how youths are processed by juvenile justice agencies.

Defining Juveniles: The Legal Perspective

Implicit in the definition of juvenile delinquency is an assumption about which youths are considered juveniles. As noted earlier, persons who are subject to the jurisdiction of the juvenile court are considered minors; however, the legal definition of a minor varies from state to state.

In the majority of states, in the District of Columbia, and in the federal system, the upper age limit for juvenile court jurisdiction is age 17 years. This means that after youths turn age 18 years, they are subject to the jurisdiction of adult courts. In a number of other states (Georgia, Illinois, Louisiana, Massachusetts, Michigan, Missouri, New Hampshire, South Carolina, Texas, and Wisconsin), youths become adults in the eyes of the criminal justice system when they become age 17 years. In three states (Connecticut, New York, and North Carolina), adult court jurisdiction begins at age 16 years.[14]

In addition to an upper age limit, some states have a lower age limit for juvenile court jurisdiction, which varies from age 6 years to age 10 years. Where a lower age limit exists, minors younger than the statutorily specified age cannot engage in delinquent behavior.

These youths are felt to be too young to have criminal intent, or ***mens rea***. The idea that children cannot commit crimes is based on the legal principle of ***doli incapax***, which holds that young children do not know right from wrong. As a result, they cannot be held liable for actions that would be considered illegal if committed by an adult.

mens rea Criminal intent.

doli incapax Not capable of criminal intention or malice; unable to formulate criminal intent (*mens rea*); not able, due to lack of maturity, to know right from wrong; not criminally responsible.

Although each state sets age limits that determine which youths are subject to juvenile court jurisdiction, some state juvenile codes allow juvenile justice agencies to continue jurisdiction over a youth for a specified period after he or she passes the maximum age for juvenile court jurisdiction. In most states, after the juvenile court takes jurisdiction over a case, it can retain that jurisdiction until the youth turns age 21 years; in four states (California, Montana, Oregon, and Wisconsin) juvenile court jurisdiction can extend until age 25 years.[15] However, after a youth who is under juvenile court jurisdiction achieves the age at which persons are considered adults, subsequent offenses fall under the jurisdiction of the adult court.

Although state juvenile codes establish the general parameters for juvenile court jurisdiction, state laws also make it possible to transfer some juveniles to criminal courts for trial. The process by which some juveniles are processed as adults is typically called **waiver**, **remand**, **transfer**, or **certification** to adult court. An in-depth examination of waiver is found in Chapter 9.

waiver, remand, transfer, or certification A legal process whereby a juvenile is sent to adult court for trial; see waiver.

Using age as a basis for waiver satisfies legal criteria that require clear-cut definitions for those juveniles who are eligible for trial in adult court; however, it suffers from at least two shortcomings. First, it flies in the face of existing knowledge about human psychological, social, and physical development. An adult is commonly defined as a person who has reached maturity, whereas a **juvenile** is commonly defined as a person who is no longer a child but not yet an adult. At what stage of life, however, does one become an adult, and what exactly is maturity? Psychologists who take a developmental approach suggest that a mature person is one who possesses certain skills that are the product of both cognitive development and the nature of the person's interactions with his or her environment. Moreover, they maintain that these skills are acquired in a developmental sequence. According to this perspective, the ability of individuals to understand their world, including how they relate to others and how their behavior affects and is affected by others, depends on their level of cognitive development and the quality of the interactions they have had with their environment. Accordingly, their

juvenile A youth who falls within an age range specified by state law. The age range varies from state to state.

FYI

Human Development

Jean Piaget argued that the ability of the individual to understand and interpret his or her world proceeds in a series of stages, beginning with the sensorimotor period, which lasts roughly from birth until age 2, and ending with the formal operations period, which lasts from roughly age 11 through adulthood. During the sensorimotor period, the cognitive abilities of the child are quite different than those of an adult. During this period, the young child learns to make sense out of visual, tactile, auditory, and other sensations and focuses on how the world responds to the child's actions. However, during the middle school years, the child enters the formal operations stage, at which point the child's cognitive abilities change greatly. During this period, the child is able to understand and interpret the world differently because of his or her ability to engage in more abstract thought. In addition, the development of the child's cognitive abilities is, to some extent, influenced by the child's environment.[17]

ability to understand the effects of their actions on others and themselves is not static, but changes over time.[16]

Building on the work of Piaget, Lawrence Kohlberg argued that an individual's ability to engage in moral reasoning develops over time and is affected by his or her cognitive development, education, and social experience.[18] At the earliest level of moral reasoning, the individual is more egocentric, focusing attention on how actions affect him or her. At higher stages of moral development, the individual considers how his or her actions affect everyone.[19] Importantly, a number of studies that have examined the relationship between levels of moral reasoning and delinquency have found that low levels of moral reasoning are related to delinquency and criminal behavior. Although these studies do not prove that low moral reasoning causes delinquency, they do suggest that low moral development, along with other psychological and social factors, appears to play an important role.[20]

A second shortcoming of relying on age as the major determinant of adulthood is that it fails to consider variations in the social and psychological development of youths. Although developmental psychologists indicate that cognitive and moral development occur in stages that typically begin and end around specific ages, they recognize that youths vary considerably in their rates of development. Also, some research indicates that boys may develop "other-oriented" reasoning, the ability to be empathic, later than girls.[21] Consequently, the fact that a youth has turned age 17 or age 18 and become subject to the jurisdiction of an adult criminal court does not mean that he or she is a mature adult capable of making adult decisions.

Given the lack of consensus about when youths become adults, it is not surprising that state laws often reflect considerable confusion over this very issue. Although a state law may mandate that youths who are age 16 or age 17 fall under the jurisdiction of adult criminal courts, other laws within that state may deny those youths full participation in adult social and political life by denying them the right to vote, possess alcohol, and enter into legal contracts until they reach age 18 or age 21. From a developmental perspective, this type of inconsistency highlights the problems associated with using age as a criterion for determining when one becomes an adult. Moreover, it is important to consider psychological and moral development issues because doing so reveals the variation that exists among youths who become the clients of both the juvenile and adult justice process. Also, knowledge of child development is needed in order to develop more effective responses to juvenile offenders.

■ The Juvenile Justice System

In referring to juvenile justice practices, it is common to use the phrase "the juvenile justice system." However, the extent to which juvenile justice constitutes a system is a matter of debate. Although state juvenile codes usually specify two main purposes of the juvenile court—to protect the best interests of children and to protect the community—daily juvenile justice operations typically result in considerable variability in practices as well as conflict between juvenile justice agencies and practitioners. In this section, we examine factors that are responsible for variability in juvenile justice practice and for conflict between agencies and practitioners.

Variability in the law is one factor that leads to variability in juvenile justice practices. Each state has its own juvenile laws, which vary regarding the age ranges that fall under the jurisdiction of the juvenile court. Consequently, a youth who is age 16 in South

MYTH VS REALITY

Juvenile Justice Goals Often Vary by Agency and Individual Practitioner

Myth—The juvenile justice system comprises police agencies, courts, and correctional institutions that work together to achieve the same common goals.

Reality—Although the various components of the juvenile justice process (i.e., police agencies, courts, and correctional institutions) and individuals who work within these different components are charged with protecting the public and serving the best interests of children, they often work at cross-purposes. Importantly, those who work in different areas of juvenile justice, such as police officers, district attorneys, defense attorneys, judges, and probation officers, have different roles and responsibilities. Moreover, those who work in juvenile justice perceive their roles differently and they are influenced by a variety of factors that result in a considerable amount of conflict over how particular cases or types of cases should be handled.

Carolina is considered a juvenile, whereas a youth who is age 16 in North Carolina falls under the jurisdiction of adult courts. Moreover, the expressed purposes of juvenile codes also vary from state to state. Some juvenile codes emphasize child welfare (e.g., the District of Columbia, Kentucky, Massachusetts, and West Virginia), while others emphasize accountability and protection of the public (e.g., Connecticut, Hawaii, North Carolina, Texas, Utah, and Wyoming).[22] Differences in the law exist within states as well, because states contain smaller jurisdictional units, such as cities, townships, and counties, that can develop local ordinances prohibiting certain juvenile behaviors. For example, the city of Charlotte, North Carolina, has a curfew ordinance that prohibits youths younger than 16 years from being out between 11:00 p.m. and 6:00 a.m. unless they are supervised by an adult. However, this ordinance does not affect youths younger than 16 years who live outside the city limits.

Differences in the social, political, and economic environments found in different communities also influence variability in juvenile justice practices. The level of juvenile crime, the range of economic opportunities for youths, the quality of the public schools, the existence of activist community groups that demand certain responses to juvenile crime, and a host of other social, political, and economic factors can influence juvenile justice practices. As a result, the response to a youth who violates the law (e.g., by shoplifting or failing to obey his or her parents) in one county may be quite different than the response to those who commit the same offense in an adjacent county in the same state.

Another factor that leads to variability in juvenile justice practice is the **discretion** that juvenile justice decision makers (police officers, district attorneys, probation officers, etc.) have when determining how to respond to youths who violate the law. Discretion is the ability to make judgments on one's own authority. However, it is important to recognize that the use of discretion itself is influenced by a number of political, social, economic, legal, and other factors. For example, juvenile justice decision makers work within a legal context that places statutory limits on their decision-making ability. Moreover, their decisions may be constrained by monetary resources and by their awareness of what the community will tolerate or support. In addition, juvenile justice decision makers have different **juvenile justice ideologies**, which is to say that they have different ideas about the causes of juvenile crime and what should be done about it.

discretion The authority to exercise judgment.

juvenile justice ideology A set of ideas about the causes of juvenile crime and what should be done about it.

The individualized nature of juvenile justice also contributes to variablilty in juvenile justice practice. Since their inception, juvenile courts, like other courts, have taken an individualized approach to dealing with offenders. This individualized approach has been reflected in efforts by juvenile justice decision makers to consider the characteristics of each offender (e.g., his or her age, maturity, mental capacity, and home environment) and the circumstances surrounding the offense (e.g., whether there was provocation) in determining the appropriate response. Finally, different juvenile justice practitioners have different roles and responsibilities in the juvenile justice process. These different roles and responsibilities inevitably lead to some degree of conflict between agencies and individuals as they strive to carry out what they feel their positions require of them.

The fact that youths are often dealt with differently, even within the same juvenile justice agency, should not be surprising because (1) the local social, economic, political, and legal context of juvenile justice practice places limits on juvenile justice decision making (2) local decision makers within juvenile justice have discretion; (3) decision makers have different views about what constitutes an appropriate response to various types of delinquent behavior; (4) the history of individualized justice given to youths, and (5) the different roles and responsibilities assumed by those who work in juvenile justice. Nor should it be surprising that there can be considerable conflict between the agencies and individuals responsible for responding to delinquent youths. In some instances, of course, juvenile justice practice involves relatively coordinated responses to juvenile offenders and relatively low levels of conflict. What makes the idea of a juvenile justice "system" open to question are the many instances of low levels of coordination and high levels of conflict between juvenile justice agencies and/or practitioners.

■ Public Perceptions of Delinquency and the Politics of Juvenile Justice

The preceding sections have addressed some important issues related to basic concepts employed in juvenile justice. In this section, we turn our attention to understanding public perceptions of delinquency. These perceptions are important because they color much of what we think about the so-called "delinquency problem" and how we should respond to it.

Interestingly, public concern over delinquency is not new. Indeed, each generation of Americans seems to believe that the country is experiencing a juvenile crime wave.[23] Concern about youth crime and waywardness in Eastern cities led reformers in the early 1800s to develop the first specialized juvenile institutions. In the late 1800s, other reformers, called the "child savers," had similar concerns and developed the first juvenile courts. During the 1950s, many members of the public were alarmed at what they perceived to be the negative influences of youth culture on adolescents. Movies about young rebels, played so well by actors such as James Dean and Marlon Brando, caused many Americans to question the influence of the media on middle-class youths, who, it was felt, were being seduced by lower-class values that spawned delinquency. As a result, various efforts were undertaken to censor movies, comic books, and other media believed to spread the wrong values.[24]

Although public concern about delinquency is hardly new, each wave of concern produces its own set of solutions to the delinquency problem. Changes in the juvenile justice response to offenders have not always coincided with actual increases in the level of delinquency nor are responses to delinquency always in line with our knowledge of effective responses to juvenile offenders. Since the late 1970s, concerns about serious, violent, and chronic juvenile offenders, sometimes referred to as juvenile "superpredators," have led to a variety of "get tough" efforts, including legislation in a number of states intended to ensure that serious juvenile offenders receive maximum terms of confinement, often by making it easier to process juveniles in adult courts.[25] Unfortunately, these policies have been driven more by the politics of juvenile justice than by our knowledge of juvenile crime or juvenile offenders. As we demonstrate in Chapter 2 and Chapter 14, various data sources on delinquency provide no evidence of a juvenile crime wave nor do they indicate that there is a growing population of violent juvenile offenders. This highlights three important facts about juvenile justice.

1. How we respond to juvenile offenders is, to a large extent, a reflection of our perception of juvenile crime.
2. Changes in our *perception* of the nature and extent of juvenile offending, irrespective of the actual level of juvenile crime, can produce changes in juvenile justice policies.
3. Responses to juvenile crime are the result of a political process in which particular sets of ideas about what constitutes a reasonable response to the delinquency problem win out over others.

Unfortunately, this process is often driven by public perceptions based on sensational and distorted media accounts of juvenile crime that influence public beliefs about the level of juvenile crime, the **etiology** of delinquency, the characteristics of juvenile offenders, and the most appropriate response to the juvenile crime problem. One result is that existing policies frequently are at odds with our current knowledge about delinquency and about the best way to deal with juvenile offenders. In fact, many of these policies are harmful to youths, their families, and the communities in which we live because they are ineffective and, in some cases, counterproductive. Such policies also have a negative impact on those who work in juvenile justice because they limit the ability of these individuals to facilitate successful client outcomes.

etiology The study of causation.

As noted earlier, public perception that a serious juvenile crime problem existed during the last two decades resulted in a variety of policies intended to "get tough" with juvenile offenders, despite a lack of evidence that the juvenile crime problem was actually getting worse. The cumulative effect of these policies was to increase reliance on incarceration as a response to delinquency, a policy that continues today in many jurisdictions, even though there is no sound evidence that incarceration has any significant effect on levels of violent juvenile crime. Nevertheless, the "get tough" approach continues to have considerable appeal because it fits well with the perception that many people have about what should be done. An important question to ask, however, is the following: What is

FYI

Public Opinion

Although increasingly more punitive responses to juvenile offenders have been developed around the country,[26] there is considerable evidence that the public still favors the traditional rehabilitative focus of the juvenile court when it comes to treating juvenile offenders.[27] Thus, efforts to treat youths more punitively may not necessarily reflect public sentiment.

the most appropriate response to juvenile crime? The information provided throughout this text is intended to assist you in formulating your own answer to this question (as well as answers to many other questions you will have).

Chapter Summary

This chapter was designed to introduce some of the basic concepts necessary for developing a clear understanding of the juvenile justice process. It began by defining two important concepts, delinquency and juvenile, and exploring a number of shortcomings of these definitions. Clearly, there are several ways to think about what constitutes delinquency and who is considered to be a delinquent (and consequently subject to the jurisdiction of juvenile justice agencies). The chapter also pointed out the broad range of behaviors that are treated as juvenile offenses and the tremendous variability in the psychological, biological, and social characteristics of the youths who engage in illegal behavior.

In addition, the chapter suggested that talk about the juvenile justice system may be best viewed as rhetorical. It examined a variety of factors that produce variability and conflict in juvenile justice practice and that make systemic responses to juvenile offenders difficult. Some commonalities in juvenile justice operations across the United States exist. Nevertheless, it is important to recognize that there is often considerable conflict between the agencies and individuals responsible for responding to juvenile offenders—conflict that sometimes impedes systemic responses to juvenile crime.

Finally, the chapter discussed the important role of public perceptions and politics in the development of juvenile justice policy. As noted, public opinion about juvenile justice is not always based on accurate knowledge of the juvenile justice process or juvenile offenders. Nevertheless, public perceptions and beliefs about the causes of delinquency and the best way to respond play critical roles in the politics of juvenile justice. Unfortunately, the politics of policy development do not always produce responses to delinquency that help youths or their families, nor do they always lead to safer communities and high success rates among juvenile justice practitioners.

Key Concepts

certification: A legal process whereby a juvenile is remanded to adult court for trial; see waiver.

CHINS: Children in need of supervision; see PINS.

comparative juvenile justice: The examination of juvenile justice history and practice in one country or culture by comparing it to the history and practice of juvenile justice in another country or culture.

delinquency: Behaviors that fall within the jurisdiction of the juvenile court and result in processing by official juvenile justice agents.

discretion: The authority to exercise judgment.

***doli incapax*:** Not capable of criminal intention or malice; unable to formulate criminal intent (*mens rea*); not able, due to lack of maturity, to know right from wrong; not criminally responsible.

etiology: The study of causation.

jurisdiction: A geographic area subject to a particular law or court.

juvenile: A youth who falls within an age range specified by state law. The age range varies from state to state.
juvenile justice ideology: A set of ideas about the causes of juvenile crime and what should be done about it.
labeling: Defining and responding to a youth as a delinquent. Labeling can lead to stigmatization, lost opportunities, and additional delinquent behavior.
***mens rea*:** Criminal intent.
PINS: Persons in need of supervision; children who engage in status offenses.
remand: A legal process whereby a juvenile is sent to adult court for trial; see waiver.
status offenses: Acts that are considered illegal when committed by a minor but not when committed by an adult (e.g., running away, school truancy, and failure to obey parents' directions).
transfer: A legal process whereby a juvenile is sent to adult court for trial; see waiver.
waiver: A legal process that enables a juvenile to be tried as an adult.

Review Questions

1. What are the potential benefits of comparing juvenile justice history and practice in the United States to the history and practice of juvenile justice in other countries or cultures?
2. What are the two broad categories of behaviors that fall within legal definitions of delinquency?
3. What are the shortcomings of legal definitions of delinquency?
4. How common is delinquent behavior in the United States?
5. How could labeling a youth as a delinquent affect his or her future behavior?
6. Are all juvenile offenders alike? Cite evidence to support your view.
7. What is the definition of a juvenile from a legal perspective?
8. What are the problems associated with using an age criterion to define a juvenile?
9. Define waiver, which is also known as transfer, remand, and certification.
10. Is there a juvenile justice "system?" Provide support for your view.
11. According to state statutes, what are the two primary functions of the juvenile courts?
12. Identify the factors that lead to variability and conflict in juvenile justice practices.
13. How do public perceptions of the delinquency problem affect responses to juvenile crime?
14. What role does politics play in juvenile justice?

Additional Readings

Bernard, T. J. (1992). *The cycle of juvenile justice.* New York: Oxford University Press.
Gilbert, J. (1986). *A cycle of outrage: America's response to the juvenile delinquent in the 1950s.* New York: Oxford University Press.
Krisberg, B. (2005). *Juvenile justice: Redeeming our children.* Thousand Oaks, CA: Sage.
Schwartz, I. M. (1989). *(In)justice for juveniles: Rethinking the best interests of the child.* Lexington, MA: Lexington Books.

Notes

1. Platt, A. M. (1977). *The child savers: The invention of delinquency*. Chicago: University of Chicago Press.
2. Dammer, H. R. & Fairchild, E. (2006). *Comparative criminal justice systems* (3rd ed.). Belmont, CA: Thompson-Wadsworth, p. 8.
3. Cloward, R. & Ohlin, L. (1960). *Delinquency and opportunity: A theory of delinquent gangs*. New York: The Free Press.
4. Eaton D. K., Kann L., Kinchen S., Ross J., Hawkins J., Harris W. A., et al., Youth risk behavior surveillance—United States, 2005. (2006, June 9). *Morbidity and Mortality Weekly Report*, 55(SS–5).
5. Garfinkel, H. (1956). Conditions of successful degradation ceremonies. *American Journal of Sociology*, 61, 420–424.
6. Frazier, C. (1976). *Theoretical approaches to deviance: An evaluation*. Columbus, OH: Merrill.
7. Chambliss, W. J. (1973). The saints and the roughnecks. *Society*, 11, 341–355.
8. Frazier, 1976.
9. Bodwitch, C. (1993). Getting rid of troublemakers: High school disciplinary procedures and the production of dropouts. *Social Problems*, 40, 493–509.
10. Schwartz, R. D. & Skolnick, J. H. (1964). Two studies of legal stigma. In Becker, H. S. (Ed.), *The other side: Perspectives on deviance*. New York: The Free Press.

    Buikhuisen, W. & Dijksterhuis, P. H. (1971). Delinquency and stigmatization. *British Journal of Criminology*, 11, 186.
11. Bernburg, J. G. & Krohn, M. D. (2003). Labeling, life chances, and crime: The direct and indirect effects of official intervention in adolescence on crime in early adulthood. *Criminology*, 4, 1287–1318.
12. Lemert, E. (1951). *Social pathology: A systematic approach to the theory of sociopathic behavior*. New York: McGraw-Hill.
13. Cavan, R. S. & Ferdinand, T. N. (1975). *Juvenile delinquency* (3rd ed.). Philadelphia: Lippincott.
14. Snyder, H. N. & Sickmund, M. (2006). *Juvenile offenders and victims: 2006 national report*. Washington, DC: Office of Juvenile Justice and Delinquency Prevention.
15. Snyder & Sickmund, 2006.
16. Kohlberg, L. (1987). *Child psychology and childhood education: A cognitive-developmental view*. New York: Longman. This resource provides an example of the developmental perspective.
17. Mayer, R. E. (1987). *Educational psychology: A cognitive approach*. Boston: Little, Brown & Co.
18. Kohlberg, 1987.
19. Bartol, C. (1991). *Criminal behavior: A psychosocial approach*. Englewood Cliffs, NJ: Prentice-Hall.
20. Andrews, D. A. & Bonta, J. (1994). *The psychology of criminal conduct*. Cincinnati: Anderson.

21. Hoffman, M. (1977). Sex differences in empathy and related behaviors. *Psychological Bulletin*, 84, 712–722.

Morash, M. (1983). An explanation of juvenile delinquency: The integration of moral-reasoning theory and sociological knowledge. In W. S. Laufer and J. M. Day (Eds.), *Personality, theory, moral development, and criminal behavior*. Lexington, MA: Lexington Books.

22. Snyder & Sickmund, 2006.

23. Bernard, T. J. (1992). *The cycle of juvenile justice*. New York: Oxford University Press.

24. Gilbert, J. (1986) *A cycle of outrage: America's reaction to the juvenile delinquent in the 1950s*. New York: Oxford University Press.

25. Krisberg, B., Schwartz, I., Litsky, P., & Austin, J. (1986). The watershed of juvenile justice reform. *Crime and Delinquency*, 32, 5–38.

26. Torbet, P. & Szymanski, L. (1998). State juvenile responses to violent juvenile crime: 1996–97 update. *Juvenile Justice Bulletin*, Washington, DC: Office of Juvenile Justice and Delinquency Prevention.

27. Moon, M. M., Sundt, J., Cullen, F., & Wright, J. (2000). Is child saving dead? Public support for juvenile rehabilitation. *Crime and Delinquency*, 46, 38–60.

Measuring the Extent of Juvenile Delinquency

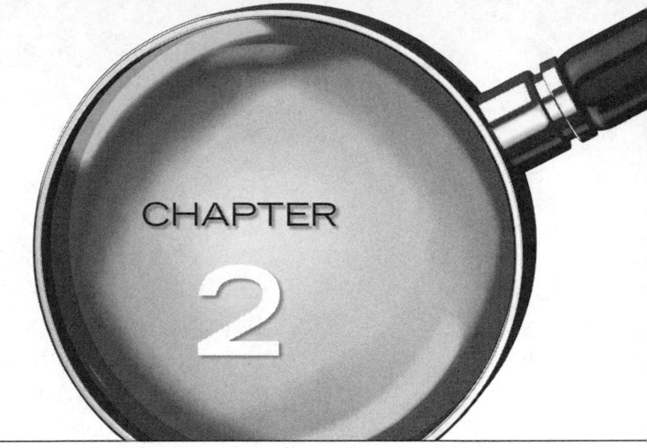

CHAPTER OBJECTIVES

After studying this chapter, you should be able to

- Explain the differences between official and unofficial data
- Describe the different types of official data sources on juvenile delinquency
- Describe the picture of delinquency presented by each of the official data sources
- Describe the shortcomings of official data
- Describe the different types of unofficial data sources on juvenile delinquency
- Describe the insights provided by and the weaknesses of unofficial data sources
- Explain why official and unofficial data sources are needed to develop a comprehensive view of delinquency
- Describe the extent of juvenile delinquency in the United States
- Describe trends in juvenile case processing over the past 20 years

CHAPTER OUTLINE

- Introduction
- Official Data Sources
- Unofficial Data Sources
- Legal Issues
- Chapter Summary
- Key Concepts
- Review Questions
- Additional Readings
- Notes

Introduction

Chapter 1 examined a variety of conceptual issues important for understanding the operation of juvenile justice in the United States. In addition, it indicated that public perceptions of juvenile crime and politics influence juvenile justice operations. However, as noted in Chapter 1, the perceptions of juvenile crime and juvenile justice held by the public and policy makers are not always based on knowledge about the actual levels of delinquency or the actual operation of juvenile justice agencies. To close the gap between perception and reality, it is important to consider several questions: How extensive is juvenile delinquency in the United States? To what degree, if at all, has the level of delinquency changed over time? How extensive is juvenile justice processing of offenders, and to what degree has such processing changed over time? Unfortunately, there are no simple answers to these questions. Moreover, the answers arrived at will depend on which data sources and measures are used to investigate the questions. To assist readers in developing an understanding of the extent of delinquency and the response to delinquency in the United States, this chapter explores various data sources and measures of delinquency. It also examines trends in juvenile offending and how the processing of juvenile offenders has changed over time. In addition, comparisons of delinquency in the United States and other industrialized countries are presented. The chapter concludes with a consideration of the questions addressed by different data sources and measures, as well as the suitability of various data sources for answering different questions about delinquency in the United States.

official data Information on juvenile delinquency collected by formal juvenile justice agencies, such as police agencies, juvenile courts, and juvenile detention and correctional facilities.

Official Data Sources

There are two types of data on juvenile delinquency: data derived from official sources and data derived from other sources. These data are collected by different agencies and individuals using different methodologies, and they often present different pictures of the extent of delinquency as well as youths who engage in delinquent behavior. **Official data** are collected by juvenile justice agencies, such as law enforcement agencies, juvenile courts, and various correctional programs that handle juveniles. These data may include information such as the names, ages, gender, race, psychological adjustment and offense history of youths involved with juvenile justice as well as information on their family backgrounds, school history, and other types of information. These data help policy makers and practitioners understand the number and characteristics of youths processed by juvenile justice agencies and the offenses they commit. Also, it is used in making decisions about individual cases.

Uniform Crime Report (UCR) The most comprehensive compilation of known crimes and arrests. The report is published each year by the Federal Bureau of Investigation under the title *Crime in the United States*.

The Uniform Crime Report

The most well-publicized source of official data on delinquency is the FBI's **Uniform Crime Report (UCR)** which compiles data from more than 17,000 law enforcement agencies and represents approximately 95% of the population of the United States.

FYI

The UCR Focuses on the Incidence of Arrests

Note that arrest data contained in the UCR refers to the incidence of arrests (the frequency of arrests) rather than the prevalence of arrests (the number of individual persons who are arrested). For instance, an individual may be arrested several or more times in one year. Each of these arrests may be counted in the UCR.

MYTH VS REALITY

Juveniles are Responsible for Most of the Violent Crime in the United States
Myth—Juveniles are responsible for much of the violent crime in the U.S.
Reality—Although people between the ages of 10 and 17 years were slightly overrepresented in arrests for Index violent offenses in 2007, they accounted for only 16% of all arrests for serious violent crimes that year.

The UCR separates crimes into two broad categories: Part I (or Crime Index) offenses and Part II crimes. Part I crimes are further separated into Crime Index violent offenses (murder and nonnegligent manslaughter, forcible rape, robbery, and aggravated assault) and Crime Index property offenses (burglary, larceny-theft, motor vehicle theft, and arson). (Note: In this chapter, all mentions of violent offenses refer to violent offenses as defined by the UCR Crime Index. Likewise, all mentions of property offenses refer to the offenses defined by the UCR Crime Index.) Part II crimes consist of all offenses (except traffic violations) that are not Index offenses. Also, the UCR contains three basic types of data: (1) the number (frequency) and kinds of offenses that are known to the police; (2) the frequency of arrests for different kinds of offenses; and (3) data on the characteristics of offenders, such as their gender, race, age, and whether they live in large or small metropolitan areas or more rural areas.[1] Because the UCR contains information on arrests by age group, it is an important source of information on juvenile crime and the responses of law enforcement agencies to the illegal behaviors of youths.

According to UCR data, juvenile crime represents a significant problem in the United States. In 2007, there were over 1.6 million arrests of persons younger than 18 years. However, as shown in Table 2-1, at least 82% of those arrests were for nonviolent crimes.[2] Indeed, over 60% of all arrests of persons younger than 18 years were for nonassaultive Part II offenses. Approximately 12% of all arrests of persons younger than 18 years were for **status offenses**, such as running away from home, curfew violations, and loitering.[3]

status offense An act that is considered a crime or legal violation when committed by a juvenile but not by an adult (e.g., running away from home, incorrigible behavior, failure to attend school, and failure to obey school rules).

Although most arrests of persons younger than 18 years were for nonviolent crimes, offenses against people still account for a substantial portion of all youth arrests. For example, in 2007, persons under the age of 18 accounted for 15% of all arrests (both Part I and Part II arrests). Furthermore, persons younger than 18 years accounted for 23% of all Crime Index arrests, 16% of all arrests for Index violent offenses, and 26% of all arrests

FYI

The Development of NIBRS
The first UCR for the United States was published in 1930. Since that time, the UCR has undergone significant changes. Calls for more comprehensive changes in the UCR program by the 1980s resulted in the development of the National Incident-Based Reporting System (NIBRS). The NIBRS system is intended to provide more comprehensive information (i.e., about the offense, victim, offender, and types and value of property involved) on each incident and arrest within 22 offense categories that are called Group A offenses. Consequently, NIBRS provides more detailed crime information than that provided by traditional UCR data.[6] By 2007, the FBI had certified 26 state programs for NIBRS participation, 12 states were in the process of testing NIBRS, and 8 states were planning on developing their NIBRS capability.[7]

TABLE 2-1 Arrests of Persons Younger Than 18 Years by Crime Type, 2007

Crime Type	Number	Percentage of Total
Index Offenses		
Index Violent Offenses		
Murder and nonnegligent manslaughter	1,011	0.1
Forcible rape	2,633	0.2
Robbery	26,324	1.6
Aggravated assault	43,459	2.6
Subtotal Index Violent Offenses	73,427	4.4
Index Property Offenses		
Burglary	61,695	3.7
Larceny-theft	229,837	13.9
Motor vehicle theft	22,266	1.3
Arson	5,427	0.3
Subtotal Index Property Offenses	319,225	19.3
Total Index Offenses	**392,652**	**23.8**
Part II Offenses		
Other assaults (including offenses against family and children)	181,378	11.0
Forgery and counterfeiting, fraud, and embezzlement	9,331	0.6
Buying, receiving, and possessing stolen property	16,889	1.0
Vandalism	84,744	5.1
Carrying and possessing weapons	33,187	2.0
Prostitution and commercialized vice	1,160	0.1
Sex offenses (except forcible rape and prostitution)	11,575	0.7
Drug abuse violations	147,382	8.9
Gambling	1,584	0.1
Offenses against family and children	4,205	0.2
Driving under the influence	13,497	0.8
Liquor law violations	106,537	6.5
Drunkenness and disorderly conduct	166,259	10.1
Curfew and loitering	109,815	6.7
Runaways	82,459	5.0
Other	287,323	17.4
Total Part II Offenses	**1,257,325**	**76.2**
Total Arrests	**1,649,977**	**100.0**

Note: Percentages may not total 100 because values were rounded.

Source: Data from Federal Bureau of Investigation (2008). *Crime in the United States 2007*, Table 38. Retrieved June 2, 2009, from http://www.fbi.gov/ucr/cius2007/data/table_38.html.

FYI

Accessing FBI Data

To access UCR data collected by the FBI go to http://www.fbi.gov/ucr/ucr.htm#cius. To access population data needed to construct juvenile arrest rates, go to http://ojjdp.ncjrs.org/ojstatbb/ezapop/asp/profile_selection.asp.

for Index property offenses.[4] These percentages take on added significance when one considers that persons between the ages of 10 and 17 years (inclusive) account for only 13% of the total U.S. population older than the age of 9 years.[5] Thus, youths are slightly *overrepresented* with respect to arrests for violent offenses; their overrepresentation in arrest data is much greater when property offense arrests are considered.

Juvenile Arrest Trends

Although there were noticeable peaks in juvenile arrests during the late 70s and early 80s and then again in the late 80s and early 90s, the number of arrests of youths younger than age 18 years has shown no noticeable upward trend over the past 30 years. In fact, there was a downward trend in arrests of persons younger than age 18 years between the mid-1990s and 2004, however, arrests of juveniles have been increasing in recent years (see Figure 2-1(a)). Arrests of persons younger than age 18 years for Index property and violent offenses have also declined since the mid-1990s, although recent increases, particularly in arrests for violent offenses, are evident in recent years (see Figures 2-1(b) and 2-1(c)).

Of particular concern to many policy makers, juvenile justice practitioners, researchers, and members of the public in recent years has been juveniles' involvement in violent crime. Indeed, the number of arrests of persons younger than age 18 years for

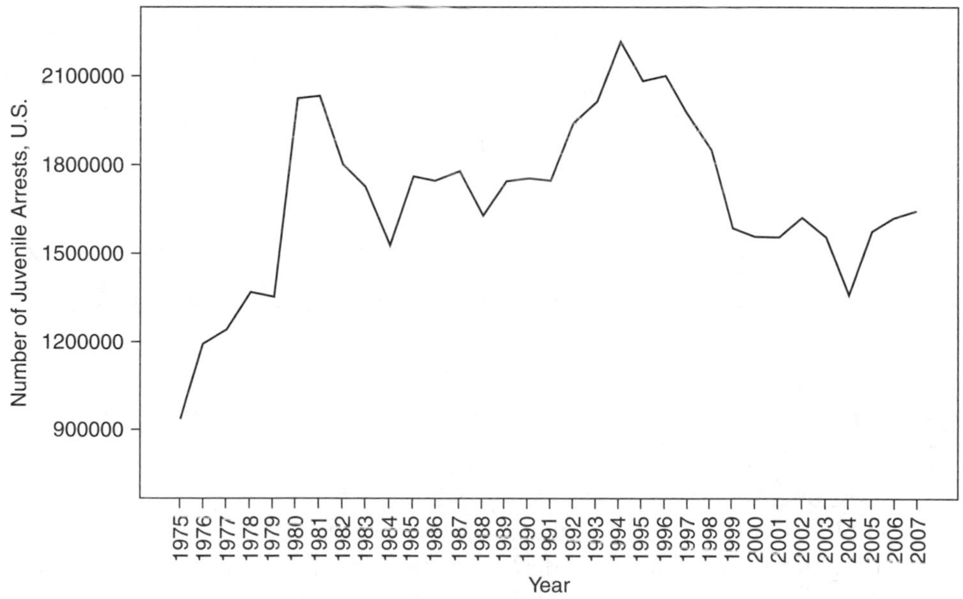

Figure 2-1 a Number of Juvenile Arrests, 1975–2007

Source: Data from the FBI. *Crime in the United States, 1975–2007.* Washington, DC: US Department of Justice.

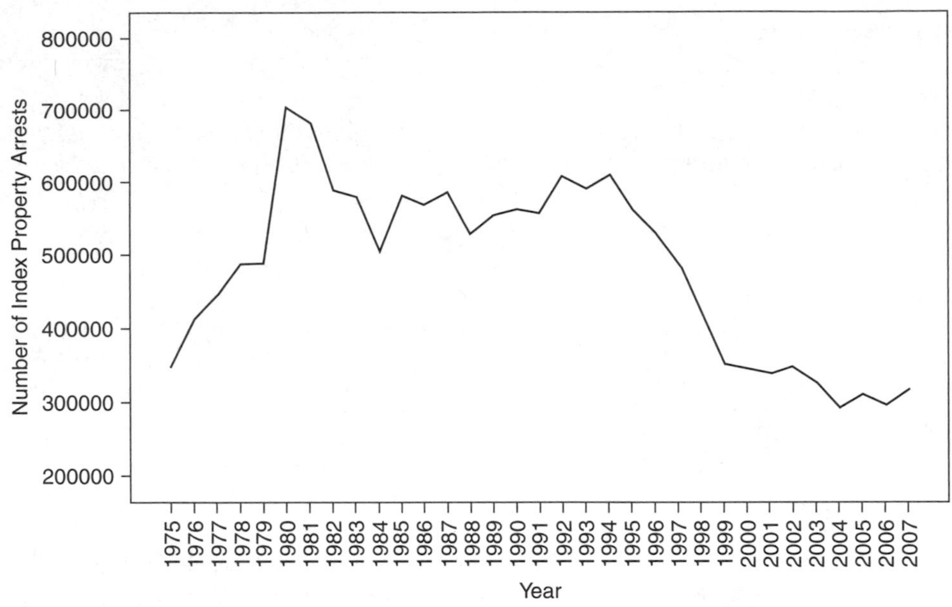

Figure 2-1 b Number of Juvenile Index Property Arrests, 1975–2007

Source: Data from the FBI. *Crime in the United States, 1975–2007.* Washington, DC: US Department of Justice.

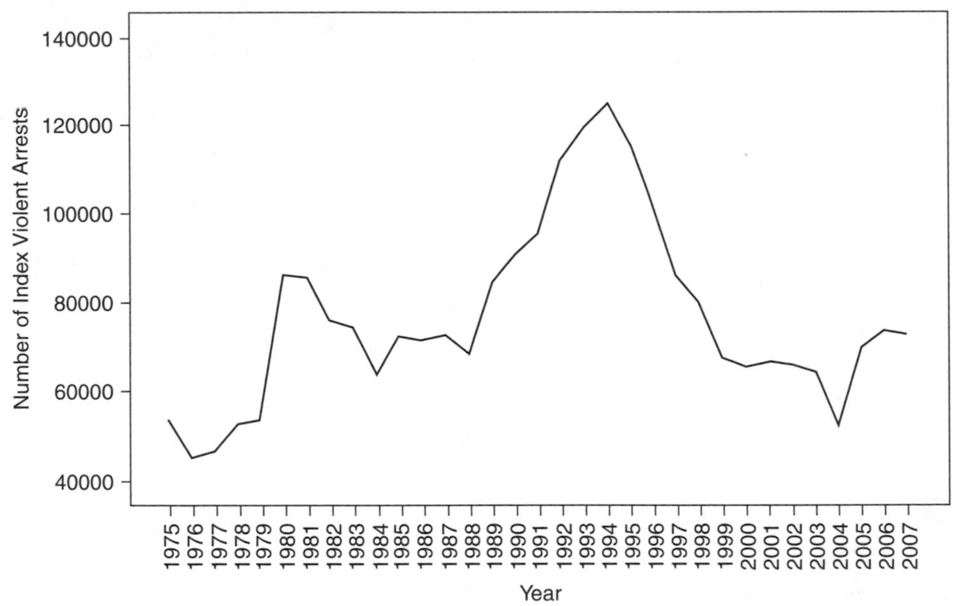

Figure 2-1c Number of Juvenile Index Violent Arrests, 1975–2007

Source: Data from the FBI. *Crime in the United States, 1975–2007.* Washington, DC: US Department of Justice.

Index violent offenses increased 48% from 1989 to 1994. However, as we noted in our discussion of juvenile arrest trends, the number of arrests of persons younger than age 18 years for violent offenses began to decline in 1995, and this general decline continued until 2004, when it began to increase again.[8] However, it is too soon to determine if this increase represents the beginning of a more sustained trend or the normal fluctuation in juvenile arrests that can be expected over time.

Population Effects on Juvenile Arrest Trends

The number of arrests of persons younger than age 18 years is an indicator of the total number of juvenile crimes committed. The number of juvenile crimes committed, however, is partially determined by the size of the juvenile population. In other words, as the size of the juvenile population (i.e., potential juvenile offenders) increases, so does the absolute level of juvenile crime (assuming the rate of juvenile crime remains constant). In fact, it has been estimated that changes in the age structure of the population may account for as much as 40% of the changes in the arrest rate.[9] Although the actual strength of the relationship between levels of juvenile crime and the size of the juvenile population is a matter of debate, population size does have some effect on levels of delinquency. Consequently, another important indicator of juvenile crime is the juvenile **arrest rate.** An examination of the juvenile arrest rate is important because it takes into consideration the size of the juvenile population.

arrest rate The number of arrests adjusted for the size of the population; often reported as arrests per 100,000 youths age 10–17 years in the population.

Because the age ranges of youths who come within the jurisdiction of the juvenile court varies from state to state, juvenile arrest rates are typically calculated by using the number of arrests of persons between the ages of 10 and 17 years taken from the UCR (youths younger than 10 years are rarely arrested relative to older youths) and dividing that number by population figures taken from census data.

Juvenile Arrest Rate Trends

An examination of juvenile arrest rate trends since 1975 indicates a similar pattern to that indicated by arrest trends. Although juvenile arrests rates vary some from year to year, juvenile crime has remained rather stable over time. For example, an examination of Figure 2-2(a) reveals an increase in juvenile arrest rates beginning in the late 1980s and continuing through the mid-1990s, followed by a steady decline until 2004, when juvenile arrest rates began to increase again. However, the juvenile arrest rate in 2007 was still below the arrest rates recorded during the 1980s.[10] Thus, there is no evidence of a continuing increase in juvenile crime over time as measured by arrest rates.

A similar conclusion can be drawn by examining Index violent and Index property offense arrest rates. As Figure 2-2(b) indicates, arrest rates for persons younger than age 18 years were quite stable during the later 1970s and through the late 1980s. The violent offense arrest rate for juveniles, however, while remaining remarkably stable during most of the 1980s, increased by more than 60% between 1988 and 1994.[11] Indeed, this increase in arrests for violent offenses between the late 1980s and the mid-1990s was the primary driving force for the increase in overall juvenile arrests during that period. However, as Figure 2-2(b) indicates, Index violent arrest rates for juveniles declined substantially between the mid-1990s and 2004 when it began to increase again. Like the overall juvenile arrest rate, the property offense arrest rate for juveniles remained quite stable between 1980 and the mid-1990s, at which time it began to decline as Figure 2-2(c) shows.

FYI

Juvenile Arrest Rate

The juvenile arrest rate per 100,000 juveniles is calculated as follows:

$$\frac{\text{Number of juvenile arrests}}{\text{Total juvenile population}} \times 100{,}000 = \text{juvenile arrest rate per 100,000 juveniles}$$

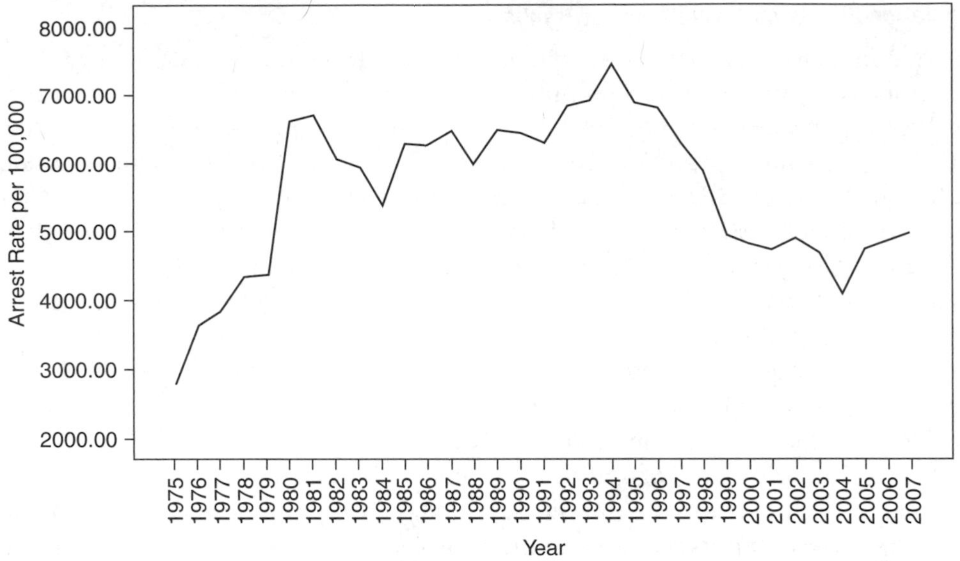

Figure 2-2 a Juvenile Arrest Rate per 100,000 (Ages 10–17), 1975–2007

Source: Data from the FBI. *Crime in the United States, 1975–2007.* Washington, DC: US Department of Justice; Bureau of the Census. (1982). *Current population estimates and projections series* (p. 25, No. 917). Washington, DC: USGPO; Bureau of the Census. (1992). *Current population reports: Population projections for the United States by age, sex, race, and hispanic origin, 1992–2050* (pp. 25–1092). Washington, DC: USGPO; Puzzanchera, C., Sladky, A., & Kang, W. (2008). Easy access to juvenile populations: 1990–2007. Retrieved June 2, 2009, from http://www.ojjdp.ncjrs.gov/ojstatbb/ezapop/.

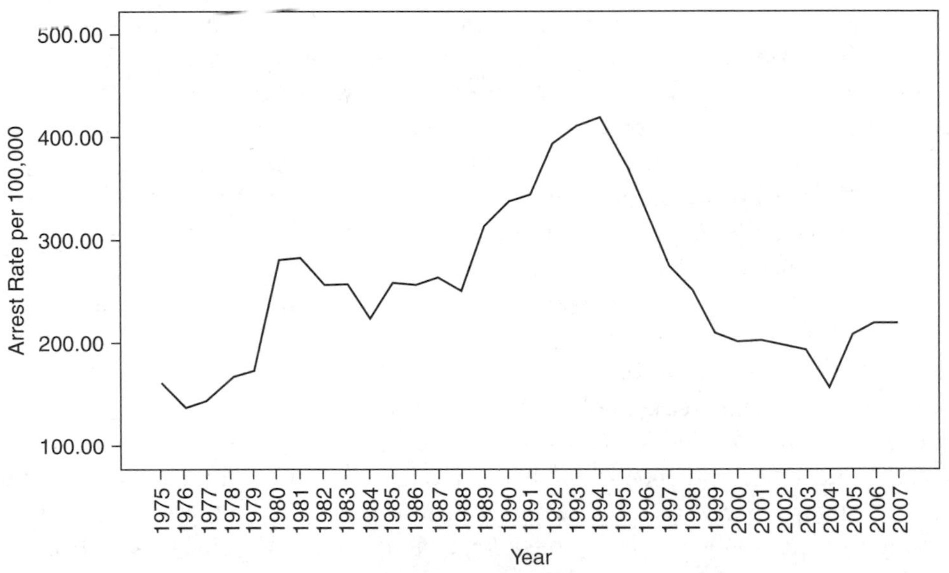

Figure 2-2 b Juvenile Index Violent Arrest Rate per 100,000 (Ages 10–17), 1975–2007

Source: Data from the FBI. *Crime in the United States, 1975–2007.* Washington, DC: US Department of Justice; Bureau of the Census. (1982). *Current population estimates and projections series* (p. 25, No. 917). Washington, DC: USGPO; Bureau of the Census. (1992). *Current population reports: Population projections for the United States by age, sex, race, and hispanic origin, 1992–2050* (pp. 25–1092). Washington, DC: USGPO; Puzzanchera, C., Sladky, A., & Kang, W. (2008). Easy access to juvenile populations: 1990–2007. Retrieved June 2, 2009, from http://www.ojjdp.ncjrs.gov/ojstatbb/ezapop/.

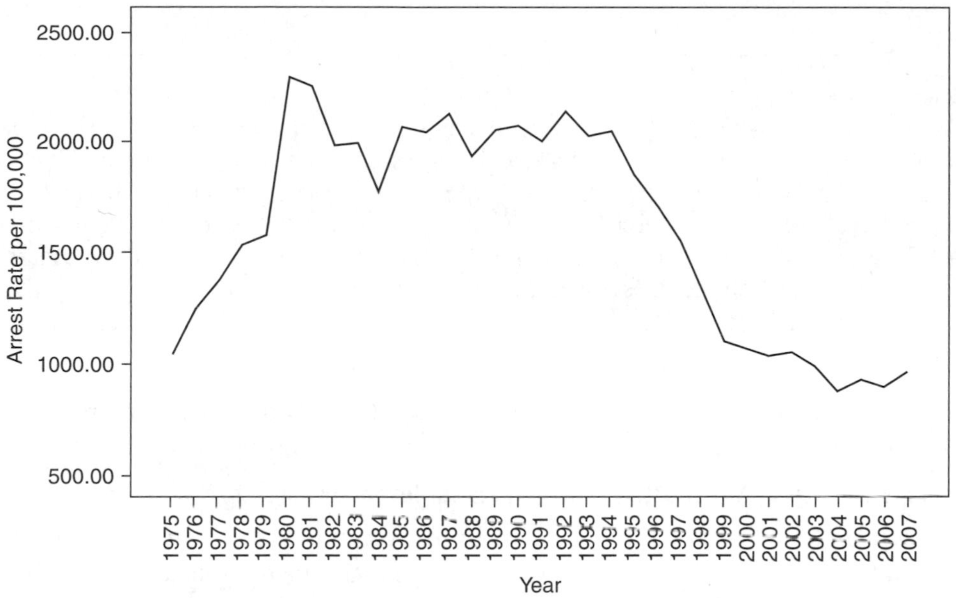

Figure 2-2 c Juvenile Index Property Offence Arrest Rate per 100,000 (Ages 10–17), 1975–2007

Source: Data from the FBI. *Crime in the United States, 1975–2007.* Washington, DC: US Department of Justice; Bureau of the Census. (1982). *Current population estimates and projections series* (p. 25, No. 917). Washington, DC: USGPO; Bureau of the Census. (1992). *Current population reports: Population projections for the United States by age, sex, race, and hispanic origin, 1992–2050* (pp. 25–1092). Washington, DC: USGPO; Puzzanchera, C., Sladky, A., & Kang, W. (2008). Easy access to juvenile populations: 1990–2007. Retrieved June 2, 2009, from http://www.ojjdp.ncjrs.gov/ojstatbb/ezapop/.

The Race and Gender of Arrested Juveniles

In addition to information about numbers of arrests, the UCR also provides information on the racial and gender characteristics of persons who are arrested. According to 2007 UCR data, Whites accounted for approximately 67% of the arrests of persons younger than age 18 years. Whites also represented approximately 66% of those younger than age 18 years who were arrested for Index property offenses and approximately 48% of those arrested for Index violent crimes. In contrast, African American youths accounted for 31% of all arrests, about 32% of Index property arrests, and almost 51% of Index violent arrests. Other racial groups accounted for the remainder of juvenile arrests.[12] However, although White youths accounted for the majority of juvenile arrests, African American youths were arrested in disproportionate numbers, because they account for only about 17% of the juvenile population between ages 10 and 17 years.[13]

The UCR data further indicate that juvenile lawbreaking is predominately the domain of males. For example, 2007 UCR data indicate that males accounted for approximately 171% of all arrests of persons younger than age 18 years. Males accounted for about 183%

FYI

Juvenile Crime Trends

An examination of both arrest trends and arrest rate trends in Figures 2-1(a) and 2-2(a) reveals almost identical patterns, which indicate that, overall, juvenile crime has not become significantly more problematic over time, despite what many people may believe.

dark figure of juvenile crime The amount of juvenile crime that is not reported.

FYI

Hidden Juvenile Crime

The amount of juvenile crime that goes unreported—the **dark figure of juvenile crime**—is not known; however, it appears that well over half of all juvenile offenses are not reported to the authorities. For instance, findings from the National Youth Survey (NYS), a large-scale self-report study; the UCR; and other sources indicate that only about 24% of juveniles who commit offenses for which they could be arrested are actually taken into custody by the police.[16] Moreover, the probability that an individual will be arrested for serious offenses is estimated to be only 2 in 100.[17]

of persons younger than 18 years arrested for Index violent offenses and approximately 165% of those younger than age 18 years arrested for Index property crimes.[14]

Problems with Using Arrest Data as an Indicator of Juvenile Crime

Although police arrests of juveniles constitute an important indicator of the extent of juvenile crime, there are a number of problems with using official data as a measure of delinquency. First, many juvenile offenses do not come to the attention of the police. Because crimes known to the police are usually reported by citizens, the ability or willingness of citizens to make complaints influences the number of crimes included in the UCR. Moreover, a variety of factors affect the ability and willingness of citizens to report crimes to the police. For example, some offenses go undetected, which prevents them from being reported. Furthermore, the seriousness of the offense, the relationship between the offender and the victim, fear of possible reprisals by the offender, the belief that nothing will be done to the offender, embarrassment, and the involvement of the complainant in criminal behavior he or she wants to keep from the police are factors that can influence a citizen's willingness to report their victimization.[15] As a result, many offenses do not come to the attention of law enforcement agencies.

Second, several factors influence the accuracy and completeness of the data that are collected and published in the UCR. For example, police must decide if a particular behavior technically qualifies as an offense. This decision can be influenced, for example, by political pressures to reduce or inflate crime rates, the quality of the interaction between the police and the offender (e.g., whether the offender is hostile toward the police), and the style of policing employed in a particular jurisdiction (e.g., the extent to which the police use a more legalistic and formal style of policing rather than a more preventive style, which encourages the informal processing of cases when possible).[18] These factors can lead to variability across jurisdictions regarding the extent to which behaviors are actually recorded as crimes, thus making comparisons across jurisdictions difficult. In addition, factors such as political pressure and changes in the style of policing can result in changes in arrest data over time that do not accurately reflect real changes in the level of crime. For example,

police discretion The authority of police to make their own judgments about which crimes or delinquent acts are subject to investigation and which juveniles are subject to arrest.

FYI

Police Discretion

It is important to note that **police discretion** plays an important role in determining whether a particular act is recorded as a crime. Moreover, police decision making is influenced by a variety of factors (see Chapter 6).

MYTH VS REALITY

Problems with the UCR

Myth—The UCR contains a comprehensive listing of crimes that are committed in the United States.

Reality—Although the media, political leaders, and policy makers often use the UCR as their primary source of national, state, and local crime data, it suffers from several shortcomings as a source of data on juvenile crime. For example, (1) it omits a sizable amount of crime that is never reported to the police; (2) it contains only limited data on the characteristics of people arrested; and (3) a number of factors influence the accuracy of the data that is recorded.

pressure exerted by a powerful interest group can lead to police crackdowns on particular types of behaviors (e.g., truancy and prostitution). This, in turn, will be reflected in an increase in the numbers of arrests for those behaviors even though their actual incidence did not change. Also, in the UCR data, not all offenses are counted when multiple offenses occur. Only the most serious offense is reported, except when arson is also involved (an act of arson is always counted, even if it is not the most serious offense).[19]

Third, arrest data say as much about the actions of the police as about the extent of crime. Because police departments make decisions to focus their attention on certain types of criminal activity (i.e., the types of offenses contained in the UCR) and these criminal activities more often occur within particular communities (i.e., lower-class neighborhoods), certain actions (e.g., robbery, vandalism, larceny, and curfew violations) are more likely to cause a response and the people in those communities (i.e., poor people and minorities) are more likely to be arrested and seen as criminals.

Using Clearance Data to Understand Juvenile Crime

Although arrest data provide important insights on juvenile crime, they suffer from one important limitation—they *overestimate* juveniles' involvement in criminal activity. This occurs because juvenile offending is more likely to involve other persons, often juveniles, than crimes committed by adults.[20] As a result, juvenile offenses are more likely than adult offenses to involve multiple arrests, even though some of those arrested are not knowing or willing participants in criminal activity.

Another way to understand the extent to which juveniles are involved in crime, which provides a more accurate estimate of juvenile involvement in offending, is to examine clearance data. An offense is cleared when at least one person is arrested for the offense, charged with an offense, or turned over to the court for prosecution. In addition, "exceptional clearances" occur in situations where the offender dies, the victim refuses to cooperate in the prosecution of the case, or the offender is denied extradition and is prosecuted in another jurisdiction.[21] Of the Crime Index offenses cleared by law enforcement in 2007, persons younger than age 18 years accounted for about 12% of violent crime clearances and approximately 18% of property crime clearances.[22] These percentages are lower (particularly that for property offenses) than the arrest percentages presented earlier for Index violent and Index property offenses. However, it is likely that these data also overestimate juvenile involvement in crime because crimes like robbery often result in multiple juvenile arrests.[23]

Trends in the proportion of violent crimes attributed to juveniles present an overall picture that is similar to that portrayed by arrest data. Although the proportion of violent crimes attributed to persons younger than age 18 years grew from approximately 9% in

COMPARATIVE FOCUS

Comparing crimes in different countries is difficult because definitions of crimes vary by country. For example, in Germany assault is only defined as a violent crime when a weapon is used.[25] Moreover, countries vary in terms of their political, social, and economic development, which also complicates cross country comparisons.[26] However, comparisons of juvenile crime in the U.S. and some other developed countries indicates that the rate of violent juvenile crime in the U.S. appears considerably higher.[27] Indeed, the juvenile homicide rate in the United States was almost twice the rate in Japan in 2000 and the robbery rate was nearly three times the Japanese rate.[28] However, some property crime rates, particularly burglary, are higher in countries like Canada, England, Wales, and the Netherlands.[29]

the late 1980s to 14% in 1994, it had decreased to 12% in 2001 and 2002, and was 12% in 2007.[24] Like arrest data, an examination of clearance data indicates that juvenile crime has remained quite stable over time and has shown some decline in recent years.

The Uses of Official Data

As noted, official data on juvenile delinquency, such as UCR arrest statistics, are used regularly by politicians, juvenile justice policy makers, and the media in their efforts to understand juvenile crime. Moreover, because much of what the public knows about juvenile crime comes from the media, politicians, and policy makers, these data play a critical role in educating the public about the nature and extent of delinquency. For example, newspapers and television news departments regularly analyze and report UCR data on juvenile arrests. Of course, the media also rely on the findings of researchers who study juvenile crime, but these researchers often depend on official data as well. Researchers investigating police responses to juvenile offending rely on official arrest data to determine how many youths are arrested by the police, the basic demographic characteristics of those youths, and the types of offenses for which youths are arrested. Similarly, researchers interested in the way youths are dealt with by juvenile courts and correctional institutions employ official court and corrections data. In addition, official arrest data has been used in countless studies that have explored the **etiology** of delinquency and the effectiveness of various types of juvenile justice programs.

etiology The study of causation.

■ Unofficial Data Sources

Besides official data sources, there are several other important sources of **unofficial data** on juvenile crime, including studies performed by researchers not connected to formal juvenile justice agencies (i.e., police, juvenile courts, and juvenile detention and corrections facilities). Among the main types of unofficial data are self-report studies of delinquency, cohort studies, developmental studies, observational studies, and victimization studies. Each of these sources is discussed in the following sections.

unofficial data Information collected by researchers not connected with formal juvenile justice agencies (i.e., police agencies, juvenile courts, and correctional institutions).

Self-Report Studies

As the name implies, a **self-report study** asks people to report on their own behavior. A primary advantage of the self-report method is that it can elicit information on offenses not known to the police. Consequently, it allows researchers to better understand the dark figure of juvenile crime. Another advantage is that a variety of data important to understanding delinquency (e.g., income, education, quality of life, work, family life, and peer group affiliations) can be collected. Unlike the UCR and other official sources of data, self-report

self-report study A study in which subjects are asked to report their involvement in illegal behavior.

studies have been conducted by a variety of researchers unaffiliated with formal juvenile justice agencies. Moreover, some long-term self-report studies have provided important information on changes in juvenile delinquency over time. A good example of an ongoing self-report study is the *Monitoring the Future* study conducted by researchers at the University of Michigan's Institute of Social Research (ISR).[30] This study, which began in 1975, collects a variety of data on the attitudes and behaviors of a national sample of high school seniors.

As one might expect, self-report studies indicate that youths engage in considerably more illegal behavior than is indicated by official data. Particularly when offenses such as school truancy, alcohol consumption, using a false ID, petty larceny, and vandalism are examined, delinquency appears to be normal rather than abnormal. Self-report studies generally indicate that most youths do not engage in serious criminal activity, a finding that mirrors the picture of delinquency presented in the official data sources. Nevertheless, some self-report studies show that a sizable percentage of juveniles are involved in serious offenses. For example, according to the *Monitoring the Future* study, in 2003, almost 12% of high school seniors indicated that during the preceding year they had hurt someone badly enough to require a doctor's attention; approximately 10% indicated that they had stolen something worth more than $50; and approximately 14% reported that they were in a serious fight at work or school during the year.[31] In another ongoing large-scale study of high school students conducted by the Centers for Disease Control and Prevention in 2007, 18% of respondents indicated that they had carried a weapon at least once in the 30 days prior to the survey, and 35.5% reported that they had engaged in a physical fight one or more times in the previous 12 months.[32]

Although self-report studies reveal that delinquency is a pervasive phenomenon in the United States, they do not indicate that juvenile crime is getting worse. There are short-term fluctuations in youths' self-reported delinquency, but an examination of long-term trends in delinquent behavior indicates that it has remained quite stable over time.[33] For instance, an examination of data collected as part of the *Monitoring the Future* study between 1990 and 2002 indicates that no clearly identifiable trends were evident for most offenses. To the extent that trends could be discerned in the data, they tended to be toward less involvement in delinquency.

Another important self-report study is the *National Youth Survey* (NYS), an ongoing study that began in 1976 and is conducted by researchers at the University of Colorado's Behavioral Research Institute. The NYS is a **panel study**, which means that a sample of youths (now adults) are surveyed each year over a number of years. Panel studies are valuable because they allow researchers to examine changes in young people's behaviors and attitudes over time and into adulthood.

panel study A study that involves the examination of the same select group over time.

The NYS indicates that youths engage in a range of criminal activity, although most are involved in minor offenses. It also reveals that the frequency of offending and the kinds of offenses youths commit change over time. Youths' illegal behaviors tend to increase in severity as they reach their late teens and move into their early adult years, but offending decreases after that period. Younger juveniles tend to engage mostly in status offenses, whereas older youths engage in more property and personal offenses.[34]

Analysis of the NYS has also produced some important findings regarding the relationship among race, social class, and delinquency. One finding of considerable interest, reported by Delbert Elliott and Suzanne Ageton, is that the levels of offending for African American youths and White youths are similar. Although the researchers found that African American youths reported slightly more involvement in serious crimes, the size of the difference was not as great as indicated by arrest data.[35] This suggests that the disparity in the arrest levels of White and African American youths is, in part, a reflection of the actions of police agencies.

The relationship between social class and delinquency found in the NYS is also of considerable interest. For example, Delbert Elliott and David Huizinga found that the prevalence of delinquency did not differ among social classes when all types of offenses were considered. In other words, the proportion of middle-class and lower-class youths who engaged in delinquency was similar. However, when the researchers examined different types of offenses, they discovered significant class differences. Middle-class youths had the highest rates of involvement in offenses such as stealing from their families, cheating on tests, cutting classes, disorderly conduct, lying about their age, and drunkenness. Lower-class youths had higher rates of involvement in more serious offenses, such as felony assault and robbery.[36]

Self-report studies also have examined the relationship between gender and delinquent behavior. Overall, these studies indicate that females engage in considerably more delinquency than is indicated by arrest data, which suggests a possible gender bias within juvenile justice. Nevertheless, females engage in less delinquency than males and tend to be involved in less serious types of delinquency, although differences in delinquent behavior between males and females are much smaller when minor offenses are examined.[37]

A very important point to note regarding recent self-report research is that it produces an overall picture of delinquency that is similar to that painted by official data. Although the extent of delinquency depicted in self-report research is considerably greater than is depicted in arrest data, the pattern of delinquency is similar. However, despite improvements in recent self-report studies, there are several problems with self-report studies that should be noted. First, there is no standard reporting format used in self-report studies. Consequently, it is difficult to compare different studies. It is particularly difficult to compare older self-report studies with more recent studies because older studies focused exclusively on less serious types of delinquent behavior and employed less sophisticated methodologies, raising questions about the accuracy of the results obtained. Second, although newer self-report studies include more serious offenses, they still do not cover all types of serious delinquency. Thus, the full range of delinquent behaviors is not covered. Third, there is some doubt as to whether present studies have been able to capture a completely representative sample of youths to study. Self-report studies that rely on school populations miss students who are not in school or who have dropped out. Even studies that do not sample from school populations may fail to include some youths. For instance, research conducted by Stephen Cernkovich, Peggy Giordano, and Meredith Pugh indicated that institutionalized chronic offenders have

MYTH VS REALITY

Self-Report Studies Provide Valuable Data

Myth—Self-report studies are not reliable because youths are unwilling to be truthful about their behavior.

Reality—Efforts to examine the validity of the self-report studies indicate that the self-report method is a valid method of measuring delinquent behavior.[39]

not been adequately represented in self-report studies.[38] Regardless of these problems, however, self-report studies are a good way to (1) estimate the dark figure of juvenile delinquency, (2) determine the extent of delinquency using more representative samples, and (3) examine factors that are believed to be related to delinquent behavior.

Cohort Studies

A **cohort study** is designed to examine specific subpopulations over a period of time. Several important cohort studies have been published since the 1970s, and these provide an important picture of delinquency within the juvenile population. Several of these studies were conducted by Marvin Wolfgang and his associates at the University of Pennsylvania. One of the best known of these studies was published in 1972 by Wolfgang, Robert Figlio, and Thorsten Sellin under the title *Delinquency in a Birth Cohort*, and it has had a profound influence on many people's thinking about delinquency.[40]

cohort study A study in which some specific group (e.g., those born during a certain year in a particular geographic location) is studied over a period of time.

Wolfgang, Figlio, and Sellin studied a cohort of 9,945 males who were born in Philadelphia in 1945 and resided there until 1963, when they became age 18 years. For each youth, the researchers collected information on formal police contacts from police records. In addition, they collected information from public, private, and parochial (Catholic) school records on academic achievement, IQ, types of schools attended, school and residential changes, and highest grade completed. They determined each youth's socioeconomic status based on the youth's place of residence.

The results of this study were important, not only because of the valuable insights provided, but also because of the policy responses the results suggested. Of the 9,945 cohort subjects, 35% had at least one formal contact with the police. Among White youths, 28.6% had some official police contact, while 50.2% of nonwhite cohort members had at least one official police contact. Among higher **socioeconomic status (SES)** youths, 26.5% had at least one official police contact, compared with 46% of the lower SES youths. Indeed, the results indicated that race and SES were the two most important variables related to official police contact, and that school-associated variables were related to delinquency. Consequently, the researchers noted that a number of variables related to delinquency seemed indicative of what they referred to as a "disadvantaged position." For example, they reported that the nonwhite youth who has official police contact is "likely to belong to the lower socioeconomic group, experience a greater number of school and residential moves (that is, be subject to the disrupting forces of intra city mobility more than the nondelinquent) and have the lowest average grade completed, the lowest achievement level, and the lowest IQ score."[41] Moreover, their analysis revealed that nonwhite and lower SES youths were more likely to be arrested than White and more affluent youths. They found that youths who had police contact at an early age often had lengthy delinquent careers. They, also, found evidence of offense escalation in instances when youths repeated an offense, which was greatest when youths had committed injury offenses.[42] In concluding their book, Wolfgang, Figlio, and Sellin noted that the juvenile justice system did an adequate job isolating the serious juvenile offender, but those who received more "punitive" responses (fines, institutional placement, and probation) were more likely to reoffend and to commit more serious offenses than youths who received less severe responses.[43]

socioeconomic status (SES) A person's position in a stratified society based on criteria such as place of residence, family income, educational background, and employment history.

The findings however, that received the most attention concerned what Wolfgang, Figlio, and Sellin termed the "chronic offender." The researchers divided the cohort into three groups: nonoffenders, one-time offenders, and **recidivists**. Of those youths who had an official police contact, 46% were one-time offenders and 54% were recidivists.

recidivist A person who continues to commit crimes despite efforts to rehabilitate or deter him or her.

Moreover, the recidivists were more likely to be in a "disadvantaged position" compared to other cohort members and to commit more serious offenses. The researchers categorized the recidivists into two additional groups: (1) those who had more than one police contact but less than five, whom they labeled "nonchronic recidivists," and (2) those who had five or more contacts, whom they labeled "chronic recidivists." Of the cohort, 1,235 youths (35.6% of all delinquents or 12.4% of the entire cohort) were nonchronic recidivists. Chronic recidivists numbered 627 youths (18% of all delinquents or 6% of the total cohort). Although the chronic offenders made up only 6% of the cohort, they accounted for a disproportionate share of the offenses attributed to the entire cohort, including serious offenses. For example, chronic offenders accounted for more than half (51.9%) of all offenses attributed to the cohort. Even more striking was their involvement in serious criminality. This 6% of the cohort accounted for 71% of the homicides, 73% of the rapes, 82% of the robberies, and 69% of the aggravated assaults attributed to the cohort.[44] Moreover, evidence of offense specialization was found for chronic offenders, although not for occasional delinquents, and this tendency increased as the number of offenses increased.[45]

Finally, Wolfgang and his associates found that nonwhite and low SES youths were more likely to be treated more harshly by juvenile justice agencies than White and higher SES youths, even when controlling for offense seriousness and prior record. Moreover, they discovered that juvenile justice responses to chronic offenders had little positive effect on their recidivism. Indeed, juvenile justice responses such as fines, probation, and incarceration did not reduce recidivism.

In a subsequent study published in 1987, Wolfgang, Figlio, and Terence Thornberry examined a 10% sample (*N* = 975) of the 1945 cohort from the time the cohort members were age 18 years until they turned age 30 years in order to examine the criminal involvement of members beyond their juvenile years.[46] The researchers supplemented Philadelphia police records with FBI data and conducted personal interviews with 58.2% of the follow-up sample in order to collect self-report and attitudinal data. The results of this study indicated that 47% of the sample had a recorded police contact by age 30 years. Also, as found in the original study, being nonwhite and being poor were strongly related to having an arrest record. The researchers also classified the offenders into three groups: (1) juvenile offenders, (2) adult offenders, and (3) persistent offenders (those who were both juvenile and adult offenders). Their analyses indicated that persistent offenders were more likely to come from disadvantaged backgrounds and have extensive delinquency histories. Similar to their original findings, most offenses committed by adults were non-Index offenses; however, unlike the earlier findings, adult offenses tended to become more serious over time. Indeed, the most serious offenses were more likely to be committed by persistent offenders.

Of course, it is possible that the findings of the 1945 cohort study conducted by Wolfgang and his associates were not representative of other cohorts. Consequently, replications of this study are needed to determine the extent to which its findings might be generalized to other populations. In 1977, this effort began, and its culmination was the publication of *Delinquency in Two Birth Cohorts* in 1985 and *Delinquency Careers in Two Birth Cohorts* in 1990.[47] In these publications, Wolfgang, Figlio, and Paul Tracy examined a cohort of all youths (males and females) who were born in Philadelphia in 1958 and resided there at least from their 10th to their 18th birthdays (a total of 27,160 youths). Again, data on police contacts were gathered as well as information from public and private school records. Finally, cohort members were assigned a socioeconomic status position derived from an analysis of census data. Analyses of the 1958 cohort data revealed that race and socioeconomic status were again related to delinquency, although the relationships were not as strong as they were in the 1945 cohort. Again, being in a disadvantaged position—indicated by being nonwhite, possessing low SES, experiencing residential instability, doing poorly in school, and failing to graduate from high school—was found to be related to delinquency. And again, chronic offenders were found to account for a disproportionate share of criminal involvement, including involvement in serious offending. In the 1958 cohort, chronic offenders made up 7.5% of the entire cohort and 23% of the delinquents. However, they accounted for 61% of all offenses attributed to the cohort, including 61% of the homicides, 75% of the rapes, 65% of the aggravated assaults, and 66% of the offenses that resulted in injury. Moreover, a greater percentage of serious offenses were committed by the 1958 cohort.[48]

Also, because information was collected on females, data on female contacts with the police were available. The data indicated that males were far more likely than females to have a police contact by age 18 years. Of cohort males, 32.8% had such contacts, compared with only 14.1% of females. Moreover, male contacts were more likely to involve serious offenses. Indeed, the overall male to female offense ratio was 4:1, but it was 9:1 for Crime Index offenses and 14:1 for violent Crime Index offenses. The data also indicated that nonwhite females were more likely to be arrested than White females and to be given more serious dispositions, such as probation and institutionalization by the juvenile court. Finally, an analysis of the dispositions of offenders found that, although differences existed in the processing of nonwhite offenders and White offenders, these differences were not as strongly related to race and socioeconomic status as they had been for youths born in 1945. Moreover, court responses to the juvenile offenders appeared to be more effective in the 1958 cohort.[49]

The disproportionate involvement in criminality of a small population of chronic offenders has been noted by other researchers as well. For instance, Lyle Shannon examined three cohorts of youths born in Racine, Wisconsin, in 1942, 1949, and 1955. In each case, he uncovered the existence of a chronic offender population.[50] According to Shannon, less than 25% of each cohort's male subjects had five or more nontraffic offenses. Nevertheless, this group accounted for 77–83% of the police contacts involving males. Moreover, 8–14% of the subjects in each cohort accounted for all of the serious felony offenses. According to Shannon, approximately 5% of each cohort was responsible for the majority of all offenses committed by the cohort and was also responsible for about 75% of all felonies committed by cohort members.

A cohort study conducted at Ohio State University called the Dangerous Offender Project focused on the violent juvenile offender. In this study, Donna Hamparian and her associates examined a cohort of 1,138 youths who were born in Franklin County (Columbus), Ohio, between 1956 and 1960 and who had been arrested for at least one violent offense before age 18 years.[51] Rather than discovering a large number of violence-prone youths, the researchers found that only about 2% of the juveniles in Columbus had been arrested for a violent offense. Moreover, although youths in the cohort averaged about four arrests during their juvenile years, few youths were involved in repetitive acts of violence.

Two federally funded studies in the 1980s produced findings similar to those reported by Hamparian. In one of these studies, over 340,000 case records submitted to the

National Juvenile Court Archive by courts in 12 states revealed that 6% of the juveniles referred to juvenile courts in those states for a law violation in 1984 were referred for a violent offense. Moreover, the researchers found that nonwhite and male juveniles were more likely to be referred to courts for violent offenses.[52] In the second study, staff at the National Center for Juvenile Justice examined the court careers of more than 69,000 juveniles dealt with by the Maricopa County (Phoenix) juvenile court. Like other studies, the researchers found that only 5% of the youths referred to the juvenile court were referred for a violent offense and less than 1% had more than one violent court referral in their career. However, although violent offenders were the least common type of juvenile offender, these juveniles were the most likely to return to court charged with a violent offense, and this finding was true for both males and females.[53]

Cohort studies have produced valuable information about youth crime and delinquent offenders. In fact, the attention they have focused on chronic and violent offenders has led to a number of policy proposals intended to better control this population, often through incarceration or transfer to adult courts for trial. Although important, such studies suffer from some weaknesses, notably their reliance on official data to measure delinquency. Indeed, research conducted by Dunford Franklyn and Delbert Elliott as part of the National Youth Survey found that a substantial amount of serious delinquency goes undetected by the police. In their study, only about 24% of serious career offenders reported that they had ever been arrested, which suggests that reliance on official data leads to an underestimation of the size of the chronic offender population.[54] Nevertheless, trends in self-report data provide no support for the contention that serious or violent juvenile crime has substantially increased over time.[55]

Developmental Studies

In recent years, other researchers have built upon and added to our understanding of the chronic and violent offender as well as our understanding of youths' patterns of offending over time. One such approach that aids our understanding of these phenomena is found in **developmental criminology**. Essentially, developmental criminology is concerned with changes in individual and group patterns in offending over time and understanding factors that influence these patterns. One developmental approach to understanding juvenile crime is being developed by Terrie Moffitt. According to Moffitt, a small group of individuals engage in various types of antisocial behavior (e.g., biting and hitting when they are age 4 years, shoplifting at age 10 years, selling drugs at age 16 years, robbery at age 22 years, and child abuse at age 30 years) throughout their lives. She refers to these individuals as *life-course-persistent offenders*. Because these antisocial behaviors begin early in the life course, Moffit argues that they are influenced by various neuropsychological factors that affect children's behavioral development, temperament, cognitive abilities, or all three. In addition, she maintains that these youths are also at greater risk of being exposed to environments that increase the likelihood of antisocial behaviors.[56] Thus, from Moffitt's point of view, life-course-persistent offending is influenced by a variety of neuropsychological and environmental factors that lead to the onset of antisocial behavior and help maintain the behavior over the life course.

developmental criminology
An area of criminology concerned with changes in individual and group patterns in offending over time, and understanding the factors that influence these patterns.

Most youths who engage in delinquency, however, are not life-course-persistent offenders. According to Moffitt, the great majority of youths who engage in delinquent behavior are *adolescence-limited offenders*. These individuals' antisocial behavior is short in duration and takes place during adolescence. Among this population, delinquent behavior can begin and end abruptly, the types and seriousness of offenses can change over time, and youths' behavior can vary in different situations. For example, these youths may engage in sporadic delinquent acts with friends, but conform to rules at home and school. According to Moffitt, the delinquency exhibited by adolescence-limited offenders is a product of social mimicry in which adolescence-limited youths model the behaviors of their life-course-persistent peers. For adolescence-limited youths, delinquent behavior can be seen as a way of exerting autonomy, taking risks, and demonstrating their maturity. As these youths mature and they take on jobs and other social responsibilities, however, their involvement in delinquency declines because antisocial behavior becomes less rewarding.[57]

Developmental studies such as those developed by Moffitt and others, like other studies that are intended to understand delinquency, suggest that certain types of interventions for juvenile offenders are necessary. From the developmental perspective, interventions with some youths should begin early, and they should be developmentally appropriate. We will return to these two issues later, when we examine correctional interventions in Chapter 11 and Chapter 12, as well as in Chapter 15 and Chapter 16 at the conclusion of the book.

Observational Studies

observational study A study in which the researcher observes and collects data on subjects in a field setting.

Chapter 1 discussed an important **observational study** done by William Chambliss that examined two groups of boys, the Saints, a group of middle-class males, and the Roughnecks, a group of lower-class males.[58] The study showed that the "outsider" status of the Roughnecks operated to reinforce a conception of them as boys headed for trouble, whereas the middle-class background of the Saints led authorities to see them as bright youths "sowing their wild oats." As a result, the Roughnecks were more likely than the Saints to be formally processed by the authorities. Such studies are important because they remind us that societal perceptions play a critical role in determining which youths become clients of the juvenile justice process.

Another important observational study was done by Herman and Julia Schwendinger and published in their highly regarded book *Adolescent Subcultures and Delinquency*.[59] The Schwendingers' work is based on more than 4 years of participant observation of peer groups in both working-class and middle-class communities in southern California (not to mention the many years they spent developing a theoretical understanding of delinquency). Through their research, the Schwendingers were able to elaborate on what they argued was a critical shortcoming of many theories of delinquency—the inability to explain the substantial amount of middle-class delinquency documented by self-report studies as well as the extensive lower-class delinquency more likely to be reflected in official data.

Rather than relying on social class as a starting point for understanding delinquency, the Schwendingers focus on adolescent peer groups (see Box 2-1). According to the Schwendingers, the particular form of capitalist development that occurred in the United States has produced a large variety of adolescent social types, peer networks, and status groups that compose what is often referred to as "youth culture." As the Schwendingers note, youth culture actually comprises various subcultures and peer networks that cut across social class lines. These subcultures are reflected in various "social types" that have their own distinct designations, such as "intellectuals," "hodads," "greasers," "homeboys," "socialites," "preppies," "athletes," and others, which are distinguished by distinctive dress and linguistic patterns that are characteristic of these groups. Moreover, they note that some of these adolescent peer networks encourage delinquency among their members and can be categorized as falling into different "stradoms" comprising distinctive social types. The Schwendingers' research reveals that youths who belong to these stradoms are more likely to engage in delinquent behaviors than youths who do not belong, regardless

BOX 2-1 Interview: Herman and Julia Schwendinger

Q: Both of you have backgrounds in social work. In what ways has this influenced your study of delinquency?

A: From the very beginning of our social work careers, we worked with adolescents and delinquent gangs in high-crime areas. Herman's own childhood and adolescence involved gang activity in a poverty-stricken New York City community, and his earliest assignments as a social group worker included field work with street gangs. In her first social work position, Julia also worked with children and teens in a poor high-crime community. Our research evolved from questions raised by our social work experience. In fact, Herman, for instance, originally attended UCLA as a part-time student to acquire knowledge that he could use as an administrator of a youth program that served delinquents as well as nondelinquents.

Q: What led to your interest in studying crime and delinquency?

A: Herman's initial UCLA experience highlighted the lack of fundamental information about delinquent groups, and his interest in developing such knowledge was strongly encouraged by sympathetic faculty. Julia's doctoral dissertation on rape went hand in hand with her work in founding the first antirape group in the country.

Q: I know that Herman has a background in math and quantitative analysis, but he has also spent a lot of time observing youth groups. How important are qualitative methods in understanding youth behaviors?

A: Qualitative work has been crucial for developing some of our most important theoretical ideas. Also, our recent work indicates that ethnographic observations combined with quantitative methods (e.g., field experiments and network analysis) can produce much more rapid advances than traditional surveys and interviews for the study of delinquency.

Q: How has this observational work informed our understanding of youth behavior?

A: We believe that our work will have considerable impact when a new generation of criminologists abandons armchair theorizing and gets out in the field. Unfortunately, today too many are stuck with "social control" and "delinquency subculture" theories that emphasize personality abnormalities and ignore the variety of adolescent subcultures in our society. Subcultures composed of jocks, preppies, homeboys, metalbangers, freaks, brains, etc., exhibit different forms and degrees of misconduct. But even the most delinquent subculture cannot be attributed to pathological traits of leaders, early childhood disorders, or the social incapacities of the members at large.

Q: What are some of the problems that you encountered in doing observational research?

A: Observational research has problems that are never encountered by researchers who ride around in patrol cars or spend time devising survey questionnaires. One has to deal with boredom from hanging around at all hours listening to teen gossip and small talk; creating trusting relationships and guaranteeing trust; staving off manipulation and being used by conflicting groups or individuals; trying to deal with ethical questions due to knowledge about crimes that have occurred and that will occur; and developing an awareness of and techniques to assure personal safety.

Q: What else would you like to tell students interested in field work?

A: Despite the myriad problems related to participant observation, we had a lot of fun. We formed warm friendships (and sometimes became "role models"), which have been maintained to this day with some of our "subjects" and their subsequent families.

Herman Schwendinger. Psychology, CCNY; MSW, Columbia; PhD, Sociology, UCLA.

Julia Schwendinger. Sociology, Queens College; MSW, Columbia; PhD, Criminology, University of California, Berkeley.

Source: H. Schwendinger (personal communication, June 15, 2004).

of their social class. For example, they note that "intellectuals" are the least delinquent youths in a peer society, whereas members of "streetcorner" stradoms are the most delinquent. They also note that as stradoms mature, the types of delinquency committed by their members change.

Participant observation is a very time-consuming form of research, and participant observation studies, like other studies, can have weaknesses. It is possible, for example, that research results gleaned from observing youths in one locality may not be easily generalized to youths living somewhere else or during a different period. However, because participant observers spend a considerable amount of time with their subjects, they are in a position to learn things about the subjects' behaviors and attitudes that cannot be easily learned through other research methods.

Victimization Studies

Another valuable source of information about juvenile offending is the **victimization study**, which involves the collection of data from crime victims through victim surveys or interviews. These studies allow researchers to gather a range of information about the experiences of crime victims, such as the level of victimization within the population, relationships between offenders and victims, circumstances surrounding the victimization (e.g., where the victimization occurred and the victim's possible contribution to the victimization), and characteristics of victims and offenders. Victimization studies also can help researchers understand the dark figure of crime. The most important source of data on youth victimizations comes from the National Crime Victimization Survey (NCVS), a large-scale national survey conducted by the Bureau of Justice Statistics, a branch of the U.S. Department of Justice, and the U.S. Census Bureau.

victimization study
A study that focuses on the crime victimization experiences of subjects.

Although the NCVS is not a good source of data for most crimes committed by juveniles, it can be used to assess offending rates for certain violent offenses because the data for these offenses contain estimates of the offender's age. As might be expected, levels of offending based on victimization data are considerably higher than those found in UCR data. In a study of trends in violent offending based on NCVS data, James Lynch found that offending rate estimates calculated from NCVS data were 9 to 15 times greater than those based on the UCR. However, he also found that the general trends in violent offending seen in NCVS data are similar to those revealed in the UCR.[60] Like the UCR data, NCVS data indicate that there has been a decline in violent offending among juveniles since the mid-1990s.[61]

Although victimization data can provide some important insights on levels of juvenile offending, they are even more valuable in aiding our understanding of youths' victimization experiences. Overall, NCVS data indicate that violent and property crime victimization has been declining since the early 1990s. However, this data also indicates

MYTH VS REALITY

Teens Have High Victimization Rates

Myth—Many people feel that the elderly are the most common victims of crime.

Reality—Teenagers are more likely to be crime victims than the elderly. For example, the average annual rate of violent crimes experienced by persons age 12–15 years in 2004 and 2005 was 46.9 per 1,000 compared to only 2.3 per 1,000 for persons age 65 years and older.[63]

FYI

School Victimization

Violent deaths at school are rare events. During the 2006–2007 school year, there were 27 homicides of students at school. This represented about one homicide for every 1.6 million students. In contrast, nonfatal victimizations of youths are common in schools. For example, students from age 12 to 18 years were more likely to be the victims of theft at school than away from school. Nevertheless, the percentage of youths reporting victimization at school declined from 10% in 1995 to 4% in 2007.[67]

that youths have high rates of victimization compared to other groups. For example, between 1993 and 2003, persons age 12 to 17 years were more than 2.5 times more likely than adults to be victims of violent offenses. In addition, some groups of teenagers (i.e., males, older teens, and those living in urban areas) have higher rates of violent crime victimization than others.[62]

The NCVS also provides information about the relationship between offenders and victims as well as information on the location and time of offenses. Most teenagers are victimized by people they are at least acquainted with, and a sizable percentage of victimizations involve offenders who are well-known to the victim; for instance, between 1993 and 2003 among teens age 12 to 14 years, the victim knew the perpetrator in 61% of reported victimizations and for youths age 15 to 17 years, the victim knew the offender 47% of the time.[64] Moreover, a sizable amount of teenage victimization occurs at school. In 2006, students between age 12 and 18 years reported 1.7 million nonfatal victimizations at school.[65] However, the peak times for violent victimization of youths is from 3:00 p.m. to 6:00 p.m. on school days, although it tends to be later (between noon and midnight) on nonschool days and during the summer.[66]

Although victimization studies are a rich source of information about crime, they are limited in their ability to describe the extent of delinquency for several reasons. Many crimes involve no contact between victim and offender; consequently, the age of the offender is unknown. Also, when there is personal contact, the victim may be unable to provide an accurate description of the offender or even determine the offender's age. Some youths simply look older than their actual age, whereas some adults look younger than they are. Despite these limitations, victim data constitute an important source of juvenile crime data. They can be used to estimate offending rates among juveniles for some violent offenses, and they provide valuable information about youths' victimization experiences.

■ Legal Issues

The Use of Arrest Versus Conviction Data

The presumption of innocence is an accepted common law principle in the United States. Any person arrested must be formally charged, arraigned, tried, and convicted beyond a reasonable doubt before he or she can be considered guilty. Therefore, arrest data, such as the information found in the UCR and other formal and informal data, may not accurately reflect the extent of adult or juvenile crime. Being arrested for an alleged offense does not mean one is guilty. Nevertheless, arrest data are often used as measures of juvenile crime because juvenile courts have varying rules of confidentiality regarding juvenile conviction records, which makes such records difficult to examine in many jurisdictions. Furthermore, police departments generally have fewer confidentiality requirements, which make arrest data more accessible and contributes to their use as a measure of juvenile crime.

Confidentiality for Juveniles

As long as there have been juvenile courts, there has been debate concerning the confidentiality of juvenile court records and proceedings. As with many legal debates, both sides of the issue have good points to make. On the side favoring the confidentiality of juvenile proceedings and records is the argument that the fundamental reason for the establishment of juvenile courts was to allow the mistakes and indiscretions of children to be dealt with away from public scrutiny so that the emphasis could be on rehabilitation, not stigmatization.

Advocates on the other side of the issue contend that allowing juvenile courts to operate outside the scrutiny of the public has led to abuses that have harmed more children than confidentiality has helped. They also point out that society has a legitimate interest in protecting itself from dangerous individuals, be they juveniles or adults, and no dangerous person should be allowed to use confidentiality to avoid public scrutiny or appropriate consequences. Indeed, this argument has gained more weight as the public has grown increasingly concerned with the rise in serious juvenile crime.

Another unintended consequence of confidentiality and closed courtrooms is that it prevents the public from seeing the difficult cases that the courts handle and the societal problems represented in juvenile court cases. Not only does this lead people to downplay the seriousness and legitimacy of juvenile courts, but it also allows some people to ignore social trends and problems that are apparent to court personnel. Unfortunately, "out of sight, out of mind" seems to apply to much of what happens in the juvenile justice process. For an excellent discussion of these public policy conflicts and balance, see *People v. Smith*, 437 Mich 293, 470 NW2d 70, 78 (1991), in which it is stated, "The purpose of the court rule, and of similar rules or statutes in other jurisdictions, is to prevent a juvenile record from becoming an obstacle to educational, social, or employment opportunities. When, however, a juvenile offender appears in court again as an adult, his juvenile offense record may be considered in imposing sentence. The law contemplates a differentiation in sentencing between first-time offenders and recidivists, juvenile or adult."[68]

Chapter Summary

Taken together, official and unofficial sources of delinquency data provide valuable insights regarding delinquency and responses to delinquency within the United States. These data sources indicate that delinquency is widespread within our society and consists of a wide range of behaviors, from serious crimes against persons to failure to obey one's parents. They also indicate that far more males than females engage in delinquent behavior and that male dominance in delinquency increases with the seriousness of the offense. Yet most delinquency, whether committed by males or females, involves minor offenses and does not come to the attention of the authorities. Further, none of the data sources suggest that juvenile crime has gotten appreciably worse over the last 20 years. Taken together, various data sources indicate that, overall, juvenile crime has remained stable or declined in recent years. Still, there is some indication that rates of violent juvenile crime in the United States may be higher than in many other developed countries, although levels of property crime may be higher in some other developed countries than in the United States.

Clearly, official and unofficial data sources provide a more comprehensive view of juvenile crime than one gets from the media, which present a distorted view of juvenile crime. However, no single data source provides a comprehensive picture of delinquency. Moreover, each method of measuring delinquency and juvenile justice processing possesses some weaknesses. Consequently, caution needs to be exercised by those who use each of the data sources in formulating policy responses to the delinquency problem.

Which data source is best and which method of measuring delinquency or the processing of juvenile offenders should one rely on? Perhaps the best answer to this question is to say that it depends on the question to be addressed. For example, if the object is to discover the number of juvenile arrests and the characteristics of those arrested, the UCR arrest data would be most helpful. On the other hand, someone interested in understanding the dark figure of delinquency and exploring a variety of factors related to delinquency would find self-report data to be more helpful. Indeed, the various sources of data on juvenile crime are probably best viewed as complementary—each tells something about the phenomena of juvenile crime and the processing of juvenile offenders.

Key Concepts

arrest rate: The number of arrests adjusted for the size of the population; often reported as arrests per 100,000 youths ages 10–17 years in the population.

cohort study: A study in which some specific group (e.g., those born during a certain year in a particular geographic location) is studied over a period of time.

dark figure of juvenile crime: The amount of juvenile crime that is not reported.

developmental criminology: An area of criminology concerned with changes in individual and group patterns in offending over time, and understanding the factors that influence these patterns.

etiology: The study of causation.

observational study: A study in which the researcher observes and collects data on subjects in a field setting.

official data: Information on juvenile delinquency collected by formal juvenile justice agencies, such as police agencies, juvenile courts, and juvenile detention and correctional facilities.

panel study: A study that involves the examination of the same select group over time.

police discretion: The authority of police to make their own judgments about which crimes or delinquent acts are subject to investigation and which juveniles are subject to arrest.

recidivist: A person who continues to commit crimes despite efforts to rehabilitate or deter him or her.

self-report study: A study in which subjects are asked to report their involvement in illegal behavior.

socioeconomic status (SES): A person's position in a stratified society based on criteria such as place of residence, family income, educational background, and employment history.

status offense: An act that is considered a crime or legal violation when committed by a juvenile but not by an adult (e.g., running away from home, incorrigible behavior, failure to attend school, and failure to obey school rules).

Uniform Crime Report (UCR): The most comprehensive compilation of known crimes and arrests. The report is published each year by the Federal Bureau of Investigation under the title *Crime in the United States.*

unofficial data: Information collected by researchers not connected with formal juvenile justice agencies (i.e., police agencies, juvenile courts, and correctional institutions).

victimization study: A study that focuses on the crime victimization experiences of subjects.

Review Questions

1. What is the distinction between official and unofficial sources of data on delinquency?
2. What types of data are contained in the UCR?
3. How much of a problem is juvenile crime according to UCR arrest data?
4. Why is it important to consider arrest rate trends when attempting to examine juvenile crime trends?
5. What does the examination of juvenile arrest rate trends tell us about juvenile crime over the last 20 years?
6. What problems are associated with using UCR arrest data as an indicator of juvenile crime?
7. What are the primary uses of official data?
8. What should one consider when comparing juvenile crime data from different countries?
9. Does the United States have more or less juvenile crime than other countries?
10. What is the picture of delinquency presented by self-report studies?
11. What are potential problems associated with self-report studies?
12. What are the major findings of cohort studies of delinquency?
13. What insights have developmental criminology provided about juvenile offending and its treatment?
14. What are the major findings of the observational studies of delinquency conducted by William Chambliss and by Herman and Julia Schwendinger?
15. In what ways are observational studies superior to official data sources for learning about delinquency?
16. What do victimization studies tell us about juvenile crime and teenage victimization?

Additional Readings

Chambliss, W. J. (1973). The Saints and the Roughnecks. *Society, 11*, 341–355.

Elliott, D. S. & Ageton, S. S. (1980). Reconciling race and class differences in self-reported and official estimates of delinquency. *American Sociological Review, 45*, 95–110.

Howell, J. C. (2003). *Preventing and reducing juvenile delinquency: A comprehensive framework.* Thousand Oaks, CA: Sage.

Lauritsen, J. L. (2003). How families and communities influence youth victimization. *Juvenile Justice Bulletin.* Washington, DC: Office of Juvenile Justice and Delinquency Prevention.

Lynch, J. P. (2002). Trends in juvenile violent offending: An analysis of victim survey data. *Juvenile Justice Bulletin*. Washington, DC: Office of Juvenile Justice and Delinquency Prevention.

Moffitt, T. E. (1993). Adolescence-limited and life-course persistent antisocial behavior: A developmental taxonomy. *Psychological Review, 100*: 674–701.

Schwendinger, H. & Schwendinger, J. S. (1985). *Adolescent subcultures and delinquency*. New York: Praeger Publishers.

Notes

1. United States Department of Justice, Federal Bureau of Investigation. (September 2008). *Crime in the United States 2007*. Retrieved July 14, 2009, from http://www.fbi.gov/ucr/07cius.htm.
2. This percentage includes arrests for carrying and possessing weapons. If these offenses are excluded from the calculations, the percentage of arrests for nonviolent offenses increases to 84.3%.
3. United States Department of Justice, Federal Bureau of Investigation, September 2008.
4. United States Department of Justice, Federal Bureau of Investigation, September 2008.
5. Puzzanchera, C., Sladky, A., & Kang, W. (2008). Easy access to juvenile populations: 1990–2007. Retrieved June 2, 2009, from http://www.ojjdp.ncjrs.gov/ojstatbb/ezapop/.
6. Federal Bureau of Investigation. (2004) *UCR uniform crime reporting handbook*. Washington, DC: Federal Bureau of Investigation.
7. Federal Bureau of Investigation. (April 19, 2007). *National incident-based reporting system (NIBRS)*. Retrieved August 8, 2007, from http://www.fbi.gov/ucr/faqs.htm.
8. Federal Bureau of Investigation. (1995–2007). *Crime in the United States*. Retrieved August 8, 2007, from http://www.fbi.gov/ucr/ucr.htm#cius.
9. Steffensmeir, D. (1987). Is the crime rate really falling? An 'aging' U.S. population and its effect on the nation's crime rate, 1980–1984. *Journal of Research in Crime and Delinquency, 24*, 23–48.
10. Cook, P. J. & Laub, J. H. (1986). The (surprising) stability of youth crime rates. *Journal of Quantitative Criminology, 2*, 265–277.

    Osgood, D., O'Malley, P. M., Bachman, J. G., Johnston, L. D. (1989). Time trends and age trends in arrests and self-reported illegal behavior. *Criminology, 27*, 389–417.
11. Snyder, H. N. & Sickmund, M. (2006). *Juvenile offenders and victims: 2006 national report*. Washington, DC: Office of Juvenile Justice and Delinquency Prevention.
12. United States Department of Justice, Federal Bureau of Investigation, September 2008.
13. Puzzanchera, Sladky, & Kang, 2008.
14. United States Department of Justice, Federal Bureau of Investigation, September 2008.

15. O'Brien, R. M. (1995). Crime and victimization data. In J. F. Sheley (Ed.), *Criminology: A contemporary handbook* (2nd ed.). Belmont, CA: Wadsworth.

16. Huizinga, D. & Elliott, D. S. (1987). Juvenile offenders: Prevalence, offender incidence, and arrest rates by race. *Crime and Delinquency, 33,* 206–223.

17. Elliott, D. S. (1995). *Lies, damn lies, and arrest statistics.* Paper presented at the annual meeting of the American Society of Criminology, Boston.

18. O'Brien, 1995.

19. O'Brien, 1995.

20. Snyder & Sickmund, 2006.

21. United States Department of Justice, Federal Bureau of Investigation. (2003). *Crime in the United States 2002.* Retrieved February 23, 2004 from http://www.fbi.gov/ucr/cius_02/pdf/3sectionthree.pdf.

22. United States Department of Justice, Federal Bureau of Investigation, September 2008.

23. Snyder & Sickmund, 2006.

24. Federal Bureau of Investigation. 1995–2007.

25. National Research Council & Institute of Medicine. (2001). *Juvenile crime, juvenile justice.* (J. McCord, C. Widom, & N. Crowell, Eds.). Washington, DC: National Academy Press.

26. Ren, X., & Friday, P. C. (2006). Different legal traditions and patterns of delinquency. In P. C. Friday & X. Ren (Eds.), *Delinquency and juvenile justice systems in the non-western world.* Monsey, NY: Criminal Justice Press.

27. National Research Council & Institute of Medicine, 2001.

28. Elrod, P. & Yokoyama, M. (2006). Juvenile justice in Japan. In P. C. Friday & X. Ren (Eds.), *Delinquency and juvenile justice systems in the non-western world.* Monsey, NY: Criminal Justice Press.

29. National Research Council & Institute of Medicine, 2001.

30. Johnston, L. D., O'Malley, P. M., & Bachman, J. (1994). *Monitoring the future.* Ann Arbor, MI: University of Michigan.

Gold, M. (1966). Undetected delinquent behavior. *Journal of Research in Crime and Delinquency, 3,* 27–46.

Gold, M. (1970). *Delinquent behavior in an American city.* Belmont, CA: Brooks/Cole.

Erickson, M. & Empey, L. (1963) Court records, undetected delinquency, and decision-making. *Journal of Criminal Law, Criminology, and Police Science, 54,* 446–469.

Short, J., & Nye, F. I. (1958). Extent of unrecorded delinquency. *Journal of Criminal Law, Criminology, and Police Science, 49,* 296–302.

31. Pastore, A. L., & Maguire, K. Maguire (Eds.). *Sourcebook of criminal justice statistics.* Retrieved June 3, 2009, from http://www.albany.edu/sourcebook/pdf/t343.pdf.

32. Eaton, D. K., Kann, L., Kinchen, S., Shanklin, S., Ross, J., Hawkins, J., et al. (June 6, 2008). Youth risk behavior surveillance—United States, 2007. *MMWR Surveillance Summaries, 57.* Retrieved July 28, 2009, from http://www.cdc.gov/HealthyYouth/yrbs/pdf/yrbss07_mmwr.pdf.

33. Osgood, O'Malley, Bachman, & Johnston, 1989.

34. Elliott, D. S., Huizinga, D., Knowles, B., & Canton, R. (1983). The prevalence and incidence of delinquent behavior: 1976–1980. *National Youth Survey Report. 26.* Boulder, CO: Behavioral Research Institute.

35. Elliott, D. S. & Ageton, S. S. (1980). Reconciling race and class differences in self-reported and official estimates of delinquency. *American Sociological Review, 45,* 95–110.

36. Elliott, D. S. & Huizinga, D. (1983). Social class and delinquent behavior in a national youth panel: 1976–1980. *Criminology, 21,* 149–177.

37. Chesney-Lind, M. & Shelden, R. G. (2004) *Girls, delinquency and juvenile justice* (3rd ed.). Belmont, CA: Thompson/Wadsworth.

38. Cernkovich, S., Giordano, P., & Pugh, M. (1985). Chronic offenders: The missing cases in self-report delinquency research. *Journal of Criminal Law and Criminology, 76,* 705–732.

39. Hindelang, M., Hirschi, T., & Weis, J. (1981). *Measuring delinquency.* Beverly Hills, CA: Sage Publications.

40. Wolfgang, M., Figlio, R., & Sellin, T. (1972). *Delinquency in a birth cohort.* Chicago: University of Chicago Press.

41. Tracy, P. E., Wolfgang, M., & Figlio, R. (1985). *Delinquency in two birth cohorts.* Washington, DC: U.S. Department of Justice.

Wolfgang, Figlio, & Sellin, 1972, p. 246.

42. Tracy, P. E. , Wolfgang, M., & Figlio, R. (1990) *Delinquency careers in two birth cohorts.* New York: Plenum Press.

43. Wolfgang, Figlio, & Sellin, 1972, p. 252.

44. Wolfgang, Figlio, & Sellin, 1972.

45. Tracy, Wolfgang, & Figlio, 1990.

46. Wolfgang, M. E., Thornberry, T., & Figlio, R. (1987). *From boy to man, from delinquency to crime.* Chicago: University of Chicago Press.

47. Tracy, Wolfgang, & Figlio, 1985. Tracy, Wolfgang, & Figlio, 1990.

48. Tracy, Wolfgang, & Figlio, 1985.

49. Tracy, Wolfgang, & Figlio, 1990.

50. Shannon, L. (1982). *Assessing the relationship of adult criminal careers to juvenile careers: A summary.* Washington, DC: U.S. Department of Justice.

51. Hamparian, D. M., Dinitz, S., & Schuster, R. (1978). *The violent few.* Lexington, MA: Lexington Books.

52. Office of Juvenile Justice and Delinquency Prevention. (1989). The juvenile court's response to violent Crime. *Office of Juvenile Justice and Delinquency Prevention Update on Statistics.* Washington, DC: U.S. Department of Justice.

53. Snyder, H., (1988). *Court careers of juvenile offenders.* Washington, DC: Office of Juvenile Justice and Delinquency Prevention.

54. Franklyn, D., & Elliott, D. (1984). Identifying career offenders using self-report data. *Journal of Research in Crime and Delinquency, 21,* 57–86.

55. Howell, J. C. (2003). *Preventing and reducing juvenile delinquency: A comprehensive framework.* Thousand Oaks, CA: Sage. This resource provides a review of research on this topic.

56. Moffitt, T. E. (1993). Adolescent-limited and life-course persistent antisocial behavior: A developmental taxonomy. *Psychological Review, 100*, 674–701.

57. Moffitt, 1993.

58. Chambliss, W. J. (1973). The saints and the roughnecks. *Society, 11*, 341–355.

59. Schwendinger, H. & Schwendinger, J. S. (1985). *Adolescent subcultures and delinquency.* New York: Praeger Publishers.

60. Lynch, J. P. (2002). Trends in juvenile violent offending: An analysis of victim survey data. *Juvenile Justice Bulletin.* Washington, DC: Office of Juvenile Justice and Delinquency Prevention.

61. Lynch, 2002. Snyder & Sickmund, 2006.

62. Baum, K. (2005). Juvenile victimization and offending: 1993–2003. *Bureau of Justice Statistics Special Report.* Washington, DC: U.S. Department of Justice.

63. Catalano, S. M. (2006) Criminal victimization, 2005. *Bureau of Justice Statistics Bulletin.* Washington, DC: U.S. Department of Justice.

64. Snyder & Sickmund, 2006.

65. Dinkes, R., Kemp, J., Baum, K., & Snyder, T. D., (2009). *Indicators of school crime and safety: 2008.* Washington, DC: U.S. Departments of Education and Justice.

66. Snyder & Sickmund, 2006.

67. Dinkes, et al. 2009.

68. *People v. Smith*, 437 Mich 293, 470 NW2d 70, 78 (1991).

Theory and Research: The Social Context of Juvenile Delinquency and Juvenile Justice

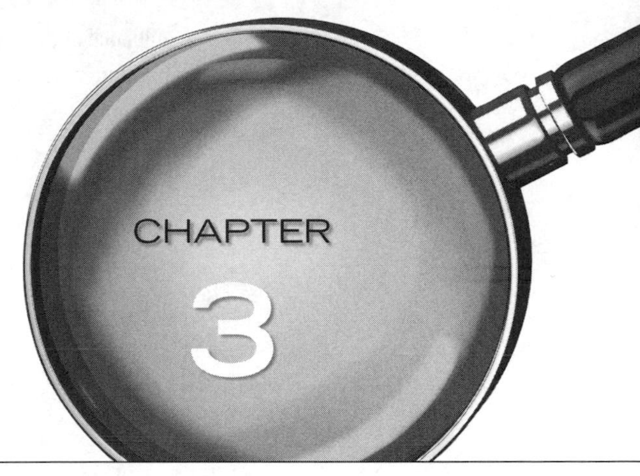

▸ ▸ CHAPTER OBJECTIVES

After studying this chapter, you should be able to

- Explain how the political economy of the United States influences delinquency
- Describe the key role that the family plays in the socialization of children and the relationship between child socialization and delinquency
- Describe the importance of parental economic or workplace power, parental criminality, and family conflict, and explain how each of these factors may influence delinquency
- Explain the connection between parental abuse, neglect of children, and delinquency
- Explain the connection between school performance and delinquency
- Describe how schools and societal factors contribute to student failure
- Describe the phenomenon of "social disorganization" and how it can encourage the creation of a delinquent "culture" in neighborhoods or urban areas
- Explain how youth peer groups and gangs differ from one another and how youths' peer associations encourage or discourage delinquency
- Describe how the development, management, and distribution of political and economic resources influence both youths and the operation of the social institutions that make up the juvenile justice process

▸ ▸ CHAPTER OUTLINE

- Introduction
- The Political and Economic Context of Juvenile Delinquency and Juvenile Justice
- The Family and Delinquency
- Schools and Delinquency
- The Community and Delinquency
- Peer Associations and Delinquency
- The Social Context of Delinquency Theories
- The Social Context of Juvenile Justice
- The Influence of the Political Economy on the Practice of Juvenile Justice

Introduction

Delinquency and the practice of juvenile justice occur, not in a vacuum, but in a social context. This does not mean that individual factors, such as biological makeup and psychological functioning, do not play a role in delinquency or the operation of juvenile justice. Nor does it imply that individuals do not make choices, often conscious choices, to engage in delinquent behaviors. However, it recognizes that individuals, and the choices they make, cannot be adequately understood without considering the social contexts within which they live and act.

As noted in Chapter 1, the social context helps shape our views of juvenile crime and the operation of juvenile justice through the portrayal of delinquency and juvenile justice in the media. In fact, much of what most people (including the makers of juvenile crime policy) know about juvenile crime and juvenile justice comes from the media. However, the social context of juvenile delinquency and juvenile justice comprises more than the media. In the United States, juvenile delinquency and juvenile justice are influenced by a variety of factors found within the political economy of the United States and within communities, families, schools, peer groups, and other important socializing institutions. How political and economic arrangements and socializing institutions, such as families, communities, and schools, influence delinquency is a primary focus of theory and research within the field of criminology. Indeed, courses in criminology, juvenile delinquency, and criminological theory focus attention on how factors such as economic inequality, school failure, residence in a high-crime neighborhood, child discipline practices, child abuse, association with criminally involved peers, and many other factors are related to delinquency. Explanations of criminal behavior that refer to such factors comprise a significant body of criminological theory. Moreover, theories are important, as Stephen Pfohl has noted, because they "provide us with an image of what something is and how we might best act toward it."[1] The development of good theories of delinquency, then, could be used to develop policies that reduce delinquency.

Although the purpose of this text is not to examine the many theoretical explanations of delinquency, it should be kept in mind that delinquency is the product of a complex set of factors and that juvenile justice exists within a complex social context that has real and profound effects on the practice of juvenile justice. Importantly, juvenile justice practitioners often spend considerable energy attempting to help their clients develop more law-abiding responses to the social context within which their clients live. Many practitioners also spend considerable energy attempting to manipulate or alter various aspects of the social context in ways that will lessen the likelihood of continued delinquent behavior.[2]

This chapter discusses some of the important components of the overall social context that influence juvenile crime and the operation of juvenile justice. It begins by considering the effects of economic and political factors on delinquency. Next, it explores a variety of important social institutions, such as families, neighborhoods, schools, and peer associations, and their relationship to youths' delinquent behavior. It then examines how the social context influences the development of delinquency theory. It closes by looking at the ways in which the social context influences the practice of juvenile justice.

The Political and Economic Context of Juvenile Delinquency and Juvenile Justice

The term **political economy** refers to the basic economic and political organization of society. The political economy determines the ways in which economic and political resources are developed, managed, and distributed. The development, management, and distribution of these resources are important because they directly affect the ability of basic socializing institutions, such as families and schools, to meet people's needs. For example, the distribution of job opportunities and the pay earned by workers influence the amount of economic resources possessed by families and the quality of life experienced by family members. As the National Commission on Children noted in a 1993 report, "Economic security is fundamental to children's well-being. Children need material support to have a better chance to grow up healthy, succeed in school, and become capable and caring adults."[3] Furthermore, decisions regarding the development, management, and distribution of economic resources do not occur in a social vacuum; rather, they are products of a political process in which different groups with conflicting concerns and varying degrees of power try to protect and further their interests.

political economy The basic political and economic organization of society, including the setting of political priorities and the management and distribution of economic resources. People's economic status largely determines the opportunities that are available to them, their feeling of political empowerment, their ability to access the political system, and their ability to influence political and economic decision making.

The Political Context

The forms of political organization found in most modern societies can be separated into two main categories: democratic forms and authoritarian forms. The United States, although relatively democratic, is not a pure democracy in which every person has an equal voice in decision making and not every political decision is intended to maximize the common good. Instead, governmental decision making is regularly influenced by powerful special interests that seek to maximize benefits for themselves at the expense of others. These **special interest groups** consist of individuals, families, corporations, unions, and various other organizations, and they use a variety of means to obtain tax breaks, favors, subsidies, and favorable rulings from congressional committees, regulatory agencies, and executive agencies.[4]

special interest group An organization consisting of persons, corporations, unions, and so on, that have common political, and often economic, purposes and goals. These individuals or groups come together to use their influence, monetarily and politically, to persuade politicians and other persons in power to act favorably toward them.

Although the government can and does act in ways that benefit the majority of Americans, it is not always neutral.[5] When government entities decide to regulate certain activities and not others, when they enforce certain laws and not others, when they select which resources to develop and how to manage and distribute these resources, their decisions and actions invariably favor some groups and interests over others. Whether to provide or withdraw government support for abortion, whether to increase or reduce regulations concerning gun ownership, whether to reduce or eliminate support for summer jobs programs for youths, or whether to make it more difficult for low-income youths to receive loans for higher education while providing a range of tax breaks and subsidies to large corporations are hardly neutral decisions. In determining what should be done, government is generally biased toward policies that benefit the wealthy, especially the big business community.[6]

power The ability to influence political and economic decisions and actions; the ability to force, coerce, or influence a person's actions or thoughts.

The ability to influence political decision making is a form of power. **Power** is the "capacity of some persons to produce intended and foreseen effects on others."[7] However, power is not equally distributed across individuals and groups in our society. Instead, it is concentrated among those who possess substantial economic resources—resources that can be used to directly and indirectly influence government decision making. The expenditure of those resources to influence government decision making is seen as a reasonable cost because the decisions reached partially determine the development, management, and distribution of resources within society and typically benefit those who possess substantial economic resources.

From a juvenile delinquency and juvenile justice standpoint, the question we must ask ourselves is this: To what extent do present political arrangements contribute to the well-being of children and to the healthy functioning of other socializing institutions, such as families and communities, that have an important impact on children's lives? We also must ask, to what extent do present political arrangements encourage delinquency, either directly or indirectly?

Decisions to invest our resources in ways that encourage the healthy development of communities, schools, families, and children are political decisions. This does not mean that families, communities, schools, and other organizations do not share responsibility for the welfare of children and for the problem of delinquency. Community decision making and school decision making are inherently political activities. However, it must be kept in mind that government decision making does have important ramifications for the quality of life experienced by many Americans and for the level of delinquency that exists in our society. The following section examines one important outcome of decision making in the United States, namely, economic inequality, as well as the relationship of economic inequality to delinquency.

capitalism An economic system based upon three fundamental principles: (1) private ownership of property, (2) competition between economic interests, and (3) personal profit as a reward for economic risk and effort.

socialism A political and economic system that is based on democratic decision making, equality of opportunity for all, collective decision making designed to further the interests of the entire community, public ownership of the means of production, and economic and social planning.

The Economic Context

There are two primary forms of economic organization in the modern world: **capitalism** and **socialism**. Although pure capitalism and pure socialism do not exist, the economic organization of each country tends toward one form or the other. In the United States, the predominant form of economic organization is capitalism. This form of economic organization is characterized by three basic principles: private ownership of personal property, personal profit, and competition. In addition, according to those who favor pure capitalism, government interference in economic life should be kept to a minimum. The basic view of capitalism's supporters is that "the profit motive, private ownership, and competition will achieve the greatest good for the greatest number in the form of individual self-fulfillment and the general material progress of society."[8]

Clearly, many people in the United States are able to achieve considerable prosperity. Yet, it is also true that many others have relatively few material resources. In 2006, it was estimated that 12.3% of the U.S. population (36.5 million people) were officially poor. Most of these poor people were white, and the greatest concentrations of poverty were found in central cities. Furthermore, 2.9 million of those persons classified as poor worked full time and another 6.3 million worked at least part time.[9] However, other estimates of the number of poor persons—estimates not based on the official poverty line—suggest that the number of people who lack an adequate income to meet their basic needs may be as high as 80 million persons.[10] Even using official data, the number of persons considered *near* poverty was estimated to be 49.7 million in 2006.[11]

FYI

Official Poverty

The poverty line is defined as three times the amount of income necessary for a minimally nutritional diet. In 2006, the official poverty line for a family of four with two children younger than 18 years was $20,444.[12] Many economists and others argue that the official poverty line underestimates the number of poor persons because those earning slightly above the poverty line still lack access to adequate shelter, diet, housing, clothing, and medical care, but they are not considered officially poor. Moreover, researchers indicate that families in some places would need double the poverty rate to meet their basic needs.[13]

It is important to note that poverty is not equally distributed within the population. Minorities, children, and families headed by women are disproportionately represented among the impoverished in the United States. In 2006, approximately 8% of whites had an income below the official poverty level, compared with approximately 24% of African Americans and 21% of Hispanics. Although most officially poor people in the United States are White, African Americans and Hispanics are disproportionately represented in poverty statistics. Similarly, 17% of persons younger than 18 years and 28% of families headed by women were below the poverty line in 2006.[14]

The gap between the wealthiest Americans and the poorest Americans has widened in recent years. Research that has examined trends in income inequality indicates that it has increased since the 1970s. The increase in income inequality appears to be related to six factors: (1) the shift from better-paid manufacturing jobs to less well-paid service jobs; (2) technological advances that have increased the demand for highly skilled workers at the expense of less skilled workers; (3) globalization, which forces American workers to compete with foreign workers, driving down American wages; (4) immigration, which drives down wages among less skilled workers; (5) a decline in unionization that decreases the ability of workers to bargain for higher wages; and (6) the segregation of poorer people into areas where their opportunities for economic well-being are lessened.[16]

For most Americans, the economic resources of the families they are born into play a critical role in their future economic status. A person born into a poor family will likely remain poor or close to the poverty level, and a person born into a wealthy family will likely remain wealthy.[19] This does not mean, however, that everyone's economic circumstances remain static over time. Even individuals from middle-class families experience some unexpected changes in their economic circumstances at some point in their

MYTH VS REALITY

Comparative Focus on Income Inequality

Myth—The United States has less income inequality than other countries.

Reality—The United States has greater income inequality than many other Western or industrialized countries. For example, countries such as Italy, Britain, Japan, Australia, Germany, France, Switzerland, Netherlands, Sweden, and Denmark have considerably less income inequality than the United States.[15] To view an interesting video on income inequality in the United States, see the PBS presentation *Income Inequality* at http://www.pbs.org/now/shows/332/index.html.

MYTH VS REALITY

Comparative Focus on the Quality of Life Experienced by Americans

Myth—The quality of life experienced by Americans is greater than the quality of life in any other country.

Reality—Although many Americans live in much better conditions than many other people in the world, some Americans are not as fortunate. On the 2007/2008 Human Development Index, which is based on life expectancy, educational attainment, and real income, the United States ranked 12th, behind Finland, France, the Netherlands, Japan, Switzerland, Sweden, Ireland, Canada, Australia, Norway, and Iceland.[17] While the United States ranks first in military spending, gross domestic product, and the numbers of millionaires and billionaires, it ranks 18th in the income gap between rich and poor children, and 22nd in infant mortality.[18]

lives, and most will experience periodic, though predictable, changes in their standard of living during their lifetimes. Divorce or separation, the death of a spouse or parent, becoming a family head or spouse, unemployment, work loss because of retirement, and a reduction in the work hours of individuals and family members can lead to financial hardship. Thus, it is not only the poor who are likely to suffer economic hardship during their lifetimes; they are merely more likely to suffer longer and more acutely because of their economic situation.

An important key to economic well-being is the ability to find and hold a job that pays a decent wage. Yet, many Americans have a hard time locating such jobs. Periodic economic recessions in the 1980s, 1990s, and the 2000s; decisions by U.S. companies to move their production operations to other countries; reductions in the workforce as a result of automation; efforts to make businesses more efficient; and a shift away from the production of goods to the provision of services not only have reduced job opportunities for the traditionally unemployed, but also have affected the wages and the future employment prospects of those who are working and those who will soon enter the workforce. Although these trends have influenced people of all racial groups and ages, they have had a profound influence on people living in urban areas, many of whom are minorities, and children.

The shift from manufacturing jobs to service jobs; the movement of entry-level jobs in manufacturing, retail sales, and customer services from urban areas to the urban periphery or overseas; and the relocation of high-tech jobs to cities have resulted in fewer job opportunities for low-skilled urban residents. Moreover, a lack of affordable rental housing in the urban periphery, coupled with racial residential segregation, prevents many poor and low-skilled workers from relocating to areas that are closer to these jobs.

FYI

Children's Health Outcomes are Directly Affected by their Parents' Economic Status

For example, compared to non-poor children, poor children are 1.6 times more likely to die in infancy, 1.9 times more likely to suffer from low birth weight, 2.7 times more likely to have no regular health care, and 8 times more likely to live in a family that has too little food a least some of the time.[20]

Further, low levels of automobile ownership and poor public transportation mean that poor low-skilled urban residents find it difficult to commute to jobs.[21] Indeed, research indicates that distance from home is a major factor that influences workforce participation for both men and women, but it is particularly strong for women, who are often hesitant to take jobs that are far from home because they do not want to be too far away from their children's caregivers.[22] Simultaneously, many new jobs that have been created in urban areas require a high degree of skill and are not well matched to the skills of many urban residents. The result has been a high degree of economic segregation characterized by concentrated pockets of poverty, particularly in urban areas,[23] at one extreme and areas of affluence at the other.

Economic resources are important because of the material goods and opportunities that such resources can procure. For those living in affluent areas, it means having a range of public services and private resources. It means well-funded schools, recreational and cultural facilities, programs for youths, and access to a range of private resources that can be used to support and respond to family and individual problems. In poor communities, it often means a lack of quality public or private resources for youths and families. It means that those who are disproportionately exposed to the stress of life have the fewest resources to help them deal with the problems that they encounter.

Adverse economic conditions, such as unemployment and declining wages, place a strain on many families. For example, research on unemployed men has found them to feel less satisfied with themselves and with their lives in general. They also tend to feel more victimized, anxious, depressed, and hostile toward others than men who are employed.[24] Financial loss appears to be related to changes in men's attitudes and parenting practices, and many fathers whose earnings decline become more tense, irritable, explosive, arbitrary, and punitive in responding to their children.[25]

Of course, women are also affected by financial hardship, and financial hardship is particularly acute for families headed by single women. In 2006, 28.3% of all female-headed families lived in poverty.[26] Moreover, life for single women who head families is often complicated by the many responsibilities that these women have. Indeed, females who head families have multiple responsibilities and demands that go beyond their own immediate needs. They are responsible for providing food, shelter, and clothing for their children; seeing that they get to school; keeping them healthy; supervising and monitoring their behavior; meeting their emotional and psychological needs; managing family finances; and meeting a wide variety of other demands that emanate from employers, relatives, and others. In addition, they must deal with issues of gender discrimination, poor labor market prospects, and welfare policies and programs that provide minimal relief and are often seen as humiliating.[27]

Economic hardship also affects the young. Poor infants are much more likely to suffer low birth weight, to die during the first year of life, and to suffer hunger or abuse while growing up. They are less likely to receive immunizations or adequate medical care, and they are less likely to undergo the type of cognitive development that will allow them to do well in school.[28] Moreover, negative psychological and social effects are associated with poverty. Long-term exposure to poverty is related to increased levels of anxiety and unhappiness among children, and current exposure to poverty is associated with behaviors such as disobedience and aggression.[29]

Economic resources clearly influence the ability of people to obtain those things that are necessary for survival and that are desired in our society. Such resources influence an

MYTH VS REALITY

Comparative Focus on the Well-Being of Children

Myth—Children in the United States enjoy a greater level of well-being than in any other country.

Reality—Out of the 21 nations in the industrialized world, the average ranking of the United States on child well-being is 20, just ahead of the United Kingdom. The countries that have the highest rankings in terms of material well-being, health and safety, education, family and peer relationships, behaviors and risks, and subjective well-being are the Netherlands, Sweden, and Denmark.[30]

individual's life chances. As Hans Gerth and C. Wright Mills noted, life chances involve "everything from the chance to stay alive during the first year after birth to the chance to view fine art, the chance to remain healthy and grow tall, and if sick to get well again quickly, the chance to avoid becoming a juvenile delinquent—and very crucially, the chance to complete an intermediary or higher educational grade."[31]

Furthermore, economic status is an important factor from a juvenile justice perspective, because it appears to be strongly related to delinquent behavior. Indeed, many theories of crime and delinquency stress the role of economic inequality in the production of crime, and these theories have received some support in the research literature. For example, ethnographic studies of low income neighborhoods,[32] large-scale self-report studies of the link between **social class** and delinquency, and research on the relationship between persistent childhood poverty and delinquency have found that poverty appears to have an effect on delinquent behavior. In a large-scale study of social class differences in delinquency, Delbert Elliott and David Huizinga found that, although class differences were almost nonexistent when minor types of delinquency were examined, lower class youths reported significantly more involvement in serious forms of delinquency.[33] In a more recent study of the relationship between persistent poverty and delinquency, Roger Jarjoura, Ruth Triplett, and Gregory Brinker found that exposure to persistent poverty is associated with delinquency. Moreover, they also found that exposure to persistent poverty during early childhood is related to an increased likelihood of delinquency later in life.[34]

social class A person's or family's economic, social, and political standing in the community. Social class can be directly related to type of employment, neighborhood, circle of contacts, and friends.

Political and economic arrangements in our society that foster inequality among Americans clearly influence the types of lives that many people, including children, experience. As noted, there is evidence that political and economic arrangements have an important effect on delinquent behavior. In addition, they provide a context that influences the operation of other important socializing institutions, such as families, schools, communities, and peer groups, that are related to delinquency. The remaining sections of this chapter examine the role of these socializing institutions in the lives of children and the research that has looked at the relationship between these socializing institutions and delinquency.

■ The Family and Delinquency

There appears to be widespread agreement among both social scientists and the general public that the **family** plays a key role in child development and socialization. The family can be a place where members love each other, care for one another, and provide a mutually beneficial environment for healthy human growth. On the other hand, the

family A collective body of any two or more persons living together in a common place under one head or management; a household; those persons who have a common ancestor or are from the same lineage; people who live together and who have mutual duties to support and care for one another.

family can be characterized by conflict, a lack of mutual support and nurturance, and violence. Like other major socializing institutions, families are profoundly influenced by the political and economic context within which they operate. As D. Stanley Eitzen, Maxine Baca Zinn, and Kelly Eitzen Smith note, a family's placement in the class system is the most important factor in determining family outcomes.[35] For example, a family's placement in the political and economic structure shapes the family's access to and interconnection with other institutions, such as work establishments, schools, churches, and voluntary associations. These institutions can function as resources for the family and can facilitate access to other resources.[36] Thus, children from wealthy families will have a definite advantage in life while children from poor families will face a variety of obstacles in their efforts to achieve the American dream.

Not only does the family determine the economic conditions within which children live, but it also plays a primary role in shaping a child's values, personality, and behavior. Not surprisingly, a variety of criminological theories suggest that the family plays a significant role in the production or prevention of delinquent behavior, and much research has been devoted to examining various aspects of the family that appear to be associated with delinquency. As Walter Gove and Robert Crutchfield note, "the evidence that the family plays a critical role in juvenile delinquency is one of the strongest and most frequently replicated findings among studies of deviance."[37] Essentially, this research has focused on two broad areas—family structure and family relations—which are felt to have a strong influence on juvenile delinquency and later criminality. Indeed, one important conclusion that can be drawn from this research to date is that the individual's experiences during infancy and early childhood influence behavior over the life course.[38]

Family Structure and Delinquency

Family structure refers to the ways in which families are constituted. For instance, are both natural parents present, is a stepparent present, is the family headed by a single parent, and how large is the family? One element appears clear: The structure of the American family has changed dramatically overtime.

The Impact of the Single-Parent Home on Delinquency

One important fact is that children today are more likely to live in single-parent households than in the past. For example, in 1970 about 12% of all children lived with one parent, typically their mother;[39] but, by 2002, the percentage of children living in single-parent homes had risen to approximately 28%, and another 4% lived in households where neither parent was present.[40] There are two primary reasons for the increase in the number of children who live in single-parent homes. First, the divorce rate has been rising over the past 20 years. The United States has the highest divorce rate in the world, with about half of all marriages ending in divorce. Moreover, the majority of all divorces involve children. A second reason why more children are living in single-parent homes is that more children are being born out of wedlock. The number of out-of-wedlock births has increased significantly since the 1960s, especially among teenagers, and the rise has contributed to the increasing number and proportion of children in single-parent homes.[41]

Of course, the high rate of marital dissolution and the growing number of out-of-wedlock births have not occurred in a social and economic vacuum. Rather, they are influenced by a complex set of social and economic factors. Although most Americans indicate they value marriage, there is considerable evidence that the importance given

FYI

Broken Homes and Delinquency

One problem with examining the relationship between broken homes and delinquency is the vagueness of the concept of a "broken home." According to the standard definition, a broken home is a home in which at least one natural (biological) parent is absent. However, this definition ignores the potential differences in types of broken homes and variations in parental absence. For instance, a number of events might cause the absence of a parent, including divorce, separation, disability, military service, and job responsibilities. The length of parental absence can vary as well. Unfortunately, the effects of different types of parental absence on youths and their relationship to delinquency are not known.

to marriage and two-parent families has eroded in recent years. The result has been a relaxation of social constraints on divorce, out-of-wedlock childbearing, and single parenthood.[42] At the same time, the economic prospects of many young men, particularly those who are minorities, have worsened. This has not only led to joblessness, but also influenced marital dissolution and made them less attractive as marriage partners who can support a family.[43]

broken home A home or family in which at least one natural (biological) parent is absent.

The relationship between single-parent families, which are sometimes referred to as **broken homes**, and juvenile delinquency has been the focus of considerable debate within the fields of criminology and juvenile justice. Indeed, some research has discovered statistically significant relationships between single-parent homes and delinquency.[44] However, research also suggests that the effects of coming from a single-parent home may not be the same for all youths. For instance, some research has indicated that coming from a single-parent home is associated, not with serious delinquency, but with status offenses like running away from home and truancy.[45] Also, some research has found that White youths, young females,[46] and youths from high-income families[47] are more likely to be affected in more adverse ways by parental absence.

Although there does appear to be a relationship between living in a single-parent home and involvement in delinquent behavior, the relationship is not very strong. Furthermore, the research uncovering such a relationship often has been based on official data. A potential problem with using official data is that authorities may treat youths from single-parent homes differently than youths from intact homes.[48] For example, police may be more inclined to formally process youths from single-parent homes than youths from two-parent homes. The existence of such a bias is suggested by the fact that studies relying on self-report data often have failed to find a strong relationship between single-parent homes and delinquency.[49]

FYI

Multiple Family Transitions and Delinquency

Many children experience family breakups several times during their lives. Research indicates that children who experience multiple family transitions during their lives are more likely to engage in delinquent behavior.[50]

Research that has not discovered a relationship between single-parent homes and delinquency suggests that what is most important is not whether youths come from a single-parent family or a two-parent family, but the quality of the relationship that exists between those parents that are present and their children.[51]

The Effect of Family Size on Delinquency

Another aspect of family structure that may influence delinquency is family size. In his classic study of delinquency, Travis Hirschi found that, even when controlling for academic performance, parental supervision, and attachment between youths and their parents, family size was related to delinquency.[52] This finding is supported by research conducted in England, although the relationship discovered was much weaker for middle-class families than for lower-class families,[53] which suggests that it may be economic resources rather than family size that is most important. Perhaps parents of larger families who have substantial economic resources may be better able to meet their children's needs in ways that reduce the probability of delinquency. Some researchers have questioned the relationship between family size and delinquency, arguing that a more important variable is having a delinquent sibling. Their line of reasoning is that having a delinquent sibling is related to delinquency and that in larger families youths are more likely to have a delinquent sibling.[54]

Family Relations and Delinquency

The term *family relations* refers to the quantity and quality of interactions and relationships among family members. Like family structure, however, family relations are also influenced by the larger social context within which families reside as well as the economic condition of the family itself. As noted earlier, families have changed in a variety of ways in recent years in response to political and economic developments. This section begins by examining how changes in the family may have influenced family socialization. It then discusses different aspects of family socialization that have been linked to delinquency.

The Influence of Women's Employment on Family Relations and Delinquency

In 1960, less than 7% of married women with children participated in the paid labor force.[55] In 2002, 67% of married women with children were employed.[56] Indeed, one of the most important changes in the American family has been the growing number of women, including mothers, who are entering the workforce. The effects of mothers' employment on children are not entirely clear, however. By working outside the home, mothers are able to improve the economic well-being of their families. Indeed, in poor families and many middle-income families, the mother must work if the children are to receive adequate care.[57] A working mother can make a significant contribution to the economic well-being of the family; however, it also means that the mother has less time to spend interacting with her children. This is a problem that single working mothers share with parents in dual-earner families, when both husband and wife work outside

FYI

Latchkey Children

In 1999, it was estimated that about 3.3 million youths, age 6 to 12 years, regularly spent time unsupervised or in the care of a young sibling.[59]

the home. In response, many families have come to rely increasingly on child-care or leave children to care for themselves for at least part of the day.[58] Children who care for themselves for part of the day are often referred to as **latchkey children**

latchkey children
Children who care for themselves at least part of the day due to parental unavailability because of employment.

Although many people feel that a decline in the amount of parent–child interaction can obstruct child development, the extent of the impact is far from apparent. Some research indicates that there is virtually no developmental difference among toddlers who attend day-care programs and children taken care of at home, although there may be some negative health outcomes for infants.[60] Moreover, quality child-care programs can facilitate children's cognitive, emotional, and social development, particularly for disadvantaged children.[61] Even self-care does not always result in poor outcomes. Many children who are left unsupervised become more independent and learn to become more responsible for themselves over time. In contrast, however, other children who care for themselves experience loneliness and isolation and may be significantly disadvantaged by a lack of adult supervision.[62] According to criminologist Travis Hirschi, a lack of parental supervision is related to an increase in the likelihood of delinquent behavior.[63] Furthermore, research has shown that when children are not involved with their parents, they are more likely to report involvement in delinquency.[64] However, it appears that parental employment is not the most significant factor in determining youths' involvement in delinquency. For example, research by Thomas Vander Ven and his colleagues found that the mothers' employment had very little effect on children's delinquent behavior when they were adequately supervised.[65] Consequently, the mothers' employment may have little impact on delinquency. What appears to be more important is whether children receive appropriate supervision and the quality of parent–child relations.

Researchers also have examined the ways in which parents' roles and experiences in the workplace influence their relationship with their children. Criminologists Mark Colvin and John Pauly argue that parents tend to reproduce at home the authority relations they experience in the workplace. The problem is that the differences in power characteristic of a capitalist economy result in workplace experiences that are authoritarian and coercive. In turn, these experiences engender coercive and authoritarian relationships within many homes—relationships that are not conducive to the establishment of intimate bonds between family members and increase the likelihood of delinquency.[66]

The relationship between the economic sphere and the family is also the focus of John Hagan's power control theory. According to Hagan, "Work relations structure family relations, particularly relations between fathers and mothers and in turn relations between parents and their children, especially mothers and their daughters."[67] Data collected by Hagan and his colleagues in Canada tend to support this hypothesis. From Hagan's perspective, the power that parents have within the workplace is typically reproduced within the family. Hagan argues that when both parents are in positions of power in the workplace, the parents share power and the family structure is egalitarian. In such families, male and female children are socialized in similar ways, which results in comparable levels of delinquency among male and female children. However, in traditional patriarchal families, in which the mother remains at home, as well as in single-parent families, daughters are more likely to become the objects of control by mothers, who socialize their daughters to avoid risk. One outcome is that in such families males tend to engage in more delinquency than females.[68]

The broader political and economic environment within which people live appears to exert important influences on the types of relationships found within families.

Moreover, a wealth of information exists indicating that various types of family relations are associated with delinquency.

Family Socialization and Delinquency

An important process through which family relations are developed is socialization. Socialization refers to the ways that a child is taught cultural roles and normal adult responsibilities, and it involves a variety of interactions, such as touching, holding, hugging, kissing, and talking to the child; listening to the child; feeding and clothing the child; and taking care of the child's need for safety, security, and love. All of these interactions convey important messages to children.

One important aspect of socialization—one that is associated with delinquent behavior—is the process by which social control is developed and implemented in the family. One form of social control consists of the bonds that children develop to the family and family members. Indeed, research indicates that youths who lack closeness to parents or caregivers, or who feel there is little family cohesiveness, are more likely to engage in delinquency.[69] In addition, some researchers have uncovered a relationship between family conflict, hostility, a lack of warmth and affection among family members, and delinquency.[70] Others have found a relationship between parental criminality and delinquency (the children of parents who are involved in criminal behavior are more likely to engage in delinquency than youths whose parents are not involved in criminality).[71]

Family crises and changes also have an effect on family relations and appear to be related to delinquency. There is some evidence that disruptions in family life, such as moving to a new residence, separation or divorce of parents, and family conflict, can produce pressures that push youths toward acting out behaviors.[72] Overall, these studies suggest that when the quality of parent–child relations is poor, and when there are significant disruptions in family life, delinquency is more likely. They also suggest the converse: Positive parent–child relations act to control delinquent behavior.

Of course, parents exert other forms of social control, including the imposition of discipline. There is considerable evidence that inconsistent discipline, as well as overly harsh or lax discipline, is related to delinquent behavior.[734] Unfortunately, some parents' responses to their children's objectionable behavior are not simply lax or overly harsh; they are neglectful or abusive.

child abuse Nonaccidental acts directed against children that result in physical, sexual, psychological, or emotional harm.

Child abuse consists of acts of commission—things done to children. Types of child abuse include physical, sexual, psychological, and emotional abuse. **Neglect** consists of acts of omission; in other words, a parent or guardian fails to meet the needs of his or her child (e.g., the need for food, shelter, medical care, clothing, education, or affection). Neglect can be physical, emotional, or psychological. However, although child abuse and neglect consist of different forms of behavior, they often occur simultaneously. That is, children who are abused are often neglected as well, and children who are neglected are often abused.

neglect The nonaccidental failure of a caregiver to meet the physical, psychological, or emotional needs of a minor or another person in the charge of the caregiver.

FYI

Actual Level of Child Maltreatment is Unknown

In 2007, an estimated 3.5 million children were the subjects of an investigation for child abuse.[74] However, because many incidents of abuse and neglect are not reported to the authorities, the actual level of child maltreatment is unknown.

FYI

A national survey of intimate partner violence conducted in the mid-1990s found that 1.3 million women and 835,000 men are assaulted each year by their partners.[77]

The sad fact is that many children are subjected to abuse and neglect by their parents. Moreover, abuse and neglect have been found to be related to a variety of health, cognitive, educational, and social difficulties, including brain injuries, mental disorders, poor school performance, fear, anger, and antisocial behavior.[75] In one of the most comprehensive studies done on the relationship among child abuse, neglect, and criminality, Cathy Spatz Widom and Michael Maxfield found that youths who were abused or neglected were more likely to engage in delinquent and adult criminal behaviors than youths who were not subjected to such treatment. Youths who were abused or neglected were 59% more likely to be arrested as a juvenile and 28% more likely to be arrested as an adult than youths with no abuse or neglect history. Moreover, youths who had been abused or neglected were 30% more likely to be arrested for a violent crime.[76] These findings support the idea of a cycle of violence (a cycle in which those who experience violence as children are more likely to engage in violence as adults).

Aside from child abuse and neglect, another indicator of the quality of family life is conflict between parents. Such conflict sometimes takes the form of domestic violence, which usually, although not always, involves males abusing their female companions. Like child abuse, the actual extent of domestic violence is not known because much of this behavior is not reported to the police. Nevertheless, domestic violence is a significant problem.

Like child abuse, domestic violence rarely occurs as an isolated incident, possibly because each act of violence tends to reduce the inhibition against violence.[78] Also like child abuse, domestic violence has negative effects on children. Research indicates that children who observe domestic violence tend to be more withdrawn and anxious; they are more likely to perform well below their peers in school performance, organized sports, and social activities; and they are more likely to exhibit aggressive and delinquent behaviors.[79] In addition, research suggests that domestic violence and child abuse are related. For example, research on the co-occurrence of domestic violence and child maltreatment indicates that there is a 30 to 60% overlap between violence directed at women and violence directed at children in the same families.[80]

There is an undeniable connection between violence in families and violent children. Parents who physically abuse their children teach them that it is okay to physically confront other people when angry and that physical violence is an acceptable form of interaction between people. Men who abuse their wives or girlfriends teach their male

FYI

A significant number of persons in the U. S. are victimized before age 18 years. Findings from the National Violence Against Women Survey indicated that 43.4% of women and 54.3% of men were the victims of rape, physical assault by a caretaker, or stalking by the time they reached age 18 years.[81]

children that it is acceptable to hit women. By their behavior, they also teach female children that it is acceptable for women to be hit by men and that women should expect and accept physical abuse by men.

In a similar manner, children who have been sexually abused, especially those victimized by a close family member or a person who has authority over them, often become perpetrators themselves. The inappropriate form of social and sexual contact inherent in sexual abuse teaches the child victim all of the wrong lessons about appropriate sexuality and, just as importantly, breaks down natural or societal taboos regarding sexual behavior.

Schools and Delinquency

School is another institution that has a profound influence on the lives of young people. Today, a much larger percentage of the youth population attends school than was true in the past. For example, in 1890 only 7% of the school-age population attended school. However, owing to the development of compulsory education laws, today over 95% of school-age youths are enrolled in school.[82] Moreover, the amount of time that youths spend in school has been increasing. For example, in 1940 only 38% of persons in the United States between the ages of 25 and 29 years had completed high school. By 2007, the percentage of persons between the ages of 25 and 29 years who had completed high school rose to 87%.[83]

Clearly, school is an important institution because it gives young people the academic skills that are critical for effective participation in today's society. Yet school is important for other reasons as well. It has become the primary socializing institution "through which … community and adult influences enter into the lives of adolescents."[84] It is in school that youths learn values, attitudes, and skills, such as punctuality and deference to authority, that are necessary for participation in economic and social life. For many families, a substantial amount of the interaction between parents and children revolves around school–related issues.[85] Indeed, school is generally acknowledged to affect "the lives of youth in ways which transcend the more obvious influences of academic knowledge acquisition."[86] It is, among other things, the place where youths develop a better sense of who they are and how they stand in relationship to others, both peers and adults.

Aside from its role as a major socializing institution, school is important because it is a primary determinant of both economic status and social status. It is, for many people, the primary avenue to economic and social success because it confers the credentials necessary for entry into well-paying jobs. Consequently, persons who complete high school, in general, earn significantly more than those who do not graduate.[87]

MYTH VS REALITY

Overall, the Benefits of Education Vary by Gender and Race

Myth—Education is a path to social and economic equality for all Americans.

Reality—Arguably, there are many benefits to receiving an education, because income tends to improve with educational attainment. However, not everyone with the same amount of education does equally well. For example, men with professional degrees can expect to earn almost $2 million more than women with equivalent degrees over the course of their lives. Also, non-Hispanic Whites, on average, earn substantially more than African Americans or Hispanics during the course of their lives at every level of educational attainment.[88]

Although clear benefits are tied to educational attainment, many youths experience both academic and social problems at school. Consider some of the following statistics cited by Richard Lerner in his book, *America's Youth in Crisis*:

- About 25% of the approximately 40 million children and adolescents enrolled in America's 82,000 public elementary and secondary schools are at risk for school failure.
- About 4.5 million children ages 10–14 years are one or more years behind in their modal grade level.
- Each year about 700,000 youths drop out of school.
- Unemployment rates for dropouts are more than double those of high school graduates.
- At any point in time, about 18% of all adults between age 18 and 24 years and 30% of dropouts between age 23 and 25 years are under the supervision of the criminal justice system. Among African Americans, the corresponding percentages are about 50% and 75%.[89]

Unfortunately, as these statistics demonstrate, school is an unpleasant experience characterized by failure and dropout for many youths. Furthermore, there is considerable evidence that failure in school and other school-related factors are related to delinquent behavior. The remainder of this section examines some of the school-related factors found to be associated with delinquency.

School Failure and Delinquency

As Gary Jensen and Dean Rojek note, "One of the most persistent findings concerning the school and delinquency is that students who are not doing well in school have higher rates of delinquency than those who are faring better."[90] After reviewing a number of studies, John Phillips and Delos Kelly found a strong relationship between school failure and delinquency. They also discovered that, opposite to the view held by some, school failure precedes delinquency and not the reverse.[91] Moreover, a review of the research in this area by Eugene Maguin and Rolf Loeber found that children who were not doing well academically were almost twice as likely to engage in delinquency as youths earning good grades.[92]

Considerable research has been done on a number of other factors found to be related to school failure and delinquency. For example, research on students' feelings of belonging, attachment, and commitment to school reveal that these factors are related to school violence, vandalism, and delinquency. Studies also have found that students who do not like their teachers or school are more likely to report involvement in delinquency than those who claim to have a strong attachment to their teachers or school.[93] Similarly, in the 1978 *Safe School Study Report* that was produced by the National Institute of Education and delivered to the U.S. Congress, student alienation was found to be an important factor linked to school violence and property loss.[94]

It is hardly surprising to find that students who are less committed to school, who are less attached to their teachers and schools, and who feel alienated are more likely to engage in disruptive or delinquent behaviors in and out of school.[95] However, a lack of attachment and commitment to school should not be seen simply as a product of individual failure, but as a product of the position students occupy in relation to others in school and as a product of the differences in opportunity available to students in the educational environment. One factor that determines the position that students occupy and the opportunities they may be exposed to is social class.

Social Class, School Performance, and Delinquency

The importance of social class to the school failure–delinquency relationship was first made explicit by Albert Cohen in his 1955 book, *Delinquent Boys*.[96] According to Cohen, school is the one place youths of all social classes come together and compete for status. However, working-class youths are at a disadvantage in this competition because they lack the necessary skills to be successful and because success is defined in middle-class terms. One tempting response available to working-class youths is to form a delinquent subculture (with its own status system) that adheres to nonconventional values and encourages delinquent behavior.

Although Cohen claims that the social class position of working-class youths acts as a stumbling block, other evidence indicates that school practices constitute the main barrier to achievement for many students.[97] Indeed, various practices make it difficult for some students to succeed academically or socially within the school environment. Moreover, a number of these practices have been found to encourage delinquent behavior.

Tracking, School Performance, and Delinquency

Tracking, the sorting of students according to ability or achievement, is one practice that fosters inequality among students and has been found to be associated with delinquent behavior. Common in American schools, tracking typically begins early in students' educational careers. Once assigned to a particular track, students tend to stay in that track.

Although tracking is common, a variety of negative consequences are associated with this practice, including delinquency. Research has found that students in college preparatory tracks get much higher grades than those in noncollege tracks.[98] Also, being placed in a noncollege track has been found to be related to a lack of participation in school activities, lowered self-esteem, school misbehavior, dropping out, and delinquency.[99]

Irrelevant Curricula, School Performance, and Delinquency

Today, school curricula are designed primarily for students who are planning to attend college, while noncollege and technical programs are frequently of inferior quality. As a result, many students have difficulty understanding how much of what they are taught will help them in the future roles they will occupy.[100] This appears to be particularly true of low-income, noncollege-track students, who often feel that school is a waste of time. Research indicates that when students feel that school is not relevant to their future job prospects, rebelliousness, school violence, property loss, and delinquency tend to increase.[101]

School Dropouts and Delinquency

For many students, dropping out is seen as a solution to the problems they face in school. Yet, dropping out has substantial negative consequences for the individual. Because of continuing technological sophistication, the skills that people need to function effectively in society have increased. Consequently, dropouts are often woefully unprepared to compete for and maintain positions requiring even basic skills. As a result, they face diminished job prospects and often experience difficulty meeting subsistence income needs.

In addition to the economic consequences of dropping out of school, there are psychological and social consequences associated with leaving school before graduation.

Research indicates that dropouts usually regret their decision to drop out,[102] and dropping out appears to be associated with further dissatisfaction with themselves and their environment.[103] In addition, dropouts typically have lower occupational aspirations than those who graduate, and they also have lower educational aspirations for their children.[104]

Not surprisingly, there appear to be important differences between youths who remain in school and those who drop out. Compared with students who stay in school, those who drop out tend to be from low socioeconomic status groups, to be members of minority groups, and to come from homes with fewer study aids and fewer opportunities for nonschool-related learning. Dropouts also are more likely to come from single-parent homes, to have mothers who work, and to receive less parental supervision. In addition, compared to those who stay in school, future dropouts are more likely to receive poor grades and low scores on achievement tests in school, are less likely to be involved in extracurricular activities, and are more likely to have school discipline problems.[105]

As noted earlier, dropping out of school clearly makes it more difficult for the individual to obtain a well-paying job in an economy that requires increasing technological sophistication on the part of workers. Not surprisingly, many people assume that there is a direct relationship between dropping out and involvement in delinquency. However, research on the relationship between dropping out and delinquency has produced conflicting results. For example, research conducted by Delbert Elliott and Harwin Voss found that youths who drop out of school engage in more delinquency than those youths who remain in school. Yet, their research also found that the level of delinquency among youths who drop out was greatest right *before* they dropped out rather than after. Also, the reasons youths gave for dropping out were directly tied to their negative schooling experience. Specifically, dropouts tended to find school alienating, they were not successful academically, and they associated with peers who were involved in delinquency.[106] The finding that delinquency tends to decrease after youths drop out suggests that negative school experiences encourage youth to engage in delinquency and leave school. After the condition for these negative experiences is eliminated, however, the motivation to engage in delinquency tends to decrease.

On the other hand, other studies have uncovered evidence that when youths drop out of school, their involvement in criminal activities tends to immediately *increase*. For instance, research conducted by Terrence Thornberry and his colleagues found that dropouts, compared with students who stayed in school, were more likely to engage in crime soon after leaving school. Furthermore, they found consistently higher arrest rates for dropouts until the two groups reached their mid-twenties.[107] Thornberry and his colleagues argue that delinquency immediately increases for school dropouts because leaving school severs ties with an important conventional socializing institution, namely school. Similar results have been noted in a more recent study by Marvin Krohn and

FYI

According to data compiled by the National Center for Educational Statistics, in 2006, 9.3% of persons between the ages of 16 and 24 years were school dropouts. However, among persons between the ages of 16 and 24 years, the dropout rate was 10.7% for Blacks and 22.1% for Hispanics.[109]

his associates, who analyzed data from the Rochester Youth Development Study. Their analysis revealed that school dropouts tended to engage in more delinquency and reported more drug use than youths who remained in school.[108]

The literature on school dropouts clearly has policy implications. Some have suggested that compulsory education laws be relaxed, based on research findings that dropping out leads to a decrease in delinquent behavior.[110] However, others are strongly opposed to such a policy. They argue that the research is not clear on the relationship between dropping out and delinquency. Their position is based on the research that indicates that dropping out of school is associated with a variety of negative outcomes, such as increased involvement in delinquency. Consequently, they maintain that the focus of policy should be on reducing the dropout rate by improving the ability and willingness of schools to meet the educational needs of *all* students.

The Community and Delinquency

As Robert Bursik and Harold Grasmick note in their book, *Neighborhoods and Crime: The Dimensions of Effective Community Control*, concern about the influence of the neighborhood on delinquency and crime is hardly new.[110] Indeed, since the development of cities in the United States, considerable emphasis has been placed on the negative influences found within some areas of the urban environment—influences believed to be related to delinquency, adult crime, and a host of other social problems, such as poverty and drunkenness.[112] However, particular attention began to be devoted to some of the more negative aspects of the urban environment around the turn of the century, and it was then that sociologists, such as Clifford Shaw and Henry McKay, started their pioneering work designed to understand the influence of the community on delinquency.

As part of their efforts, Shaw and McKay mapped areas of Chicago where official delinquents lived. The maps indicated that the highest rates of delinquency were located in deteriorating inner-city areas characterized by decreasing population, a high percentage of foreign-born persons and African American households, low levels of home ownership, low rental values, close proximity to industrial and commercial establishments, and an absence of agencies designed to promote community well-being.[113] Also, they discovered that, despite changes in the ethnic composition of these high delinquency areas over time, the delinquency rates remained relatively constant.[114] This convinced Shaw and McKay that the high delinquency rates in certain areas could not be attributed to residents' individual pathologies, but resulted from a set of conditions that added up to, in their terminology, "**social disorganization**." They believed that these transitional neighborhoods typically suffered a breakdown in social control characterized by a lack of community cohesiveness, common values, and institutions that prevented delinquency.[115] They also thought that, in these neighborhoods, gangs and delinquent groups formed in which delinquent traditions were passed from one generation to the next (i.e., through "cultural transmission"). According to Shaw and McKay, cultural transmission accounted for high rates of delinquency in these areas, despite changes in their ethnic composition over time.[116]

A number of more recent research efforts have documented the importance of community influences on delinquency. Ora Simcha-Fagan and Joseph Schwartz examined the effects of both community and individual factors on delinquency in a study of 12 New York City neighborhoods and found that communities characterized by low organizational participation by residents and the existence of a criminal subculture are likely

social disorganization The state of a neighborhood in which it is unable to exert control over its members. Social disorganization can result from a decrease in population, a large transient population, the existence of groups with differing values, little home ownership, low property values, proximity to industrial and commercial establishments, and a lack of agencies designed to promote a sense of community.

FYI

The Chicago Area Project Attempted to Turn Theory into Practice

Not only did Shaw and McKay want to understand the ways in which conditions in communities influence levels of delinquency, they also were interested in creating programs designed to help disorganized communities respond to delinquency. In 1932, Clifford Shaw initiated the Chicago Area Project in three high-delinquency neighborhoods; the project eventually grew to include more neighborhoods and lasted for approximately 30 years. The intent of the project was to get local residents to respond to local problems, including delinquency, by developing social programs, such as sports and recreation, gang intervention, counseling, discussion groups, and community improvement campaigns.[117] Unfortunately, however, no systematic evaluation of the effectiveness of the Chicago Area Project was conducted. Although some researchers claim that the project likely had little real impact on delinquency because it failed to address the underlying causes of community disorganization,[118] others have suggested that it may have increased community cohesion and led to reductions in delinquency in some areas.[119]

to experience high levels of delinquency.[120] Robert Bursik examined delinquency rates, and a variety of other variables, for 74 communities in Chicago from 1930 to 1970.[121] In many of the communities studied, the results were similar to those found by Shaw and McKay: Delinquency rates remained high even though the racial and ethnic composition of the communities changed. Some communities did not exhibit this pattern, however. In one community, an increase in delinquency rates appeared to be influenced by rapid changes in adjacent neighborhoods. This finding suggests the importance of examining the ways in which communities are linked together as well as possible causal factors outside the community. Still other research has found that neighborhoods go though cycles of change.[122] Although some of the changes, such as neighborhood deterioration, have been found to be related to increases in delinquency, other changes, such as neighborhood revitalization, may be associated with decreases.[123]

Research on the relationship between community characteristics and delinquency has found that communities play an important role in the encouragement of delinquent behavior. Communities that are economically deprived appear to be particularly susceptible to high levels of criminality. Indeed, economic deprivation appears to interact with a variety of other community and family characteristics that produce increased levels of delinquency, and this appears to be true regardless of the racial makeup of community members.[124] Poor, physically deteriorating communities, where drugs are readily available to youths, where residents avoid involvement in community organizations, and where criminal subcultures exist are likely to experience high rates of delinquent behavior.[125] In such communities, there appears to be a lack of close personal ties between residents and a variety of criminal role models that result in a lessening of restraints on illegal behavior.

Peer Associations and Delinquency

Concern about delinquency as a group phenomenon is hardly new. In fact, concern about youth groups that threatened citizens was a major impetus for the establishment of the first juvenile courts in the United States, which occurred around the beginning of the 20th century. Furthermore, early efforts to study delinquency using a sociological approach typically focused on the group nature of delinquent behavior. The early research of Shaw and McKay in Chicago, as well as some of the more popular theoretical work on delinquency through the mid-1960s, focused on gangs and other types of delinquent subcultures. There is considerable evidence that youths' peers exert a strong influence on their behavior, and such influence seems apparent when groups engage in delinquent behaviors. There is some evidence that the larger a youth's **accomplice network** (the pool of potential co-offenders a youth associates with), the more likely a youth is to engage in delinquency.[126] Moreover, other research on the relationship between peers and delinquency indicates that many, though not all, youths pass through a progression from no delinquency to more serious delinquency that involves the following steps: (1) youths interact with mildly delinquent peers before the onset of delinquency; (2) minor delinquency results from this association; (3) involvement in minor delinquency leads to interactions with more delinquent peers; (4) this interaction leads to involvement in more serious forms of delinquency.[127] However, peer influence may also be present when an individual commits a lone act of deviance. Conversely, simply because two or more youths in the same location are engaging in delinquent acts does not always mean that their illegal behavior is a product of some group dynamic.[128]

accomplice network
The pool of potential co-offenders with whom a youth interacts.

One type of peer group that has received attention in both the research literature and the popular press is the gang. However, there is little agreement among researchers regarding the proper definition of the term *gang*. Sometimes, the term is used to describe any congregation of youths who have joined together to commit a delinquent act. At other times, it is used to refer to more structured ongoing groups that hold or defend a particular territory. Indeed, some communities do have highly structured groups that fit the popular conception of a gang. Nevertheless, it is worth noting that research on gangs has discovered that gangs vary in a number of ways, including their involvement in delinquent behavior.[130]

Other researchers have noted that some groups actually have very little organization and cohesion. Lewis Yablonsky has argued that a more accurate description of many youth groups involved in crime would be "near group."[131] Such groups are characterized by (1) diffuse role definitions, (2) limited cohesion, (3) impermanence, (4) minimal consensus on norms, (5) shifting membership, (6) disturbed leadership, and (7) limited definition of membership expectations. Yablonsky draws our attention to the fact that not all youth groups that engage in delinquent activities are highly organized and cohesive.

FYI

Most Delinquency Occurs in a Group Context

Studies that have examined the group context of delinquency suggest that between 62% and 93% of delinquent activity occurs in group settings. A recent study that examined data from a nationally representative sample of youths discovered that 73% of their delinquent offenses were committed in the company of others. This included 91% of all burglaries and alcohol violations, 79% of all drug violations, 71% of all assaults, 60% of all acts of vandalism, and 44% of all thefts.[129] Interestingly, the percentage of delinquency committed in groups is higher in official arrest data than the percentage shown in self-report data. The discrepancy is possibly explained by the **group hazard hypothesis**, which holds that delinquency committed in groups is especially likely to be detected and to result in a formal response. If youths who engage in delinquency in groups are more likely to be arrested than lone delinquents, then official arrest data would overestimate group delinquency rates.

group hazard hypothesis
The hypothesis that delinquency committed in groups is more likely to be detected and to result in a formal response by the authorities.

This is not to say that organized gangs are not a significant problem in many communities around the country. However, although gang membership appears to be common in some areas, only a minority of youths belong to gangs at any point in time, even in neighborhoods where gangs exist.[132] Nevertheless, gang membership is linked to drug dealing, vandalism, violent crime, and a variety of other illegal activities, although different gangs may favor different types of criminal activities. Moreover, illegal gang activity is a significant problem in many large cities and even in some smaller cities and rural counties, despite the fact that more gang members are being arrested, prosecuted, and incarcerated than ever before.[133] There is also considerable evidence that many gangs are becoming more oriented toward violence than in the past. Early studies of gangs found that violent activity was not common in these groups. Furthermore, when gangs did engage in violence, they rarely used firearms.[134] Today, gang activity is more likely to be violent and lethal because of the availability and possession of sophisticated weapons by many gang members and the types of violent behaviors they exhibit (e.g., drive-by shootings).[135]

A variety of factors are associated with youth involvement in gangs. Gang formation appears to be facilitated by a social context characterized by poverty, inequality, social disorganization, easy access to drugs, and an absence of well-paying jobs.[137] Many youths who join gangs are marginalized within their community. Such youths face a variety of stressful conditions: They have few legitimate opportunities for earning money, and they have few strong bonds to conventional institutions, such as school and family. In many instances, gang members come from destitute and troubled families where parents exhibit poor parenting practices. Moreover, many gang members have family members who are involved in gangs, and they have few positive educational or vocational role models.[138] Research also indicates that learning disabilities, poor academic performance, having friends who engage in problem behaviors, early use of drugs, and involvement in violence at a young age are strong predictors of gang involvement.[139] For many marginalized youths, gangs hold out the promise of economic and social opportunities.[140] Gangs also provide youths with a sense of belonging and status[141] as well as protection from other gangs and a means for dealing with a socioeconomic environment that fosters aggression and violence.[142]

Youth involvement in gangs appears to be linked to increases in delinquent behavior. For example, several studies have found that, prior to joining gangs, gang members' involvement in delinquency was similar to that of nongang youths. When these youths joined a gang, however, their involvement in delinquency, particularly violent delinquency and drug sales, increased. However, after youths left gangs, with the exceptions of drug sales, their involvement in delinquency decreased.[143]

FYI

Gang Involvement Is Associated with Criminality

Gang members have higher rates of delinquent involvement than nongang youths. Moreover, in areas where there are gang problems, gang members account for a majority of delinquent acts that are reported; this is particularly true for serious offenses. For example, one study that examined a sample of Denver youths found that 14% were gang members. However, these youths accounted for 79% of serious violent crimes, 71% of serious property offenses, and 87% of drug sales.[136]

Organized gangs are one type of peer group that seems to be related to increased delinquency, but researchers have noted that other peer groups apparently encourage delinquent behavior as well. As noted earlier, Yablonsky argues that many youth groups involved in delinquency lack the type of organization and cohesion often associated with gangs. In studying delinquency in Flint, Michigan, in the 1960s, Martin Gold found that a considerable amount of delinquent behavior occurred spontaneously in rather loosely structured youth groups.[144] Gold concluded that delinquency of this type resembles a "pickup game" in which opportunities for delinquent behavior present themselves to ordinary peer groups, leading to delinquent behavior.

The research of Herman and Julia Schwendinger has done the most to highlight the complexity of peer groups and how various peer groups contribute to delinquency.[145] Their book, *Adolescent Subcultures and Delinquency*, focuses on the complexity of youth culture and the variety of peer groups that exist. The Schwendingers' observational studies of youth culture in southern California communities reveal that youth culture is far more complex than many assume, a finding supported by other researchers.[146] Rather than being a monolithic entity or strictly based on social class, youth culture comprises a variety of subcultures and peer networks that cut across class lines. The various peer formations have their own designations (e.g., "intellectuals," "greasers," "homeboys," "socialites," and "athletes") and have their own distinctive dress and linguistic patterns. They also are accorded differing degrees of status and prestige by their members as well as by other youths and adults.

Although the Schwendingers identified a variety of adolescent subcultures, they noted that there were three persistent types: streetcorner, socialite, and intellectual groups. Moreover, they found that delinquency tended to vary between these types. For example, delinquency is less common among intellectuals, who focus on academic or technical interests (e.g., computers, mathematics, electronics, and physics), have little interest in adolescent fashion, and often spend considerable time doing homework or participating in adult-sponsored activities. In contrast, streetcorner groups are more likely to consist of youths who are economically and politically disadvantaged and who engage in delinquency, including serious delinquency. Falling between intellectual and streetcorner groups, as regards involvement in serious delinquency, are socialite groups, which typically consist of youths from economically and politically advantaged families. Although members of these groups engage in less serious delinquency than members of streetcorner groups, they nevertheless engage in a considerable amount of garden variety delinquency, such as driving violations, vandalism, drinking, petty theft, truancy, gambling, and sexual promiscuity. The Schwendingers' studies not only help us understand the complexity of youth culture, but also indicate the necessity of carefully examining the ways in which youth subcultures encourage or inhibit delinquency among their members.

This chapter has highlighted the contextual nature of delinquent behavior and pointed out some of the ways in which economic, political, family, school, neighborhood, and peer contexts influence juvenile offending. The fact that delinquent behavior is affected by factors that lie outside the individual does not imply, of course, that youths are incapable of exercising some degree of free will. Clearly, youths who engage in delinquent behavior make a choice to act as they do, although the decision-making processes that they use could not always be described as particularly mature or rational from the point of view of those who are not involved in illegal behavior. However, given the social context in which many youths live, delinquent behavior is often seen as normal, even expected, behavior.

Moreover, in order to effectively respond to youths' delinquent behavior, policy makers and juvenile justice practitioners must address the economic, political, and social factors that make delinquency appear to be a viable choice for so many youths.

The Social Context of Delinquency Theories

theory A statement or set of statements that is designed to explain how two or more phenomena are related to one another.

Social, economic, and political contexts affect not only delinquent behavior, but also the explanations, or theories, of delinquency that gain prominence during particular historical periods. A **theory** of delinquency is a statement or a set of statements that is designed to explain how one or more events or factors are related to delinquency.[147] Such theories are important for two primary reasons. First, they help us make sense of delinquency and understand why it occurs. Second, they guide us in our attempts to reduce crime. Importantly, hypotheses about why delinquency occurs suggest actions we might take in order to reduce it.

Theories of delinquency, like other theories, can be in sharp conflict. Some of them are based on the assumption that political and economic conditions play a crucial role in generating delinquency within American society, whereas others treat delinquency as primarily the product of rational choices made by individual youths. Our object here is not to review these theories (an object more fitting for a book on criminological theory or juvenile delinquency) but to emphasize their role in helping us understand how and why particular factors appear to be related to juvenile crime and what their role is in guiding responses to the problem of delinquency.

Theories of delinquency, like delinquency itself, are products of a particular historical context. Most people, including those who study delinquency, have ideas about why youths engage in delinquent behavior, ideas that are influenced by their particular life experiences.[148] For example, people in medieval Europe favored explanations that treated deviant behavior as a product of otherworldly spirits. Such explanations made perfectly good sense to people within that particular historical context. Today, influenced by ideas derived from the social and behavioral sciences and by popular notions about human behavior, people are much less inclined to explain delinquency in ways that would make sense to medieval Europeans.

This does not mean that, at present, there is general agreement over the causes of delinquency. Indeed, the historical context within which we live is conducive to the promulgation of a variety of theoretical perspectives on delinquency and considerable debate over its causes. Moreover, the differing theoretical perspectives on delinquency lead to differing, sometimes opposing, responses to delinquent behavior. A theory based on the idea that delinquency is the product of choices made by rational actors leads to policies that stress punishment as a logical response. In contrast, a theory based on the idea that delinquency is the product of the oppression of youths calls forth a different type of policy response,[149] as would a theory that views delinquency as the product of abnormal thinking patterns. In short, social context influences theoretical explanations of delinquency, which in turn suggest various juvenile justice responses to the delinquency problem.

The Social Context of Juvenile Justice

The previous sections of this chapter examined the influence of social context on delinquency and on the ways we explain delinquency. Social context also influences the practice of juvenile justice. Indeed, as noted previously, different explanations of delinquency are associated with different juvenile justice responses. Other aspects of the sociopolitical environment influence the practice of juvenile justice as well, and some of these are examined in the remainder of this chapter.

■ The Influence of the Political Economy on the Practice of Juvenile Justice

The development, management, and distribution of political and economic resources not only has an impact on the behavior of young people in our society, but also influences the operation of important social institutions, including those institutions that comprise the so-called juvenile justice system. The economic resources that are available to law enforcement agencies, juvenile courts, and correctional agencies have a profound effect on the level of staffing, the types of programs operated, and the support (both personnel and materiel) provided to the individuals and agencies charged with responding to delinquency. For example, an increase in the funding for a probation department might allow more staff to be hired and additional types of interventions to be developed. In turn, this may lead to better services given to youths and their families and a reduction in juvenile offending. Similarly, reductions in funding levels can result in staff reductions, increased caseloads, additional stress for juvenile services workers, a reduction in services and monitoring of clients, and higher levels of recidivism.

Although the effects of changing levels of funding on the operation of juvenile justice agencies are easy to imagine, it should be kept in mind that such changes take place within a highly political environment. In other words, funding allocation decisions are the result of a political process in which various ideas and interests vie for supremacy. Moreover, this political process is found at the state, local, and agency level. At the state level, legislative as well as executive decisions can have a profound effect on the levels of funding available to juvenile justice agencies for correctional programs, staff salaries, staff hiring and promotions, and staff training. State-level political decision making plays a particularly important role in those states where juvenile courts and correctional agencies are state operated. However, even when juvenile justice agencies are county or city run, state-level political decisions can affect their operation. For example, political decisions made at the state level can determine the level of monetary and human resources, as well as the types of institutions and programs (and their operation) available to local courts for the treatment of youths.

Local political decision making also can have a profound impact on the operation of juvenile justice agencies. For example, when juvenile justice agencies are county run, political decisions made by county governments play a major role in determining the level of monetary and human resources available to the courts as well as the types of programs operated at the local level. Similarly, political decisions made within juvenile justice agencies also can affect the allocation of monetary and human resources, the amount and types of training given to staff, as well as the types of programs that are operated by those agencies.

Political decision making at the state, local, or agency level is often characterized by considerable conflict between groups possessing different ideologies and interests. State legislatures, county and city governments, executive agencies at the state and local level, and local juvenile justice agencies typically encompass groups and individuals that have conflicting views about delinquency and juvenile justice. As a result, conflict is a common element of juvenile justice practice.

The Influence of the Local Community and the Media on the Practice of Juvenile Justice

The local community and the media play a substantial role in determining juvenile justice practice. Many communities possess at least one group that seeks to influence the local response to juvenile delinquency. Indeed, the practice of juvenile justice, like the practice of criminal justice in general, is often a highly political endeavor. In addition, if juvenile justice decision makers, such as prosecuting attorneys and judges, are elected officials, they can be quite sensitive to public perceptions of their performance. As a result, support for various juvenile justice programs and practices can be extended or withdrawn depending on perceptions of "what the public wants."

The decisions made by prosecuting attorneys and judges also can influence the actions of other important decision makers. For example, in some states, court personnel (e.g., court administrators) and corrections staff (e.g., detention unit, probation, and other casework personnel) are, in effect, employees of the judge. As a result, a judge in one of these states can have tremendous influence over the allocation of juvenile justice resources. Even in jurisdictions where judges do not directly control probation and other juvenile justice staff, they are often able to exert considerable influence over other components of the juvenile justice process.

Courts that deal with juveniles are also political entities. Most juvenile court judges are elected officials who not only are accountable to the electorate, but also must contend with other elected officials for coveted tax dollars. Competition for local dollars among local officials and law enforcement or law-related agencies, such as prosecutors, sheriffs, and the courts, can force courts to make decisions about juvenile services based on economic considerations rather than the best interests of youths.

When juvenile courts are faced with budget limits restricting the number of probation staff or the amount of money available for out-of-home placements, certain actions by those courts are predictable. First, judges will limit out-of-home placements by increasing their tolerance for delinquent behavior and/or probation violations before removal. Second, many courts will try to develop dispositional alternatives, such as day treatment programs, expanded use of local detention facilities, and an increase in the number of foster home or group home beds. Although these strategies are fiscally responsible, they may allow more serious delinquents to remain in the community. As a result, courts can be caught between the public outcry to get dangerous delinquents off the streets and budgetary restraints that prevent them from heeding the public's demand for protection.

media-controlled perceptions People's perceptions of some phenomenon (or area of interest) that are substantially shaped by the media's coverage of that phenomenon. The public's perceptions of juvenile justice are determined largely by the media because the public lacks information from other sources about the daily operation of juvenile justice agencies.

The public influences nonelected juvenile justice decision makers as well as elected ones. Court administrators, middle management personnel such as chief probation officers, and other juvenile justice personnel are often sensitive to public perceptions and demands. For example, complaints made by organized community groups or perceptions held by local political leaders, including judges and prosecuting attorneys, that juvenile justice practices are at odds with community preferences can produce changes in these practices. Within the community, the media often play an important role in framing public perceptions of delinquency and the operation of juvenile justice. Indeed, much of what the public knows about juvenile justice practices is a reflection of **media-controlled perceptions**. This is not surprising, because the juvenile justice process has historically been a closed process intended to prevent the stigmatization of children who come before the court. However, the media may not present the most balanced view of juvenile justice practices to the public, preferring instead to focus on the most sensational cases and the most obvious failings. Whether media coverage is balanced or not, the public relies on the media for information about juvenile justice practices, and the information received by the public can be used by organized groups in their efforts to influence juvenile justice practices.

Chapter Summary

This chapter examined the context of juvenile justice, including juvenile justice theory and juvenile justice practice. More specifically, it looked at the political, economic, and social contexts within which delinquency occurs and described the ways that a variety of political, economic, and social factors (e.g., community conditions; family structure, relations, and socialization; school experiences; and peer relations) influence youths' involvement in illegal behavior. In addition, the chapter noted that the political, economic, and social contexts within which we live shape our ideas about why youths engage in delinquency and influence the ways in which we respond to delinquent behavior. Thus, these contexts exert considerable influence on the operation of juvenile justice. Consequently, recognizing and understanding their impact on delinquency, theory development, and juvenile justice practice is a requisite for developing more effective juvenile justice practices.

Key Concepts

accomplice network: The pool of potential co-offenders with whom a youth interacts.

broken home: A home or family in which at least one natural (biological) parent is absent.

capitalism: An economic system based upon three fundamental principles: (1) private ownership of property, (2) competition between economic interests, and (3) personal profit as a reward for economic risk and effort.

child abuse: Nonaccidental acts directed against children that result in physical, sexual, psychological, or emotional harm.

family: A collective body of any two or more persons living together in a common place under one head or management; a household; those persons who have a common ancestor or are from the same lineage; people who live together and who have mutual duties to support and care for one another.

group hazard hypothesis: The hypothesis that delinquency committed in groups is more likely to be detected and to result in a formal response by the authorities.

latchkey children: Children who care for themselves at least part of the day due to parental unavailability because of employment.

media-controlled perceptions: People's perceptions of some phenomenon (or area of interest) that are substantially shaped by the media's coverage of that phenomenon. The public's perceptions of juvenile justice are determined largely by the media because the public lacks information from other sources about the daily operation of juvenile justice agencies.

neglect: The nonaccidental failure of a caregiver to meet the physical, psychological, or emotional needs of a minor or another person in the charge of the caregiver.

political economy: The basic political and economic organization of society, including the setting of political priorities and the management and distribution of economic resources. People's economic status largely determines the opportunities that are available to them, their feeling of political empowerment, their ability to access the political system, and their ability to influence political and economic decision making.

power: The ability to influence political and economic decisions and actions; the ability to force, coerce, or influence a person's actions or thoughts.

social class: A person's or family's economic, social, and political standing in the community. Social class can be directly related to type of employment, neighborhood, circle of contacts, and friends.

social disorganization: The state of a neighborhood in which it is unable to exert control over its members. Social disorganization can result from a decrease in population, a large transient population, the existence of groups with differing values, little home ownership, low property values, proximity to industrial and commercial establishments, and a lack of agencies designed to promote a sense of community.

socialism: A political and economic system that is based on democratic decision making, equality of opportunity for all, collective decision making designed to further the interests of the entire community, public ownership of the means of production, and economic and social planning.

special interest group: An organization consisting of persons, corporations, unions, and so on, that have common political, and often economic, purposes and goals. These individuals or groups come together to use their influence, monetarily and politically, to persuade politicians and other persons in power to act favorably toward them.

theory: A statement or set of statements that is designed to explain how two or more phenomena are related to one another.

Review Questions

1. Why is delinquency theory important in juvenile justice?
2. How does the political economy of the United States affect children?
3. What are the essential elements of a capitalist economy?
4. How do the economic circumstances of adults and children affect the quality of their lives?
5. How is family structure related to delinquent behavior?
6. What types of family relationships are related to delinquency?
7. What kinds of violence are found in American families and how are they related to delinquent behavior?
8. What are the characteristics of school that encourage both school failure and delinquency?
9. What community characteristics are related to high levels of delinquent behavior?
10. What are the different types of youth peer groups that exist, and how do these groups encourage delinquency?
11. How do economic conditions influence the operation of juvenile justice agencies?
12. How do politics affect the operation of juvenile justice?

Additional Readings

Bowles, S. & Gintis, H. (1976). *Schooling in capitalist America: Educational reform and the contradictions of economic life.* New York: Basic Books.

Edwards, R. C., Reich, M., & Weisskopf, T. (Eds.). (1978). *The capitalist system: A radical analysis of American society* (2nd ed.). Englewood Cliffs, NJ: Prentice-Hall.

Ehrenreich, B. (2001). *Nickel and dimed.* New York: Henry Holt.

Howell, J. C. (2003). *Preventing and reducing juvenile delinquency: A comprehensive framework.* Thousand Oaks, CA: Sage.

McCord, J. (Ed.). (1997). *Violence and childhood in the inner city.* New York: Cambridge University Press.

McWhirter, J., McWhirter, B., McWhirter, A., & McWhirter, E. (1993). *At-risk youth: A comprehensive response.* Pacific Grove, CA: Brooks/Cole.

National Commission on Children. (1991). *Beyond rhetoric, A new American agenda for children and families. Final report of the National Commission on Children.* Washington, DC: U.S. Government Printing Office.

Schwendinger, H. & Schwendinger, J. S. (1985). *Adolescent subcultures and delinquency.* New York: Praeger Publishers.

Shulman, B. (2003). *The betrayal of work: How low-wage jobs fail 30 million Americans.* New York: The New Press.

Simons, R. L., Simons, L., & Wallace, L. (2004). *Families, delinquency, and crime: Linking society's most basic institution to antisocial behavior.* Los Angeles, CA: Roxbury Publishing Company.

Wilson, W. J. (1987). *The truly disadvantaged: The inner city, the underclass, and public policy.* Chicago: University of Chicago Press.

Notes

1. Pfohl, S. J. (1985). *Images of deviance and social control* (p. 9). New York: McGraw-Hill.
2. Jacobs, M. D. (1990). *Screwing the system and making it work: Juvenile justice in the no-fault society.* Chicago: University of Chicago Press. See this resource for an interesting account of how juvenile justice practitioners may work to manipulate the "system" in order to help clients.
3. National Commission on Children. (1993). *Ensuring income security.* Washington, DC: National Commission on Children.
4. Domhoff, W. G. (1978). *The powers that be: Processes of ruling class domination in America.* New York: Random House.
5. Parenti, M. (1978). *Power and the powerless* (2nd ed.). New York: St. Martin's Press.
6. Eitzen, D. S., Zinn, M. B., & Smith, K. E. (2009). *Social problems* (11th ed.). Boston: Allyn & Bacon.
7. Wrong, D. (1979). *Power: Its forms, bases, and uses* (p.2). New York: Harper & Row.
8. Wrong, 1979, p. 23.
9. DeNavas-Walt, C., Proctor, B. D., & Smith, J. (2007) *Income, Poverty, and Health Insurance Coverage in the United States: 2006* (U.S. Census Bureau, Current Population Reports, P60-233). Washington, DC: U.S. Government Printing Office.

10. Bernstein, J., Brocht, C., & Spade-Aguilar, M. (2000). *How much is enough? Basic family budgets for working families.* Economic Policy Institute. Available from http://www.epi.org/content.cfm/books_howmuch.

Edleman, P. (2001, August 1). The question now isn't just poverty: For many, it is survival. *Washington Spectator, 27,* 1–3.

11. DeNavas-Walt, Proctor, & Smith, 2007.

12. DeNavas-Walt, Proctor, & Smith, 2007.

13. Bernstein, Brocht, & Spade-Aguilar, 2000.

Shulman, B. (2003). *The betrayal of work: How low-wage jobs fail 30 million Americans.* New York: The New Press.

14. DeNavas-Walt, Proctor, & Smith, 2007.

15. The Economist (2006, June 17). Inequality and the American dream. *The Economist.*

16. Swanstrom, T., Dreier, P., & Molenkopf, J. (2002). Economic inequality and public policy. *City and Community, 1,* 349–372.

17. To view the United Nations Human Development Index rankings, see http://hdr.undp.org/en/statistics/.

18. The Children's Defense Fund. (2001). *The state of America's children, 2001.* Washington, DC: Children's Defense Fund.

19. Harrington, M. (1963). *The other America: Poverty in the United States.* Baltimore: Penguin Books.

Harrington, M. (1985). *The new American poverty.* Baltimore: Penguin Books.
Ryan, W. (1981). *Equality.* New York: Vintage Books.

20. The Children's Defense Fund. (2004). *The state of America's children, 2004.* Washington, DC: Children's Defense Fund.

21. Jaret, C., Reid, L., & Adelman, R. (2003). Black-white income inequality and metropolitan socioeconomic structure. *Journal of Urban Affairs, 25,* 305–333.
Swanstrom, Dreier, & Molenkopf, 2002.

22. Swanstrom, Dreier, & Molenkopf, 2002.

23. Swanstrom, Dreier, & Molenkopf, 2002.

24. Buss, T. & Reddman, F. S. (1983). *Mass unemployment: Plant closings and community mental health.* Beverly Hills, CA: Sage.

Gary, L. E. (1985). Correlates of depressive symptoms among a select population of black men. *American Journal of Public Health, 75,* 1220–1222.

25. Galambos, N., & Silbereisen, R. (1987). Income change, parental life outlook, and adolescent expectations for job success. *Journal of Marriage and the Family, 49,* 141–149.

Lempers, J., Clark-Lempers, D., & Simons, R. L. (1989). Economic hardship, parenting, and distress in adolescence. *Child Development, 60,* 25–39.

26. Institute for Research on Poverty, University of Wisconsin. (2008). *Is poverty different for different groups in the population?* University of Wisconsin-Madison. Retrieved from http://www.irp.wisc.edu/faqs/faq3.htm.

27. Shulman, 2003.

28. Duncan, G. J., & Brooks-Gunn, J. (1997). Consequences of growing up poor. New York: Russell Sage Foundation.

National Commission on Children. (1991). *Beyond rhetoric, A new American agenda for children and families. Final report of the National Commission on Children*. Washington, DC: U.S. Government Printing Office.

29. McLeod, J. D. & Shanahan, M. J. (1993). Poverty, parenting, and children's mental health. *American Sociological Review, 58,* 351–366.

30. UNICEF. (2007). Child poverty in perspective: An overview of child well-being in rich countries. *Innocenti Report Card 7.* Florence, Italy: UNICEF Innocenti Research Centre.

31. Gerth, H. & Mills, C. W. (1953). *Character and social structure: The psychology of social institutions.* (p. 313). New York: Harcourt, Brace & World.

32. Anderson, E. (1990). *Streetwise: Race, class, and change in an urban community.* Chicago: University of Chicago Press.

Anderson, E. (1997). Violence and the inner city street code. In J. McCord (Ed.), *Violence and childhood in the inner city*. New York: Cambridge University Press.

Sanchez-Jankowski, M. (1995). Ethnography, inequality, and crime in the low income community. In J. Hagan & R. D. Peterson (Eds.). *Crime and inequality.* Stanford, CA: Stanford University Press.

33. Elliott, D. S. & Huizinga, D. (1983). Social class and delinquent behavior in a national youth panel: 1976–1980. *Criminology, 21,* 149–177.

34. Jarjoura, G. R., Triplett, R. & Brinker, P. (2002). Growing up poor: Examining the link between persistent childhood poverty and delinquency. *Journal of Quantitative Criminology, 18,* 159–187.

35. Eitzen, Zinn, & Smith, 2009, p. 329.

36. Rapp, R. (1982). Family and class in contemporary America. In B. Thorne & M. Yalom (Eds.), *Rethinking the family: Some feminist questions* (pp. 168–187). New York: Longman.

37. Gove, W. & Crutchfield, R. (1982). The family and juvenile delinquency. *Sociological Quarterly, 23,* 301–319.

38. Wright, K. N. & Wright, K. E. (1994). *Family life, delinquency, and crime: A policymaker's guide, research summary.* Washington, DC: Office of Juvenile Justice and Delinquency Prevention.

39. National Commission on Children, 1991.

40. Fields, J. (2003). *Children's living arrangements and characteristics: March 2002* (Current Population Reports, P20-547). Washington, DC: U.S. Census Bureau.

41. Fields, J. & Casper, L. M. (2001). *America's families and living arrangements: March 2000* (Current Population Reports, P20-537). Washington, DC: U.S. Census Bureau.

National Commission on Children, 1991.

42. National Commission on Children, 1991.

43. Wilson, W. J. (1987). *The truly disadvantaged: The inner city, the underclass, and public policy*. Chicago: University of Chicago Press.

44. Canter, R. (1982). Family correlates of male and female delinquency. *Criminology, 20*, 149–167.

Glueck, S. & Glueck, E. (1950). *Unraveling juvenile delinquency.* Cambridge, MA: Harvard University Press.

Wells, L. E. & Rankin, J. (1991). Families and delinquency: A meta-analysis of the impact of broken homes. *Social Problems, 38*, 71–90.

Wilson, J. Q. & Herrnstein, R. (1985). *Crime and human nature.* New York: Simon & Schuster.

45. Wells & Rankin, 1991.

46. Toby, J. (1957). The differential impact of family disorganization. *American Sociological Review, 22*, 505–512.

47. Chilton, R. & Markle, G. (1972). Family disruption, delinquent conduct, and the effects of subclassification. *American Sociological Review, 37*, 93–99.

48. Johnson, R. E. (1986). Family structure and delinquency: General patterns and gender differences. *Criminology, 24.*

49. Gove and Crutchfield, 1982.

50. Thornberry, T. P., Smith, C. A., Rivera, C., Huizinga, D., & Stouthamer-Loeber, M. (1999). Family disruption and delinquency. *Juvenile Justice Bulletin.* Washington, DC: Office of Juvenile Justice and Delinquency Prevention.

51. Cernkovich, S. A. & Giordano, P. C. (1987). Family relationships and delinquency. *Criminology, 25*, 295–321.

Laub, J. H. & Sampson, R. J. (1988). Unraveling families and delinquency: A reanalysis of the Gluecks' data. *Criminology, 26*, 355–380.

52. Hirschi, T. (1969). *Causes of delinquency.* Berkeley, CA: University of California Press.

53. Rutter, M. & Giller, H. (1984). *Juvenile delinquency: Trends and perspectives.* New York: Gilford Press.

54. Loeber, R. & Stouthammer-Loeber, M. (1986). Family factors as correlates and predictors of juvenile conduct problems and delinquency. In M. Tonry & N. Morris (Eds.). *Crime and justice: An annual review of research* (Vol. 7). Chicago: University of Chicago Press.

55. Curran, D. J. & Renzetti, C. M. (2000). *Social problems: Society in crisis* (5th ed.). Boston: Allyn & Bacon.

56. Bureau of Labor Statistics. (2003). *Employment characteristics of families in 2002.* Washington, DC: U.S. Department of Labor.

57. National Commission on Children. (1993). *Just the facts: A summary of recent information on America's children and their families.* Washington, DC: National Commission on Children.

58. National Commission on Children, 1991.

59. Vandivere, S., Tout, K., Capizzano, J., & Zaslow, M. (2003). *Left unsupervised: A look at the most vulnerable children* (Child Trends Research Brief Publication No. 2003-05). Washington, DC: Child Trends.

60. Ehrle, J., Adams, G., & Tout, K. (2001). *Who's caring for our youngest children? Child care patterns of infants and toddlers* (Occasional Paper No. 42). Washington, DC: The Urban Institute.

Galtry, J. (2002). Child health: An underplayed variable in parental leave policy debates? *Community, Work and Family, 5*, 257–278.

61. Kagan, S. L. & Neuman, M. J. (1997). Defining and implementing school readiness: Challenges for families, early care and education, and schools. In R. Weissberg, T. Gullotta, R. Hampton, B. Ryan, & G. Adams (Eds.), *Healthy children 2010: Establishing preventive services*. Thousand Oaks, CA: Sage.
62. Vandivere, Adams, & Tout, 2003.
63. Hirschi, T. (1985). Crime and family policy. In R. A. Weisheit & R. G. Culbertson (Eds.), *Juvenile delinquency: A justice perspective*. Prospect Heights, IL: Waveland Press.
64. Laub and Sampson, 1988.
65. Vander Ven, T., Cullen, F., Carrozza, M., & Wright, J. (2001). Home alone: The impact of maternal employment on delinquency. *Social Problems, 48*, 236–257.
66. Colvin, M. & Pauly, J. (1983). A critique of criminology: Toward an integrated structural-Marxist theory of delinquency production. *American Journal of Sociology, 90*, 513–551.
67. Hagan, J. (1989). *Structural criminology* (p. 13). New Brunswick, NJ: Rutgers University Press.
68. Hagan, J., Gillis, A. R., & Simpson, J. (1985). The class structure of gender and delinquency: Toward a power-control theory of common delinquent behavior. *American Journal of Sociology, 90*, 1151–1178.

    Hagan, J., Simpson, J., & Gillis, A. R. (1987). Class in the household: A power-control theory of gender and delinquency. *American Journal of Sociology, 92*, 788–816.

    Hagan, J., Gillis, A. R., & Simpson, J. (1990). Clarifying and extending power-control theory. *American Journal of Sociology, 95*, 1024–1037.

    Blackwell, B. S. (2000). Perceived sanction threats, gender, and crime: A test and elaboration of power-control theory. *Criminology, 38*, 439–488.
69. Hanson, C. (1984). Demographic, individual, and familial relationship correlates of serious and repeated crime among adolescents and their siblings. *Journal of Counseling and Clinical Psychology, 52*, 528–538. Hirschi, 1969.

    Laub & Sampson, 1988.

    Rankin, J. H. & Kern, R. (1994). Parental attachments and delinquency. *Criminology, 32*, 495–515.

    Smith, R. & Walters, J. (1978). Delinquent and nondelinquent males' perceptions of their fathers. *Adolescence, 13*, 21–28.
70. Farrington, D. P. (1995). The development of offending and antisocial behaviour from childhood: Key findings from the Cambridge study in delinquent development. *Journal of Child Psychology and Psychiatry, 360*, 929–964.

    Farrington, D. P. (1996). The explanation and prevention of youthful offending. In J. D. Hawkins (Ed.), *Delinquency and crime: Current theories*. Cambridge, England: Cambridge University Press.

    Farrington, D. P. (1996). Criminological psychology: Individual and family factors in the explanation and prevention of offending. In C. R. Hollin (Ed.), *Working with offenders*. Chichester, England: John Wiley.

    Henggeler, S. (1989). *Delinquency in adolescence*. Newbury Park, CA: Sage.

Sampson, R. J. & Laub, J. H. (1993). *Crime in the making.* Cambridge, MA: Harvard University Press.

Wright and Wright, 1994.

71. Laub & Sampson, 1988.

72. Amato, P. & Keith, B. (1991). Parental divorce and the well-being of children: A meta-analysis. *Psychological Bulletin, 110*, 26–46.

Grych, J. H. & Fincham, F. D. (1990). Marital conflict and children's adjustment: A cognitive–contextual framework. *Psychological Bulletin, 108*, 267–290.

Hershorn, M. & Rosenbaum, A. (1985). Children of marital violence: A closer look at the unintended victims. *American Journal of Orthopsychiatry, 55*, 260–266.

Jaffe, P., Wlofe. D., Wilson, S., & Zak, L. (1986). Similarities in behavior and social maladjustment among child victims and witnesses to family violence. *American Journal of Orthopsychiatry, 56*, 142–146.

Thornberry, Smith, Rivera, Huizinga, & Stouthamer-Loeber, (1999). Wright and Wright, 1994.

73. Agnew, R. (1983). Physical punishment and delinquency: A research note. *Youth and Society, 15*, 225–236.

Laub & Sampson, 1988.

Patterson, G. R., DeBaryshe, B., & Ramsey, E. (1989). A developmental perspective on antisocial behavior. *American Psychologist, 44*, 329–335. Wright and Wright, 1994.

74. U.S. Department of Health and Human Services, Administration on Children, Youth and Families (2009). *Child maltreatment 2007.* Washington, DC: U.S. Government Printing Office.

75. Chalk, R., Gibbons, A., & Scarupa, H. (2002). *The multiple dimensions of child abuse and neglect: New insights into an old problem.* (Child Trends Research Brief). Washington, DC: Child Trends.

76. Widom, C. S. & Maxfield, M. G. (2001). An update on the cycle of violence. *National Institute of Justice Research in Brief.* Washington, DC: U.S. Department of Justice.

77. Tjaden, P. & Thoennes, N. (2000). *Full report of the prevalence, incidence, and consequences of violence: Findings from the National Violence Against Women Survey.* Washington, DC: National Institute of Justice.

78. Gelles, R. J. & Cornell, C. P. (1985). *Intimate violence in families.* Beverly Hills, CA: Sage.

79. Kolbo, J. R. (1996). Risk and resilience among children exposed to family violence. *Violence and Victims, 11*, 113–128.

80. National Clearinghouse on Child Abuse and Neglect Information. (2003). *In harm's way: Domestic violence and child maltreatment.* Washington, DC: National Clearinghouse on Child Abuse and Neglect Information.

81. Tjaden & Thoennes, 2000.

82. U. S. Census Bureau, American Fact Finder. *School Enrollment.* Retrieved July 18, 2009 from http://factfinder.census.gov/servlet/STTable?_bm=y&-geo_id=01000US&-qr_name=ACS_2006_EST_G00_S1401&-ds_name=ACS_2006_EST_G00_.

83. U.S. Office of Education. (1969). *Digest of educational statistics.* Washington, DC: U.S. Government Printing Office.

National Center for Education Statistics. (2008). *The condition of education 2008.* Washington, DC: Department of Education.

84. Polk, K. & Schafer, W. E. (Eds.). (1972). *Schools and delinquency.* Englewood Cliffs, NJ: Prentice Hall.

85. Johnson, G., Bird, T., & Little, J. W. (1979). *Delinquency prevention: Theories and strategies.* Washington, DC: Office of Juvenile Justice and Delinquency Prevention.

86. Elrod, H. P. & Friday, P. C. (1986, October). *Delinquency reduction through school organizational change: Some thoughts on the relationship between theory, process and outcomes.* Paper presented at the American Society of Criminology, Atlanta.

87. National Center for Education Statistics. (1999). *Annual earnings of young adults, by educational attainment* (Indicator of the Month). Washington, DC: National Center for Education Statistics.

88. Day, J. C. & Newburger, E. C. (2002). *The big payoff: Educational attainment and synthetic estimates of work–life earnings* (Current Population Reports). Washington, DC: U. S. Census Bureau.

89. Lerner, R. M. (1995). *America's youth in crisis.* (pp. 4–5). Thousand Oaks, CA: Sage.

90. Jensen, G. & Rojek, D. (1998). *Delinquency and youth crime* (3rd ed.). (p. 273). Prospect Heights, IL: Waveland.

91. Phillips, C. & Kelly, D. H. (1979). School failure and delinquency: Which causes which? *Criminology, 17*, 194–207.

92. Maguin, E. & Loeber, R. (1996). Academic performance and delinquency. In M. Tonry (Ed.), *Crime and justice: A review of research* (Vol. 20). Chicago: University of Chicago Press.

93. Agnew, R. (1985). A revised strain theory of delinquency. *Social Forces, 64*, 151–167.

Hirschi, 1969.

Hindelang, M. J. (1973). Causes of delinquency: A partial replication. *Social Problems, 21*, 471–487.

94. Boesel, D., Crain, R., Dunteman, G., Ianni, F., Martinolich, M., Moles, O., et al. (1978). *Violent schools—Safe schools: The safe schools study report to the Congress* (Vol. 1) Washington, DC: U.S. Department of Health, Education and Welfare.

95. Agnew, 1979.

Elrod, P., Soderstrom, I., & May, D. (2009). Theoretical predictors of delinquency in and out of school among a sample of rural public school youth. *Southern Rural Sociology, 23*, 131–156.

Hirschi, 1969.

96. Cohen, A. K. (1955). *Delinquent boys.* New York: The Free Press.

97. Bowles, S. & Gintis, H. (1976). *Schooling in capitalist America: Educational reform and the contradiction of economic life.* New York: Basic Books.

Gold, M. (1978). School experiences, self-esteem, and delinquent behavior. *Crime and Delinquency, 24*, 290–308.

Polk & Schafer, (Eds.), 1972.

98. Schafer, W. E., Olexa, C., & Polk, K. (1972). Programmed for social class tracking in high school. In K. Polk & W. E. Schafer (Eds.), *School and delinquency.* Englewood Cliffs, NJ: Prentice Hall.

99. Schafer, Olexa, & Polk, 1972.

Oakes, J. (1985). *Keeping track: How schools structure inequality.* New Haven, CT: Yale University Press.

100. Boesel, D., Crain, R., Dunteman, G., Ianni, F., Martinolich, M., Moles, O., et al.1978. Polk & Schafer, (Eds.), 1972.

Wertleib, E. L. (1982). Juvenile delinquency and the schools: A review of the literature. *Juvenile and Family Court Journal, 33,* 15–24.

101. Boesel, Crain, R., Dunteman, G., Ianni, F., Martinolich, M., Moles, O., et al. 1978. Polk & Schafer, (Eds.), 1972.

Stinchcombe, A. (1964). *Rebellion in a high school.* Chicago: Quadrangle Press.

102. Peng, S. S. & Takai, R. T. (1983). *High school dropouts: Descriptive information from high school and beyond* (ERIC No. ED 236 3666). Washington, DC: National Center for Education Statistics.

103. Ekstrom, R. B., Goertz, M., Pollack, J., & Rock, D. (1986). Who drops out of high school and why? Findings from a national study. *Teacher's College Record, 87,* 356–373.

104. McWhirter, J., McWhirter, B., McWhirter, A., & McWhirter, E. (1993). *At-risk youth: A comprehensive response.* Pacific Grove, CA: Brooks/Cole.

105. Ekstrom, Goertz, Pllack, & Rock, 1986.

Kaufman, P., Alt, M., & Chapman, C. (2001). *Dropout rates in the United States: 2000.* Washington, DC: National Center for Education Statistics.

106. Elliott, D. S. & Voss, H. L. (1974). *Delinquency and dropout.* Lexington, MA: Lexington Books.

107. Thornberry, T. P., Moore, M., & Christenson, R. L. (1985). The effect of dropping out of high school on subsequent criminal behavior. *Criminology, 23,* 3–18.

108. Krohn, M. T., Thornberry, T., Collins-Hall, L. & Lizorre, A. (1995). School dropout, delinquent behavior, and drug use. In H. Kaplan (Ed.), *Drugs, crime, and other deviant adaptations: Longitudinal studies.* New York: Plenum Press.

109. National Center for Education Statistics, 2008.

110. Toby, J. (1983). *Violence in schools.* Washington, DC: National Institute of Justice.

111. Bursik, R. J., Jr. & Grasmick, H. G. (1993). *Neighborhoods and crime: The dimensions of effective community control.* New York: Lexington Books.

112. Bernard, T. J. (1992). *The cycle of juvenile justice.* New York: Oxford University Press.

Rothman, D. J. (1971). *The discovery of the asylum: Social order and disorder in the New Republic.* Boston: Little, Brown & Co.

113. Shaw, C. R. & McKay, H. D. (1931). Social factors in juvenile delinquency. In *National commission on law observance and enforcement: Report on the causes of crime* (Vol. 2). (Publication No. 13).Washington, DC: U.S. Government Printing Office.

114. Shaw, C. R. & McKay, H. D. (1972). *Juvenile delinquency in urban areas* (rev. ed.). Chicago: University of Chicago Press.

115. Shaw & McKay, 1972.

116. Shaw & McKay, 1972.

117. Lundman, R. J. (1993). *Prevention and control of juvenile delinquency* (3rd ed.). New York: Oxford University Press.

118. Snodgrass, J. (1982). *The jackroller at 70.* Chicago: University of Chicago Press.

119. Schlossman, S. & Sedlack, M. (1983). *The Chicago Area Project revisited.* Santa Monica, CA: Rand.

Sorrentino, A. (1977). *Organizing against crime: Redeveloping the neighborhood.* New York: Human Sciences Press.

120. Simcha-Fagan, O. & Schwartz, J. E. (1986). Neighborhood and delinquency: An assessment of contextual effects. *Criminology, 24,* 667–703.

121. Bursik, R. J., Jr. (1986). Ecological stability and the dynamics of delinquency. In A. J. Reiss, Jr., & M. Tonry (Eds.), *Communities and crime.* Chicago: University of Chicago Press.

122. Schuerman, L. & Kobrin, S. (1986). Community careers in crime. In A. J. Reiss, Jr., & M. Tonry (Eds.). *Communities and crime.* Chicago: University Of Chicago Press.

123. McDonald, S. C. (1986). Does gentrification affect crime rates? In A. J. Reiss, Jr., & M. Tonry (Eds.), *Communities and crime.* Chicago: University of Chicago Press.

124. Hawkins, J. D., Herrenkohl, T., Farrington, D. P., Brewer, D., Catalano, R. F., & Harachi, T. W. (1998). A review of predictors of youth violence. In R. Loeber & D. P. Farrington (Eds.), *Serious and violent juvenile offenders.* Thousand Oaks, CA: Sage.

Krivo, L. J. & Petersen, R. D. (1996). Extremely disadvantaged neighborhoods and urban crime. *Social Forces, 75,* 619–650.

Peeples, F. & Loeber, R. (1994). Do individual factors and neighborhood context explain ethnic differences in juvenile delinquency? *Journal of Quantitative Criminology, 10,* 141–157.

125. Harries, K. & Powell, A. (1994). Juvenile gun crime and social stress: Baltimore, 1980–1990. *Urban Geography, 15,* 45–63.

Hawkins, Herrenkohl, Farrington, Brewer, Catalano, & Harachi, 1998.

Spelman, W. (1993). Abandoned buildings: Magnets for crime? *Journal of Criminal Justice, 21,* 481–493.

126. Warr, M. (1996). Organization and instigation in delinquent groups. *Criminology, 34,* 11–37.

Haynie, D. L. (2002). Friendship networks and delinquency: The relative nature of peer delinquency. *Journal of Quantitative Criminology, 18,* 99–134.

127. Elliott, D. & Menard, S. (1996). Delinquent friends and delinquent behavior: Temporal and developmental patterns. In J. D. Hawkins (Ed.), *Delinquency and crime: Current theories.* Cambridge, England: Cambridge University Press.

128. Klein, M. W. (1969). On the group context of delinquency. *Sociology and Social Research, 54,* 63–71.

129. Warr, 1996.

130. Fagan, J. (1989). The social organization of drug use and drug dealing among urban gangs.

Criminology, 27, 633–669.

Huff, C. R. (1989). Youth gangs and public policy. *Crime and Delinquency, 35*, 524–537.

Yablonsky, L. (1959). The delinquent gang as a near group. *Social Problems, 7*, 108–117.

131. Yablonsky, 1959.

132. Esbensen, F. A., Huizinga, D., & Weiher, A. W. (1993). Gang and non-gang youth: Differences in explanatory variables. *Journal of Contemporary Criminal Justice, 9*, 94–116.

Hill, K. G. Lui, C., & Hawkins, J. D. (2001). Early precursors of gang membership: A study of Seattle youth. *Juvenile Justice Bulletin*. Washington, DC: Office of Juvenile Justice and Delinquency Prevention.

Thornberry, T. P., Krohn, M., Lizotte, A., & Chard-Wierschem, D. (1993). The role of gangs in facilitating delinquent behavior. *Journal of Research in Crime and Delinquency, 30*, 55–87.

133. Egley, A. & O'Donnell, C. E. (2008). *Highlights of the 2006 National Youth Gang Survey* (OJJDP Fact Sheet). Washington, DC: Office of Juvenile Justice and Delinquency Prevention.

Thornberry, T. P. (1998). Membership in youth gangs and involvement in serious and violent offending. In R. Loeber & D. P. Farrington (Eds.), *Serious and violent juvenile offenders*. Thousand Oaks, CA: Sage.

Weisel, D. L. (2002). The evolution of street gangs: An examination of form and variation. In W. L. Reed and S. H. Decker (Eds.), *Responding to gangs, evaluation and research*. Washington, DC: National Institute of Justice.

134. Egley, A. & Major, A. K. (2003). *Highlights of the 2001 National Youth Gang Survey* (OJJDP Fact Sheet). Washington, DC: Office of Juvenile Justice and Delinquency Prevention.

Miller, W. B., Geertz, H., & Cutter, H. S. G. (1961). Aggression in a boy's street-corner group. *Psychiatry, 24*, 283–298.

Miller, W. B. (1975). *Violence by youth gangs.* Washington, DC: U.S. Government Printing Office.

135. Klein, M. W., Maxson, C., & Miller, J. (Eds.). (1995). *The modern gang reader.* Los Angeles: Roxbury.

136. Thornberry, 1998.

137. Curry, G. D. & Spergel, I. A. (1992). Gang involvement and delinquency among Hispanic and African-American adolescent males. *Journal of Research in Crime and Delinquency,* 29, 273–291.

Jackson, P. I. (1991). Crime, youth gangs, and urban transition: The social dislocations of postindustrial economic development. *Justice Quarterly*, *8*, 379–397.

Vigil, J. D. (1988). *Barrio gangs.* Austin, TX: Texas University Press.

138. Curry & Spergel, 1992.

Klein, M. W. (1971). *Street gangs and street workers.* Englewood Cliffs, NJ: Prentice Hall.

Thornberry, 1998. Vigil, 1998.

139. Hill, Lui, & Hawkins, 2001.

Thornberry, 1998.

140. Jankowski, M. S. (1991). *Islands in the street: Gangs and American urban society.* Berkeley, CA: University of California Press.

141. Vigil, 1998.

Miller, W. B. (1958). Lower class culture as a generating milieu of gang delinquency. *Journal of Social Issues, 14*, 5–19.

142. Jankowski, 1991.

143. Thornberry, 1998.

144. Gold, M. (1970). *Delinquent behavior in an American city.* Belmont, CA: Brooks/Cole.

145. Schwendinger, H. & J. S. Schwendinger. (1985). *Adolescent subcultures and delinquency*. New York: Praeger Publishers.

146. Berger, R. J. (1991). Adolescent subcultures, social type metaphors, and group delinquency. In R. J. Berger (Ed.), *The sociology of juvenile delinquency*. Chicago: Nelson-Hall.

Schwartz, G. & Merten, D. (1967). The language of adolescence: An anthropological approach to the youth culture. *American Journal of Sociology, 72*, 453–468.

Warr, 1996 and Haynie, 2002 contain complimentary studies.

147. Curran, D. J. & Renzetti, C. M. (2001). *Theories of crime.* Boston: Allyn & Bacon.

148. Lilly, J. R., Cullen, F. T., & Ball, R. A. (2007). *Criminological theory: Context and consequences* (4th ed.). Thousand Oaks, CA: Sage.

149. Regoli, R. M. & Hewitt, J. D. (2003). *Delinquency in society.* New York: McGraw-Hill. This resource contains an explication of a theory focusing on child oppression.

Early Juvenile Justice: Before the Juvenile Court

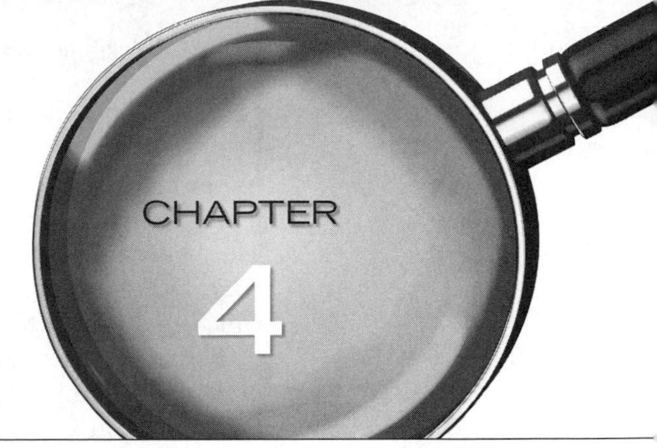

CHAPTER OBJECTIVES

After studying this chapter, you should be able to

- Describe how the young have historically been viewed in Western societies
- Describe the characteristics of colonial villages that encouraged conformity among children and adults
- Explain how the social context of the early 1800s led to the development of the first specialized correctional institutions for youths in the United States
- Describe the original goals of the houses of refuge and the mechanisms they used to reform youths
- Describe the shortcomings of the houses of refuge
- Describe the development and operation of placing out
- Describe the shortcomings of placing out
- Describe the role that John Augustus played in the development of probation
- Describe the focus of early juvenile probation programs
- Describe the rationale for the development of reform schools, industrial schools, and training schools
- Describe the operation of early reform schools, industrial schools, and training schools
- Describe the development and operation of early institutions for females

CHAPTER OUTLINE

- Introduction
- Families and Children in Developing European Societies: The Medieval View of Childhood
- Early Social Control: The Role of the Family
- Children in the New World
- The Social Control of Children in the Colonies
- Families and Children in the 1800s
- The Development of Early Juvenile Correctional Institutions in the United States
- Institutional Care: A Legacy of Poor and Often Inhumane Treatment
- Legal Issues
- Chapter Summary
- Key Concepts
- Review Questions
- Additional Readings
- Notes

Introduction

From a historical perspective, the notion of juvenile delinquency and the establishment of separate formal social control institutions for youths are recent developments. Throughout most of recorded history, childhood did not enjoy the special status we now accord it. Our modern view that childhood and adolescence are special times during which the young need nurturing and guidance for their healthy development did not exist until the later part of the Middle Ages. Before this, the young were seen either as property or as miniature adults who were expected to assume the same responsibilities as other adults by age 5 or 6 years.[1] Because people did not recognize childhood as a distinct period in human development, they did not see a need to create a separate legal process to deal with youths who violated community norms or laws.

As the ideas of childhood and adolescence developed, however, the development of institutions designed for children began to be seen as a logical response to problem youths. As a result, during the 1800s, several institutions designed to control problem youths and change their behavior were established in the United States. However, the development of these early institutions, like the development of contemporary correctional responses to youths, is best understood by examining the historical context within which these developments occurred. This chapter begins by examining the position of youths in early European societies and the role that families played in social control. Next, it examines the colonial period and the types of social control mechanisms that pervaded colonial life. The final section looks at the social context that led to the development of the first specialized correctional institutions for youths in the United States.

Families and Children in Developing European Societies: The Medieval View of Childhood

How adults have responded to children throughout history has depended on existing conceptions of the young—conceptions that are best understood within their historical context. During the Middle Ages, life was difficult for most people. During the fifth century, the collapse of the Western Roman Empire, which had dominated Western civilization for centuries, created many uncertainties for people in western Europe. Although our knowledge of children's lives during this time is limited, there are some indications that the young often fared poorly. For example, there is evidence that infanticide (i.e., the practice of killing children), which had been common since antiquity, continued to be practiced during the Middle Ages and beyond.[2] Historical evidence from this period indicates that mothers sometimes deliberately suffocated their offspring or abandoned them in the streets or latrines. According to one priest in 1527, "The latrines resounded with the cries of children who have been plunged into them."[3]

FYI

Throughout History, Youths Have Sometimes Been Spared the Harshest Punishments

Although a separate legal process for the young did not exist at this time, there is evidence that youths were sometimes spared the harshest punishments because of their age.

MYTH VS REALITY

Contemporary Conceptions of Childhood Are a Recent Historical Development

Myth—The concept of childhood as a period when the young need and deserve nurturing and protection has always existed.

Reality—Contemporary ideas about childhood are historically recent. During early times, children were often treated as property and were expected to act like adults from a very young age.

Infants who were born deformed or were felt to be too burdensome were particularly vulnerable. In other instances, destitute parents abandoned their unwanted children or took them to monasteries to be raised by monks.[4]

Life in Europe during the Middle Ages was very different than it is today. Although the nobility, some members of the clergy, and skilled craftsmen were relatively well off, many people led a precarious existence. There were wars, rebellions, disagreements between citizens, and fires (most structures were made of wood), all of which posed regular threats to personal safety. Homelessness, drunkenness, and crime were problems in the cities. In addition, sanitation practices represented a serious threat to public health. Refuse was discarded in the streets, which contained open sewers. The refuse heaps attracted rats and other rodents and served as a breeding ground for disease.[5]

Life tended to be short. The young were particularly at risk from various threats, including plagues and famine. Moreover, common child-rearing practices led to the premature death of many children. Swaddling (i.e., wrapping a child tightly in a long bandage) was a common practice. This prevented the child from wandering away, crawling into a sewer, or knocking over candles, which was a major concern because homes and buildings were usually constructed of wood and they contained a variety of highly flammable materials.[6] Swaddling also made the child easier to handle and protected it from air, sunlight, and soap, which were felt to present a threat to healthy growth.[7] Other common practices included feeding children from a horn and wrapping food in rags for them to suck. Such practices seem appalling by today's standards, but they were accepted methods of caring for the young during this time.[8]

Although the Middle Ages is often seen as a period of extreme hardship during which little social and intellectual progress occurred, by the later part of the Middle Ages Europe experienced a revival in trade and town life.[9] Moreover, by this time the concepts of childhood and adolescence had emerged, at least among some of the wealthier members of medieval society. Although historians are not in agreement as to the exact period during which childhood and adolescence began to be recognized as distinct stages of life, there is some evidence that many people began to see the young differently by the 1400s. For example, accounts of village life in southern France during the 1300s indicate that parents displayed love for their children and many grieved when their children died.[10] In addition, medical and scientific texts, literary works, and folk terminology used during the later Middle Ages in England indicate that many adults were concerned with helping children and youths make the transition to adulthood.[11] Less frequently were the young seen as miniature adults or as simply the property of their parents. Childhood was beginning to be accepted as a unique period in an individual's development, a period during which the individual should be molded and guided in order to become a moral and productive member of the community.

feudalism A system of land tenure governed by relations between nobles and lesser nobles whereby land (fief) was given by an overlord to a lesser noble (vassal) to hold (not to own) in return for military service. The land could then be given to still lesser nobles. This system depended on the agricultural production of the peasants (serfs) who worked the land.

Changes in the conception of childhood coincided with a variety of other changes in social, political, and economic life. **Feudalism**, a form of social organization characterized by relationships between nobles (i.e., knights and greater lords), the exploitation of the peasantry by the nobility, and subsistence farming, dominated political and economic life in Europe for centuries. However, substantial increases in the size of the population, as well as the use of agricultural methods that depleted the land, strained the ability of the feudal agriculture system to feed everyone, particularly when harvests were poor.[12] Furthermore, famine and disease claimed many lives, particularly the lives of children. It is estimated that as much as two-thirds of the population died before age 20 years, and the average life expectancy was about 30 years.[13]

Despite high mortality rates, however, the number of people in Europe grew during the 1500s and 1600s, which placed considerable pressure on both towns and rural areas. Simultaneously, political and economic power began to shift away from the church to a few monarchs who were molding strong centralized states and to a growing merchant and capitalist class. As a result, feudal forms of political and economic organization, which had been based on a subsistence system of agriculture, began to give way to a trade-oriented capitalistic system based on cash crops and the consolidation of land.[14]

Because of these broad changes in the economic and political organization of European states, large numbers of peasants were displaced from the land as landowners developed a capitalistic pasturage system geared more toward the production of cash crops. A result was the migration of peasants into the cities that strained the cities' resources and produced conflict between established workers and artisans. This conflict occurred because the influx of peasants threatened to drive down city wages and disrupt the status quo. The result was rising levels of public disorder and crime.[15]

In order to control the threat of the "dangerous classes," towns and cities passed legislation intended to restrict or discourage the immigration of peasants. For example, "poor laws" were enacted that were intended to prevent peasants from obtaining citizenship. The laws restricted their ability to join guilds, and, in some cases, barred them from entering the cities. In England, the Statute of Artificers was passed in 1562, which restricted access to certain trades and compelled many of the rural young to remain in the countryside.[16] Despite the efforts of city dwellers and the wealthy to restrict movement into the cities, however, there was no real alternative for many of the rural poor. As a result, cities continued to grow, and urban institutions, such as the guilds and the courts, came under increasing pressure.[17]

Early Social Control: The Role of the Family

family A social unit responsible for, among other things, the raising and socialization of children. Historically, the family in its various forms has been society's tool for teaching children how to act, what to believe, and what values to hold (see also Chapter 3).

During the Middle Ages and through the 1500s and 1600s, the **family** served as the primary unit of social control. Within the family, power was typically vested in the father, and the mother and the children were expected to obey his commands.[18] However, the nuclear family of today was relatively rare in the Middle Ages. Rather, a typical family consisted of an extended group of related people and other individuals who were not related. The father, particularly if he was a merchant, might spend considerable time away from the home. High mortality rates meant that either parent might have died and been replaced by a stepparent. In addition, servants, apprentices, and journeymen sometimes lived in the household, and neighbors were often present.[19] Consequently, children were exposed to a large number of adults who played a role in their socialization and control.

> **FYI**
>
> **The Bridewell Was an Early Correctional Institution for Youths and Adults**
> The first Bridewell was opened by private businessmen under a contractual agreement with the local government. It was designed to control a range of people, both young and old, who were poor, idle, and involved in petty crimes, prostitution, and lewd conduct. These institutions stressed work and, in some instances, children were committed to the Bridewell by their parents, who believed these institutions would help the child learn the value of labor.

Because of the **binding out** and **apprenticeship** systems, the family unit played a major role in the socialization and control of not only those born into the family, but also other youths. The practice of binding out young people consisted of placing them in the service of others for a specific period of time, from a week to many years. Although some adults became servants when they were elderly and some children became servants when they were as young as age 7 years, most who entered servitude did so in their teens. Binding out presented a way for poor youths to earn and save money, obtain room and board, learn a skill, prepare for adulthood, and meet a potential wife or husband.[20]

> **binding out** The placing of a youth or adult in the service of others for a specific period of time, from a week to many years. Although children as young as age 7 years and elderly adults could be placed as servants, the practice was mostly restricted to children in their teens. Binding out was a way for a poor youth to earn and save money, obtain room and board, learn a skill, prepare for adulthood, and meet a potential wife or husband.

Although binding out was common among the lower class, apprenticeships were usually reserved for the wealthy. Typically, an apprenticeship began when a youth was age 14 to 18 years and lasted 10–12 years. A distinguishing characteristic of apprenticeships, that set them apart from binding out, was that the master was obligated to teach the apprentice a trade as well as control their behavior.[21]

> **apprenticeship** A period, set by contract, during which a novice worked for a master in return for instruction in the master's trade. Apprenticeships were typically reserved for middle-class or wealthy youths. They usually began when youths were ages 14 to 18 years and lasted 10–12 years.

Despite the existence of the binding out and apprenticeship systems and the recognized sovereignty of adults—particularly the sovereignty of adult males over their children, apprentices, and servants—undisciplined and destitute youths, social and political unrest, and crime were common problems. Indeed, the number of poor and destitute people (many of whom were children) engaged in begging, prostitution, and petty crime in London had become so large that in 1556 an institution specifically designed to control these people, the Bridewell, was established.[22] However, youth crime and beggary were not confined to London. Consequently, Parliament passed a law in 1576 that created similar institutions in each county in England.[23]

Institutions built along the Bridewell model also spread to the continent. In 1595, the best known of these, the Amsterdam House of Corrections, was established to deal, in part, with the youth crime problem of the day. It and similar institutions were intended to instill discipline and teach youths to be hard-working and productive members of society.[24]

Early youth correctional institutions accepted a wide range of children—the destitute, the infirm, the needy, and serious offenders. Some youths were placed in these institutions by parents who believed that, through institutionalization, their children would learn to be more obedient. Whether youths benefited from such placement is debatable. What is not debatable is that the founders of these correctional institutions saw them as a source of cheap labor for local industries.[25]

Children in the New World

Population pressures, high mortality rates, the desire for land, political unrest, religious persecution, and the desire for greater wealth and influence led to a period of exploration and colonization after 1500. Children were counted on to play an important role in the expansion of European influence and became an integral part of plans to colonize the New World. Indeed, for some people, the colonization of the New World was viewed as an opportunity for children to be involved in productive work. Merchants saw children as potential providers of the labor that would be needed to produce the many goods necessary for survival in the colonies. Some people felt that work was good for children because it kept them occupied and insulated them from the temptations of the street. For others, such as the Puritans, the New World represented an environment where their religion might prosper free of religious persecution and where the souls of their children might be saved.[26]

Initially, the mid-Atlantic colonies were settled primarily by individuals—farmers, artisans, and indentured servants—instead of families. However, the need for child labor in the colonies, along with worsening conditions in Europe, soon led to the development of a variety of mechanisms by which youths could make their way to the New World. In Europe, "spirits" (commission agents of shipowners and merchants) were responsible for signing up young men, young women, and even children for the voyage to America. Sometimes the spirits bribed, tricked, or coerced people into making the voyage. In other cases, they found desperate people who dreamed of a better life. In exchange for their passage to the colonies, those who signed with the spirits agreed to a period of indenture (contract), usually 4 years, as a way of paying for their passage.[27] To ensure that they did not renege on their contract, people were often imprisoned until they could be transported to the colonies. Furthermore, laws were passed in the colonies that allowed the recording and enforcement of these contracts.[28] Indeed, it is believed that the majority of people who colonized the Chesapeake region during the 1600s came over as indentured servants.[29]

Although many youths came to the New World as servants or apprentices, others came as part of a family. This was particularly true in New England, because members of Puritan families typically sailed together.[31] In other instances, poor, destitute, and wayward children were transported from European countries in order to reduce the costs of providing relief to the poor or incarcerating them and trying to correct their behavior. In addition, the colonies needed workers, and children were seen as a cheap source of labor.[32]

The great demand for labor in the colonies led not only to an influx of European children, but also to the forced transportation of large numbers of African children. In 1619, the same year that the colony of Virginia obtained an agreement for the regular shipment of orphans and destitute children from England, the first African slaves arrived in that colony. Most of these slaves were children. Slave traders thought that children would bring higher prices, and more of them could be transported in the limited cargo space of ships. Furthermore, the slave traders encouraged childbearing in order to increase their capital. As Barry Krisberg and James Austin note, "African babies were a commodity to be exploited just as one might exploit the land or the natural resources of a plantation, and young slave women were often used strictly for breeding."[33]

FYI

Conditions in Europe Encouraged Immigration to the New World

It has been estimated that about 80,000 people, or 2% of the population, left England between 1620 and 1642. Of the people who left, about 58,000 are believed to have gone to North America.[30]

The Social Control of Children in the Colonies

Those who settled in the New World brought a variety of ideas with them about childhood as well as European mechanisms for responding to those who violated social and legal rules. Like the Old World, the discipline of children in the New World was stern. Parents believed that corporal punishment was the appropriate method of teaching children an appreciation for correct behavior, sound judgment, and respect for their elders. The maxim of parents in colonial America was "spare the rod and spoil the child."[34]

In addition to stern child-rearing practices, several other features of colonial life encouraged conformity among young and old. Even though the colonists came from varied social class backgrounds, they generally shared beliefs and values.[36] Also, most of them resided in small towns, where there was regular interaction between people intimately familiar with one another. Because the necessities for survival were produced locally, children spent much of their time involved in economic production and under the watchful eyes of adult members of the community.[37] In short, the organization of community life was a major factor in the establishment of conformity among community members.

Two other institutions dominated colonial social life: the family and the church.[38] During the colonial period, the family was the basic unit of economic production[39] and the primary mechanism through which social control was exerted.[40] Survival of the family and the community depended on the ability of all able-bodied persons to contribute to the production of needed commodities. Consequently, a primary responsibility of the family was to oversee the moral training and discipline of the young and perpetuate the values that supported existing social institutions and economic production. This responsibility was so important that parents who failed to instill respect for community values in their young were subject to punishment by the authorities.[41] According to the Laws and Liberties of Massachusetts (1648 edition):

> The selectmen of every town are required to keep a vigilant eye on the inhabitants to the end that the fathers shall teach their children knowledge of the English tongue and of the capital laws, and knowledge of the catechism, and shall instruct them in some honest lawful calling, labor, or employment. If parents do not do this, the children shall be taken away and placed (boys until twenty-one, girls until eighteen) with masters who will teach and instruct them.[42]

MYTH VS REALITY

Several Factors Encouraged Conformity Among Colonial Youths

Myth—The stern child-rearing practices found in colonial towns and villages prevented youths from engaging in deviant behavior.

Reality—Low levels of serious misbehavior on the part of youths were most likely the product of several features of colonial life. Nevertheless, youthful misconduct was a concern of many adults. In writing about the youths of his town in the 1700s, religious reformer Jonathan Edwards wrote that "licentiousness for some years greatly prevailed among the youth of the town," many of whom were "very much addicted to night walking, and frequenting the tavern, and lewd practices, wherein some, by their example exceedingly corrupt others." The youths often gathered "in conventions of both sexes, for mirth and jollity, which they called frolics" and in which they spent "the greater part of the night" and "indeed family government did too much fail in the town."[35]

The family, along with the binding out and apprenticeship system, played a critical role in the training and control of children in colonial America. The binding out and apprenticeship systems were responsible for the immigration of many youths and served as a primary means by which colonial youths learned a skill, earned and saved money, and prepared themselves for adulthood. They were also mechanisms of social control. Children who were difficult to handle or needed supervision were often bound over to masters for care. Masters were responsible for the discipline of those within the household, including servants and apprentices. Under the binding out system, masters were not required to teach servants a trade, and boys were often given farming tasks while girls were assigned domestic duties. In contrast, apprenticeships were typically reserved for wealthy youths, and masters were obligated to teach their apprentices a trade.[43]

Religion, particularly in New England, was another powerful force shaping social life in the colonies. Contemporary concerns about the separation of church and state were nonexistent. Regular church attendance was expected,[44] and religious beliefs dominated ideas about appropriate behavior. Little differentiation was made between sin and crime.[45] What was believed to be immoral was also unlawful and subject to punishment by the authorities.

In colonial towns, children had few incentives or opportunities to act in ways that deviated from family and community expectations.[46] Gossip, ridicule, stern discipline by parents, and work, as well as regular supervision by parents, masters, and others in the community, ensured that most children did not stray too far from community norms. When children committed minor rule violations, their parents or masters were expected to punish them. In other instances, youths who violated community rules might be sent to the town minister for a stern lecture and a warning to avoid further infractions. However, rule violations that, by today's standards, would be considered minor were seen as serious by people in colonial times. For example, in some colonies a child who rebelled against his or her parents could be put to death.[47] Colonial codes contained a long list of capital offenses, such as murder, horse stealing, arson, robbery, burglary, and sodomy. According to the Massachusetts Bay Colony Laws of 1660, for sodomy, a capital crime, children younger than age 14 years were to be "severely punished" but not executed; for cursing and smiting parents, a capital crime, only those "above sixteen years old, and of sufficient understanding" could be put to death. Those who were older than age 16 years could be executed for arson, being a stubborn or rebellious son, and "denying the Scriptures to be the infallible word of God," all of which were capital crimes.[48]

Banishment was another method used to permanently eliminate offenders from the community.[49] On the frontier, however, banishment could be tantamount to a death sentence. In addition, punishments such as fines, whipping, branding, and placement in stocks or a pillory served as reminders to both young and old that violations of community norms were serious matters.[50] Incarceration was also used during this period. Some offenders were placed in the small jails that existed in many towns, but incarceration was used primarily to hold debtors and other offenders awaiting trial. The use of incarceration as a punishment for offenders did not become popular until later.[51]

Despite the relative homogeneity of the population of the colonies and the variety of social control mechanisms in place, social unrest still occurred,[52] including servant revolts, slave revolts, strikes, demands for political representation, and discontent among the poor. As the colonies grew, the antagonism between the poor and wealthy intensified. Although some servants became landowners after their period of servitude was over, most continued to be poor, often becoming tenants on large plantations and providing the owners with a cheap source of labor. Moreover, the towns were typically run by wealthy elites who maintained their power through intermarriage and other forms of alliance between families.

In colonial America, the majority of people were poor, and many of the poor resented their treatment by the wealthy. Even during the Revolutionary War, those who supported independence were concerned about the possibility of mutiny and the lack of support for their cause. This was particularly true in the South, where many poor people felt that a victory by the colonies would simply mean changing one master for another. Indeed, the problems of poverty and social unrest prompted the growing cities to establish poorhouses to provide for and control the elderly, widows, the physically challenged, orphans, the unemployed, war veterans, and new immigrants. A letter to Peter Zinger's *New York Journal* in 1737 described poor street children in New York during this period: "an Object in Human Shape, half starv'd with Cold, with Cloathes out at the Elbows, Knees through the Breeches, Hair standing on end.... From the age about four to Fourteen they spend their Days in the Streets ... then they are put out as Apprentices, perhaps four, five, or six years."[53]

Families and Children in the 1800s

By the early 1800s, colonial social organization was changing as a result of continuing immigration and economic and social developments occurring throughout the new country. The diversity of the population had increased, and English settlers had been joined by Scots-Irish, German, Irish, and French immigrants.[54] Family-based production, which had characterized colonial social life, was giving way to a factory-based system of production in the growing towns. Moreover, the factory-based system was beginning to supplant the binding out system, which had been the primary means by which children entered the labor force during the colonial period.[55] As a result, the factory and the factory boss gradually took the place of the master as agents of socialization and control.

As more parents, particularly fathers, and children began to leave the home for work in factories, fundamental changes occurred in the relationships among family members and in the role of the family in controlling the behavior of children. Before the Industrial Revolution, parents were involved in making occupational choices for their children, typically when the children were age 10 to 12 years. Moreover, there were important class differences between children. Poor children were bound out or placed in apprenticeships, whereas wealthier children were sent to schools or began preparing for careers in the military, the government, or medicine by their late teens. Even within the lower classes, there were various distinctions between youths, because those who were apprenticed in the better trades would be assured of higher incomes. And, of course, there were tremendous differences in the occupational opportunities available to males and females. As a result, children developed specific identities based on their gender and class position. The advent of industrialization, along with increases in the population, led to the erosion of these traditions, however. The professions became more difficult to get into because of the larger number of people vying for entry. More often, youths became employees rather than apprentices. Finally, young people began to remain at home longer, and the period of transition into full adulthood lengthened.[56]

Another change that accompanied industrialization was the decline of the large extended family. A typical family of the colonial period comprised parents, children, relatives, and perhaps servants and apprentices; but, this type of family began to be replaced by the nuclear family. Also, the work that people performed for pay began to be carried out in factories and other job sites away from the home. Not only did men work outside the home, but in many cases women and children began to work outside the home as well, as a result of the development of industrial machinery. For example, Samuel Slater, the "father of American manufacturers," initially employed a workforce of nine boys in his Rhode Island factory. By 1801, this workforce had expanded to 100 youths between the ages of 4 and 10 years.[57] As apprenticeships became less important as a way of learning a trade and earning money and as the extended family declined in importance, childcare became much less of a collective responsibility (shared by family members and the community) and more of a responsibility of the mother.

Although industrialization brought prosperity to some, it was accompanied by growing concern about social unrest and crime. There were increasing numbers of poor people as well as people suffering from mental illness. In the smaller, more homogeneous towns and villages of colonial times, people had accepted these problems and attempted to devise a community response. Although some offenders and those considered rogues or vagabonds were banished or barred from communities, no systematic attempts were made to isolate those who were deviant or the dependent. By the early 1800s, however, notions of dependency and deviancy had changed dramatically. Deviant children were more often seen as products of pauperism and many people believed that problem children could be transformed into productive hard-working adults.[58] It was within this context that new institutions for the control of children were developed.[59]

The Development of Early Juvenile Correctional Institutions in the United States

The United States grew significantly during the late 1700s and early 1800s. With the growth in population came an increase in industry and the development of larger towns and cities. The population also became more diverse, and the conflict that had characterized colonial times continued to be a feature of American life. There was antagonism not only between the elite, who had political and social advantages, and the poor, but also between the citizens of the new country, Native Americans, and African Americans.[60]

The population of the United States grew rapidly after the revolution. In 1790, there were fewer than 4 million Americans, and the majority of them lived within 50 miles of the Atlantic Ocean.[61] Colonial America was rural; even the largest cities were small. When Washington became president, "only two hundred thousand Americans lived in towns with more than twenty-five hundred people. ... In 1790, no American city had more than fifty thousand residents."[62] However, by 1820 the population of New York City was about 120,000 and growing rapidly as a result of immigration.[63] Indeed, by 1830, the United States was home to approximately 13 million people, and more than 4.5 million lived west of the Appalachian Mountains. Moreover, immigration and slavery changed the ethnic composition of many communities, making them less homogeneous than during colonial times. In contrast to the influx of immigrants, the population of Native Americans living east of the Mississippi was reduced from 120,000 to approximately 30,000 in 1820. They were either killed or forced to leave their land.[64]

MYTH VS REALITY

The Problem Behaviors of Youths Have Been an Ongoing Concern
Myth—In the good old days, people did not worry about youth crime because it was not a problem.
Reality—The criminal behavior of young people has been a recurring concern in American cities since at least the 1800s.

Although the size of the United States grew considerably between 1790 and 1820, the population increased even more dramatically after 1820. By 1860, the population of the United States had grown to approximately 31 million people. Moreover, the diversity of the population continued to expand. Increasingly, new arrivals came from the Scandinavian countries and from Ireland and Germany.[65]

The United States presented tremendous opportunities for people and many prospered. For others, however, the American dream was elusive. The New World had always had its share of poor people. Many of those who arrived during the colonial period came as servants, and many of the immigrants to the United States were poor people seeking a better life. Both before and after the American Revolution, the poor outnumbered the wealthy. Moreover, poor and affluent alike faced a variety of insecurities. The wealthy worried about the threat of crime, protests, riots, periodic slave revolts, and various forms of political resistance to their leadership.[66] In addition, periodic economic downturns threatened the livelihood of many people and left many out of work.[67] Furthermore, there was no safety net to rely on in difficult times. There was no minimum wage, health care plan, social security, or pension program to protect workers' interests if they lost their jobs, became ill, or became too old to work.

The Houses of Refuge

Accompanying the changes in the social and economic life of the growing cities were a host of social problems, such as vagrancy, drunkenness, and crime, including crimes committed by children. In response to the growth in juvenile crime, the first correctional institution specifically for youths in the United States was developed. This institution, the House of Refuge, was established in New York City in 1825. Soon other **houses of refuge** were founded in other cities, such as Boston (1826) and Philadelphia (1828).[68] In the 1840s, houses of refuge opened in Rochester, Cincinnati, and New Orleans; in the 1850s, they opened in Providence, Baltimore, Pittsburgh, Chicago, and Saint Louis. By 1857, the refuge movement had expanded to the point that a national convention of refuge superintendents was held in New York. According to its committee on statistics, 17 institutions for youths were in operation in the United States. These institutions had land and buildings worth almost $2 million and total annual expenditures of approximately $330,000.[69]

house of refuge
A privately operated child-care institution developed in the United States for children who engaged in illegal behavior or were felt to be at risk of such behavior. Although the initial goal was to educate and train youths for productive lives, houses of refuge became little more than prisons characterized by harsh discipline and hard work.

The development of the houses of refuge represented a new approach to problem children. This new approach relied on formal child-care institutions as opposed to families, the church, and informal community controls.[70] The houses of refuge were championed by wealthy reformers who saw youth crime and waywardness as a natural outgrowth of the pauperism prevalent in cities like New York. These reformers were mostly men from established families, and they intended to oversee the moral well-being of the community and to develop policies that would protect their way of life from the threat posed by the poor.[71] These reformers decried the unwillingness of the criminal courts to deal with children who committed minor offenses. The incarceration of youths who committed any type of criminal offense was possible because children fell under the jurisdiction of the criminal courts, but the reformers realized that adult correctional institutions did nothing to reform youths.[72] According to one of these reformers, Cadwallader Colden, who was both the mayor of New York and the presiding judge of the city's municipal court, "The penitentiary cannot but be a fruitful source of pauperism, a nursery of new vices and crimes, a college for the perfection of adepts in guilt."[73] What was needed, according to the reformers, were separate institutions for youths that could shield them from the corrupting influences of adult institutions, control them, and transform them into hard-working citizens.

In order to accomplish the reformers' mission, youths were placed in the houses of refuge for indeterminate periods of time or until they reached age 18 or 21 years. Placement did not require a court hearing,[74] and a child could be committed to a refuge by a constable or a parent or on the order of a city alderman.[75]

Despite the name and the lofty goals touted by the reformers, the houses of refuge were run more like prisons and were built for the secure confinement of their inmates.[76] Indeed, the daily operation of these institutions focused more on control than reform.[77] While there, children were exposed to a strict daily regimen that included hard work, military drill, enforced silence, and both religious and academic training.[78] Another common practice intended to support the upkeep of these institutions was to operate shops within the refuge that were run by outside contractors. In these shops, children labored 8 hours a day producing goods such as shoes, brass nails, and furniture.[79] In return, the refuge was paid 10 to 15 cents per youth per day. This arrangement allowed the refuge to cover a substantial percentage of its daily operating expenses[80] and supplement the private and public funds that it received.[81] However, the child workers were often punished when they failed to meet production quotas.[82]

When children rebelled against their treatment, discipline was often severe. Minor rule violations were typically dealt with by loss of supper and an early bedtime. More serious infractions resulted in a variety of punishments, including bread and water diets, placement in solitary confinement (sometimes in a "sweat box"), manacling with a ball and chain, placement in a straightjacket, hanging by the thumbs, or whipping with a cat-o'-nine-tails.[83] When "reformed," boys were often indentured to farmers or tradesmen,[84] and girls were placed in domestic service.[85] Boys who were more difficult to manage were sometimes sent on extended whaling voyages as a punishment for their difficult behavior.[86]

Despite the initial optimism concerning the ability of the houses of refuge to control and reform youths, a less sanguine view of these institutions soon developed among

FYI

Since the establishment of correctional institutions for youths in the United States, many institutions have asserted that their primary mission is to treat and rehabilitate youth, while they actually have focused on the punishment of youths.

some reformers. Indeed, the houses of refuge faced several problems that contributed to their failure to achieve their main goals. For one thing, they admitted a range of youths, many of whom had committed no criminal offense. Thus, children who had no means of support, those who were neglected, and those who had committed criminal offenses were confined together.[87] Indeed, the diversity of youths placed in the houses of refuge made it impossible for them to meet the needs of their residents.

The length of placements also created problems. Because many children were placed for long periods of time, the houses of refuge soon became overcrowded. The number of children requiring placement always surpassed the number of available beds. Moreover, the harsh treatment inflicted on children by staff in these institutions and by the masters to whom they were indentured led many children to run away.[88] Indeed, the treatment youths received was hardly conducive to reform and was thought by some reformers to compound the problems that the houses of refuge were initially intended to solve.

placing out The placement of children on farms and ranches in the Midwest and West. Placing out, which was popularized by private charitable organizations in the mid-1800s (and lasted until 1929), was an attempt to remove vagrant and wayward children from the corrupting influences of the cities and locate them in what was believed to be a more wholesome environment. The idea of using the frontier as a geographical "cure" for societal ills was consistent with the views of Frederick Jackson Turner, a 19th-century historian who believed that the American experience of the frontier helped shape the American character by cultivating individualism, industry, and self-reliance—the very values that the social reformers wanted all children to acquire.

Placing Out

Although institutional responses to juvenile waywardness and delinquency spread during the last half of the 1800s, many reformers recognized the inability of the houses of refuge to accommodate the large numbers of children needing placement. For example, Charles Loring Brace, director of the Children's Aid Society in New York, noted that the harsh punishments and military regimen found in these institutions had failed to transform their residents into law-abiding citizens. According to Brace, the longer a youth remained in a house of refuge, "the less likely he is to do well in outside life."[89]

One early attempted remedy for the problems of the houses of refuge was **placing out**, which involved the placement of children on farms in the West and Midwest (see Exhibit 4-1). Initially, placing out was used to assist youths in finding employment when they were ready to leave the houses of refuge.[90] However, placing out picked up momentum by the mid-1800s. Faced with continuing juvenile delinquency problems exacerbated by population growth, the dislocations produced by the Civil War, and what many thought was the failure of the houses of refuge to achieve their goals, some reformers came to view placing out as a solution to the growing number of problem children.[91] Agencies such as the Children's Aid Society of New York, the Boston Children's Aid Society, and other child relief organizations began to promote placing out as the ideal way to deal with problem children, and they hired agents to help place these children with farm and ranch families in the West.[92]

Advocates of placing out claimed that it had several advantages over the houses of refuge. In their view, it removed children from the corrupting influences of the cities,

FYI

Placing Out Was One Response to Problem Children

The Children's Aid Society of New York placed approximately 60,000 youths into western homes in the 30 years after its founding in 1853.[93] The idea of using the frontier as a geographical "cure" for societal ills was consistent with the views of Frederick Jackson Turner,[94] a 19th-century historian who believed that the American frontier experience helped shape the American character by cultivating individualism, industry, and self-reliance—the very values that the social reformers wanted all children to acquire.

Homeless . .
. . Children.
THE CHILDRENS' HOME SOCIETY
HAS PROVIDED
2990 Children With Homes, in Families.

All children received under the care of this Association are of **SPECIAL PROMISE** in intelligence and health, and are in age from one month to twelve years, and are sent **FREE** to those receiving them, on ninety days trial, **UNLESS** a special contract is otherwise made.

Homes are wanted for the following children:

8 BOYS Ages, 10, 6 and 4. Brothers, all fine, healthy, good looks. Of good parentage. Brothers 6 and 4 years; English parents, blondes. Very promising, 2 years old, blonde, fine looking, healthy, American; has had his foot straightened. Walks now O. K. Six years old, dark hair and eyes, good looking and intelligent, American.

10 BABES Boys and girls from one month to three months. One boy baby, has fine head and face, black eyes and hair, fat and pretty; three months old. Send two stamps.

REV. M. B. V. VAN ARSDALE,
General Superintendent.
Room 48, 280 La Salle Street, CHICAGO.

Exhibit 4-1

Source: Nebraska State Historical Society.

which they saw as breeding grounds for idleness and crime. In addition, they pointed to the failure of the houses of refuge to control or rehabilitate youths[95] and suggested the rural countryside was an ideal environment for instilling in children the values the reformers cherished—discipline, hard work, and piety.

Yet, the wholesome characteristics of the country life depicted by the advocates of placing out were a far cry from the realities faced by many of the children who were sent to the West. Although some children were put into caring homes and were treated as members of the family, others were not so lucky. Considerable evidence exists that many of the children placed out were required to work hard for their keep, were abused, or were never fully accepted into the family.[96] Moreover, there was little systematic follow-up to determine if the children were treated humanely by those responsible for their care.[97]

Even when it was popular, placing out was not without its critics. Some pointed to the fact that children who were placed out engaged in delinquency after their placement.[98] Others, including supporters of juvenile institutions, made the accusation that the practice was contaminating the West with unfit youths, an accusation that struck a chord with many people in the areas to which the children were being shipped.[99] Despite these concerns, however, placing out continued to be used until after the turn of the century.

Probation

Another effort to deal with troubled children was initiated by a Boston shoemaker, John Augustus. Augustus spent considerable time observing the operation of the Boston Police Court and became convinced that many minor offenders could be reformed. In 1841, as a result of his concern and willingness to work with offenders, Augustus was permitted to provide bail for his first **probation** client, a drunkard who showed remarkable improvement during the period he was supervised. The court was impressed with Augustus's work and permitted him to supervise other minor offenders, including children. During Augustus's lifetime, he provided supervision to over 2,000 people, very few of whom violated the terms of their release.[100]

After Augustus died, the Boston Children's Aid Society and other volunteers carried on his work.[101] Eventually, in 1869, the state of Massachusetts formalized the existing volunteer probation system by authorizing visiting probation agents to work with adults and child offenders who showed promise. Under a probation arrangement, youths were allowed to return home to their parents, provided the youths obeyed the law.[102] In addition to the supervision of youths, the visiting probation agents were charged with investigating children's cases, making recommendations to the criminal courts, and arranging for the placement of children. In 1878, a law was passed that provided for paid probation officers in Boston.[103] Subsequently, several other states authorized the appointment of probation officers. However, it was not until after the turn of the century and the development of the first juvenile court that probation gained widespread acceptance.[104]

probation The supervised release of an individual by a court. Initially, probation was promoted by John Augustus, a Boston shoemaker. Today children are usually returned to their parents by courts on conditions of release called "rules of probation," and court or state employees are the supervisors or monitors of the people on probation. In the 1800s, the monitoring of youths on probation was largely done by private citizen volunteers or volunteer societies.

Reform Schools, Industrial Schools, and Training Schools

Although reformers like Charles Loring Brace were skeptical of the ability of institutions to reform or control youths, the role of institutions in providing social control continued to grow during the latter half of the 1800s. The dislocations produced by the Civil War placed tremendous strain on the existing houses of refuge and the placing out system, and the number of problem youths continued to grow. In response, **reform schools**, also known as industrial schools and training schools, proliferated during the postwar period.

The development of these schools was made possible by the willingness of state and city governments to play a larger role in the administration of institutions for wayward and criminal youths. In some instances, state and city governments took over the administration of existing institutions; in other instances, they built new institutions.[105] Although called reform schools, industrial schools, or training schools,[106]

reform school An institution for the control and reform of youths, it was funded by local or state government (unlike the privately funded or self-sustaining houses of refuge). Reform schools (also known as industrial schools and training schools) were often built in rural areas to remove youths from the corrupting influences of the city. Like in the houses of refuge, however, children were often subjected to inhumane treatment in these institutions.

cottage reformatory
A type of juvenile institution that was developed in the mid-1800s and intended to simulate family life. In a cottage reformatory, youths were separated into smaller groups than usual, were supervised by cottage parents, and had mentors or "big brothers" who served as normative models.

institutional reformatory
A large juvenile correctional institution with hundreds of beds that eventually was characterized, like the houses of refuge, by hard work, long hours, and exploitive conditions.

these institutions were mainly of two types: **cottage reformatories** and **institutional reformatories.**

The cottage reformatories were usually located in rural areas in order to avoid the negative influences of the urban environment. In addition, they were intended to closely parallel family life. Typical of these institutions was the Ohio Reform School, which was established in 1857. Youths in this institution were divided into separate cottage families. According to a report of the commissioners of the Ohio Reform School, the boys lived on a farm with "elder brothers," who were selected because of their ability to inspire good behavior and a love of the country in others.[107] The reformatory cottages typically contained 20 to 40 youths[108] supervised by cottage parents charged with the task of overseeing their training and education.[109]

In addition to cottage reformatories, institutional reformatories also were developed in many states. Like the cottage reformatories, these institutional reformatories were usually located in rural areas in an effort to remove youths from the criminogenic influences of city life. However, the institutions were often large and overcrowded. For example, the Elmira Reformatory in New York was intended to be a model reformatory; it contained 500 cells, but by the late 1890s it held approximately 1,500 residents.[110]

Another development that occurred in the mid-1800s was the establishment of separate institutions for females, such as the Massachusetts State Industrial School for Girls, which was established in 1856.[111] Prior to this time, girls had been committed to the same institutions as boys, although there was strict gender segregation. More often than boys, girls were committed by parents or relatives for moral as opposed to criminal offenses. These moral offenses included "vagrancy, beggary, stubbornness, deceitfulness, idle and vicious behavior, wanton and lewd conduct, and running away."[112]

The expressed goal of the specialized institutions for females was to prepare them to be good housewives and mothers. Yet these institutions, like those for boys, differed little from the prisons of the day. The daily regimen revolved around the teaching of basic domestic skills, the methods of treatment were based on coercion, and life in the institution was essentially a form of punitive custody.[113]

Although they placed more emphasis on formal education, the reform, industrial, and training schools, were in many respects similar to the houses of refuge. Indeed, they confronted many of the same problems. As a result, the conditions in the reform, industrial, and training schools were generally no better than those that existed in the houses of refuge, and in many cases were actually worse.

COMPARATIVE FOCUS

Comparative Focus

The development of ideas about the causes of delinquency as well as institutional responses to delinquency in the United States and in Europe were influenced by writing and practices on both sides of the Atlantic. For example, Americans were influenced by European theorists such as Herbert Spencer, Cesare Lombroso, Emile Durkheim and others. Moreover, European theorists played a major role in developing the idea that delinquents could be treated and turned into law-abiding citizens. This idea was put into practice in American institutions for youths.[114]

Institutional Care: A Legacy of Poor and Often Inhumane Treatment

Although the terms used to refer to institutions developed to control and reform children changed periodically,[115] the treatment of children in these institutions did not. Children in these institutions continued to be seen as a source of cheap labor that could be exploited by businesses and could help defray the costs of operating the institutions. Beatings, whippings, and other forms of corporal punishment; bread and water rations; solitary confinement; and straitjackets continued to be the standard tools to respond to recalcitrant youths or to deter would-be troublemakers.[116] Furthermore, youths continued to be placed into apprenticeships without careful follow-up to determine if their treatment was humane. Both boys and girls were subjected to beatings and other mistreatment by their masters, and the sexual abuse of females was not unheard of.[117]

Juvenile correctional institutions continued to grow in popularity as a way of controlling problem youths throughout the latter half of the 1800s, despite the economic hardships that resulted from the Civil War. The amount of state funds going to these institutions during the war had declined, which forced them to rely more heavily on contract labor to meet operating expenses. The use of child labor led to charges that the institutions were more concerned with the economic exploitation of children than their reformation. Indeed, reports of the exploitation of youths in the institutions surfaced periodically. For example, in 1871 a New York Commission on Prison Labor found that refuge youths were paid 30 cents per day for work that would command $4 per day in the community. In addition, workingmen's associations complained about the unfair competition of child inmate labor.[118]

By the end of the 1800s, a variety of institutions had been developed in response to problem children. Yet the many problems presented by children who were believed to be in need of correctional treatment—problems such as homelessness, neglect, abuse, waywardness, and criminal behavior—proved difficult to solve. Nevertheless, during the late 1800s a new group of reformers, the Child Savers, began to advocate for a new institution for dealing with problem children: the juvenile court.

Legal Issues

To the extent that adults in the past viewed children as property, they failed to recognize that children had legal rights as well. During colonial times and throughout the early 1800s, children were put into houses of refuge and reformatories and placed out to rural families without any kind of due process or legal recourse. As bad as it must have been for some families and children, however, slave families and children suffered more and had even less ability to protect themselves. It has only been within the last half century that the concept of children as legal entities with legal rights has been accepted by large segments of society.

Chapter Summary

Historically, childhood is a recent invention. For much of recorded history, the young have been regarded as "miniature adults" and expected to participate in family and community life from a very early age. On the one hand, such participation helped youths earn a place in the community and helped ensure their survival. On the other hand, it placed the young in situations where they were subject to exploitation and abuse by those who were older.

Youths were seen as both a blessing and a curse. They were seen as a blessing because they helped ensure the continuation of the family and the survival of the community. They were seen as a curse when they were too young to contribute to economic survival. Children who were physically or mentally incapable of contributing economically were viewed as a burden, and many of these children were killed or abandoned.

In the past, life was often precarious for young and old alike. Disease, famine, natural disasters, and wars claimed the lives of many. Although there is clear evidence that many parents in Europe and elsewhere demonstrated care and concern for their children, there is also evidence that very little thought was given to the needs of many others. Large numbers of youths lived harsh lives characterized by abuse and exploitation, perhaps because the difficult and precarious nature of life produced a culture that downplayed the importance of children. Whatever the reason, children's lives were very different than those experienced by many children in the United States today.

Although there were many controls on the behaviors of the young, the idea that the young were always obedient in the "good old days" is a myth. Indeed, each generation of youths appears to have presented challenges to their elders, and there is clear evidence that juvenile crime has long been seen as a problem, particularly by urban dwellers in Europe and the United States. However, unlike today, only one criminal justice system existed for handling illegal behavior. Consequently, youths who violated the law were frequently treated like adults, although on other occasions they were spared the harshest punishments because of their age.

By the 1800s, several factors made the creation of separate institutions for problem youths possible, including the growing acceptance by many that childhood and adolescence were distinct developmental periods during which the young needed guidance, protection, and correction. This view of childhood, along with a number of broad economic, social, and cultural changes, provided the foundation for juvenile correctional institutions and juvenile courts.

The 1800s were a time of enormous change in the United States. Many eastern towns were growing into crowded urban areas, the country was becoming more culturally diverse, and industrialization was well under way. Although many people benefitted from these changes and the new opportunities that became available, many others were not so fortunate. Poverty, homelessness, mental health problems, disease, alcoholism, racism, a lack of social services, and crime (including juvenile crime) were reminders that not everything was perfect. The problem of juvenile crime, together with the idea that children were impressionable, led to the development of the first specialized correctional institutions for youths in the United States. These institutions, the houses of refuge, were touted as havens for youths who needed protection and guidance and as places where they could become hardworking, productive, God-fearing citizens. Yet, like other correctional institutions, they were designed to control those who were seen as problems, and eventually control became their primary focus. The result was that many of these early correctional institutions for youths were no better than adult jails and prisons.

Because of the failure of the houses of refuge, the growing number of problem youths needing control and reform, and increasing worry that the cities were breeding grounds for crime, several additional correctional responses aimed at youths were developed during the 1800s. These included placing out; probation; and publicly operated reform schools, which were also known as industrial and training schools. Each approach faced obstacles that prevented it from effectively controlling or reforming youths. In the case of placing out, little effort was made to determine the appropriateness of placements, and some Westerners began to complain that Easterners were simply attempting to dump their problems on Western communities. Early probation officers were typically volunteers (often former police officers) who had little training and lacked the personal and community resources needed to provide assistance to many of their clients. The newer correctional institutions faced many of the same problems that had plagued the houses of refuge. Overcrowding, mixing of youths of various ages and delinquent histories, inhumane treatment, and a failure to provide follow-up services to youths after their release prevented these institutions from achieving their correctional goals.

By the end of the 1800s, despite the existence of varied correctional responses, the problem of juvenile crime had, if anything, grown worse. As a result, a new group of reformers, the Child Savers, began to advocate for the development of a new institution to coordinate efforts to control and reform youth. This new institution, the juvenile court, is the topic of the following chapter.

Key Concepts

apprenticeship: A period, set by contract, during which a novice worked for a master in return for instruction in the master's trade. Apprenticeships were typically reserved for middle-class or wealthy youths. They usually began when youths were ages 14 to 18 years and lasted 10–12 years.

binding out: The placing of a youth or adult in the service of others for a specific period of time, from a week to many years. Although children as young as age 7 years and elderly adults could be placed as servants, the practice was mostly restricted to children in their teens. Binding out was a way for a poor youth to earn and save money, obtain room and board, learn a skill, prepare for adulthood, and meet a potential wife or husband.

cottage reformatory: A type of juvenile institution that was developed in the mid-1800s and intended to simulate family life. In a cottage reformatory, youths were separated into small groups, were supervised by cottage parents, and had mentors or "big brothers" who served as normative models.

family: A social unit responsible for, among other things, the raising and socialization of children. Historically, the family in its various forms has been society's tool for teaching children how to act, what to believe, and what values to hold (see also Chapter 3).

feudalism: A system of land tenure governed by relations between nobles and lesser nobles whereby land (fief) was given by an overlord to a lesser noble (vassal) to hold (not to own) in return for military service. The land could then be given to still lesser nobles. This system depended on the agricultural production of the peasants (serfs) who worked the land.

house of refuge: A privately operated child-care institution developed in the United States for children who engaged in illegal behavior or were felt to be at risk of such behavior. Although the initial goal was to educate and train youths for productive lives, houses of refuge became little more than prisons characterized by harsh discipline and hard work.

institutional reformatory: A large juvenile correctional institution with hundreds of beds that eventually was characterized, like the houses of refuge, by hard work, long hours, and exploitive conditions.

placing out: The placement of children on farms and ranches in the Midwest and West. Placing out, which was popularized by private charitable organizations in the mid-1800s (and lasted until 1929), was an attempt to remove vagrant and wayward children from the corrupting influences of the cities and locate them in what was believed to be a more wholesome environment.

probation: The supervised release of an individual by a court. Initially, probation was promoted by John Augustus, a Boston shoemaker. Today children are usually returned to their parents by courts on conditions of release called "rules of probation," and court or state employees are the supervisors or monitors of the people on probation. In the 1800s, the monitoring of youths on probation was largely done by private citizen volunteers or volunteer societies.

reform school: An institution for the control and reform of youths, it was funded by local or state government (unlike the privately funded or self-sustaining houses of refuge). Reform schools (also known as **industrial schools** and **training schools**) were often built in rural areas to remove youths from the corrupting influences of the city. Like in the houses of refuge, however, children were often subjected to inhumane treatment in these institutions.

Review Questions

1. What were the social conditions faced by children in the Middle Ages?
2. What role did the family play in the socialization and control of children before the 1800s?
3. What role did children play in the colonization of the New World?
4. What characteristics of colonial life encouraged conformity among both young and old?
5. What social conditions during the early 1800s contributed to the development of the houses of refuge?
6. In what ways did the houses of refuge fall short of their initial goals?
7. What was placing out?
8. What were the shortcomings of placing out?
9. Why is John Augustus called the "father of probation"?
10. What were the basic features of the reform, industrial, and training schools that were developed during the last half of the 1800s?
11. What were the distinguishing characteristics of the specialized female training schools?

Additional Readings

Bernard, T. J. (1992). *The cycle of juvenile justice.* New York: Oxford.

Bremner, R. H. (Ed.). (1970). *Children and youth in America: A documentary history* (Vols. 1–3). Cambridge, MA: Harvard University Press.

Rothman, D. J. (1971). *The discovery of the asylum: Social order and disorder in the New Republic.* Boston: Little, Brown & Co.

Schlossman, S. & Wallach, S. (1998). The crime of precocious sexuality: Female delinquency in the Progressive Era. In P. M. Sharp & B. W. Hancock (Eds.), *Juvenile delinquency: Historical, theoretical, and societal reactions to youth.* Upper Saddle River, NJ: Prentice Hall.

Sommerville, C. J. (1990). *The rise and fall of childhood.* New York: Vintage Books.

Zinn, H. (1995). *A people's history of the United States 1492–present.* New York: HarperCollins.

Notes

1. Aries, P. (1962). *Centuries of childhood: A social history of family life.* (R. Baldick, Trans.). New York: Random House. (Original work published 1960).
2. Sommerville, C. J. (1990). *The rise and fall of childhood.* New York: Vintage Books.
3. DeMause, L. (1974). The evolution of childhood. In L. deMause (Ed.), *The history of childhood* (p. 29). New York: Psychohistory Press.
4. Boswell, J. E. (1984). Exposito and oblatio: The abandonment of children and the ancient and medieval family. *American Historical Review, 89,* 10–33.
5. Hanawalt, B. A. (1993). *Growing up in medieval London: The experience of childhood in history.* New York: Oxford University Press.
6. Hanawalt, 1993.
7. Robertson, P. (1974). Home as a nest: Middle-class childhood in nineteenth century Europe. In L. deMause (Ed.), *The history of childhood.* New York: Psychohistory Press.
8. Hanawalt, 1993.
9. Sommerville, 1990.
10. Ladurie, E. L. R. (1983). Parents and children. In B. Tierney (Ed.), *Readings in medieval history: Vol. 2. The Middle Ages* (3rd ed.). New York: Knopf.
11. Hanawalt, 1993.
12. Postan, M. M. (1983). Land and population. In B. Tierney (Ed.), *Readings in medieval history: Vol. 2. The Middle Ages* (3rd ed.). New York: Knopf.
13. Gillis, J. R. (1974). *Youth and history.* New York: Academic Press.
14. Krisberg, B. & Austin, J. F. (1993). *Reinventing juvenile justice* (p. 8). Newbury Park, CA: Sage.
15. Krisberg & Austin, 1993.
16. Krisberg & Austin, 1993.
17. Krisberg & Austin, 1993.
18. Krisberg & Austin, 1993.
19. Hanawalt, 1993.
20. Hanawalt, 1993.
21. Hanawalt, 1993.
22. Silverman, I. J. (2001). *Corrections: A comprehensive view.* Belmont, CA: Wadsworth.

23. Krisberg & Austin, 1993.

24. Krisberg & Austin, 1993.

25. Krisberg & Austin, 1993.

26. Bremner, R. H. (Ed.). (1970). *Children and youth in America: A documentary history* (Vol. 1), (p. 3). Cambridge, MA: Harvard University Press.

27. Bremner, 1970.

28. Zinn, H. (1995). *A people's history of the United States 1492–present.* New York: HarperCollins.

29. Bremner, 1970.

30. Bremner, 1970, (p. 6). This resource cites Bridenbaugh, C. (1968). *Vexed and troubled Englishmen, 1590–1642.* New York: Oxford University Press.

31. Bremner, 1970.

32. Bremner, 1970.
Sommerville, 1990.

33. Krisberg & Austin, 1993.

34. Hanawalt, 1993.

35. Beales, R. W. (1985). In search of the historical child: Miniature adulthood and youth in colonial New England. In R. Hiner & J. M. Hawes (Eds.), *Growing up in America: Children in historical perspective* (p. 21). Chicago: University of Illinois Press.

36. Michalowski, R. (1985). *Order, law and crime.* New York: Random House.

37. Waegel, W. B. (1989). *Delinquency and juvenile control: A sociological perspective* (p. 4). Englewood Cliffs, NJ: Prentice Hall.

38. Waegel, 1989, p. 4.

39. Zaretsky, E. (1976). Capitalism, the family and personal life. In R. C. Edwards, M. Reich, & T. Weisskopf (Eds.), *The capitalist system.* Englewood Cliffs, NJ: Prentice Hall.

40. Krisberg & Austin, 1993.

41. Demos, J. (1970). *A little commonwealth: Famly life in Plymouth Colony.* New York: Oxford University Press.
Farber, B. (1972). *Guardians of virtue: Salem families in 1800.* New York: Basic Books. This resource contains additional information.

42. Simonsen, C. E. & Gordon, M. S. (1982). *Juvenile justice in America* (p. 19). New York: Macmillan. This resource cites the Laws and Liberties of Massachusetts (1648).

43. Krisberg & Austin, 1993.

44. Waegel, 1989, p. 4.

45. Rothman, D. J. (1971). *The discovery of the asylum: Social order and disorder in the New Republic.* Boston: Little, Brown & Co.

46. Waegel, 1989.

47. Bremner, 1970.

48. Powers, E. (1966). *Crime and punishment in early Massachusetts* (p. 442). Boston: Beacon Press.

49. Rothman, 1971.

50. Barnes, H. E. (1972). *The story of punishment* (2nd ed.). Montclair, NJ: Patterson Smith.

51. Rothman, 1971.

52. Zinn, 1995.

53. Zinn, 1995, p. 49.

54. Zinn, 1995.

55. Krisberg & Austin, 1993.

56. Sommerville, 1990.

57. Bremner, 1970.

58. Bernard, T. J. (1992). *The cycle of juvenile justice.* New York: Oxford University Press.

59. Rothman, 1971.

60. Zinn, 1995.

61. Zinn, 1995.

62. Rothman, 1971, p. 57.

63. Bernard, 1992.

64. Zinn, 1995.

65. Zinn, 1995.

66. Zinn, 1995.

67. Krisberg & Austin, 1993.

68. Bremner, 1970.

69. Rothman, 1971.

70. Rothman, 1971.

71. Krisberg & Austin, 1993.

72. Bernard, 1992.

73. Peirce, B. K. (1869). *A half century with juvenile delinquents: The New York House of Refuge and its times* (p. 40). New York: D. Appleton and Company.

74. Bernard, 1992.

75. Bernard, 1992.
Rothman, 1971.

76. Roberts, A. R. (1998). *Juvenile justice: Policies, programs, and services.* Chicago: Nelson-Hall.
Rothman, 1971.

77. Rothman, 1971.

78. Rothman, 1971.

79. Bremner, 1970.

80. Mennel, R. M. (1973). *Thorns and thistles.* Hanover, NH: University Press of New England.

81. Bremner, 1970.

82. Schlossman, S. L. (1977). *Love and the American delinquent.* Chicago: University of Chicago Press.

83. Rothman, 1971. Pisciotta, A. W. (1982). Saving the children: The promise and practice of parens patriae, 1838–1898. *Crime and Delinquency, 28*, 410–425.

84. Pisciotta, A. (1985). Treatment on trial: The rhetoric and reality of the New York House of Refuge, 1857–1935. *American Journal of Legal History, 29*, 151–181. Rothman, 1971, p. 231.

85. Krisberg & Austin, 1993.

86. Roberts, 1998.

87. Krisberg & Austin, 1993.

88. Mennel, 1973.

89. Roberts, 1998, p. 98.

90. Bernard, 1992, p. 65.

91. Mennel, 1973.
Rothman, 1971.

92. Krisberg & Austin, 1993.

93. Sommerville, 1990.

94. Turner, F. J. (1920). *The frontier in American history*. New York: Holt, Rinehart, & Winston.

95. Rothman, 1971.

96. Bremner, 1970.

97. Krisberg & Austin, 1993.

98. Sommerville, 1990.

99. Krisberg & Austin, 1993.

100. McCarthy, B. R. & McCarthy, B. J. (1991). *Community-based corrections* (2nd ed.), (p. 98). Pacific Grove, CA: Brooks/Cole.

101. Empey, L. T. & Stafford, M. C. (1991). *American delinquency: Its meaning and construction* (3rd ed.). Belmont, CA: Wadsworth.

102. Bartollas, C. & Miller, S. J. (1994). *Juvenile justice in America*. Englewood Cliffs, NJ: Regents/ Prentice Hall.

103. Empey & Stafford, 1991.

104. McCarthy & McCarthy, 1991.

105. Krisberg & Austin, 1993.

106. Hart, H. (1910). *Preventive treatment of neglected children*. New York: Russell Sage.

107. Bremner, 1970.

108. Bartollas & Miller, 1994.

109. Whitehead, J. T. & Lab, S. P. (1990). *Juvenile justice: An introduction*. Cincinnati, OH: Anderson.

110. Roberts, 1990.

111. Brenzel, B. (1983). *Daughters of the state*. Cambridge, MA: MIT Press.

112. Brenzel, 1983, p. 81.

113. Chesney-Lind, M. & Shelden, R. G. (2004). *Girls, delinquency and juvenile justice* (3rd ed.). Belmont, CA: Thompson/Wadsworth.

114. Platt, A. M. (1977). *The child savers: The invention of delinquency* (2nd ed.). Chicago: University of Chicago Press.

115. Hart, 1910.

116. Pisciotta, 1982.

117. Pisciotta, 1982.

118. Krisberg & Austin, 1993.

The Development of the Juvenile Court

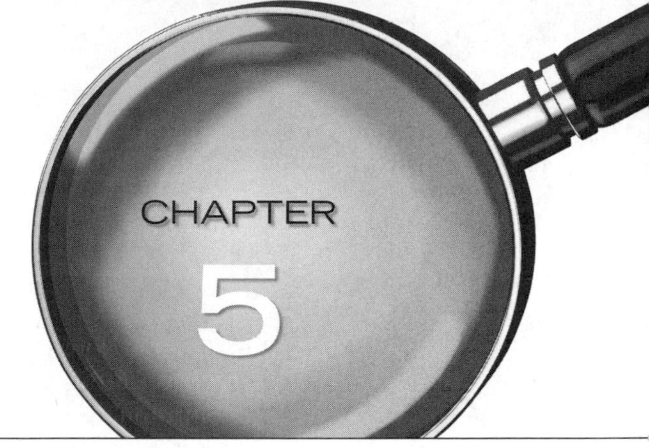

CHAPTER OBJECTIVES

After studying this chapter, you should be able to

- Describe how the middle and upper classes in the late 1800s viewed the poor and the working class
- Describe how the elite's view of the poor and working class influenced the types of social control mechanisms that were developed
- Describe who the "child savers" were and why they were interested in poor, wayward youths
- Explain how the inability or unwillingness of criminal courts to deal with young offenders encouraged the development of the juvenile courts
- Describe the ways in which the social context of the late 1800s and early 1900s contributed to the development of the juvenile courts
- Describe the legal doctrine of *parens patriae*, and indicate how it was used as a justification for state intervention into family life
- Explain why concern about the lack of due process in juvenile courts grew during the middle of the 1900s
- Describe the significance of *Kent v. United States, In re Gault, In re Winship*, and *McKeiver v. Pennsylvania* and how these cases influenced juvenile justice practice

CHAPTER OUTLINE

- Introduction
- The Social Context of the Juvenile Court
- The Legal Context of the Juvenile Court
- The Operation of the Early Juvenile Courts
- The Legal Reform Years
- Legal Issues
- Chapter Summary
- Key Concepts
- Review Questions
- Additional Readings
- Cases Cited
- Notes

Introduction

This chapter examines the development of the juvenile court in the United States. It describes the social context influencing the creation of the first juvenile court, established in Chicago in 1899, and the rapid spread of juvenile courts to other jurisdictions. It explores the legal developments that formed the underlying basis of juvenile court practice. It also looks at early juvenile court operation and important landmark cases that have affected contemporary practice. The chapter closes with an examination of the extent to which important legal cases actually influence the practice of juvenile justice.

The Social Context of the Juvenile Court

Progressive Era
A period in American history, lasting roughly from 1890 to 1924, during which a variety of economic, social, and political reforms and reform movements occurred, including women's suffrage; trust busting; Prohibition; reduction in working hours; elimination of child labor; adoption of social welfare benefits; and the popular voting measures of initiative, recall, and referendum. The reforms were primarily concerned with responding to popular unrest and problems uncovered in the operation of economic, social, and political institutions.

The period from 1880 to 1920, referred to as the **Progressive Era** by historians, was a time of major change in the United States. The pace of industrialization, urbanization, and immigration quickened during this period, and the population became more diverse. The stream of Irish and German immigrants that had begun during the early 1800s trailed off at the end of the century, but they were replaced by Italians, Russians, Jews, Greeks, and other immigrants from Eastern and Southern Europe—people who came from cultural backgrounds that were alien to native-born Americans.[1]

The dream of a better life drew many people to American cities during the Progressive Era. This dream, however, frequently remained unrealized. Wealth was becoming increasingly concentrated in the hands of business elites who sought to dominate economic life, workers were forced to work long hours in unsafe work settings for low wages and no benefits, and housing was often inadequate. Periodic economic depressions ruined many people financially and made life particularly difficult for the poor. Not surprisingly, urban areas faced many social problems, including poverty, labor unrest, alcoholism, disease, racial and ethnic prejudice and discrimination, and crime.

In response, members of the middle and upper classes attempted to implement reforms to alleviate the worst conditions faced by the poorest citizens. Of course, poverty, labor unrest, alcoholism, and racial and ethnic prejudice were hardly new. However, as the size and diversity of the population increased, these problems had grown by the end of the 1800s. Furthermore, workers' movements had become more national in scope and were clearly seen as more threatening to political and economic elites. As a result, these elites sought to devise better mechanisms of control through which their interests might be protected.[4]

Problems that continued to cause concern in the developing urban areas were youth waywardness and crime. Thousands of indigent children roamed the streets in the larger cities,[6] and many of these children engaged in immoral or illegal behavior that threatened

FYI

Urban Growth Was Only One Important Change During the Progressive Era

The city of Chicago provides a good example of the changes experienced during the Progressive Era. Between 1890 and 1910, the population of Chicago doubled, growing in those two decades from 1 million to 2 million people. Between 1880 and 1890, the number of factories nearly tripled. Furthermore, by 1889, nearly 70% of the inhabitants of the city were immigrants.[2]

FYI

American History Is One of Conflicting Interests

The United States has seen several conflicts between opposing interests, often the wealthy and the poor. One of the earliest of these was Shays's Rebellion, which began in western Massachusetts during the summer of 1786. The uprising represented an attempt by debt-ridden farmers to seize courthouses in order to forestall foreclosure proceedings against them. In 1839, an anti-rent movement spread throughout the Hudson Valley; thousands of renters attempted to make their voices heard and to resist the efforts of wealthy landlords to collect rents and taxes from poor farmers. All through the later 1800s, workers engaged in strikes and other efforts in order to win better wages and improve working conditions. The conflict between workers and employers sometimes turned violent. For example, the great railroad strikes of 1877, which involved approximately 100,000 strikers, resulted in 1,000 arrests and 100 deaths.[3]

the tranquility of city life. Of course, criminal laws could be used to prosecute youths who committed crimes, but if the offenses were minor, the courts were often reluctant to do more than give the culprits a stern lecture. Furthermore, the laws were ineffective against those who had not committed a criminal offense. As a result, reformers sought to develop better ways of controlling youths.

Those who led the reform effort, known as **child savers**, were primarily well-educated Protestant, middle-class women of Anglo-Saxon descent, but some were the daughters or wives of the most influential men within their community.[7] They also tended to be conservative in their thinking. For example, they believed in traditional roles for women, and like other affluent people, they took for granted their natural superiority over the poor. From the child savers' perspective, homemaking and child-care were women's primary responsibilities. However, because these traditional duties were performed in their

child savers A group of reformers, mostly well-educated, middle-class Protestant women of Anglo-Saxon descent, who had the time and resources to fight for improved conditions for delinquent youths in jails and reformatories and who eventually played an important role in the creation of the juvenile courts. On the positive side, their work led to wider acceptance of adolescence as an important developmental stage of life, and they correctly believed that the adult criminal justice system was harmful to children or at best ineffective. On the negative side, their conservative ideas on the family and their "noblesse oblige" attitude toward the poor resulted in reforms that focused more on the social control of children and the protection of the interests of the wealthy than on attacking the underlying causes of problem youth behaviors.

FYI

The Progressive Era Was a Time of Change

The Progressive Era was characterized by a political revolt against the social and economic evils of the Industrial Revolution and a belief that government intervention, even on a national scale, was necessary to remedy these evils. Politically, it began with a series of urban and state reform movements directed against corrupt and boss-ridden local governments.

The Progressive Era also was characterized by the belief that American ingenuity and spirit could solve social problems and make life better for people. Tremendous technological advances were taking place, including the development of the automobile, the airplane, moving pictures, the phonograph, and the electric light bulb, just to name a few. Many Americans believed that advances also could be made in social, political, and economic life.

The American Constitution: Its Origin and Development by A. H. Kelly and W. A. Harbison describes the Progressives as follows: "Like the Jeffersonians of a century before, the Progressives had an abiding faith in the intelligence and good will of the American people. Fundamentally, their remedy for the failures of democracy was more democracy. Let the will of the people really reach into the Congress, the courts, the state legislatures, and America could then solve its problems."[5]

own homes by servants, they had few outlets for their energies. Child saving provided a mechanism for these women to perform an important public service—the control and care of less fortunate children.[8]

The child savers were informed by the latest thinking on adolescence and youth crime. Adolescence was seen as a problem period in the child's development,[9] a time when youths were subject to a variety of negative influences. Indeed, youths who engaged in crime were felt to suffer from poor parenting and weak morals and to be unusually susceptible to the temptations of the streets.[10]

Initially, the child savers sought to improve jail and reformatory conditions.[11] However, their attention soon turned toward the development of more effective means of controlling the growing population of problem youths. Indeed, the primary focus of the child-saving movement was on extending government control over children through stricter supervision and improved legal mechanisms designed to regulate their behavior.[12] In short, the child savers felt that intrusion into the lives of children was necessary in order to prevent them from leading immoral and criminal lives—and, equally important, to protect the state and the interests of the wealthy.

Unlike earlier reformers, the child savers thought that the interventions needed to control and uplift children should be performed by government agencies, such as the police and courts, with the assistance of local charitable organizations. According to Julia Lathrop, an influential child saver in Chicago who later became head of the Federal Children's Bureau:

> There are at the present moment in the State of Illinois, especially in the city of Chicago, thousands of children in need of active intervention for their preservation from physical, mental and moral destruction. Such intervention is demanded, not only by sympathetic consideration for their well-being, but also in the name of the commonwealth, for the preservation of the State. If the child is the material out of which men and women are made, the neglected child is the material out of which paupers and criminals are made.[13]

The Legal Context of the Juvenile Court

Although the social conditions that existed in urban areas led reformers to advocate for changes in how communities responded to youth crime and waywardness, the reformers also recognized that existing legal mechanisms for controlling and protecting children were inadequate. One immediate concern of the child savers was the placement of children in jail. For example, a study of the county jail system in Illinois in 1869 discovered 98 children younger than 16 years in 40 different jails. In Michigan, 377 boys and 100 girls younger than 18 years were placed in county jails in 1873; 182 boys and 29 girls were placed in jails in Ohio in 1871; and more than 2,000 youths, 231 of whom were younger than 15 years, were placed in jails in Massachusetts in 1870. The jailing of children was a common practice.[14] However, as the Board of State Charities in Illinois noted, the county jails in the state contained cells that were "filthy and full of vermin," and they were claimed to be "moral plague spots" where children were turned into "great criminals."[15]

The child savers, concerned as they were about the conditions children were exposed to in adult jails and lockups, condemned the fact that adult courts failed to mete out appropriate punishments for many of the transgressions of the young. Unless their offenses were severe, criminal court judges often treated youths leniently. From the child savers' perspective, children either were treated too severely, by being placed in adult facilities where they were corrupted by older and more hardened offenders, or were let go without receiving any assistance.[16] Thus, according to the child savers, children were neither controlled nor helped.

By the late 1800s, legal mechanisms for treating children separately from adults had been in existence for some time. For example, laws establishing the minimum age at which a child could be considered legally responsible for criminal behavior as well as age limits for placement in adult penitentiaries were enacted during the first half of the century.[17] As noted in Chapter 4, the first special institution dealing with youths, the House of Refuge, was established in New York City in 1825.

The legal justification for state intervention in the lives of children was based on the doctrine of ***parens patriae*** (the state as parent), which was given legal standing in an important case, *Ex parte Crouse*, in 1838. Mary Ann Crouse was committed to the Philadelphia House of Refuge by her mother, but against her father's wishes. The father questioned Mary Ann's placement, arguing that she was being punished without having committed a criminal offense. The Pennsylvania Supreme Court ruled that Mary Ann's placement was legal because the purpose of the house of refuge was to reform youths, not punish them; that formal **due process** protections afforded to adults in criminal trials were not necessary because Mary Ann was not being punished; and that when parents were unwilling or unable to protect their children, the state had a legal obligation to do so.[18]

The right of the state to intervene in the lives of children did not go unchallenged, however. In another important case, *People v. Turner* (1870), the Illinois Supreme Court ruled that Daniel O'Connell, who was committed to the Chicago House of Refuge against his parents' wishes, was being punished, not helped by his placement. In some respects this case was similar to *Crouse*. Daniel was institutionalized even though he had committed no criminal offense, and he was perceived to be in danger of becoming a pauper or criminal. In other respects, however, the two cases were decidedly different. Both parents had objected to Daniel's placement, and, even more important, the court ruled that his placement was harmful, not helpful. Furthermore, the court decided that because placement in the house of refuge was actually a punishment, due process protections were necessary.[19]

People v. Turner was an important case because it was seen by reformers as an obstacle to their efforts to help and control youths. The Illinois Supreme Court's ruling in *People v. Turner* that required due process protections prior youths' placement, along with the increasing concern over the unwillingness of the criminal courts to sentence youths, led reformers in Chicago to consider other mechanisms by which their aims might be achieved. What they finally did was create the first juvenile court, which was established in Cook County (Chicago) in 1899.[20]

The juvenile court allowed reformers to achieve their goals of assisting and controlling children's behavior without undue interference from the adult courts and without undue concern for the due process protections afforded adults. This was accomplished by setting up the court as a civil or **chancery court** intended to serve the "best interests" of children (as opposed to a criminal court, which focuses on the punishment of offenders). Because the new court was not a criminal court and its goal was not to punish but to help children, the need for formal due process protections was obviated.[21]

parens patriae The obligation of the government to take responsibility for the welfare of children. The doctrine of *parens patriae* recognizes the moral obligation of the government to take care of children when there is no family available or if the family is not suitable.

due process A course established for legal proceedings intended to ensure the protection of the private rights of the litigants. The essential elements of due process are (1) proper notice as to the nature of the legal proceedings, (2) a meaningful hearing in which the individual has an opportunity to be heard and/or defend him- or herself, and (3) the objectivity of the tribunal before which the proceedings take place.

chancery court English court that was primarily concerned with property rights. Chancery courts, first established during the Middle Ages, were partially based on the idea that children, particularly children who owned or were in a position to inherit property, should fall under the protective control of the king. Indeed, it was in these courts that the concept of *parens patriae* was developed. Over time, the chancery courts became more involved in the general welfare of families and children. Because of its role in protecting the interests of children, it served as a model for the juvenile court.

The Operation of the Early Juvenile Courts

dependent children Children who are dependent on others for financial support or who do not have the support of a parent or guardian.

neglected children Children who are not given appropriate care by the parent(s) or a guardian.

The Juvenile Court Act of 1899 gave the new juvenile court in Illinois broad jurisdictional powers over people younger than 16 years who were delinquent children, **dependent children**, or **neglected children**.[22] In addition, it required that the court be overseen by a special judge, that hearings be held in a separate courtroom, and that separate records be kept of juvenile hearings.[23] It also made probation a major component of the juvenile court's response to offenders and emphasized the use of informal procedures at each stage of the juvenile court process.[24]

In practice, the informality of the juvenile court meant that complaints against children could be made by almost anyone in the community. It meant that juvenile court hearings, which were initially open hearings like those in adult court, would become closed hearings, often held in offices as opposed to traditional courtrooms. In these closed hearings, the only people present were the judge, the parents, the child, and the probation officer. The informality also meant that few, if any, records were kept of hearings, that proof of guilt was not necessary for the court to intervene in children's lives, and that little or no concern for due process protections existed. Finally, it meant that judges exercised wide discretion regarding the actions they took, which could include anything from a stern warning to placement of a child in an institution.[25]

The juvenile courts were successfully implemented, in large measure, because they served a variety of interests. They served the interests of reformers, who sought to help children on humanitarian grounds and the interests of those who were concerned primarily with the control of lower-class, immigrant children whose behaviors threatened urban tranquility.[28] They served the interests of the criminal courts because they removed children from criminal court jurisdiction, freeing up time for the trying of adult offenders. Finally, they served the interests of the economically and politically powerful, because they did not require the alteration of existing political and economic arrangements.[29] Although some undoubtedly saw the new juvenile court as an instrument of change, many others saw them as instruments of containment.

Despite the growing popularity of the juvenile courts, they did not go unchallenged. In another important court case, *Commonwealth v. Fisher* (1905), the juvenile court's mission, its right to intervene in family life, and the lack of due process protections afforded children were examined by the Pennsylvania Supreme Court. In this case, Frank Fisher, a 14-year-old male, was indicted for larceny and committed to a house of refuge (the same house of refuge that Mary Ann Crouse had been committed to over 60 years earlier) until his 21st birthday. Frank's father objected to his placement and filed a suit that argued that Frank's 7-year sentence for a minor offense was more severe than he would have received in a criminal court.[30]

In its ruling, the Pennsylvania Supreme Court upheld the idea of a juvenile court and in many respects repeated the arguments made by the court in the *Crouse* decision.

FYI

Informality Has Been a Hallmark of the Juvenile Court
Informality, a lack of concern with strict due process, has characterized the operation of many juvenile courts since their inception.

FYI

Politics Has Always Played a Role in the Juvenile Court

The first juvenile court hearing held by the Cook County Juvenile Court was a public event because the institutional lobby in Illinois that represented child care institutions in that state had successfully opposed initial provisions of the legislation that called for closed hearings. In addition, the institutional lobby also was successful in ensuring that the initial legislation that was passed in Illinois gave jurisdiction over cases to the institutions after a juvenile was placed. As a result of the efforts of the institutional lobby, the placement of children in institutional settings in Illinois increased after the juvenile court was established.[26] This is a good example of the role of politics in juvenile justice.

The idea of using juvenile courts to deal with youth crime spread rapidly after the passage of the Illinois legislation. Within 10 years, 10 states had established special courts for children, and by 1925 all but 2 states had juvenile courts.[27] These courts followed closely the model developed in Chicago: They were procedurally informal and intended to serve the best interests of children.

The court found that the state may intervene in families when the parents are unable or unwilling to prevent their children from engaging in crime and that Frank was being helped by his placement in the house of refuge. Further, it ruled that due process protections were unnecessary when the state acts under its *parens patriae* powers.[31]

Fisher set the legal tone for juvenile courts from the time they began until the mid-1960s, when new legal challenges to the courts began to be mounted. These legal challenges primarily concerned expansion of juveniles' due process protections. Critics of the juvenile courts recognized that, despite their expressed goal of serving the best interests of children, the established institutions of juvenile justice often did the opposite. Although the courts were intended to help children, they did not always act as wise and benevolent parents. Indeed, their use of coercive powers to deal with a wide range of behaviors, many of which were not criminal in nature, in an informal setting without due process protections created the potential for abuse. Moreover, while reformers had looked to the juvenile courts as a mechanism for both helping and controlling delinquent and wayward children, little attention was devoted to improving other juvenile justice institutions. Various mechanisms intended to assist the court in its mission, such as probation, relied heavily on untrained volunteers.[32] Further, the institutions used by the juvenile courts for the placement of children were the same institutions used before the establishment of the courts, and children continued to be subjected to inhumane treatment.[33]

COMPARATIVE FOCUS

Legislation Establishing a Separate Juvenile Justice Process Was Enacted in Other Western Countries During the Early 1900

For example, in 1908 the Federal Juvenile Delinquents Act (JDA), which stressed a treatment-oriented philosophy, was passed in Canada,[34] and the Children Act was passed in Britain, which established juvenile courts in England, Wales, Scotland, and Ireland.[35] In Germany, the Juvenile Welfare Act (JWA) of 1922 (with amendments in 1923) was passed to establish legal procedures for dealing with the rehabilitation and institutionalization of youths.[36]

■ The Legal Reform Years

The 1960s and 1970s were a time when American institutions, including juvenile justice institutions, came under intense scrutiny. During this period, the U.S. Supreme Court heard a number of cases that altered the operation of the juvenile courts. The most important of these was *In re Gault* (1967), which expanded the number of due process protections afforded juveniles. This section discusses this case and several others, including the first of these important cases, *Kent v. United States* (1966).

Morris Kent was accused of committing break-ins and robberies in the District of Columbia. One robbery victim was raped, but the principal evidence against Kent was a latent fingerprint left at the scene of the robbery and rape. Kent was on probation at the time of these crimes, and after his arrest he was interrogated over a 7-hour period and confessed to several house break-ins. Without a hearing or any formal notice, Kent's case was transferred to the criminal court. His attorney tried to get the case dismissed from the adult court and moved for a psychiatric evaluation and for receipt of all social reports in the juvenile court's possession. The motions by Kent's attorney were denied, however. At his trial in criminal court, Kent was convicted by a jury for robbery and housebreaking, and he was sentenced to 30–90 years in prison.

The matter was appealed on the jurisdictional issue of the waiver from juvenile court to adult court. It was contended that the waiver was defective on the following grounds:

- No waiver hearing was held.
- No indication was given as to why the waiver was ordered.
- Counsel was denied access to the social file and social reports that were reportedly used by the judge to determine the waiver.

In this case, the Supreme Court ruled that for Kent's waiver to be valid, certain due process requirements were necessary. Specifically, the Court held that Kent was entitled to representation by an attorney; to a meaningful hearing, even if informal; to access to any social reports, records, reports of probation, and so on, that would be considered by the Court in deciding the waiver; and to be apprised of the reasons for the waiver decision.

Kent was an important case for several reasons. It resulted in the first major ruling by the U.S. Supreme Court that scrutinized the operation of the juvenile courts. After more than 60 years of informal *parens patriae* procedures, the appropriateness of these procedures was being questioned, even if only narrowly, in the limited area of waivers to adult court. *Kent* also made explicit the need for due process protections for juveniles who were being transferred to adult courts for trial. The Court noted that, even though a hearing to consider transfer to adult court is far less formal than a trial, juveniles are still entitled to some due process protections.

In its decision, the Court made numerous references to the need for due process protections, stating that in a juvenile court a child may receive "the worst of both worlds: that he gets neither the protections afforded to adults nor the solicitous care and regenerative treatment postulated for children."[37] After the Court began to look at juvenile court processes and procedures, subsequent cases like *Gault*, *McKeiver*, and *Winship* became inevitable.

Having given notice that it would review the operation of the juvenile courts, within a year of the *Kent* decision the Supreme Court heard another landmark case. This case,

In re Gault (1967), went far beyond *Kent* in its examination of juvenile court practice and extended a variety of due process protections to juveniles. The facts of the case clearly demonstrate the potential for abuse found in the informal procedures of the traditional juvenile court, and consequently this case is discussed in detail.

Gerald Gault was 15 years old when he and a friend were taken into custody by the Gila County (Arizona) Sheriff's Department for allegedly making an obscene phone call to a neighbor, Ms. Cook. At the time of his arrest, Gerald was on 6 months' probation—the result of being with another friend, who had stolen a wallet from a purse. Gerald was taken into custody on the verbal complaint of Ms. Cook and was taken to the local detention unit. His mother was not notified of this by the police, but she learned about it later that day when she returned home and, not finding Gerald present, sent a sibling to search for him.

Upon learning that Gerald was in custody, Ms. Gault went to the detention facility and was told by the superintendent that a juvenile court hearing would be held the next day. On the following day, Gerald's mother, the police officer who had taken Gerald into custody and filed a petition alleging that Gerald was delinquent, and Gerald appeared before the juvenile court judge in chambers. Ms. Cook, the complainant, was not present. Gerald was questioned about the telephone call and was sent back to detention. No record was made of the hearing, no one was sworn to tell the truth, nor was any specific charge made, other than an allegation that Gerald was delinquent. At the conclusion of the hearing, the judge said he would "think about it." Gerald was released a few days later, although no reasons were given for his detention or release.

On the day of Gerald's release, Ms. Gault received a letter indicating another hearing would be held regarding Gerald's delinquency a few days later. A hearing was held, and again the complainant was not present, and no transcript or recording was made of the proceedings (later, what was said was disputed by the parties). Neither Gerald nor his mother was advised of any right to remain silent, of Gerald's right to be represented by counsel, or of any other constitutional rights. At the conclusion of the hearing, Gerald was found to be a delinquent and was committed to the state industrial school until age 21 years, unless released earlier by the court. This meant that Gerald received a 6-year sentence for an offense that, if committed by an adult, could be punished by no more than 2 months in jail and a $50 fine.[38]

In *Gault*, the Court ruled that the special circumstances that gave rise to the informal process of juvenile courts and the broad discretion of juvenile court judges did not justify the denial of fundamental due process rights for juveniles. The majority opinion stated, "As we shall discuss, the observance of due process standards, intelligently and ruthlessly administered, will not compel the states to abandon or displace any of the substantive benefits of the juvenile process." The justices went on to argue:

The constitutional and theoretical basis for this particular system is—to say the least—debatable. And in practice, as we remarked in the Kent case, supra, the results have not been entirely satisfactory. Juvenile Court history has again demonstrated that unbridled discretion, however benevolently motivated, is frequently a poor substitute for principle and procedures.... The absence of substantive standards had not necessarily meant that children receive careful, compassionate, individualized treatment. The absence of procedural rules based upon constitutional principle has not always produced fair, efficient, and effective procedures. Departures from established principles of due process have frequently resulted not in enlightened procedure, but in arbitrariness.... Failure to observe the fundamental requirements of due process has resulted in instances ... of unfairness to individuals and inadequate or inaccurate findings of fact and unfortunate prescriptions of remedy. Due process of law is the primary and indispensable foundation of individual freedom.... Under our Constitution, the condition of being a boy does not justify a kangaroo court.[39]

The Court made it clear that juveniles are to be afforded fundamental due process rights that measure up to the essentials of fair treatment. The following rights are included:

- the right to reasonable notice of the charges
- the right to counsel (either retained or appointed)
- the right to confrontation and cross-examination of witnesses
- the right against self-incrimination

However, the Court limited the application of these rights to proceedings involving a determination of delinquency that may result in commitment to an institution.

Fundamental changes were required in juvenile courts after *Gault*. Many of them carried high costs. For example, the right to appointed counsel has become a major budget concern for juvenile courts, especially because this right has been expanded several times in subsequent court decisions. The right to confrontation requires the processing of subpoenas and the costs of service. Finally, due process requirements have resulted in more adversarial hearings, which take longer to complete and thus push up costs.

Although *Gault* was a landmark juvenile law case, it was not the last Supreme Court decision that influenced juvenile court procedures. The Court further expanded protections for juveniles 3 years after *Gault*. In *In re Winship* (1970), it addressed the level of proof needed for a conviction of delinquency. The case involved a 12-year-old male who was found guilty of stealing $112 from a woman's purse. As a result, he was placed in a New York training school for a minimum period of 18 months, although the juvenile court indicated that the term of the sentence could be extended to the youth's 18th birthday. The judge who heard the case admitted that proof "beyond a reasonable doubt" was not established at trial, but held that this level of proof was not required.

A majority of the Supreme Court justices held that proof of guilt beyond a reasonable doubt was an essential element of due process in delinquency cases. The Court indicated that to allow a lesser standard would seriously harm the confidence of the community in the fairness of the adjudicative process. The Court went on to talk about the "moral force" of

exclusionary rule
A procedural rule that prohibits evidence that is obtained by illegal means or in bad faith from being used in a criminal trial.

***Miranda* warnings**
Specific warnings given to a suspect prior to questioning that informs them that statements they make to law enforcement agents can be used against them in court.

bail The money or bond used to secure the release of a person charged with a crime.

FYI

Gault Raised Important Questions About Juveniles' Rights

Gault raised several questions at the time of its release. Would procedural rules, such as the **exclusionary rule**, be applicable to juveniles? Do juveniles have the right to ***Miranda* warnings** before being interrogated by the police? Do juveniles have the right to jury trials? Do juveniles have the right to **bail**? At a hearing to determine a finding of delinquency, does a juvenile have to be found "guilty beyond a reasonable doubt"? Since Gault, many, but not all, of these questions have been answered in the affirmative, further prolonging juvenile proceedings and increasing court costs, but giving youths important protections.

the criminal law and how this force would be diluted if the standard of proof was lowered. According to the Court, a lower standard would lead the community to wonder whether innocent juveniles were being convicted and incarcerated.

The *Winship* decision added due process protections to those established by *Kent* and *Gault*. The result was that the concept of "proof beyond a reasonable doubt," which had long been accepted in adult criminal cases, was now applied to adjudications in which juveniles were at risk of institutional placement.

The Supreme Court's willingness to extend due process protections to juveniles came into question the following year, however. It heard two cases jointly, one from Pennsylvania and one from North Carolina, that concerned whether juveniles should be entitled to jury trials at the trial or adjudicative stage of the juvenile court process. In the Pennsylvania case, *McKeiver v. Pennsylvania* (1971), the court used procedures that were similar to those used in adult criminal courts (plea bargaining, motions to suppress evidence, and so on). Moreover, one possible outcome was incarceration in a prison-like facility until the juvenile's age of majority. The North Carolina case was heard in a court using much less formalized procedures than those in Pennsylvania, but the argument was made that the supposed benefits of the juvenile court system—discretionary intake, diversion, flexible sentencing, and a focus on rehabilitation—would not be hindered by the use of juries.

Justice Blackmun, writing for the majority, reviewed the historical reasons for a separate and distinct juvenile court and concluded by stating:

> The arguments necessarily equate the juvenile proceeding—or at least the adjudicative phase of it—with the criminal trial, whether they should be so equated is our issue.... If the formalities of the criminal adjudicative process are to be superimposed upon the juvenile court system, there is little need for its separate existence. Perhaps that ultimate disillusionment will come one day; but for the moment, we are disinclined to give impetus to it.[10]

Consequently, the Supreme Court declined to extend the constitutional right to jury trial to the juvenile system.

McKeiver is important because it made clear that the Supreme Court was unwilling to give juveniles all of the due process protections available to adults. Nevertheless, many states have extended the right to a jury trial to youths at the adjudicative phase of the juvenile justice process. The *McKeiver* decision was not unanimous, however. Justice Brennan, in his dissent in *McKeiver*, focused on the individual state procedure and whether it afforded sufficient protections to the juvenile from any government overreaching and from the "biased or eccentric" judge. He thought that one crucial factor that needed to be taken into account in deciding whether to mandate a jury trial was the

MYTH VS REALITY

Juveniles Do Not Have All of the Legal Protections Available to Adults

Myth—Today, juveniles have all of the due process protections afforded adults in criminal courts.

Reality—In some states, juveniles still lack some of the due process rights, such as right to a jury trial and a right to bail, given to adults.

public "openness" of trial proceedings and the ability of the jury to act as both a finder of fact and the community's conscience, preventing the state from accusing juveniles for political purposes.[41]

Although one of the objections to jury trials in juvenile proceedings was that they would cause a backlog of cases and hamper the functioning of the juvenile court, experience has not shown that such trials seriously impeded the juvenile justice process. The other issue related to efficiency is whether juries can consist of less than 12 citizens. In many states, a jury of 6 is allowed for delinquency adjudications.

A collateral benefit of allowing jury trials is that they may aid rehabilitation. A juvenile who believes that "the system" has treated him or her unfairly may be less defensive if tried by an objective jury. A perception of fair treatment may go a long way toward fostering acceptance of responsibility on the part of the juvenile, to say nothing of its promotion of feelings of self-esteem. Juveniles who believe that the system treats them fairly and accepts them as significant people whose rights must be protected may overcome feelings that they were treated unjustly.

Another point that is sometimes made about the use of juries in the juvenile system is that, unlike in adult courts, the jurors are not peers of those on trial. In some instances, the juvenile may benefit from this fact, because the adults on the jury may remember their own youthful mistakes and indiscretions and feel sympathy. If they are parents, jurors also may realize that, under other circumstances, one of their children could be at the defense table.

In reality, jury trials in delinquency proceedings have not proved to be docket cumbersome, inefficient, or exceedingly expensive. Moreover, they make a statement to the community and the juvenile regarding the juvenile court's concern for fundamental due process fairness.

Despite the due process protections extended to juveniles through the *Kent*, *Gault*, and *Winship* decisions, much of the informality of the juvenile court remains intact in practice. Indeed, one complaint is that changes in the legal procedures that supposedly govern the juvenile courts have not always resulted in fundamental changes in the daily operation of juvenile justice.[43] Many critics contend that juveniles often are denied basic protections within the juvenile justice process and that the continued informality of the juvenile courts fails to serve either the juveniles' best interests or the best interests of the community.

pretrial discovery
Efforts of a party to a lawsuit to obtain information prior to trial. The theory behind discovery is that all parties should go to trial with as much information as possible.

bill of particulars
In a lawsuit, a written itemization of claims or charges provided by the plaintiff at the defendant's request; the document can serve as the factual basis of the allegations against the defendant.

FYI

Juvenile vs. Criminal Courts

Looked at broadly, *Kent, Gault, McKeiver*, and *Winship* concern the degree to which adult criminal practice should be extended into the juvenile courts. *Kent, Gault*, and *Winship* made clear that certain due process protections that have long been part of adult court practice also should be given to youths being tried in juvenile courts. *McKeiver,* however, indicated that there were limits to how far the Supreme Court was willing to go in this area. It also showed that the Court was unwilling to completely dismantle the juvenile justice apparatus and that it wished to maintain some of the traditional informality of juvenile courts.[42] These cases left unresolved, however, whether many other adult court practices should apply to the juvenile courts, such as a defendant's right to bail, **pretrial discovery**, and **bill of particulars**.

FYI

Some States Allow Jury Trials for Juveniles

Many states have required more due process protections for juveniles than mandated by the Supreme Court. For example, many state laws specify that juveniles have a right to a jury trial (adjudication). In fact, at least one state, Texas, requires that all adjudications be heard by a jury. Nevertheless, jury trials are rare in states that have this right, including Texas. This is because the right is not often exercised (or, as in the case of Texas, it can be and often is waived). In practice, jury trials are frequently discouraged because they are felt to be time-consuming and costly or because they impinge on the power of the juvenile court judge.

It is one thing to be afforded various rights through Supreme Court or state court rulings and state statutes. It is another thing to ensure that those involved in the juvenile justice process know all of their rights and that they feel comfortable exercising those rights. However, several studies raise serious doubts about youths' access to counsel in delinquency proceedings and the quality of representation that occurs. A study by the American Bar Association Juvenile Justice Center, the Youth Law Center, and the Juvenile Law Center raised serious concerns about the quality of legal representation given youths in many juvenile courts.[44] This research found that, in many instances, juveniles are not represented by attorneys when they appear in juvenile court; attorneys who represent youths often have high caseloads and, in some instances, lack the proper training and experience in juvenile court to provide effective representation; and youths and their parents or guardians often fail to have a clear understanding of the legal process. As a result, youths are frequently in a vulnerable position, particularly in the early stages of the juvenile justice process, and this can lead to more severe dispositions than are warranted. These findings mirror those of other studies that have examined attorney representation in juvenile court proceedings.[45] For example, a study that examined statewide data in six states (California, Minnesota, Nebraska, New York, North Dakota, and Pennsylvania) found that in three of those states (Minnesota, Nebraska, and North Dakota) the defendant was represented by counsel in approximately half of the cases in which a petition was filed. In addition, this study found that youths who received assistance from counsel were more likely to receive a more severe disposition, even when the seriousness of the charges and the youths' delinquent history were taken into account.[46] Such findings indicate that not only is representation absent in many instances, but the quality of the representation provided to juveniles is often inadequate. Moreover, the failure to receive effective representation can result in harm to juveniles involved in the juvenile court.

The continued informality of the juvenile court may also explain why very few youths contest the charges against them.[47] Moreover, many others who are not petitioned and a sizable number who go to adjudications but are not found guilty are still placed on some form of probation. Data collected by the National Center for Juvenile Justice indicates that in 2005, 22% of youths who were referred to juvenile court but were not petitioned were still placed on some form of probation, 6% of youths who went to an adjudication but were not adjudicated were placed on probation, and 38% were given some other sanction.[48]

Today, juveniles have been granted many, but not all, of the due process protections given to adults in criminal trials. However, the extent to which court-mandated changes in juvenile justice procedures have actually influenced the traditional informality of the juvenile courts is open to question. Juvenile court procedures in many jurisdictions are still characterized by an informality that would be considered unacceptable for adults brought before a criminal court.[49] Is the informality necessary for the courts to carry out their mandate to serve the best interests of children and protect the community, as supporters of the traditional court procedures argue? Or does the courts' failure to adhere to stricter due process standards lead to abuses and harm many children, as critics assert? These issues continue to be the focus of an important policy debate within the field of juvenile justice.

Legal Issues

Kent, *Gault*, and *Winship* extended due process protections to juveniles when juveniles are at risk of commitment. These cases also raised the issue of the rights of juveniles who are involuntarily committed by their parents to mental health facilities, drug treatment centers, and other types of treatment programs. Should children have due process rights in these circumstances? Also, what about instances when the state or the federal government seeks treatment or commitment?

Chapter Summary

The juvenile courts were developed by progressive reformers (the child savers) in the late 1800s as a new means of controlling wayward and problem youths. They were characterized by a focus on rehabilitation, procedural informality, an individualized approach to cases, and the separation of juveniles from adult offenders.[50] The presumption of the child savers was that juvenile delinquents could be, and should be, controlled and induced to abandon their youthful waywardness. Although the child savers were concerned about the increasing number of problem children who posed a threat to community life, they also viewed "delinquents" as children who had lost their way and who needed control, guidance, and nurturing. They believed that children should feel "protected" by the state and that procedural formalities, such as those found in adult criminal courts, would only serve to intimidate children. They had in mind an image of a judge with his arm around a child, not a judge behind a bench with a gavel in his hand looking down on a child from on high. Their hope was that the juvenile court would form an understanding of each child's social and family history and use it to develop an appropriate "treatment."

Although many child savers were concerned about the well-being of children and sought to create institutions capable of serving their best interests, others were threatened by the many poor and immigrant children who lived in the rapidly growing urban areas and represented a threat to traditional institutions. From the point of view of the latter reformers, the main goal was simply to prevent delinquent and other problem behaviors, and the possibility that prevention could be achieved through the juvenile courts and correctional institutions was reason enough to support them.

By the early 1900s, juvenile courts had spread throughout the country. Clearly, some children were helped by their existence, but the juvenile courts did not always act in the kind and benevolent manner their supporters had envisioned. Nevertheless, juvenile court practice was not closely scrutinized until the 1960s.

During the 1960s, however, some of the shortcomings of existing juvenile court practice began to receive attention by the U.S. Supreme Court. In a series of important cases, the Supreme Court extended due process protections to juveniles. The essential thrust of the rulings was that juveniles had a right to some protection from the coercive powers of the juvenile court. However, whether all jurisdictions adhere firmly to the principle that juveniles are entitled to due process protections is in doubt. Indeed, there is considerable evidence that many juvenile courts still operate in an informal manner, which in effect circumvents the due process protections supposedly available to juveniles.

Key Concepts

bail: The money or bond used to secure the release of a person charged with a crime.

bill of particulars: In a lawsuit, a written itemization of claims or charges provided by the plaintiff at the defendant's request; the document can serve as the factual basis of the allegations against the defendant.

chancery court: English court that was primarily concerned with property rights. Chancery courts, first established during the Middle Ages, were partially based on the idea that children, particularly children who owned or were in a position to inherit property, should fall under the protective control of the king. Indeed, it was in these courts that the concept of *parens patriae* was developed. Over time, the chancery courts became more involved in the general welfare of families and children. Because of its role in protecting the interests of children, it served as a model for the juvenile court.

child savers: A group of reformers, mostly well-educated, middle-class Protestant women of Anglo-Saxon descent, who had the time and resources to fight for improved conditions for delinquent youths in jails and reformatories and who eventually played an important role in the creation of the juvenile courts. On the positive side, their work led to wider acceptance of adolescence as an important developmental stage of life, and they correctly believed that the adult criminal justice system was harmful to children or at best ineffective. On the negative side, their conservative ideas on the family and their "noblesse oblige" attitude toward the poor resulted in reforms that focused more on the social control of children and the protection of the interests of the wealthy than on attacking the underlying causes of problem youth behaviors.

dependent children: Children who are dependent on others for financial support or who do not have the support of a parent or guardian.

due process: A course established for legal proceedings intended to ensure the protection of the private rights of the litigants. The essential elements of due process are (1) proper notice as to the nature of the legal proceedings, (2) a meaningful hearing in which the individual has an opportunity to be heard and/or defend him- or herself, and (3) the objectivity of the tribunal before which the proceedings take place.[51]

exclusionary rule: A procedural rule that prohibits evidence that is obtained by illegal means or in bad faith from being used in a criminal trial.

***Miranda* warnings:** Specific warnings given to a suspect prior to questioning that informs them that statements they make to law enforcement agents can be used against them in court.

neglected children: Children who are not given appropriate care by the parent(s) or a guardian.

***parens patriae*:** The obligation of the government to take responsibility for the welfare of children. The doctrine of *parens patriae* recognizes the moral obligation of the government to take care of children when there is no family available or if the family is not suitable.

pretrial discovery: Efforts of a party to a lawsuit to obtain information prior to trial. The theory behind discovery is that all parties should go to trial with as much information as possible.

Progressive Era: A period in American history, lasting roughly from 1890 to 1924, during which a variety of economic, social, and political reforms and reform movements occurred, including women's suffrage; trust–busting; Prohibition; reduction in working hours; elimination of child labor; adoption of social welfare benefits; and the popular voting measures of initiative, recall, and referendum. The reforms were primarily concerned with responding to popular unrest and problems uncovered in the operation of economic, social, and political institutions.

Review Questions

1. How did the social context of the 1800s influence the development of the "child saving" movement?
2. Who were the child savers, and what was their approach to saving children?
3. According to the child savers, what caused youth crime and waywardness?
4. What legal mechanisms existed during the early and mid-1800s to respond to children who engaged in crime or other problem behaviors?
5. What were the facts of the *Crouse* case, what were the findings of the Pennsylvania Supreme Court, and why is this case significant?
6. What were the facts of the *People v. Turner* case, what were the findings of the Illinois Supreme Court, and how did this case influence the development of the juvenile court in Chicago?
7. What social and legal factors contributed to the development of the first statutorily recognized juvenile court?
8. What were the essential characteristics of the first juvenile court?
9. What were the facts of the *Fisher* case, what were the findings of the Pennsylvania Supreme Court, and why is this case significant?
10. What were the essential features of early juvenile court practice?
11. What issues were addressed by the U.S. Supreme Court in the *Kent*, *Gault*, *Winship*, and *McKeiver* cases? What were the important Supreme Court rulings in these cases?
12. How did the *Kent*, *Gault*, *Winship*, and *McKeiver* cases influence the operation of the juvenile courts?
13. Explain how the continuing informality of many juvenile courts circumvents the due process protections extended to juveniles by the Supreme Court.

Additional Readings

Bernard, T. J. (1992). *The cycle of juvenile justice.* New York: Oxford University Press.

Feld, B. C. (1999). *Readings in juvenile justice administration.* New York: Oxford University Press.

Hemmens, C., Steiner, B., & Mueller, D. (2004). *Significant cases in juvenile justice*. Los Angeles, CA: Roxbury Publishing Company.

Krisberg, B. & Austin, J. F. (1993). *Reinventing juvenile justice*. Newbury Park, CA: Sage.

Platt, A. M. (1977). *The child savers: The invention of delinquency* (2nd ed.). Chicago: University of Chicago Press.

Tanenhaus, D. S. (2004). *Juvenile justice in the making*. New York: Oxford University Press.

Cases Cited

Commonwealth v. Fisher, 213 Pa. 48 (1905).

Ex parte Crouse, 4 Whart. 9 (Pa. 1838).

In re Gault, 387 U.S. 1 (1967).

In re Winship, 397 U.S. 358 (1970).

Kent v. United States, 383 U.S. 541 (1966).

McKeiver v. Pennsylvania, 403 U.S. 528 (1971).

People v. Turner, 55 Ill. 280 (1870).

Notes

1. Zinn, H. (1995). *A people's history of the United States 1492–present*. New York: Harper Perennial.
2. Finestone, H. (1976). *Victims of change*. Westport, CT: Greenwood Press.
3. Zinn, 1995.
4. Zinn, 1995.
5. Kelly, A. H. & Harbison, W. A. (1955). *The American Constitution: Its origin and development*. (p. 612). New York: W.W. Norton.
6. Clement, P .F. (1985). Families and foster care: Philadelphia in the late nineteenth century. In R. Hiner & J. M. Hawes (Eds.), *Growing up in America: Children in historical perspective*. Chicago: University of Illinois Press.
7. Platt, A. M. (1977). *The child savers: The invention of delinquency* (2nd ed.). Chicago: University of Chicago Press.
8. Bernard, T. J. (1992). *The cycle of juvenile justice*. New York: Oxford University Press. Platt, 1977.
9. Sommerville, C. J. (1990). *The rise and fall of childhood*. New York: Vintage Books.
10. Bernard, 1992.
11. Platt, 1977.
12. Platt, 1977.
13. Mennel, R. M. (1973). *Thorns and thistles: Juvenile delinquents in the United States 1825–1940*. (p. 129). Hanover, NH: University Press of New England.
14. Platt, 1977.
15. Platt, 1977, p. 119.
16. Bernard, 1992.
17. Platt, 1977.

18. Bernard, 1992.

19. Bernard, 1992.

20. Platt, 1977.
Bernard, 1992.

21. Platt, 1977.
Bernard, 1992.

22. Waegel, W. B. (1989). *Delinquency and juvenile control: A sociological perspective.* Englewood Cliffs, NJ: Prentice Hall.

23. Empey, L. T., Stafford, M. C., & Hay, C. H. (1999). *American delinquency: Its meaning and construction* (4th ed.). Belmont, CA: Wadsworth Publishing.

24. Roberts, A. R. (1998). The emergence of the juvenile court and probation services. In A. R. Roberts (Ed.), *Juvenile justice: Policies, programs, and services* (2nd ed.). Chicago: Nelson-Hall.
Empey, Stafford, & Hay, 1999.

25. Bordelaise, C. & Miller, S. J. (1998). *Juvenile justice in America* (2nd ed.). Englewood Cliffs, NJ: Prentice Hall.
Empey, Stafford, & Hay, 1999.

26. Tanenhaus, D. S. (2004). *Juvenile justice in the making.* New York: Oxford University Press.

27. Krisberg, B. & Austin, J. F. (1993). *Reinventing juvenile justice.* Newbury Park, CA: Sage.

28. Platt, 1977.

29. Bernard, 1992.

30. Bernard, 1992.

31. Bernard, 1992.

32. Lindner, C. & Savages, M. R. (1984). The evolution of probation. *Federal Probation, 48*, 3–10.
Roberts, 1998.

33. Platt, 1977.

34. Hacker, J. C. (1984). Canada. In M. W. Klein (Ed.), *Western systems of juvenile justice.* Beverly Hills, CA: Sage.

35. Farrington, D. P. (1984). England and Wales. In M. W. Klein (Ed.), *Western systems of juvenile justice.* Beverly Hills, CA: Sage.

36. Keener, H. & Weitekamp, E. (1984). The Federal Republic of Germany. In M. W. Klein (Ed.), *Western systems of juvenile justice.* Beverly Hills, CA: Sage.

37. Bernard, 1992, p. 113.

38. Senna, J. J. & Siegel, L. J. (1992). *Juvenile law: Cases and comments* (2nd ed.). St. Paul, MN: West Publishing Co.

39. Bortner, M. A. (1988). *Delinquency and justice: An age of crisis.* (p. 62). New York: McGraw-Hill.

40. *McKeiver v. Pennsylvania*, 403 U.S. 528 (1971).

41. *McKeiver v. Pennsylvania* (1971).

42. Bortner, 1988.

Bernard, 1992.

43. Bernard, 1992.

44. Puritz, P., Burrell, S., Schwartz, R., Soler, M., & Warboys, L. (1995). *A call for justice: An assessment of access to counsel and quality of representation in delinquency proceedings.* Washington, DC: American Bar Association.

45. Feld, B. C. (1991). Justice by geography: Urban, suburban, and rural variations in juvenile justice administration. *Journal of Criminal Law and Criminology, 82*, 156–210.

Feld, B. C. (1989). The right to counsel in juvenile court: An empirical study of when lawyers appear and the difference they make. *Journal of Criminal Law and Criminology, 79*, 1185–1346.

Feld, B. C. (1988). In re Gault revisited: A cross-state comparison of the right to counsel in juvenile court. *Crime and Delinquency, 34*, 393–424.

46. Feld, 1988.

47. Bernard, 1992.

48. Sickmund, M. (2009). *Delinquency cases in juvenile court, 2005.* (OJJDP Factsheet). Washington, DC: Office of Juvenile Justice and Delinquency Prevention.

49. Jacobs, M. D. (1990). *Screwing the system and making it work: Juvenile justice in the no-fault society.* Chicago: University of Chicago Press.

50. Mnookin, R. H. & Weisberg, D. K. (1989). *Child, family, and state: Problems and materials on children and the law* (2nd ed.). Boston: Little, Brown & Co.

51. Black, H. C. (1968). *Black's law dictionary* (4th ed.). St. Paul, MN: West Publishing Co.

Herman v. Chrysler Corp., 106 Mich App 709 (1981).

Public and Police Responses to Juvenile Offenders

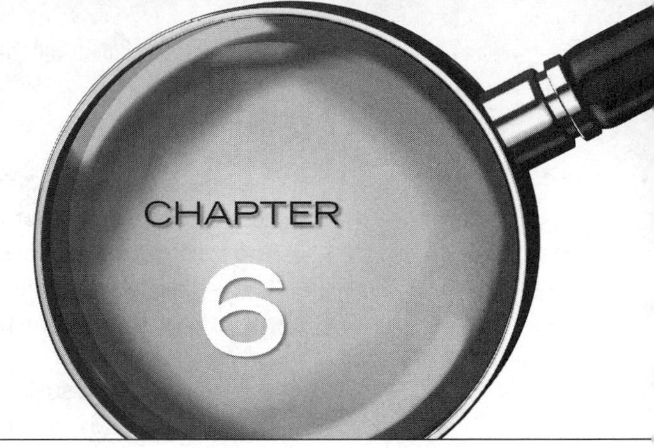

CHAPTER OBJECTIVES

After studying this chapter, you should be able to

- Describe the role that individual citizens play in both the informal and formal juvenile justice process
- Describe the evolution of police responses to youths from colonial America to the present
- Explain the three fundamental duties of the police and how these duties influence police responses to delinquent behavior
- Explain why conflict is an inherent feature of policing and how this conflict influences how the police interact with juveniles and their families
- Describe the factors that influence police decisions to refer youths to the juvenile courts
- Define the concept of police diversion of juveniles, describe its history, and indicate how it is presently used in juvenile justice
- Explain the impact of court decisions, such as *Miranda v. Arizona* and *Schall v. Martin*, on police interaction with juveniles, their families, and the courts
- Describe recent trends in police responses to juvenile offending
- Describe the focus and effectiveness of policing programs and strategies such as DARE, community policing, and youth-oriented community policing in dealing with delinquent behavior

CHAPTER OUTLINE

Introduction

This chapter examines citizen and police responses to juvenile offenders. First, it explores the role that the public plays in dealing with juvenile crime by describing the operation of an informal juvenile justice process—one that consists of efforts by citizens to handle instances of juvenile crime that do not involve the police or other formal agents of juvenile justice. Next, it examines police responses to juvenile offenders and presents a general overview of the history of youth-oriented policing. It then examines contemporary issues regarding the policing of juveniles and concludes with a discussion of recent trends in youth-oriented policing.

Youths and the Public: The Informal Juvenile Justice Process

Examinations of the juvenile justice process usually begin with the police. The police do play a critical role in juvenile justice; in fact, they are the primary gatekeepers of the formal juvenile justice process. The public also is involved, however, because police contact with juveniles is usually the result of a citizen complaint or request for assistance. Thus, an examination of the juvenile justice system should arguably begin with the public.

As noted in Chapter 2, most juveniles engage in illegal behaviors at some time during their childhood. Although some juvenile delinquency consists of serious offenses committed against people, the great majority of delinquent acts are nonviolent. Moreover, people respond to this delinquency in a variety of ways. Some of these responses are informal, which means that they consist of the actions taken by members of the public without reliance on official juvenile justice agencies. Formal responses, in contrast, involve official agencies such as the police and the juvenile courts.

When citizens, whether adults or other juveniles, encounter or observe youths whom they believe are engaging in illegal behavior, they may exercise any of a variety of options. One option is to ignore the offender. Another is to confront the offender. For example, a store clerk who observes a youth taking merchandise from a store shelf and attempting to leave the store may stop the youth and demand that the goods be returned, may lecture the youth and ban him or her from the store, may contact the youth's parents and request that the parents take some action, may call the police, or may decide to take some other action. Importantly, citizens use **discretion**. Discretion is the power to act on one's own authority.

discretion The power to act on one's own authority.

Citizens regularly encounter juveniles involved in alleged violations of the law. Yet, in many instances, they do not contact the police. As just noted, an individual may choose to ignore the youth's actions or deal with the youth in some way that does not involve the police or other formal agents of the juvenile justice process. The actions taken by parents, neighbors, business owners, teachers, and others constitute an **informal juvenile justice process**. Citizens regularly encounter youths engaged in illegal behaviors and decide to handle those situations themselves or seek the assistance of others outside the formal juvenile justice process. This informal juvenile justice process is one, potentially powerful, mechanism that operates to control youths' behavior. Moreover, the more that citizens rely on informal control mechanisms and the more effective they are, the less necessary the formal processing of juveniles becomes.

informal juvenile justice process Actions to combat juvenile delinquency taken by parents, neighbors, business owners, teachers, and others who are not part of the formal juvenile justice apparatus.

Although many delinquent actions are handled informally, in other instances members of the public seek the assistance of formal juvenile justice agencies, such as the police or the juvenile courts, and most youths become involved in the formal juvenile justice process when citizens select this option. In short, in many instances, it is a combination of the public and the police that determines who the clientele of the formal juvenile justice process will be.

Police Responses to Juvenile Offenders

The police are clearly an important component of the juvenile justice process. Indeed, the police usually determine who becomes formal juvenile justice clients. When citizens believe that juveniles are engaging in illegal or problem behaviors and seek the assistance of a formal juvenile justice agency, they usually turn to the police. This is because (1) the police are the most visible symbol of the juvenile justice process within the community and (2) they have the primary responsibility for identifying and processing youths who break the law. Consequently, how the police respond to juvenile offenders influences how many people view the juvenile justice process, and it influences who becomes the clients of the juvenile court.

Like the responses of other institutions to juvenile offenders, police responses are best understood by examining the historical development of policing, particularly as it relates to young people. Consequently, this section begins with a brief historical overview of youth-oriented policing, then examines more contemporary approaches to policing youths.

The History of Policing Youths

The British colonists who settled the New World brought with them a variety of criminal justice institutions and practices, including English common law, a court system, punishments for crimes, and policing practices. However, colonial law enforcement had a number of shortcomings, including inefficiency, corruption, and political interference.[1] Those who had formal policing responsibility, such as constables, sheriffs, and members of the night watch, engaged in a variety of activities besides law enforcement. For example, sheriffs' and other early law enforcement officers collected fees for engaging in particular activities such as collecting taxes, conducting elections, and maintaining bridges and roads.[2] This system encouraged early law enforcement officials to devote more time to civil functions that were safer and offered more regular income than law enforcement activities. As a result, law enforcement typically ranked among their less important areas of responsibility.[3]

Policing during the colonial period was primarily reactive and often ineffective. Those who had formal law enforcement duties responded to complaints brought to their attention, but they had limited resources to investigate offenses and engaged in few efforts to prevent crime. Indeed, most criminal cases in colonial America were initiated

MYTH VS REALITY

During the Early Years of the United States, Informal Controls Were the Primary Means of Social Control

Myth—Early colonial settlements established full-time police forces to maintain order.

Reality—Full-time police forces did not exist in colonial times. Indeed, the first full-time police forces in the United States were not developed until the early 1800s.[5]

MYTH VS REALITY

A Lack of Respect by Many Citizens Was a Problem for Early Police Officers
Myth—During the late 1800s and early 1900s, most city residents had a great deal of respect for the friendly neighborhood patrol officer.
Reality—Many people lacked respect for the police, whom they saw as political hacks willing to use brutality to protect their bosses' interests.[7]

by citizens who brought complaints before a judge; rare was the case in which a citizen contacted the police in order to initiate an investigation.[4] Rather than relying on formal mechanisms to control youth and adult behaviors, residents in colonial settlements relied more heavily on the family and the church.

By the early 1800s, industrialization, urbanization, population growth, and associated social problems, such as alcoholism, homelessness, riots, ethnic clashes, and lawlessness, led to the development of more formal methods of policing. Full-time police forces were created in larger cities during the 1830s and 1840s. By the 1870s, all the major cities had full-time forces and many smaller ones employed part-time police officers.

In these early days, policing in the United States was community based. Police officers came from the communities in which they worked and were paid more than many other workers. They walked the beat in the neighborhoods they patrolled, usually spending from 2 to 6 hours of a 12-hour shift on foot patrol. However, police forces were small, beat officers had little contact with supervisors while on patrol, and turnover was high.[6]

A major problem that faced early policing organizations was corruption. Because the police were paid fees for the jobs they performed, they often saw themselves as private entrepreneurs rather than public servants. Also, policing was closely tied to local politics. Police jobs were a primary form of political patronage and were controlled by local political bosses. Police officers owed their allegiance to these political bosses and the police officials who awarded them their positions rather than to the public.[8] As a consequence, they were careful to act in ways that protected the politically powerful, sometimes by ignoring illegal activities, such as gambling, prostitution, and after-hours liquor sales, and sometimes by participating in such activities themselves.[9]

Although some officers were undoubtedly respected by people in their communities, many citizens saw the police as servants of corrupt political machines.[10] It was common to view the police as primarily engaged in rigging elections, controlling ethnic and racial minorities, breaking strikes, and protecting the interests of wealthy business owners.[11] Police officers often faced considerable hostility from both young and old, and juvenile gangs often made a sport of throwing rocks at the police and taunting them.[12]

The feeling was mutual. Many early police officers exhibited contempt for problem youths as well as concern about those youths who engaged in illegal behavior or who were felt to be potential offenders. For example, in his 1849 semiannual report to the mayor of New York, the Chief of Police George W. Matsell noted the following:

I deem it to be my duty to call the attention of your Honor to a deplorable and growing evil which exists amid this community.... I allude to the constantly increasing numbers of vagrant, idle, vicious children of both sexes, who infest our public thoroughfares, hotels, docks.... Children who are growing up in ignorance and profligacy, only destined to a life of misery, shame and crime, and ultimately to a felon's doom.... Left, in many instances to roam day and night ... a large portion of these juvenile vagrants are in the daily practice of pilfering wherever opportunity offers, and begging where they cannot steal. In addition to which, the female portion of the youngest class, those who have only seen some eight or twelve summers, are addicted to immoralities of the most loathsome description. Each year makes fearful additions to the ranks of these prospective recruits of infamy and sin, and from this corrupt and festering fountain flows on a ceaseless stream to our lowest brothels—to the Penitentiary and the State Prison![13]

In dealing with youth crime and waywardness, the police employed a variety of tactics. Some of these tactics involved formal responses to youth crime, such as **arrest**. Indeed, police officers during this era had almost unlimited authority. They could place a youth in a house of refuge, seek the detention of a youth in a jail, or refer a youth to a criminal court (which could place the youth in a jail or prison). However, police officers also employed a variety of informal, not always legal, tactics in dealing with the young. Some officers developed close working relationships with both youths and adults in their community; this helped the officers respond to problem children. In other cases, officers warned and counseled children. On still other occasions, police dispensed street justice by harassing, verbally abusing, or beating up problem youths.[14]

arrest Legally authorized deprivation of a person's liberty. As a general rule, a person is under arrest when he or she is not free to walk away.

American policing underwent important and fundamental changes during the 1900s. Essentially, these changes resulted from two factors: the development of professional policing standards and the improvement of communications technology, which led to changes in the nature of police work and police administration.[15]

The push toward police professionalism grew during the Progressive Era (roughly 1880 to 1920). This period saw a variety of social movements that focused attention on problems such as economic abuses, social welfare problems, and political and police corruption. Two of the primary architects of reform were Richard Sylvester, superintendent of the District of Columbia police force from 1898 to 1915 and president of the International Association of Chiefs of Police (IACP) from 1901 to 1915, and August Vollmer, chief of police in Berkeley, California, from 1905 to 1932. Sylvester was instrumental in turning the IACP into an effective national police association that advocated for progressive police reform. Vollmer was an important innovator who developed modern police administrative methods and championed increased education for police officers. However, Vollmer did more than just advocate for these reforms. He hired graduates of the University of California to work on the Berkeley force and organized the first police science courses at the university in 1916.[16]

The **reform agenda for police** pushed by Vollmer, Sylvester, and others had a number of goals: "eliminating political influence, appointing qualified chief executives, establishing a mission of nonpartisan public service, raising personnel standards, introducing principles of scientific management, emphasizing military-style discipline, and developing specialized units."[17] The development of specialized units in police departments was an important event in youth-oriented policing in the United States. In the early 1900s, one type of specialized police unit that began to appear was the juvenile unit (sometimes called a youth aid bureau, juvenile bureau, juvenile control bureau, juvenile division, or crime prevention bureau). The development of specialized juvenile units spread during the early 1900s as professionally oriented departments began to think more about delinquency and crime prevention.[18]

reform agenda for police Agenda consisting of items such as the following: eliminating political influence and corruption, appointing qualified leaders, establishing the ideal of nonpartisan public service, raising personnel standards, implementing scientific management principles, instituting military-style discipline, and creating "specialized" units.

One significant aspect of early delinquency prevention efforts was that they became the mechanism by which women initially entered policing. The first woman to be hired as a police officer was Mary Owens, who was given a job by the Chicago Police Department in 1893 after her husband, a Chicago police officer, was killed in the line of duty. Owens made visits to the juvenile court and helped detectives on cases involving women and girls.[19] In 1905, Lola Baldwin was hired by the Portland, Oregon, Police Department. Baldwin had a background in social work, and her primary responsibility was to protect young girls and women and to prevent their involvement in crime. Indeed, women were felt to be particularly suited to dealing with the needs of children, which led several departments around the United States to follow Portland's lead by hiring women to act as police social workers.[20]

The concept of delinquency prevention spread to a number of large police departments during the early 1900s, and by 1924 approximately 90% of the nation's largest cities had specialized juvenile programs that placed welfare officers in high-crime neighborhoods, and assigned police officers to juvenile courts. Other innovative police programs, such as specialized juvenile crime units, relief programs that gave toys to children at Christmas, speakers' bureaus that gave presentations to youth organizations (e.g., the Boy Scouts and Camp Fire Girls), and police athletic leagues, were also developed.[21]

In the 1950s, two additional events highlighted the growing importance of juvenile officers in police departments around the country. In 1955, the Central States Juvenile Officers Association was formed, followed by the International Juvenile Officer's Association in 1957. Both of these organizations worked toward the development of professional standards and procedures for handling juveniles.[22]

During the 1960s and 1970s, the importance of the police role in the handling of juvenile offenders was given additional support through the Law Enforcement Assistance Administration (LEAA). LEAA provided federal funding for the development of hundreds of new police programs, a number of which targeted youths.[23] One of these programs, police diversion, became popular in a number of jurisdictions. Police diversion programs were intended to steer youths away from the formal juvenile justice process by providing various services to youths and their families. Although the services varied from jurisdiction to jurisdiction, they often included individual counseling, recreational programs, social skills training, and the development of parenting skills.[24] Other programs developed during this period included leadership and moral training programs, job assistance programs, and programs intended to reduce school truancy.[25]

In the late 1970s and 1980s, many police departments, particularly small ones, faced budget cuts and rethought the operation of specialized youth divisions. These departments did away with or reduced the size of specialized juvenile units.[26] Nevertheless, others developed still more specialized units that focused on family violence and gangs (issues that were becoming more problematic for police, especially in large jurisdictions). Many departments began to focus increased attention on juvenile crime during the late 1980s and the early 1990s due to concerns about increases in juvenile violence during that period. Today, a diverse range of police programs and responses to juvenile crime can be found throughout the country.

The Role of the Police in Modern Society: Implications for Policing Juveniles

In order to understand the role of police in modern society, it is necessary to consider the "rights and duties" as well as "the normatively approved patterns of behavior" for those in law enforcement.[27] Roles have both sociological and psychological dimensions. Sociologically, the role of police consists of the position that policing has within the social structure and the activities of those who perform policing functions. Obviously, police have considerable power and responsibility and they perform a variety of duties as part of their job. Psychologically, the police role is composed of the attitudes and beliefs that people (including the police themselves) have about those in policing. Importantly, the public and the police have certain expectations for those who work in policing, and these expectations are related to police performance.[28]

One thing is quite clear: The role of police in contemporary society is complex and often characterized by conflict.[29] Some scholars who have studied the police argue that the police role is comprised of three basic functions: law enforcement, service, and order maintenance.[30] The law enforcement role of the police is directed toward the detection, apprehension, and prevention of illegal behavior, and the collection of evidence that can be used in the prosecution of cases in court. Police efforts to enforce legal statutes, including traffic, juvenile, and criminal codes as well as investigating crimes, chasing and arresting suspects, transporting suspects to jail, enforcing traffic laws, engaging in routine patrol, and appearing in court are all examples of the law enforcement role performed by the police.

The service role of the police encompasses efforts by the police to assist citizens in a variety of ways. Indeed, police play a central role in providing a range of services to members of the community who, because of personal, financial, economic, social, or other circumstances, need assistance.[31] Service activities include assisting motorists with stalled vehicles, giving directions to motorists and others, providing various types of information to citizens, giving first aid to injured or ill people, escorting ambulances or fire trucks, responding to the needs of people with mental health problems, and providing a variety of other services to members of the public.

The order maintenance role of the police encompasses their efforts to intervene in situations that threaten to disturb the peace or that involve face-to-face conflicts between two or more individuals. Examples of order maintenance include resolving a dispute between a tenant and a landlord or between a store clerk and a customer, dealing with a noisy drunk, and responding to a group of rowdy college students at a fraternity party. Drunkenness and rowdiness are behaviors disapproved of by some members of the public, and those who disapprove of these behaviors often ask the police to control or stop them.[32]

The order maintenance role of the police is important for several reasons. First, order maintenance activities account for as much as 80% of police activity.[33] Second, actions that disrupt the public order have the potential to erupt into violence. Third, police have considerable discretion in handling public order situations.[34] Fourth, these situations often involve juveniles and their families. Police regularly encounter individual youths, as well as youths in large and small groups, who may be perceived by individual police officers and some members of the community as potentially threatening to the public order.

The most appropriate police response to situations where there is a perceived threat to public order is not always clear, however. Police feel obligated to assist people who have complaints about the behavior of juveniles, yet behaviors that members of the public find threatening or annoying are not necessarily illegal. Police officers may feel that a group of youths on a street corner represents a potentially problematic situation that demands intervention, but the youths have a constitutional right to congregate and express themselves, a right that the police are obligated to protect. As a result, the police may find themselves in a no-win situation. If they fail to disperse a rowdy group of teenagers, citizens may complain that the police are ineffectual or do not care. If they break up the group, they open themselves to a charge of police harassment. The attitudes

MYTH VS REALITY

Policing Is Characterized by Role Conflict

Myth—The public expects the police to enforce all laws.

Reality—The public expects the police to engage in selective law enforcement. For example, few citizens want the police to strictly enforce all traffic laws. Indeed, the police recognize that strict enforcement of all traffic laws would result in considerable conflict between themselves and citizens.

and behaviors exhibited by both police and juveniles in these encounters can play a large role in shaping the attitudes that police and juveniles have of one another.

police role conflict
Clash between the duty of police to investigate crimes and prevent individuals from committing crimes and the duty of police to protect the rights of offenders, crime victims, and other citizens.

As this example indicates, police often experience **police role conflict** caused by their different responsibilities. They are expected to prevent crimes and respond to the illegal behaviors of individuals, including juveniles, but they also are expected to protect the rights of all citizens. In addition, the police recognize that the public does not want them to enforce all laws and that full enforcement of the law is actually counterproductive. For example, many police recognize that the arrest of a juvenile for a minor illegal activity, such as pushing another child down on a community playground (an assault), might result in more harm than good and could be better dealt with in a more informal manner.

Because police encounters with juveniles are frequent, they often occur under stressful conditions, and they play an important role in shaping the attitudes that police and juveniles have toward one another. Therefore, it is imperative that police be carefully trained in dealing with youths. Police departments have used two basic training approaches. One approach involves training individual officers in how to approach juveniles. This approach is more appropriate for small departments that have little specialization; officers in small departments may receive little training that focuses specifically on dealing with juveniles. The second approach involves developing specialized units to deal with more serious juvenile cases and is more likely to be found in large departments with specialized units. However, the quality of training given to individual officers in handling juvenile matters, and in some cases the training of specialized juvenile officers, varies considerably across departments.[35] Some rural and small police departments do not have officers who are specially trained to deal with juveniles.[36] In contrast, other departments have given extensive training to officers to help them respond to juvenile crime more effectively.[37]

The Police: The Gatekeepers of the Formal Juvenile Justice Process

Clearly, police work is characterized by complexity and conflict. Nevertheless, police play a crucial role in juvenile justice because they act as primary gatekeepers to the formal juvenile justice process. For example, in 2002, 82% of the delinquency cases referred to the juvenile courts came from the police. In addition, police agencies were the primary referral source for 55% of court referrals involving running away from home and 92% of the referrals involving liquor law violations.[38] Like citizens, however, the police also exercise discretion in the handling of juvenile cases. Among other options, police officers may select one of the following choices:

- Warn and release the juvenile.
- Refer the juvenile to his or her parents.

- Refer the juvenile to a diversionary program operated by the police or a community agency.
- Refer the juvenile to the local juvenile court.

Factors That Influence Police Decisions to Arrest Juveniles

Describing the typical police response to juveniles is difficult, because there is considerable variability in the options available within different jurisdictions and in how individual police officers approach juvenile offenders. The factors that influence police decision making include (1) the seriousness of the offense; (2) the police organization; (3) the community, including the resources available within the community for responding to the juvenile offender; (4) the wishes of complainants; (5) the demeanor of the offender; (6) the gender of the offender; and (7) the race and social class of the offender.

Offense Seriousness

The most important factor influencing police–juvenile encounters is the seriousness of the offense. Regardless of the other factors, as the seriousness of the offense increases, so does the likelihood of arrest. Indeed, most police–juvenile encounters involving felony offenses result in an arrest.[39] Most police–juvenile encounters, however, involve minor offenses. Only 5 to 10% involve felonies.[40] Consequently, in the majority of police–juvenile interactions, other factors play an important role in determining how police react.

Police Organization and Culture

Police departments, like other types of organizations, develop their own particular styles of operation. They also develop a variety of formal and informal policies and procedures for handling juveniles involved in various types of illegal behaviors.[41] As a result, police departments vary in how they respond to juveniles. For example, one study that categorized police departments according to the extent to which they employed a legalistic style of policing (characterized by a high degree of professionalism and bureaucratic structure) found that the more legalistic departments were more likely to arrest juvenile suspects than less legalistic departments.[42] A study of four Pennsylvania communities found that the percentages of juveniles referred to court varied considerably from one community to another. In one community, only 9% of the juveniles who had police contacts were referred to court, whereas, in another community, 71% were referred.[43] Similarly, a study of 48 departments in southern California found that virtually all juveniles arrested by one department were referred to court, but that in another department, the great majority of juveniles were counseled and released.[44]

Clearly, there is considerable variability in how police departments respond to juvenile offenders, and this variability is partially a product of differences in department organizational characteristics and department policies and procedures. However, police departments do not operate in a political and social vacuum. They are influenced by the communities within which they operate. Consequently, community characteristics and attitudes toward the police also can influence police responses to juvenile offenders.

The Community Influence on Policing

One factor that can influence how police handle juvenile suspects is whether community programs for handling juvenile offenders exist. In some communities, officers may have a variety of options, including programs that provide individual, group, or family counseling; assist with conflict resolution; or provide restitution to victims. In other communities, available options are more limited.

Communities also influence policing in a variety of other ways. Through their interaction with the local community, police develop assumptions about the community, the people who reside and work there, and the ability and willingness of community members to assist in responding to crime and delinquency. When a police officer encounters a member of the community—whether a suspect, a complainant, or a witness to an alleged offense—the officer's interaction with that person is colored by his or her attitude toward the community and the subgroups that compose it. Consequently, in order to understand how communities influence policing, it is necessary to consider the attitudes police and community members have toward one another.

Research suggests that police departments operate differently in lower-class communities than in wealthier communities. The behavior of police in a specific community seems to be a reflection of what police expect as well as the realities of policing that community. Police expect lower-class communities to have higher crime rates, they are aware that more arrests occur in those communities than in wealthier communities, and they are aware that lower-class communities have few resources for informally responding to the wide array of problems experienced in those communities. Consequently, when police come into contact with juveniles in lower-class areas, formal responses are likely because informal responses are felt to be unrealistic. Indeed, research on the effect of neighborhood socioeconomic status has found that as the socioeconomic status of the neighborhood increases, the likelihood of a police–juvenile encounter ending in arrest declines.[45]

Community attitudes toward the police also exert some influence on police actions. Although the police often feel that the public is hostile toward them, the reality is that most citizens have a positive attitude toward the police. For example, citizens have more confidence in the police than they have in a number of other American institutions, including the presidency, public schools, television news, the U.S. Supreme Court, big business, banks, organized labor, church and organized religion, and the medical system.[46]

It is true, however, that certain subgroups have less favorable views of the police than people in general. In a 2007 Gallup poll, 22% of Blacks, compared with 60% of Whites, reported a great deal of confidence in the police; and 34% of Blacks, compared with 9% of Whites, indicated they had little or no confidence in the police.[48] Also, some research suggests that adolescents' views of the police become more negative as they become older.[49] An early study that examined the attitudes of almost 1,000 junior high students in Cincinnati found that hostility toward the police increased as the students moved through their junior high years. The same study found that lower-class youths were less likely to have positive attitudes toward the police than upper-class youths.[50]

MYTH VS REALITY

A Majority of Americans Respect the Police

Myth—The public lacks respect for police officers.

Reality—Gallup polls consistently report that the majority of citizens have a great deal of respect for the police. For example, in a 2007 Gallup Poll, 54% of the respondents indicated they had a "great deal" or "quite a lot" of confidence in the police. In that same poll, 25% indicated that they had a "great deal" or "quite a lot" of confidence in the presidency, 18% reported that they had confidence in big business, and 46% indicated that they had confidence in church or organized religion.[47]

Similar findings have been seen in several more recent studies as well as in public opinion polls that examine public confidence in American institutions, including the police. This research indicates that youths in high crime and minority neighborhoods often hold negative views of the police.[51] Such attitudes are important in a police–juvenile encounter because they can affect how the encounter proceeds and the extent to which the parties are satisfied with the outcome.

The Wishes of Complainants

As noted earlier, many police–juvenile interactions result from citizen complaints. Furthermore, what police do in a complaint situation typically depends on whether the complainant is present and what the complainant would like the police to do. For example, one study of police–juvenile encounters found that when the complainant requested informal action on the part of the police, police officers always did what the complainant wanted. If the complainant asked that an arrest be made, officers complied with this request about 60% of the time.[52] In a replication of this study done 8 years later, other researchers reached similar conclusions. The results of this study indicated that when citizens and suspects were both present and when citizens indicated clear preferences regarding how they wanted the police to respond, police usually complied with these requests. The researchers concluded, based on these findings as well as on the findings of the earlier research, that citizens "largely determine official delinquency rates" because their wishes are accorded considerable weight by the police.[53]

The Demeanor of the Offender

As one might expect, how youths behave toward the police can affect whether police officers decide to make an arrest. Interestingly, youths who are unusually antagonistic or unusually polite are more likely to be arrested. In contrast, youths who are moderately respectful are less likely to be arrested—as long as the offense is not serious.[54] Apparently, police officers have basic expectations regarding the behavior of those they have interaction with. They expect suspects to be neither overly hostile nor unusually polite. When suspects' behaviors fall outside the boundaries of those expectations, the likelihood of arrest increases.[55]

The Gender of the Offender

Gender also appears to influence arrest decisions, particularly when **status offenses** are involved. However, the research in this area has produced mixed results. Some studies that examined the relationship between gender and the decision to arrest found that girls are less likely to be arrested for criminal offenses than boys, even when prior criminal record and offense seriousness are taken into account.[56] However, other studies indicate that any gender bias that existed in the past has diminished or disappeared over time,[57] although other research indicates that gender continues to influence arrest decisions in some locations.[58] Still other research indicates that the relationship between gender and arrest decisions is more complex than it first appears. For example, one study that examined police–suspect encounters in 24 police departments found that arrest rates for males and females (both juveniles and adults) were similar. However, the researchers also discovered that police officers used different arrest criteria for males and females, which resulted in younger females receiving harsher treatment than older females. No differences between younger and older males were found. This suggests that, at least in those departments, police took a more paternalistic stance toward younger females, increasing the likelihood of formal processing.[59] Overall, this research suggests that although gender may not play a significant role in police decision making in some jurisdictions,

status offense
An act considered to be an offense partly because of the status of the person who performed the act. For example, juveniles (i.e., individuals who have a juvenile status) must obey the reasonable rules of their parents, attend school, and live at home. If they fail to do any of these things, they could be arrested and brought into the court system. Adults do not have to do any of those things and cannot be arrested for refusing to do them.

in combination with other factors such as the type and seriousness of delinquent offense and girls' demeanor, it continues to influence practices in other locations.

Unlike the research on the relationship between gender and arrests for criminal offenses, the research on the relationship between gender and the processing of status offense cases is generally consistent. A number of studies have found that female status offenders are more likely to be formally processed than male status offenders.[60] Even studies that found that differentials in the processing of criminal offenses have declined over time report gender differentials in the processing of status offenses.[61]

The Race and Social Class of the Offender

Strong evidence exists that the race and the social class of juveniles also influence police decision making. As noted in Chapter 2, minority and poor youth are disproportionately represented in arrest statistics. However, the relationship between race or social class and police decision making is complex, which makes straightforward descriptions of this relationship difficult.

After reading the following true story, list the factors that you believe contributed to the behavior of the police. Do you feel the police handled this situation appropriately? How do incidents such as this influence police community relations?

These events took place in 1975. A friend and I had just completed playing a high school basketball game. We had just left our high school, Pontiac Northern, on our way to meet with other members of the team. I was driving, and my friend was in the passenger seat. As we passed a major intersection, we saw police lights flashing, although they were still some distance away. As we approached the next major intersection, we saw four police cars, one at each corner of the intersection. As I entered the intersection, each of the police cars pulled into the intersection and cut off the car I was driving. The officers exited their cars with their guns drawn and demanded that we get out of the car. We asked what we had done. Did we do anything wrong? They told us to shut up and get out of the car. As we got out of the car, they grabbed us and threw us to the ground with their guns pushed against the backs of our necks. While on the ground, they asked us questions about a robbery that had taken place. We said we didn't know anything about it. We told them that we had just finished playing a high school basketball game and were on our way to a party. They told us we were lying. They said a gas station in the neighborhood had been held up, and we fit the description. After we were on the ground for a short time, they said we could go. We asked, "What's going on?" Their response was that they had a line on who the perpetrators were. We argued with them about how we had been treated and asked why they had stopped us. They said we fit the description. What description was that, we asked? Their response was, "Two African American males, about 5' 11", short hair." I remember laughing about it at the time because it fit the description of just about every other black male I knew. They didn't like us questioning their treatment of us. They then searched our car and told us that if we weren't quiet they would take us to the station. The way I saw it was that just being black makes you a suspect in the eyes of many police. Prior to this, I had no personal experiences with the police, although I had witnessed a number of negative encounters between the police and other people in my neighborhood. After this incident, my trust and faith in the police were greatly diminished. (Felix Brooks, Probation Officer, MA, Political Science)

Research on the relationship among race, social class, and delinquency has found that lower-class minority youths are especially likely to be arrested, even when taking into account both prior criminal records and offense seriousness.[62] Such findings indicate that police decisions to arrest are biased against minority and poor youths. They may be even more biased against minority and poor youths who act in an unusual manner or appear to be unconventional in some other way.

As criminologist James Q. Wilson notes:

> The patrolman believes with considerable justification that teenagers, Negroes, and lower-income people commit a disproportionate share of all reported crimes; being in those population categories at all makes one, statistically, more suspect than other people; but to be in those categories and to behave unconventionally is to make oneself a prime suspect. Patrolmen believe they would be derelict in their duty if they did not treat such people with suspicion, routinely question them on the street, and detain them for longer questioning if a crime has occurred in the area.[63]

In contrast, a number of other studies have failed to find strong evidence of police bias,[64] while admitting that a number of factors make interpreting the relationship among race, class, and arrest difficult. For example, the effect of race on the decision to arrest could be complicated by other factors, such as police decisions to focus their attention on poor and minority areas, the wishes of the complainants, and the demeanor of the offender.

Based on the available evidence, it appears that there are differences in how police departments treat poor and minority youths in some communities.[65] It also appears that in urban areas, where minority populations tend to be large, police are more likely to engage in discriminatory practices,[66] although more research on police–youth encounters in small town and rural areas is needed in order to make definitive conclusions about police–youth contacts in urban and nonurban areas.[67]

The discriminatory treatment of minority and poor youths should not be interpreted to mean that police in large urban areas are blatantly racist or hold biased attitudes toward the poor. Although some police officers, like people in other occupations, are racist or strongly biased against the poor, the disproportionately high arrest rate for lower-class and minority youths may be better understood as an outgrowth of **institutional racism and bias**—that is, the ways in which basic social institutions operate to keep minorities and poor people in subordinate positions.

institutional racism and bias The ways that basic social institutions operate to keep minorities or poor people in subordinate positions. If police believe that minority communities foster more crime, and believe that minority group members are more hostile to the police, it increases the likelihood that police will focus more crime-fighting resources in minority communities, resulting in increases in crime, police–citizen hostility, and neighborhood problems.

Police decisions to concentrate their surveillance activities in lower-class or minority communities, for example, increase the likelihood that they will uncover delinquent behaviors in those communities. High rates of delinquency in such communities may be partial products of long-term patterns of discrimination. Further, when police encounter youths in a lower-class or minority community, some youths will present a hostile demeanor caused by past unpleasant encounters with the police, which, in turn, increases the probability that the police will respond in a more forceful way. Such interactions reinforce police perceptions that more forceful responses are justified but, at the same time, they encourage more negative community perceptions and responses toward the police. The result is a cycle of action and reaction that results in negative police–juvenile encounters and increased numbers of arrests.

Police Processing of Juvenile Offenders

Two types of police units have the most contact with juveniles: patrol units and specialized juvenile units (although not all departments have specialized juvenile units).

COMPARATIVE FOCUS

Countries Vary Widely in Police Responses to Youths

In Thailand, youths are not considered to be criminally responsible for their behaviors. Consequently, they are treated differently than adults and no youth younger than the age of 14 years can be punished. Moreover, arrests of youths are only possible when a youth has committed a flagrant offense, an injured person has insisted on an arrest, or a warrant is issued under the code of criminal procedure.[68] In contrast, in Saudi Arabia, there is no defined age of responsibility and young males may be detained for offenses such as eating in restaurants with girls or making lewd comments to women in shopping malls. Moreover, the punishment for youths is the same as the punishment for adults. For minor offenses, such as those previously mentioned, flogging may be used.[69]

Of these two, patrol units have the most contact with juveniles.[70] Specialized juvenile units and youth divisions typically serve as referral units that accept juvenile cases from other departmental divisions.[71] Juvenile units are usually small,[72] but they often conduct juvenile investigations on their own, particularly investigations of serious juvenile crimes. There is no agreement among police officials regarding the best response to juveniles who engage in minor types of delinquent behavior. Some maintain that police should adopt a strict law enforcement approach that focuses on enforcing laws and making juvenile arrests when law violations occur. In contrast, others argue that a crime prevention approach that favors diversion, except in the case of serious offenses, is preferable.[73] Many patrol officers and their supervisors support the crime prevention approach for nonserious juvenile offenses. This approach also is supported by labeling theorists, who call for radical nonintervention. Radical nonintervention involves avoiding formal action whenever possible in order to avoid the labeling and stigmatization of youths that may lead to additional delinquent behavior. However, as you will see later in this chapter, overall, police have become more formal in their processing of juvenile cases by referring a greater percentage of juvenile cases to the courts.

When a police officer encounters a juvenile who has committed an illegal act, the officer must make a decision regarding the best way to handle the case. As noted earlier, the decision the officer makes can be influenced by a variety of factors. One option, of course, is to make an arrest (in some jurisdictions this is referred to as "taking into custody"). For practical purposes, an arrest is considered to have taken place if the youth is not free to walk away.

probable cause
Grounds sufficient to convince a reasonably competent person that a crime was committed and that the suspect committed it.

As a general rule, the basis for arresting a juvenile is the same as the basis for arresting an adult. The officer needs to have **probable cause**, which means that the officer has reason to believe that an offense has been committed and that the youth to be arrested committed the offense. The U.S. Supreme Court rulings in *Gault* and *Kent* as well as subsequent court cases raise the question of whether special standards should apply to the arrest of a juvenile and whether the police can deal with juveniles in the same way they deal with adults.

One of the key issues surrounding the arrest of juveniles is the interrogation of juvenile suspects. However, before determining whether special standards apply to the interrogation of juveniles, we need to look at the standards that apply to the interrogation of adults. The most important case regarding this issue is *Miranda v. Arizona*.[74] Ernesto A. Miranda had been arrested for kidnapping and rape and was in the custody of the

Phoenix police. He was identified by the complaining witness and was interrogated for 2 hours without being notified of his right to have an attorney present. Finally, the police produced a signed confession, on which was typed a statement indicating that the confession was voluntary and "with full knowledge of my legal rights."[75]

In its decision, the U.S. Supreme Court indicated the following:

> We hold that when an individual is taken into custody or otherwise deprived of his freedom by the authorities in any significant way and is subject to questioning, the privilege against self-incrimination is jeopardized. Procedural safeguards must be employed to protect the privilege ... He must be warned prior to any questioning that he has the right to remain silent, that anything he says can be used against him in a court of law, that he has the right to the presence of an attorney, and that if he cannot afford an attorney one will be appointed for him prior to any questioning if he so desires ... unless and until such warnings and waiver are demonstrated by the prosecution at trial. No evidence obtained as a result of interrogation can be used against him.[76]

In its decision, the court clearly stated that adults cannot be held in custody for long periods of time and cannot be questioned without being specifically advised of rights regarding an attorney in order to protect the Fifth Amendment right to be free from self-incrimination. If these circumstances that Ernesto Miranda found himself in were so coercive as to make his confession inadmissible, what types of circumstances would make the confession of a juvenile inadmissible?

The U.S. Supreme Court addressed this issue in 1979, in the case of *Fare v. Michael C.*[77] Michael C. was 16 years old at the time of his arrest on suspicion of murder. He was taken to the police station and, before questioning, was fully advised of his rights per *Miranda*. He was on probation to the juvenile court, and he asked to see his probation officer. The police denied this request. He proceeded to talk to the police without an attorney, making statements and drawing sketches that implicated him in the murder.

In this case the court indicated the following: "Thus the determination whether statements obtained during custodial interrogation are admissible against the accused is to be made upon an inquiry into the *totality of the circumstances* surrounding the interrogation, to ascertain whether the accused, in fact, knowingly and voluntarily decided to forgo his rights to remain silent and to have assistance of counsel."[78] The court cited age, past contacts and experience with the police, whether the juvenile was involved with the juvenile court, and the youth's intelligence as factors that were to be weighed as part of the "totality of circumstances." The court also mentioned whether the juvenile was "worn down" by improper interrogation tactics employed by the police or tricked by them as additional factors to review. In *Fare v. Michael C.*, the confession was determined to be admissible by the court.

The "totality of the circumstances" test was not a new one; it had been articulated by the Supreme Court of California in *People v. Lara*.[79] In this case, the court held that the factors to be reviewed in order to determine that a confession by a juvenile is voluntary included the juvenile's age, intelligence, education, experience, and ability to comprehend the meaning and effect of the statements made. This holding is interesting, because in *Fare v. Michael C.*, a California case, the Supreme Court of California ruled the juvenile's confession inadmissible, but the U.S. Supreme Court, adopting a similar test, ruled it admissible. As this example makes evident, courts looking at the same set of facts can interpret or construe them differently.

Fare v. Michael C. is important because it required courts to examine a variety of factors in determining whether information obtained from juveniles by the police could be used in court. It is also important because it indicated that parents, guardians, custodians, other significant adults, or attorneys do *not* have to be present when juveniles are interrogated by the police. Although the "totality of circumstances" test may sound reasonable, courts look at police conduct with hindsight that sometimes falls below 20/20 acuity. In many areas of society, children are recognized as being fundamentally different from adults. For example, minors cannot own property, sign business contracts, or engage in other adult behaviors because they are considered minors. They can, however, by themselves, confess to a murder. A strong argument can be made that this is illogical.

The issue of self-incrimination by juveniles also was given some attention by the Supreme Court of New Hampshire in the case of *State v. Benoit*. Here the court stated, "Courts employing the totality of the circumstances test do so under the belief that juvenile courts are equipped with the expertise and experience to make competent evaluations of the special circumstances surrounding the waiver of rights by juveniles."[80] The court went on to list 15 circumstances that juvenile courts should consider. Juvenile courts, however, may not routinely perform the appropriate evaluations. The expertise and experience of the court that is charged with the responsibility of protecting a juvenile's rights is a key factor in determining whether those rights are, in fact, protected. Unfortunately, the experience and expertise of many juvenile courts in this area are questionable.

The *Benoit* decision also articulated the main alternative to the "totality of the circumstances" rule, namely, the "interested adult" rule. According to this rule, adopted by a few jurisdictions, no juvenile can waive the privilege against self-incrimination without having had the opportunity to consult with and have present an adult who is "friendly" to the juvenile and who understands the juvenile's rights.

The effect of the *Benoit* holding was to increase the totality factors by adding three more factors:

1. The juvenile must be informed of his or her rights in language understandable to him or her.
2. The juvenile must be made aware of the possibility of adult criminal prosecution or waiver, if applicable.
3. When the juvenile is arrested, the officer in charge must immediately secure from the juvenile the name of a "friendly adult" the juvenile could consult.[81]

Thus, the New Hampshire Supreme Court attempted to blend the "interested adult" test into the "totality of the circumstances" test by making the securing of a friendly adult one of the voluntary factors to be considered.

While the preceding legal cases are concerned with tests to determine when evidence obtained by the police is admissible in court, little is known about actual interrogation practices employed by the police when they encounter juvenile suspects. What is known, however, indicates that police likely use the same interrogation tactics with juveniles as they do with adults—tactics such as psychological coercion, trickery, and deceit. Moreover, while police may understand that youths lack the comprehension abilities of adults, they may not employ their understanding in the interrogation of juveniles. This is particularly troubling because these types of interrogation techniques appear to increase the likelihood of false confessions among youths.[82]

One important difference in the handling of juveniles and adults at the arrest stage is that juveniles are more likely to be detained pending trial than adults who have committed similar offenses. The U.S. Supreme Court, in *Schall v. Martin*, ruled that preventive detention of juveniles is permissible.[83] This case concerned Gregory Martin, age 14 years, who was arrested on December 13, 1977, and charged with first-degree robbery, second-degree assault, and criminal possession of a weapon. At 11:30 p.m., he and two other juveniles had hit a youth on the head with a loaded gun and stolen his jacket and sneakers. When arrested, Martin had the gun in his possession and lied to the police about where and with whom he lived. He appeared in New York Family Court on December 14 with his grandmother and was ordered detained because he had been in possession of a loaded gun and he had lied to the police about his home address as well as the late hour of the incident. A probable cause hearing was held on December 19, and probable cause was found on all charges. At a fact-finding hearing held December 27 through December 29, he was found guilty. He had been detained a total of 15 days.

In this case, the court ruled that "Children, by definition, are not assumed to have the capacity to take care of themselves. They are assumed to be subject to the control of their parents; and, if parental control falters, the state must play its part as *parens patriae* ... In this respect, the juvenile's liberty interest may, in appropriate circumstances, be subordinated to the state's parens patriae interest in preserving and promoting the welfare of the child."[84] The court went on to point out that every state and the District of Columbia permit the preventive detention of juveniles:

> The fact that a practice is followed by a large number of states is not conclusive ... as to whether that practice accords with due process, but it is plainly worth considering in determining whether the practice "offends some principle of justice so rooted in the traditions and conscience of our people as to be ranked fundamental" ... we conclude that the practice serves a legitimate regulatory purpose compatible with the "fundamental fairness" demanded by the Due Process Clause in juvenile proceedings.[85]

There are several ramifications of the *Schall* decision. *Parens patriae*, an idea associated with the juvenile court, played an important role in this decision. In fact, the idea that a juvenile should and must be under someone's control is a long-standing principle in U.S. law. As long as juvenile courts have time guidelines for the setting of meaningful hearings so that juveniles do not languish in detention, the constitutional right to be free from "excessive bail," found in the Eighth Amendment, takes a back seat to society's interest in the supervision and control of juveniles. Nevertheless, the punitive nature of some detention facilities, the possibility of a child's victimization in detention, and the widespread use of detention in some jurisdictions raise serious questions about the practice of detaining juveniles prior to adjudication.

MYTH VS REALITY

Many Youths Are Arrested and Detained for Minor Offenses

Myth—Most youths detained by police have committed serious offenses and pose a threat to community safety.

Reality—Most youths detained in many jurisdictions have not committed serious offenses against people. Moreover, in many jurisdictions the great majority of youths who are detained pending a court hearing are released at the hearing because they are felt to pose no threat to community safety.

The exact procedures that police must follow when taking a juvenile into custody vary from state to state and are spelled out in state juvenile codes and in police department policies. However, these codes and policies typically require officers to notify the juvenile's parents that the juvenile is in custody when an arrest is made. Often, police request that parents come to the police station or the officer transports the youth home prior to any questioning. If the officer feels that detention of the juvenile is appropriate, the juvenile is transported to a juvenile detention facility or, in some jurisdictions, to an adult jail if a juvenile detention facility is unavailable. If the juvenile is released to his or her parents, the juvenile and parents are informed that they will be contacted by the court at a later date about their case. After the release of the juvenile, the officer completes the complaint by collecting any additional information needed to establish the offense. The completed complaint is then forwarded to the next stage of the juvenile justice process for further action. If the juvenile is detained in a juvenile detention facility or adult jail, the processing of the complaint is expedited because juvenile codes require that a detention hearing be held to determine the appropriateness of detention and the complaint will be needed at the hearing.

The interrogation and detention of juvenile suspects are sources of considerable controversy. At the heart of the interrogation controversy is a debate over the age at which juveniles are mature enough to fully understand the importance of their rights and the potential consequences associated with their waiver. Research indicates that many youths do not understand their rights when arrested. In one study, about one-third of a sample of institutionalized youths incorrectly believed that they *had* to talk with the police. The study also found that about one-third of the parents of these youths would advise them to confess to the police.[86] In another study, over half of the youths examined lacked a full understanding of the Miranda warnings.[87] Moreover, the evidence from developmental psychology indicates that youths are psychologically less mature than adults.[88] Such findings raise concerns about the vulnerability of many youths when they are questioned by the police and the reality that youths receive fewer protections than those given to adults. Concern over the detention of juveniles revolves around the fact that preventive detention amounts to punishment of someone before he or she has been found guilty of an offense. Moreover, the conditions youths are sometimes exposed to in detention units and adult jails raise additional worries. (See Chapter 12 for a more complete discussion of the placement of juveniles in detention facilities and jails.)

Trends in Police Processing of Juveniles

In 1973, approximately 50% of all juveniles taken into police custody were referred to juvenile court, approximately 45% were handled within the police department and released, and slightly more than 1% were referred to adult or criminal courts. However, the percentage of youths taken into custody and referred to juvenile courts has been increasing over time.[91] By 2007, approximately 70% of youths taken into custody by the

> **FYI**
>
> **Juveniles Are Sometimes Placed in Adult Jails**
> On June 30, 2004, it was estimated that there were 7,083 juveniles detained in adult jails in the United States.[89] This figure, however, may underestimate the number of juveniles confined in such facilities.[90]

TABLE 6-1 Percent Distribution of Juveniles Taken into Police Custody, by Method of Disposition, 1973–2007

Year	Referred to juvenile court jurisdiction	Handled within department and released	Referred to criminal or adult court	Referred to other police agency	Referred to welfare agency
1973	49.5	45.2	1.5	2.3	1.4
1975	52.7	41.6	2.3	1.9	1.4
1977	53.2	38.1	3.9	1.8	3.0
1979	57.3	34.6	4.8	1.7	1.6
1981	58.0	33.8	5.1	1.6	1.5
1983	57.5	32.8	4.8	1.7	3.1
1985	61.8	30.7	4.4	1.2	1.9
1987	62.0	30.3	5.2	1.0	1.4
1989	63.9	28.7	4.5	1.2	1.7
1991	64.2	28.1	5.0	1.0	1.7
1993	67.3	25.6	4.8	.9	1.5
1995	65.7	28.4	3.3	.9	1.7
1997	66.9	24.6	6.6	.8	1.1
1999	69.2	22.5	6.4	1.0	.8
2001	72.4	19.0	6.5	1.4	.7
2003	71.0	20.1	7.1	1.2	.6
2005	70.7	20.2	7.4	1.3	.4
2007	69.6	19.5	9.4	1.2	.4

Source: Data from United States Department of Justice, Federal Bureau of Investigation. (September 2008). Table 68. *Crime in the United States 2007.* Retrieved from www.fbi.gov/ucr/cius2007/data/table_68.html; Pastore, A. L. & Maguire, K. (Eds.). (2006). *Sourcebook of criminal justice statistics, Table 4.26.* Retrieved from http://www.albany.edu/sourcebook/pdf/t4262006.pdf.

police were referred to juvenile court, 20% were handled within the police department and released, and 9% were referred to adult or criminal courts (see Table 6-1).[92] These data indicate the existence of three important trends in police processing of juveniles: (1) the referral of more youths to juvenile court, (2) the handling of fewer cases within police departments, and (3) the referral of more youths to adult courts.

The Police and Delinquency Prevention

Although police are becoming more formal in their processing of juvenile cases, many police departments also devote resources to delinquency prevention. Indeed, such efforts have a number of advantages. For example, by preventing youths from engaging in delinquent activities, police spare youths the stigma associated with formal juvenile justice processing. Delinquency prevention also spares juvenile courts the time and costs associated with processing cases and allows the police and the courts to focus more time on serious juvenile offenses and adult crime.

One of the most widely used prevention programs is the D.A.R.E. program. (D.A.R.E. is an acronym for drug abuse resistance education.) This program, which involves cooperation between schools and police departments, typically targets upper elementary school children and attempts to provide them with the skills necessary for resisting peer pressure to take drugs. However, D.A.R.E. curriculum materials have been developed for youths in kindergarten through the 12th grade and for parents. The program is unique because it uses uniformed police officers, who present a structured curriculum designed to produce the following results:

- provide students with accurate information about drugs, including alcohol and tobacco, and the consequences of taking drugs
- teach students specific skills that will allow them to resist peer pressure to experiment with drugs
- help students develop respect for police officers
- give students alternatives to drug use
- help build students' self-esteem

More recently, the D.A.R.E. curriculum has been updated and renamed "D.A.R.E. to Resist Drugs and Violence." The new curriculum, which includes information about the use of tobacco and inhalants, attempts to teach youths conflict resolution and violence prevention skills and calls for closer cooperation between D.A.R.E. officers and classroom teachers.[93] Although the D.A.R.E. program reaches approximately 26 million children annually and receives more than $200 million in funding, there is little evidence that it is effective at reducing drug use.[94] Some research indicates that exposure to the D.A.R.E. program can increase students' knowledge about drugs and enhance their social skills. In addition, it appears to influence students' attitudes about drugs, their perceptions of the police, and their self-esteem. However, the positive effects associated with D.A.R.E. dissipate rapidly after the program. Moreover, with the exception of tobacco, D.A.R.E.'s ability to influence drug use has not been demonstrated. Importantly, the finding that DARE has little, if any, influence on drug use has been replicated in a number of studies.[95]

Another popular prevention program initially developed by the Phoenix Police Department and sponsored by the Bureau of Alcohol, Tobacco, and Firearms is G.R.E.A.T. (Gang Resistance Education and Training). Like D.A.R.E., G.R.E.A.T. consists of a curriculum that is taught by police officers in elementary and middle schools and is found in almost every state. G.R.E.A.T. also has undergone a curriculum revision that was designed to make the program more interactive and to help youths better understand the relationship between gangs and violence, learn what can be done about gangs, set goals, and develop a variety of social competencies including communication and conflict resolution skills.[96] Importantly, evaluations of the G.R.E.A.T. program have produced some positive results, although a major goal of the program, to prevent youths from participating in gangs, may not be achieved. Nevertheless, G.R.E.A.T has been associated with the development of more prosocial attitudes (including more favorable attitudes toward the police), less risk-taking behavior, less victimization, and less favorable attitudes toward gangs. Moreover, the evidence suggests that police officers can have a positive impact on youths.[97]

community policing
A policing strategy in which police attempt to identify and understand the social context of delinquent and criminal behavior and then work with the local community to rectify the problems causing crime rather than simply reacting to crime by arresting and incarcerating offenders.

Youth-Oriented Community Policing

Another significant trend in policing in recent years has been the development of problem-oriented and **community policing** strategies. In problem-oriented policing, police attempt to identify problems that encourage criminal behavior and take steps to remedy these problems.[99] In community policing, police rely on the community to identify problems that can be cooperatively addressed by the police and the community.[100] Both of these approaches are based on the realization that, in order to reduce crime, police

COMPARATIVE FOCUS

Japanese Police Play an Important Role in Delinquency Prevention

In Japan, the police have developed crime prevention associations and conduct company–police conferences and school–police conferences designed to prevent youth crime. In addition, they have established police boxes (*Koban*) in urban areas where youths congregate and police houses (*Chuzaisho*) in rural areas that are intended to help police respond to and prevent problems in the community. Moreover, Japanese police stations contain departments of community safety that are intended to coordinate local delinquency prevention efforts and they contain police counselors who are responsible for counseling youths and their parents, referring clients to local services, and supporting juvenile crime victims. Finally, the police encourage the development of a variety of voluntary associations and support the efforts of various individuals involved in providing prevention services to youths.[98]

need to take a more proactive approach in dealing with crime-generating community problems instead of simply reacting to citizen complaints after crimes occur.

Community policing employs a number of strategies, including the following examples:

- moving officers from their patrol cars to positions of direct contact with community residents (intended to give officers information about problems that exist in the community and ideas for solutions)
- changing the officers' main mode of behavior from reactive (responding to crimes committed) to proactive (preventing crimes)
- making police operations more visible to the public and increasing police accountability
- decentralizing police operations, thereby allowing individual officers to develop greater familiarity with the needs of the communities they police and to better respond to those needs
- encouraging officers to view citizens as partners in dealing with community problems related to crime, thereby improving relations between police and citizens
- placing more decision-making power in the hands of community policing officers, who know the community, its problems, and its expectations
- developing relationships between police and citizens that foster public initiatives aimed at preventing and solving crimes.[101]

A recently developed variant of community policing is known as youth-oriented community policing. Youth-oriented community policing programs have multiple goals: providing multi-agency responses to children and their families in order to prevent future delinquent behavior and adult criminality; working with youths who are already involved with the juvenile courts in order to reduce the likelihood of recidivism; involving neighborhood residents in efforts to improve their communities; and developing closer relations between community residents and public officials, including police officers. The strategies employed to achieve these goals, although varied, often include the provision of specialized training to police officers and the development of cooperative multi-agency teams to address community problems. The specialized training given to police officers might focus on topics such as child development, the psychological impact of family violence on children, assisting victims of crime, fostering effective collaboration between citizens and police, and handling mental health issues. The multi-agency teams are typically staffed by representatives of mental health agencies, schools, social service agencies, health departments, universities, and neighborhood associations, among others. These teams focus on prevention by identifying family and community problems that negatively affect youths and lead to delinquent behavior.[102]

One thing that makes community policing stand out from other policing strategies is its focus on the social context within which delinquency takes place. Rather than simply reacting to juvenile crime, youth-oriented community policing requires law enforcement agencies to identify societal factors related to delinquency and to take positive steps to address those factors in order to prevent delinquency. As a result, youth-oriented community policing is more theory and research oriented than traditional policing strategies. By having a clear understanding of factors that encourage delinquency within the community, those using a youth-oriented community policing approach are in a better position to prevent as well as respond to delinquent behavior.

Legal Issues

Although the relationship between police and citizens is a long-standing concern in U.S. law, the relationship between children and police has also received attention. Many contend that children, because of their lack of maturity, should have special protections when faced with arrest or interrogation by police. Exactly how far the police should go in protecting juveniles and which specific protections should be given to juveniles are, however, matters of debate. What is your perspective on this issue?

Chapter Summary

This chapter examined police and public responses to juvenile crime. As noted at the beginning of the chapter, citizens make up an informal juvenile justice process that responds in a variety of ways to youthful misbehavior. The ability of this informal juvenile justice process to handle juvenile misbehavior determines the extent to which formal components of the juvenile justice process become involved in the lives of youths.

Although many youth offenses are handled by citizens without official assistance, citizens frequently choose to involve the police, and in some cases the police initiate contact with youths on their own. The police play a critical role in juvenile justice because they serve as the primary gatekeepers of the formal juvenile justice process. Of particular significance is the fact that the police exercise discretion in how they handle alleged instances of juvenile delinquency. How the police decide to handle a juvenile case can have important ramifications for community safety as well as for the child and his or her family. In addition, how the police handle juveniles raises several legal issues regarding the relations between juveniles and the police and how far the police should go in protecting juveniles' rights. These issues have received some attention by the courts in cases such as *Fare v. Michael C.*, *People v. Lara*, *State v. Benoit*, and *Schall v. Martin*. However, many people are still concerned that youths may be denied important legal protections when confronted by the police, and this lack of protection can result in youths being harmed because of their involvement with juvenile justice agencies.

Data on the police processing of juveniles indicate that the police are becoming more formal in their responses to youths. Nevertheless, some police departments have also developed a variety of new strategies to deal with juveniles, such as youth-oriented community policing. To some extent, these strategies reflect the recognition by the police that their effectiveness as crime fighters requires good relationships with all members of the community, including youths. In addition, they appear to be tied to the increased emphasis on crime prevention and the growing popularity of community policing found in some police departments. However, the extent to which police departments will embrace these strategies and their effectiveness remains to be seen.

Key Concepts

arrest: Legally authorized deprivation of a person's liberty. As a general rule, a person is under arrest when he or she is not free to walk away.

community policing: A policing strategy in which police attempt to identify and understand the social context of delinquent and criminal behavior and then work with the local community to rectify the problems causing crime rather than simply reacting to crime by arresting and incarcerating offenders.

discretion: The power to act on one's own authority.

informal juvenile justice process: Actions to combat juvenile delinquency taken by parents, neighbors, business owners, teachers, and others who are not part of the formal juvenile justice apparatus.

institutional racism and bias: The ways that basic social institutions operate to keep minorities or poor people in subordinate positions. If police believe that minorities or poor people have fewer resources to respond to youth problems, believe that minority communities foster more crime, and believe that minority group members are more hostile to the police, it increases the likelihood that police will focus more crime-fighting resources in minority communities, resulting in increases in crime, police–citizen hostility, and neighborhood problems.

police role conflict: Clash between the duty of police to investigate crimes and prevent individuals from committing crimes and the duty of police to protect the rights of offenders, crime victims, and other citizens.

probable cause: Grounds sufficient to convince a reasonably competent person that a crime was committed and that the suspect committed it.

reform agenda for police: Agenda consisting of items such as the following: eliminating political influence and corruption, appointing qualified leaders, establishing the ideal of nonpartisan public service, raising personnel standards, implementing scientific management principles, instituting military-style discipline, and creating "specialized" units.

status offense: An act considered to be an offense partly because of the status of the person who performed the act. For example, juveniles (i.e., individuals who have a juvenile status) must obey the reasonable rules of their parents, attend school, and live at home. If they fail to do any of these things, they could be arrested and brought into the court system. Adults do not have to do any of those things and cannot be arrested for refusing to do them.

Review Questions

1. What is the informal juvenile justice process, and how is it related to formal juvenile justice agencies?
2. What are the typical responses available to the public when they come into contact with juvenile offenders?

3. What were the basic characteristics of policing during the colonial period?
4. What problems in policing did the progressive reforms of the late 1800s and early 1900s attempt to address?
5. What innovations in youth-oriented policing occurred between the early 1900s and the 1970s?
6. What are the basic roles played by the police and how do these roles influence police–juvenile interactions?
7. What options do police officers have when dealing with youths who have allegedly broken the law?
8. What factors influence police discretion?
9. What procedures do the police follow when taking a youth into custody?
10. What tactics do police use in the interrogation of juveniles and what are potential problems with the use of these tactics?
11. What are the important court cases concerning the interrogation of juveniles by the police, and what do these cases require the police to do?
12. In what ways have trends in the police processing of juveniles changed over the last 20 years?
13. What recent strategies or programs have been developed by the police to help them prevent or more effectively respond to delinquency? What do these strategies or programs attempt to do?

Additional Readings

Bazemore, G. & Senjo, S. (1997). Police encounters with juveniles revisited: An exploratory study of themes and styles in community policing. *Policing: An International Journal of Police Strategy and Management, 20*, 60–82.

Chesney-Lind, M. & Shelden, R. G. (2004) *Girls, delinquency and juvenile justice* (3rd ed.). Belmont, CA: Thompson/Wadsworth.

Krisberg, B. (2005). *Juvenile justice: Redeeming our children*. Thousand Oaks, CA: Sage.

Lundman, R. J., Sykes, R. E., & Clark, J. P. (1978). Police control of juveniles: A replication. *Journal of Research in Crime and Delinquency, 15*(1), 74–91.

Meyer, J. R. & Reppucci, N. D. (2007). Police practices and perceptions regarding juvenile interrogation and interrogative suggestibility. *Behavioral Sciences and the Law, 25*, 757–780.

Senna, J. J. & Siegel, L. J. (1992). *Juvenile law: Cases and comments* (2nd ed.). St. Paul, MN: West Publishing Co.

Walker, S. & Katz, C. M. (2008). *The police in America: An introduction* (6th ed.). New York: McGraw-Hill.

Cases Cited

Fare v. Michael C., 442 U.S. 707 (1979).
Miranda v. Arizona, 384 U.S. 436 (1966).
People v. Lara, 67 Cal.2d 365 (1967).
Schall v. Martin, 467 U.S. 253 (1984).
State v. Benoit, 490 A.2d 295 (NH 1985).

Notes

1. Walker, S. (1992). *The police in America: An introduction.* New York: McGraw-Hill.
2. Walker, 1992.
3. Walker, 1992.

   Berg, B. L. (1992). *Law enforcement: An introduction to police in society.* Boston: Allyn & Bacon.
4. Walker, 1992.
5. Klockars, C. B. (1985). The idea of police: Vol. 3. *Law and criminal justice series.* Newbury Park, CA: Sage.
6. Uchida, C. (1989). The development of the American police: An historical overview. In R. G. Dunham & G. P. Alpert (Eds.), *Critical issues in policing: Contemporary readings.* Prospect Heights, IL: Waveland Press.

   Walker, 1992.
7. Walker, 1992.
8. Uchida, 1989.
9. Walker, 1992.
10. Walker, 1992.
11. Trojanowicz, R. & Bucqueroux, B. (1990). *Community policing: A contemporary perspective.* Cincinnati, OH: Anderson.

    Walker, 1992.
12. Walker, 1992.
13. Bremner, R. H. (Ed.). (1970). *Children and youth in America: A documentary history* (Vols. 1–3), (p. 755). Cambridge, MA: Harvard University Press. This resource cites Matsell, G. W. (1849). *Semi-annual report, May 31–October 31, 1849, New York City Police Department.*
14. Bartollas, C. & Miller, S. J. (2008). *Juvenile justice in America* (5th ed.). Upper Saddle River, NJ: Pearson-Prentice Hall.
15. Walker, 1992.
16. Walker, 1992.
17. Walker, 1992, p. 13.
18. Fogelson, R. M. (1977). *Big-city police.* Cambridge, MA: Harvard University Press.
19. Owings, C. (1969). *Women police: A study of the development and staus of the women police movement.* Montclair, NJ: Patterson Smith.

    Schultz, D. M. (1995). *From social worker to crime fighter: Women in United States municipal policing.* Westport, CT: Praeger.
20. Walker, S. A. (1977). *A critical history of police reform.* Lexington, MA: Lexington Books.
21. Fogelson, 1977.
22. Bartollas & Miller, 2008.
23. Siegel, L. J., Welsh, B. C., & Senna, J. J. (2003). *Juvenile delinquency: Theory, practice, and law* (8th ed.). Belmont, CA: Wadsworth.

24. Roberts, A. R. (1998). The emergence and proliferation of juvenile diversion programs. In A. R. Roberts (Ed.), *Juvenile justice: Policies, programs, and services* (2nd ed.). Chicago: Nelson-Hall.

25. Bartollas & Miller, 2008.

26. Bartollas & Miller, 2008.

27. Walker, 1992, p. 63. This resource cites Yinger, M. (1965). *Toward a field theory of behavior: Personality and social structure* (pp. 99–100). New York: McGraw-Hill.

28. Walker, 1992.

29. Walker, 1992.

30. Wilson, J. Q. (1973). *Varieties of police behavior: The management of law and order in eight communities.* New York: Atheneum.

31. Reiss, A. J., Jr. (1971). *The police and the public.* New Haven, CT: Yale University Press.

32. Wilson, J. Q. (1968). *Varieties of police behavior: The management of law and order in eight communities.* Cambridge, MA: Harvard University Press.

33. Lab, S. P. (1984). Police productivity: The other eighty percent. *Journal of Police Science and Administration, 12*, 297–302.

Wilson, 1973.

34. Whitehead, J. T. & Lab, S. P. (1990). *Juvenile justice: An introduction*. Cincinnati, OH: Anderson.

35. Muraskin, R. (1998). Police work and juveniles. In A. R. Roberts (Ed.), *Juvenile justice: Policies, programs, and service* (2nd ed.). Chicago: Nelson-Hall.

36. Cox, S. M., Conrad, J., & Allen, J. (2002). *Juvenile justice: A guide to practice and theory* (5th ed.). New York: McGraw-Hill.

37. Office of Juvenile Justice and Delinquency Prevention & Training Resource Center, Eastern Kentucky University (1995). Participant's resource packet for the youth-oriented community policing national satellite teleconference. Richmond, Kentucky: Author.

38. Snyder, H. N. & Sickmund, M. (2006). *Juvenile offenders and victims: 2006 national report.* Washington, DC: Office of Juvenile Justice and Delinquency Prevention.

39. Black, D. & Reiss, A. J., Jr. (1970). Police control of juveniles. *American Sociological Review, 35*, 63–77.

Lundman, R. J., Sykes, R. E., & Clark, J. P. (1990). Police control of juveniles: A replication. In R. Weisheit and R. G. Culbertson (Eds.), *Juvenile delinquency: A justice perspective* (2nd ed.). Prospect Heights, IL: Waveland Press.

Piliavan, I. & Briar, S. (1964). Police encounters with juveniles. *American Journal of Sociology, 70*, 206–214.

40. Black & Reiss, 1970.

Lundman, Sykes, & Clark, 1990.

Piliavan & Briar, 1964.

41. Lundman, Sykes, & Clark, 1990.

Piliavan & Briar, 1964.

42. Smith, D. A. (1984). The organizational context of legal control. *Criminology, 22*, 19–38.

43. Goldman, N. (1969). The differential selection of juvenile offenders for court appearance. In W. J. Chambliss (Ed.), *Crime and the legal process.* New York: McGraw-Hill.

44. Klein, M. W. (1970). *Police processing of juvenile offenders: Toward the development of juvenile system rates* (Part 3). Los Angeles: Los Angeles County Subregional Board, California Council on Juvenile Justice.

45. Sampson, R. J. (1986). Effects of socioeconomic context on official reaction to juvenile delinquency. *American Sociological Review, 51*, 876–885.

Cicourel, A. (1968). *The social organization of juvenile justice.* New York: Wiley. Piliavan & Briar, 1964.

46. Gallup, Inc. (2007). *The Gallup Poll.* Retrieved July 12, 2007, from http://www.galluppoll.com/.

47. Gallup, Inc., 2007.

48. Pastore, A. L. & Maguire, K. (Eds.). (2007). Table 2.12.2007. In *Sourcebook of criminal justice statistics.* Retrieved July 12, 2007, from http://www.albany.edu/sourcebook/pdf/t2122007.pdf.

49. Decker, S. H. (1981). Citizen attitudes toward the police: A review of past findings and suggestions for future policy. *Journal of Police Science and Administration, 9*, 80–87.

50. Portune, R. (1971). *Changing adolescent attitudes toward the police.* Cincinnati, OH: Anderson.

51. Bouma, D. H. (1969). *Kids and cops.* Grand Rapids, MI: William E. Eerdman. Carr, P. J., Napolitano, L., & Keating, J. (2007). We never call the cops and here is why: A qualitative examination of legal cynicism in three Philadelphia neighborhoods. *Criminology, 45*, 445–480.

Decker, 1981.

Pastore, & Maguire (Eds.), 2007.

Murty, K. S., Roebuck, J. B., & Smith, J. D. (1990). The image of police in black Atlanta communities. *Journal of Police Science and Administration, 17*, 250–257.

52. Black & Reiss, 1970.

53. Lundman, Sykes, & Clark, 1990, p. 114.

54. Allen, T. T. (2005). Taking a juvenile into custody: Situational factors that influence police officers' decisions. *Journal of Sociology and Social Welfare, 32*, 121–129.

Black & Reiss, 1970.

Lundman, Sykes, & Clark, 1990.

55. Lundman, Sykes, & Clark, 1990.

56. Krohn, M. D., Curry, J. P., & Nelson-Kilger, S. (1983). Is chivalry dead? An analysis of changes in police dispositions of males and females. *Criminology, 21*, 417–437.

Elliott, D. S. & Voss, H. L. (1974). *Delinquency and dropout.* Lexington, MA: Lexington Books.

Morash, M. (1984). Establishment of a juvenile police record: The influence of individual and peer group characteristics. *Criminology, 22*, 97–111.

57. Dannefer, D. & Schutt, R. K. (1982). Race and juvenile justice processing in court and police agencies. *American Journal of Sociology, 87*, 1113–1132. Sampson, R. J. (1985). Sex differences in self-reported delinquency and official

records: A multiple-group structural modeling approach. *Journal of Quantitative Criminology, 1*, 345–367.

58. Chesney-Lind, M. & Shelden, R. G. (2004) *Girls, delinquency and juvenile justice* (3rd ed.). Belmont, CA: Thompson/Wadsworth. This resource contains a review.

59. Visher, C. (1983). Gender, police arrest decisions, and notions of chivalry. *Criminology, 21*, 5–28.

60. Krohn, Curry, & Nelson-Kilger, 1983.

Teilmann, K. & Landry, P. H. (1981). Gender bias in juvenile justice. *Journal of Research in Crime and Delinquency, 18*, 47–80.

Chesney-Lind, M. (1977). Judicial paternalism and the female status offender: Training women to know their place. *Crime and Delinquency, 23*, 121–130.

Staples, W. G. (1987). Law and social control in juvenile justice dispositions. *Journal of Research in Crime and Delinquency, 24*, 7–22.

61. Krohn, Curry, & Nelson-Kilger, 1983.

62. Dannefer & Schutt, 1982.

Piliavan & Briar, 1964.

Fagan, J., Slaughter, E., & Hartstone, E. (1987). Blind justice? The impact of race on the juvenile justice process. *Crime and Delinquency, 33*, 224–258.

Sampson, 1986.

Staples, 1987.

National Council on Crime and Delinquency. 1992. *The over-representation of minority youth in the California juvenile justice system.* San Francisco: Arthur. Thornberry, T. P. (1973). Race, socioeconomic status and sentencing in the juvenile justice system. *Journal of Criminal Law and Criminology, 64*, 90–98. Thornberry, T. P. (1979). Sentencing disparities in the juvenile justice system. *Journal of Criminal Law and Criminology, 70*, 164–171.

63. Wilson, 1968, pp. 40–41.

64. Black & Reiss, 1970.

Lundman, Sykes, & Clark, 1990.

Morash, 1984.

Terry, R. M. (1967). Discrimination in the handling of juvenile offenders by social-control agencies. *Journal of Research in Crime and Delinquency, 4*, 218–230.

Terry, R. M. (1967). The screening of juvenile offenders. *Journal of Criminal Law, Criminology and Police Science, 58*, 173–181.

Weitzer, R. (1996). Racial discrimination in the criminal justice system: Findings and problems in the literature. *Journal of Criminal Justice, 24*, 309–322.

65. Gibbons, D. C. & Krohn, M. D. (1991). *Delinquent behavior* (5th ed.). Englewood Cliffs, NJ: Prentice Hall.

66. Dannefer & Schutt, 1982.

67. Liederbach, J. (2007). Controlling suburban and small-town hoods: An examination of police encounters with juveniles. *Youth Violence and Juvenile Justice, 5*, 107–124.

68. Narkvichetr, K. (2006). Juvenile justice in Thailand. In P. Friday & X. Ren (Eds.), *Delinquency and juvenile justice systems in the non-Western World.* Monsey, NY: Criminal Justice Press.

69. Gilani, S. N. (2006). Juvenile justice in Saudi Arabia. In P. Friday & X. Ren (Eds.), *Delinquency and juvenile justice systems in the non-Western World.* Monsey, NY: Criminal Justice Press.

70. Walker, 1992.

71. Muraskin, 1998.

72. Walker, 1992.

73. National Institute for Juvenile Justice and Delinquency Prevention. (n.d.). *Police–juvenile operations: A comparative analysis of standards and practices: Vol. 2.* Washington, DC: U. S. Government Printing Office.

74. *Miranda v. Arizona*, 384 U.S. 436 (1966).

75. *Miranda v. Arizona* (1966).

76. *Miranda v. Arizona* (1966).

77. *Fare v. Michael C.*, 442 U.S. 707 (1979).

78. *Fare v. Michael C.* (1979). Italics added.

79. *People v. Lara*, 67 Cal.2d 365 (1967).

80. *State v. Benoit*, 490 A.2d 295 (NH 1985).

81. *State v. Benoit* (1985).

82. Meyer, J. R. & Reppucci, N. D. (2007). Police practices and perceptions regarding juvenile interrogation and interrogative suggestibility. *Behavioral Sciences and the Law, 25*, 757–780.

83. *Schall v. Martin*, 467 U.S. 253 (1984).

84. *Schall v. Martin* (1984).

85. *Schall v. Martin* (1984).

86. Robin, G. D. (1982). Juvenile interrogation and confessions. *Journal of Police Science and Administration, 10*, 224–228.

87. Holtz, L. E. (1987). Miranda in a juvenile setting: A child's right to silence. *Journal of Criminal Law and Criminology, 78*, 534–556.

88. Scott, E. S. & Grisso, T. (1997). The evolution of adolescence: A developmental perspective on juvenile justice reform. *The Journal of Criminal Law and Criminology, 88*, 137–189.

89. Snyder & Sickmund, 2006.

90. Schwartz, I. M. (1989). *(In)justice for juveniles: Rethinking the best interests of the child* (p. 77). Lexington, MA: Lexington Books.

91. Pastore, A. L. & Maguire, K. (Eds.). (2007). Table 4.26.2006. In *Sourcebook of criminal justice statistics.* Retrieved July 12, 2007, from http://www.albany.edu/sourcebook/pdf/t4262006.pdf.

92. United States Department of Justice, Federal Bureau of Investigation. (September 2008). Table 68. *Crime in the United States 2007.* Retrieved from www.fbi.gov/ucr/cius2007/data/table_68.html.

93. Ringwalt, C. L., Greene, J. M., Ennett, S. T., Iachan, R., Clayton, R. R., & Leukefeld, C. G. (1994). *Past and future directions of the D. A. R. E. program: An evaluation review*. Research Triangle Park, NC: Research Triangle Institute. This resource cites personal communication with C. Dunn, June 22, 1993.

94. Rosenbaum, D. P. (2007). Just say no to D.A.R.E. *Criminology and Public Policy, 6*, 815–824.

95. Clayton, R. R., Cattarello, A., & Walden, K. P. (1991). Sensation seeking as a potential mediating variable for school-based prevention interventions: A two-year follow-up of DARE. *Journal of Health Communications, 3*, 229–239.

Ringwalt, Greene, Ennett, Iachan, Clayton, & Leukefeld, 1994.

Rosenbaum, D. P., Flewelling, R. L., Bailey, S. L., Ringwalt, C. L., & Wilkinson, D. L. (1994). Cops in the classroom: A longitudinal evaluation of Drug Abuse Resistance Education (DARE). *Journal of Research in Crime and Delinquency, 31*, 3–31. Rosenbaum, 2007.

96. For information on the G.R.E.A.T curriculum, see http://www.great-online.org/.

97. Esbensen, F. (2002). *National evaluation of the Gang Resistance Education and Training (G.R.E.A.T) Program, final report*. Washington, DC: U.S. Department of Justice. Esbensen, F., Peterson, D., Taylor, T. J., Freng, A., & Osgood, D. W. (2004). Gang prevention: A case study of a primary prevention program. In F. Esbensen, S. G. Tibbetts, & L. Gaines (Eds.), *American youth gangs at the millennium* (pp. 351–374). Long Grove, IL: Waveland Press, Inc.

98. Elrod, P. & Yokoyama, M. (2006). Juvenile justice in Japan. In P. C. Friday & X. Ren (Eds.), *Delinquency and juvenile justice systems in the non-Western World*. Monsey, NY: Criminal Justice Press.

99. Goldstein, H. (1990). *Problem-oriented policing*. New York: McGraw-Hill.

100. Trojanowicz & Bucqueroux, 1990.

101. National Institute of Justice. (1992). Community policing in the 1990s. *National Institute of Justice Journal, 225*, 2–8.

102. Office of Juvenile Justice and Delinquency Prevention. (1996). Participant's resource packet for the Youth-Oriented Community Policing National Satellite Teleconference. Washington, DC: Office of Juvenile Justice and Delinquency Prevention.

Juvenile Diversion

▸ ▸ CHAPTER OBJECTIVES

After studying this chapter, you should be able to

- Define the concept of diversion and describe how it is used in dealing with youthful offenders
- Describe the social context of the 1960s and 1970s and indicate how it encouraged the development of diversion programs
- Explain why juvenile justice might be seen as a type of diversion from the criminal justice process
- Describe the theoretical underpinnings of the concept of diversion
- Describe specific diversion programs and whether or not they are effective
- Describe common problems with diversion programs and how these problems impact youths and families

▸ ▸ CHAPTER OUTLINE

- Introduction
- Early Efforts at Diversion
- The Theoretical Foundation of Diversion
- The Social Context of the 1960s and the Popularization of Diversion
- The Policy Implications of Societal Reaction Theory
- Other Rationales for Diversion Programs
- The Spread of Diversion Programs Since the 1960s
- Contemporary Diversion Strategies and Programs
- The Effectiveness of Diversion
- Legal Issues
- Chapter Summary
- Key Concepts
- Review Questions
- Additional Readings
- Notes

Introduction

diversion The processing of juvenile offenders in ways that avoid the formal juvenile justice process.

Although there is a lack of agreement on how to define **diversion**, the intent of diversion is to respond to delinquent youths in ways that keep them out of the formal juvenile justice process.[1] Diversion is based on the fact that formal responses to youths who violate the law, such as arrest and adjudication, do not always protect the best interests of children, nor do such responses necessarily protect the community. Indeed, some formal responses may be harmful to many youths and increase the likelihood of future delinquent behavior. This is because formal processing may cause a youth to develop a negative or delinquent self-image,[2] may stigmatize the youth in the eyes of significant others,[3] or may subject the youth to inhumane treatment. In addition, formal processing may restrict the youth's opportunities to associate with law-abiding peers or to engage in conventional activities, thereby increasing the chances of future delinquent activity. Consequently, efforts to divert youths from the juvenile justice process (e.g., by warning and releasing) as well as efforts to divert youth to specific diversionary programs (e.g., counseling and community service programs) have long been a part of juvenile justice practice. In addition, in recent years, diversion has been touted as a way to reduce the problem of minority overrepresentation in juvenile justice.[4]

This chapter provides an overview of the history of diversion programming. In addition, it examines the theoretical rationale for diversion and explores various types of diversion programs and their effectiveness. The chapter concludes by discussing a number of potential problems that confront diversion programs as well as issues that should be considered prior to the development and implementation of these interventions.

Early Efforts at Diversion

Efforts to divert children from normal criminal justice processing have a long history. As noted in Chapter 4, prior to the development of specialized correctional institutions for youths in the 1800s, children were subject to the same laws and the same criminal justice process as adults. Nevertheless, there is considerable evidence that they were often spared the harshest penalties because of their age. Police officers, sheriffs, constables, and others responsible for law enforcement have at times decided to ignore the illegal actions of youths. Historically, law enforcement officers have often handled matters on their own, by warning youths, returning them to their parents for punishment, or meting out punishment themselves. To some extent, efforts to spare youths from the most severe punishments or to let them escape punishment altogether were based on the recognition that the young were, in important ways, different from adults and that formal or severe punishments would rarely benefit the offender or the community.

The development of the houses of refuge in the early 1800s and the juvenile court movement in the late 1800s are often cited as early examples of efforts to divert youths from the adult criminal justice process.[5] Indeed, the houses of refuge were established, in part, to divert children from the harsh conditions of adult correctional institutions, which were seen by some reformers as harmful to youths and likely to produce more dangerous offenders.[6] The child savers of the late 1800s and their supporters helped establish juvenile courts in an effort to divert children from adult jails and prisons and from adult courts, which were often reluctant to sanction and control wayward children.[7]

Although diversion has a long history, the development of routine diversionary strategies and specialized diversion programs has substantially increased since the late

FYI

Juvenile Courts Were Created to Divert Children from Criminal Courts

The creation of specialized juvenile correctional institutions as well as the juvenile courts was based on the recognized need to respond to children and adults differently. It is the height of irony that today, in legislatures across the country, the age for the transfer of juvenile cases to adult courts is being lowered and the automatic waiver to adult courts for certain offenses is becoming law. Nowadays, more and more children are being "diverted" to the adult system—a system that is not particularly effective in dealing with many adults and is hardly capable of meeting the needs of youths.

1960s. The decade of the 1960s was a time of widespread social unrest, increases in juvenile arrests, and critical scrutiny of basic institutions, including juvenile justice. Concern about the effectiveness of formal responses to juvenile offenses was clearly reflected in recommendations made by the 1967 President's Commission on Law Enforcement and the Administration of Justice. These recommendations, formulated for the purpose of improving juvenile justice practice, served as a catalyst for the development of diversion programs during the late 1960s and 1970s. According to the commission:

> The formal sanctioning system and pronouncement of delinquency should be used only as a last resort. In place of the formal system, dispositional alternatives to adjudication must be developed for dealing with juveniles, including agencies to provide and coordinate services and procedures to achieve necessary control without unnecessary stigma.... The range of conduct for which court intervention is authorized should be narrowed, with greater emphasis upon consensual and informal means of meeting the problems of difficult children.[8]

In order to facilitate the number of diversionary responses available to communities, the commission called for the establishment of youth services bureaus. These bureaus were intended to supplement existing community agencies that dealt with children and to coordinate programs and services for both delinquent and nondelinquent youths. They also were intended to serve as an alternative to juvenile court processing, allowing, it was hoped, substantial numbers of youths to be diverted from the formal juvenile justice process each year.[9] Through federal grants, primarily from the Law Enforcement Assistance Administration (LEAA) and the Office of Youth Development and Delinquency Prevention, and smaller matching grants provided by local and state governments, **youth services bureaus** were established in communities around the country.[10] Typically, they had five basic goals:

youth services bureau
An agency that provides counseling, tutoring, mentoring, advocacy, and job referrals for youths in order to keep them out of the formal juvenile justice process.

1. Divert juveniles from the juvenile justice system.
2. Fill gaps in service by advocating for youths and developing services for youths and their families.
3. Provide case coordination and program coordination.
4. Provide modification of systems of youth services.
5. Involve youths in the decision-making process.[11]

Evaluation studies found that the typical youth services bureau was staffed by 5 or 6 full-time staff members and from 1 to 50 volunteers. Types of services offered included tutoring; recreational programming; individual, group, and family counseling; and job referral. The typical bureau operated in a low socioeconomic status urban neighborhood where there were high rates of crime and unemployment and limited resources for youths. Each year the typical bureau served approximately 350 youths who were self-referred or referred by the police, schools, and parents, among other sources.[12]

Although youth services bureaus were popular in a number of jurisdictions, they experienced some problems. One difficult issue they had to deal with was the question of voluntary participation. The ideal was to provide services to youths and possibly their families on a voluntary basis. However, if a youth or a family member failed to follow through on treatment plans developed by a bureau, a referral to the appropriate juvenile court was typically made, even if the youth avoided further delinquent behavior. In addition, if a bureau determined that a youth was not likely to benefit from its services, the youth could be referred back to the original referral source.[13] Such practices, of course, could result in formal processing—hardly what diversion programs are designed to accomplish. In addition to funding youth services bureaus, LEAA helped fund a variety of other diversion programs in communities around the country. These programs included "alternative schools, job development and training programs, police social work programs, and family counseling programs for youths referred by the police, schools, and court intake personnel."[14]

In the late 1970s, however, LEAA began to alter its priorities and shift funds away from prevention and diversion programs and into law enforcement and rehabilitation programs. The initial LEAA funding for the youth services bureaus was intended to be seed money to help establish the programs, but after a certain period, usually 2 years, communities were expected to continue funding these programs on their own. In some cases, communities did assume full support of local youth services bureaus, but by 1982 a number of bureaus developed, at least in part, with LEAA seed money were discontinued because of a lack of funds.[15] Nevertheless, a perceived need for diversion programs continued to motivate the development of such programs in communities around the country.

The Theoretical Foundation of Diversion

Like other programs for youths, diversion programs are supported by beliefs and theoretical concepts. As noted earlier, for a long time many law enforcement officials, juvenile justice reformers, and students of crime and punishment have doubted the reformative potential of adult courts, jails, and prisons and have sought to spare youths from criminal processing. Questions regarding the value of criminal processing for juveniles were heightened during the political and social unrest that characterized much of the 1960s and 1970s. In addition, a well-developed theoretical rationale for diversion became popularized during this time. These two factors, questions about the value of formal processing for many youths and a theoretical rationale for diverting youths, helped spur the development of diversion program beginning in the late 1960s and early 1970s.

The Social Context of the 1960s and the Popularization of Diversion

The 1960s and early 1970s proved to be a fruitful period for the development and proliferation of diversionary responses to juvenile offenders. For one thing, it was marked by considerable social unrest, and a substantial portion of the protests that occurred during this period were initiated and led by youths. Furthermore, many Americans, both young and old, seriously questioned the operation of basic institutions, including criminal and juvenile justice agencies. In addition, there was a marked increase in the juvenile arrest rates during this period. Not only were increasing numbers of young persons being arrested and processed by juvenile courts for violent and property offenses, but more youths were being arrested and processed for status offenses as well. Indeed, according to the President's Commission on Law Enforcement and the Administration of Justice, more than 25% of the youths in juvenile courts and institutions were status offenders.[16]

It was also during the 1960s and 1970s that the **societal reaction perspective** or **labeling,** began to play a more important role in criminological thinking.[17] This perspective focused on three interrelated topics: "(1) the historical development of deviant labels, (2) the process by which labels are applied, and (3) the consequences of being labeled."[18]

societal reaction perspective The point of view according to which social responses, particularly formal social responses, can contribute to subsequent delinquent behavior. This perspective favors the diversion and deinstitutionalization of juvenile offenders when possible and questions the wisdom of responding through formal means to each act of youth misbehavior (e.g., by arresting and trying the offender).

The first topic encompasses two sub-topics: how deviant categories, such as delinquent, are produced, and how social control mechanisms, such as the juvenile court, are established. A good example of a work focused on this topic is Anthony Platt's classic book, *The Child Savers*, which examines the "invention of delinquency" and the development of the first juvenile court in Chicago in 1899, a court that was primarily designed to control problem youths.[19]

The second topic includes the way in which control agents apply deviant labels to others and the factors that influence an individual's efforts to resist or accept these labels.[20] Societal reaction researchers in the field of juvenile justice might examine how police, judges, prosecutors, probation officers, and other control agents attempt to apply labels such as "delinquent," "status offender," and "chronic offender" to youths. In addition, they might explore a range of factors—such as youths' perceptions of themselves, family influences, peer reactions, and interactions between youths and others—that tend to reinforce a deviant label or help youths resist such a label.

The topic that has received the most attention within juvenile justice circles is the third—the consequences of being labeled. According to societal reaction theorists, labels such as "delinquent," "chronic offender," "thief," "doper," and the like can have a variety of negative consequences for those who are given such labels. For example, a label can lead others to assume things about the individual labeled that may not be true. A youth who has gotten into trouble in the past and has been labeled as a delinquent may be incorrectly treated as untrustworthy or may be treated more punitively as a result of being labeled. Indeed, the perceptions that social control agents have of others appear to play an important role in how youths are treated. If, through being labeled, youths are believed to possess undesirable characteristics associated in the public mind with criminality or potential criminality, they are more likely to be formally processed by police and other control agents and to be avoided by law-abiding individuals.

labeling The process by which a derogatory or otherwise negative term comes to be associated with a person. Being labeled with a term like ***delinquent*** or ***chronic offender*** may cause a person to develop negative self-perceptions and cause others to respond to the person based on the label. Furthermore, labeling can limit the opportunities available to the person and can lead him or her to develop a deviant identity, increasing the likelihood of subsequent offending.

Another problem noted by societal reaction theorists is that youths saddled with a negative label often have fewer opportunities for involvement in normal law-abiding activities. Without such opportunities, they are more likely to associate with those in similar circumstances, thus increasing the likelihood of further deviance.[21] Furthermore, responding to youths as if they were a lesser form of human beings may lead youths to see themselves in a more negative light.[22]

In an interesting study, Charles Frazier described the case of a young man, named Ken, who was tried and "branded" a criminal in a small town. Frazier noted that labeling Ken as a criminal led his former friends and associates to see him differently. Rejected by his former friends, Ken had fewer opportunities for engaging in conventional activities and began to see himself as a criminal.[23] According to societal reaction theorists, being treated as different and denied opportunities to associate with law-abiding persons may cause an individual to adopt a delinquent or criminal identity that becomes a master status—one that becomes the person's primary public identity and that overrides other statuses the individual may enjoy.[24]

A critical issue raised by societal reaction theorists is whether responses to deviant behavior, such as delinquency, can increase the likelihood of additional deviance. This possibility was spelled out by sociologist Edwin Lemert in 1951, when he distinguished between *primary* and *secondary deviance*. According to Lemert:

> Primary deviance is assumed to arise in a wide variety of social, cultural, and psychological contexts, and at best has only marginal implications for the psychic structure of the individual; it does not lead to symbolic reorganization at the level of self-regulating attitudes and social roles. Secondary deviation is deviant behavior, or social roles based upon it, which becomes a means of defense, attack, or adaptation to overt and covert problems created by the societal reaction to primary deviation. In effect, the original "causes" of the deviation recede and give way to the central importance of the disapproving, degradational, and isolating reactions of society.[25]

The Policy Implications of Societal Reaction Theory

Societal reaction theory has had a profound impact on social policy.[26] Yet, its policy implications are rather different than those of other theoretical approaches. In contrast to those who argue that punishment of some sort needs to be imposed in order to deter youths from further delinquent activity, societal reaction theorists call for minimal or no intervention whenever possible in order to avoid the negative consequences of labeling. For example, Edwin Lemert made a case for the use of "judicious non-intervention,"[27] arguing that much delinquent behavior is normal. Similarly, sociologist Edwin Schur argued for a policy of **radical nonintervention** (i.e., leaving children alone whenever possible) in order to avoid the negative consequences associated with involvement in the juvenile justice process.[28]

radical nonintervention The strategy of leaving youth offenders alone (i.e., not placing them in the juvenile justice process) if possible.

Societal reaction theory had an important influence on those who looked critically at the juvenile justice process in the 1960s and 1970s. In addition to serving as the theoretical rationale for diversion, it played a significant role in the development of three other juvenile justice reforms implemented during this period: decriminalization, increased emphasis on due process, and deinstitutionalization. Each of these reforms stressed keeping juveniles out of the formal juvenile justice process whenever possible.[29]

Decriminalization

Advocates of the societal reaction perspective pointed out that the criminalization of some behaviors often produces more harm than good. For example, behaviors like running away from home and not going to school are undesirable in many instances, but treating them as crimes does not necessarily benefit the youths who engage in them. On the contrary, treating truants and runaways as offenders may actually have a number of undesirable consequences. In addition, formal responses to status offenses, according to critics, were expensive and ineffective. As a result, societal reaction theorists favored the **decriminalization** (i.e., redefining status offenses as social problems to be dealt with by welfare agencies as opposed to criminal actions to be handled by juvenile courts) of status offenses, and this policy was adopted in a number of jurisdictions around the country.

decriminalization The act of redefining status offenses as social problems to be dealt with by welfare agencies as opposed to criminal actions to be handled by juvenile courts.

Increased Emphasis on Due Process

Societal reaction theorists also were wary of the discretionary powers available to juvenile justice officials, including police and judges. Many of these theorists claimed that juvenile justice officials often abused their authority and acted in ways that were harmful to those under their care. As a solution, Edwin Schur argued for "a return to the rule of law."[30] According to this approach, constitutional safeguards should be extended to youths to protect them from the power of the state. Moreover, punishments should be spelled out in law and should be determinate, preventing capricious actions on the part of officials inclined to extend punishments indefinitely.

Societal reaction theory has been cited as an important contributor to the movement to extend due process protections to juveniles. Today juveniles enjoy more due process protections than they did before the mid-1960s, but the extent to which the juvenile justice process has become more humane in its treatment of youths is not clear.[31] Simply extending legal protections to youths does not mean that those protections will be implemented in practice, and there is considerable evidence that the extension of due process protections to juveniles has had less of an impact on juvenile court proceedings than some have claimed.[32]

Deinstitutionalization

Another reform supported by those influenced by the societal reaction perspective was deinstitutionalization—the removal of juveniles from correctional facilities. Like diversion, deinstitutionalization was intended to protect youths from the harmful effects of incarceration. Indeed, criminologists from diverse theoretical positions argued that the reform of delinquents required efforts to "improve attachments to family and school, increase academic skills, open up legitimate opportunities, and reduce association with delinquent peers."[33] The incarceration of youths accomplishes none of these objectives. As the President's Commission stated in a 1967 report, *The Challenge of Crime in a Free Society*, "Institutions tend to isolate offenders from society, both physically and psychologically, cutting them off from schools, jobs, families, and other supportive influences and increasing the probability that the label of criminal will be indelibly impressed upon them."[34]

Several states, including Massachusetts, Maryland, Missouri, North Dakota, South Dakota, Oregon, and Utah, have made efforts to implement community-based correctional strategies,[35] and others have articulated the importance of community-based programs. The most ambitious of these efforts was led by Jerome Miller, the Commissioner of the Department of Youth Services in Massachusetts during the 1970s. Under Miller's direction, Massachusetts dramatically reduced its institutional population by closing training schools and placing youths in community-based programs.[36] In addition, a number of states have made considerable progress in removing status offenders from correctional facilities.[37]

■ Other Rationales for Diversion Programs

One argument offered in support of diversion is that it can reduce the stigma and other negative consequences of being arrested for delinquent behavior and thus reduce the probability of recidivism. Other supporting reasons and arguments also have been advanced. For example, diversion programs are believed to allow some youths to receive assistance who would not otherwise be helped, because juvenile corrections programs

MYTH VS REALITY

Punishment Is Not Always Needed to Correct Youths

Myth—If youths are not punished for their delinquent behavior, they not only will continue to engage in illegal behavior, but also engage in more serious types of delinquency.

Reality—Although self-report studies indicate that most youths engage in at least minor types of delinquent behavior,[40] most youths are not apprehended and processed by formal juvenile justice agencies.[41] Despite this, most of these youths will not commit serious offenses or live lives of crime when they become adults.[42]

often fail to provide needed services to youths. Advocates also point out that diversion programs allow juvenile justice decision makers to use discretion and be more flexible than many courts in responding to youths.[38] Another reason people favor diversion is that it reduces the burden placed on juvenile court resources by decreasing the number of youths referred to the courts. Furthermore, diversion programs promise the added benefit of freeing up scarce resources, which could then be devoted to dealing with more serious offenders.[39]

A final argument offered in support of diversion is based on the fact that many youths who engage in delinquent behavior are not identified and punished, and they cease offending without formal intervention. This fact suggests that formal intervention is, in many cases, unnecessary. Given the potential negative consequences of formal processing (according to societal reaction theorists), subjecting youths to formal processing can cause them great harm (and the community great harm, to the extent that these youths become more inclined to engage in further delinquency).

■ The Spread of Diversion Programs Since the 1960s

As already noted, a number of diversion programs have been developed since the late 1960s. Among the most popular have been Scared Straight programs; family crisis intervention programs for status offenders; limited individual, family, and group counseling programs for status and criminal offenders; runaway shelters; individual and family counseling programs coupled with educational, employment, and recreational services; basic casework and counseling programs; dispute resolution programs involving restitution and community service; and, more recently, teen court and restorative justice programs. Unfortunately, many of these programs have not been carefully evaluated, and their effectiveness is thus unknown. The following sections examine some of the programs that have been evaluated.

Scared Straight or Deterrence-Based Programs

Scared Straight programs, developed in the 1970s, were popularized by two films about the **Juvenile Awareness Project**, which began in September 1976 at Rahway State Prison in New Jersey, a maximum-security institution. The first film, *Scared Straight!*, received considerable publicity and presented testimonial evidence that suggested the program was a tremendous success. Indeed, the film won both an Emmy and an Academy Award; it was shown to countless youth groups and school classes, and it aired on national television in 1979. The second film, essentially a 10-year follow-up of the initial documentary, made similar claims regarding the effectiveness of the Juvenile Awareness Project.

Juvenile Awareness Project A program in which long-term prison inmates meet with juveniles in counseling, educational, or confrontational sessions for the purpose of exposing the youths to the harsh realities of prison life. Scared Straight! is an example of this type of program.

In California, "scaring kids straight" became so popular that legislation was introduced that required the busing of 15,000 juveniles to state prisons for confrontation sessions similar to those at Rahway. In many other communities, tours of prisons and jails and confrontations with inmates were arranged in order to scare youths straight.[43]

The Juvenile Awareness Project was developed by a group of inmates called the "lifers" who were serving long prison terms at Rahway. The goal of the lifers was to expose youths to the harsh realities of prison life. In order to accomplish their goal, the lifers did more than just describe conditions in Rahway. The basic intervention employed consisted of a confrontation in which verbal and, at times, physical abuse and verbal and physical intimidation were used to convey the brutality and human indignities characteristic of life in a prison such as Rahway. The inmates, who developed the program, believed that making youths understand where the consequences of their delinquent behavior might lead would act as a **deterrent** to subsequent delinquency.

deterrent Refers to actions taken to influence a person's behavior by imposing or threatening to impose a legal sanction such as probation, incarceration, a fine, etc.

Although the *Scared Straight!* films indicated that these programs were successful, an evaluation of the Juvenile Awareness Project at Rahway by James Finckenauer was far less encouraging.[45] Finckenauer found that the youths who typically attended the Juvenile Awareness Project at Rahway were less delinquent than the films suggested. More disturbing, his research indicated that youths who went through the Rahway Project were more likely to engage in subsequent delinquency than youths who had not attended the program.[46] There were, however, some problems with Finckenauer's evaluation. The evaluation design had called for the random assignment of youths to the experimental condition (the Rahway program) or to a control condition (no treatment). However, some of the agencies that selected youths did not select them on a completely random basis. Consequently, it is possible that the two groups may not be comparable. Even so, evaluations of programs similar to the Rahway program have produced results much like those reported by Finckenauer.

In a review of the evaluations of two other programs designed to scare youths straight, Richard Lundman found no evidence that those programs were effective.[47] In one program that was similar to the program at Rahway because it used a confrontational style of interaction between youths and inmates, Lundman found there were no differences between experimental and control subjects over a six-month follow-up period. In the other program, which consisted of a tour of the facility and meetings between inmates and youths, he found that youths who went through the program were *more* likely to engage in subsequent delinquency than youths in a control group (who did not participate in the program). This finding mirrors that of the most extensive examination of these types of programs to date. This research, conducted by Anthony Petrosino, Carolyn Turpin-Petrosino, and John Buehler, consisted of an extensive review of the existing

MYTH VS REALITY

Evaluations of Scared Straight Programs Fail to Produce Positive Results

Myth—After youths understand what prison is like, they will think twice about committing delinquent acts.

Reality—Evaluations of scared straight programs fail to demonstrate that these programs are effective.[44]

COMPARATIVE FOCUS

Scared Straight in Norway

A program based on the Juvenile Awareness Program at Rahway was started at Ullersmo Prison in the early 1990s as part of an effort by child welfare agencies and the police. However, the program was abandoned in the late 1990s because it was attacked on ethical grounds that it exposed youths to inhumane treatment and because there was no scientific evidence that it was effective.[50]

literature on programs like Scared Straight! as well as a meta-analyses (a data analysis technique that makes possible the examination of treatment effects across different studies) of the existing literature. This research, as well as a meta-analysis conducted by James Finckenauer and Patricia Gavin, concluded that these programs are ineffective.[48] In discussing the results of their research on juvenile awareness programs, Petrosino and his colleagues noted the following results:

> These randomized trials, conducted over a 25-year period in eight different jurisdictions, provide evidence that "Scared Straight" and other "juvenile awareness" programs are not effective as a stand-alone crime prevention strategy. More importantly, they provide empirical evidence—under experimental conditions—that these programs likely increase the odds that children exposed to them will commit offenses in the future. Despite the variability in the type of intervention used, ranging from harsh, confrontational interactions to tours of the facility converge on the same result: an increase in criminality in the experimental group when compared to a no-treatment control. Doing nothing would have been better than exposing juveniles to the program.[49]

Family Crisis Intervention Programs

Another type of diversion program involves providing crisis intervention services to children and families in hopes of preventing subsequent problems and formal court involvement. A good example of this type of diversion program was developed in Sacramento, California, in 1969. This program, known as the Sacramento 601 Diversion Project, was intended to divert status offenders from detention and formal court processing by having specially trained probation officers provide short-term crisis intervention and treatment services to the youths and their families.[51]

This program focused on status offenders who were referred to the juvenile court by police, parents, and school officials. The youths referred to this program were typically white, poor, female, and younger than 15 years of age. Also, three-quarters of the

FYI

Efforts to "Scare Kids Straight" Continue to Be Found Around the Country

Despite evidence that the Scared Straight program at Rahway may be harmful to youths, it is still in existence, and similar programs continue to operate around the country. There is even an MTV version that is shown periodically.

referred youths had no previous contact with the court. The usual reason for referral was the existence of family problems.[52] When a youth was referred to the court, a specially trained probation officer would contact the youth's parents and request that the parents and youth come to Juvenile Hall for an immediate counseling session. When they arrived at Juvenile Hall, the probation officer would read the Miranda warning and explain that participation was voluntary. The officer indicated, however, that if they did not wish to participate in an immediate counseling session, their case would be referred to the court. The youth and parents were also told that if they agreed to participate in a counseling session, they could return for a limited number of sessions (the limit was five) if they chose to do so. Most parents and their children agreed to participate.[53] After a juvenile waived his or her rights, a counseling session began in which the probation officer used family intervention techniques. Essentially, these techniques were intended to get the family to look at the problem as a family problem that should be addressed by the entire family rather than by incarcerating the youth.[54]

An evaluation of the Sacramento 601 Diversion program, which employed an **experimental design**, indicated that it produced some promising results. Only about 14% of those youths handled by the specially trained project probation officers were detained and only about 4% were petitioned to the court. In comparison, of the youths handled by regular probation officers, 55% were detained and about 20% were petitioned to court.[55]

experimental design Research design in which subjects are randomly assigned to experimental and control groups.

In addition, the program evaluation found that the project reduced recidivism. A follow-up study of youths 12 months after their involvement in the program found that 46% of those who received the specialized diversion services had another court referral and 22% were referred to the court for a criminal offense. In comparison, 54% of those handled by regular probation officers were subsequently referred and about 30% were referred for a criminal offense.[56]

Diversion Programs That Employ Individual, Family, and Group Counseling Services

Given the apparent success of the Sacramento County 601 Diversion Project, 11 other jurisdictions in California developed similar programs and began to conduct evaluations of those projects in the mid-1970s. These projects were developed in Compton, El Centro, Fremont, Irvine, La Colonia, Mendocino, Simi Valley, Stockton, Vacaville, and Vallejo. They were similar to the Sacramento 601 Diversion Program in a number of ways. For example, they substituted short-term family and individual counseling for detention and referral to juvenile court.[57]

There were, however, important differences between these projects and the Sacramento project. For example, more of the youths involved in these projects were male, Hispanic, and African American, and a number were diverted after committing criminal offenses. In these projects, juveniles were referred by either the police or by probation officers. Youths received individual, family, and group counseling for 4 to 6 weeks on average, and the counseling included nearly 6 hours of contact services.[58]

The evaluators of these projects employed a **quasi-experimental design**. That is, they did not use random assignment of youth to experimental and control conditions, but rather attempted to approximate equivalence of the groups by comparing youths serviced by the diversion programs with youths who had similar race, age, sex, ethnicity, prior arrest, and referral source characteristics but were not serviced by the programs. Arrests of the individuals in the two groups were then compared 6 months after the arrest that brought them to the attention of the authorities.[59]

quasi-experimental design Research designs that can be used to explore possible cause-and-effect relationships when true experimental designs are not possible. What distinguishes these designs from experimental designs is that they do not involve random assignment of subjects to experimental and control groups.

A comparison of the two groups revealed that youths serviced by the diversion programs were arrested slightly less often during the 6-month follow-up period than youths in the other group (about 25% compared with about 31%).[60] These results suggest that diversion programs may produce modest reductions in recidivism. The researchers also found that diversion seems to have some benefits for males and females and for both criminal and status offenders.

Diversion Programs That Employ Individual and Family Counseling in Conjunction with Employment, Educational, and Recreational Services

As the number of diversion programs increased in the United States, researchers raised questions about their effectiveness. Among other things, they looked at the types of evaluations that had been done. In most instances, the diversion research in California was based on quasi-experimental designs. Consequently, it was difficult to tell if it was the programs themselves that had reduced recidivism or some unknown factor. Furthermore, it was unclear if diversion could be used effectively with criminal offenders, although there was some indication that it could. Finally, none of the evaluations of diversion programs had employed a "no treatment" group. Consequently, it was difficult to determine what the benefits of diversion programs were compared with doing nothing—a radical nonintervention approach.[61] In response to these concerns and others, the Office of Juvenile Justice and Delinquency Prevention (OJJDP) made $10 million available for the National Evaluation of Diversion Programs in the 1970s. As a result, four jurisdictions—Kansas City, Missouri; Memphis, Tennessee; Orange County, Florida; and New York City—developed experimental diversion projects. Each project excluded status offenders and included radical nonintervention as one of the experimental conditions. Experimental conditions consisted of (1) assigning youths to the diversion project, (2) doing nothing, and (3) referring the youths back to the original referral source for normal handling. This meant that some, but not all, of these youths penetrated the juvenile justice system.[62]

Diverted juveniles in each of the four projects received individual and family counseling, and educational services were available to youths. However, each of the projects tended to emphasize different services. For example, in Orange County, recreational services were emphasized, whereas in Kansas City, some diverted juveniles were assigned adult counselors who acted as advocates in dealings with public agencies and organizations, such as schools.[63]

An evaluation of these projects indicated that there was no difference in re-arrests among the three groups 6 and 12 months following the intervention. At the 6-month point, 22% of the youths who had been enrolled in a diversion program, 22% of those released (radical nonintervention), and 22% of those handled in the usual way had experienced at least one re-arrest. After 12 months, 31% of those diverted, 30% of those released, and 32% of those handled in the usual way had experienced at least one re-arrest.[64] The finding of "no difference" in the outcomes of the different types of diversion is important for at least two reasons: It suggests that criminal offenders may benefit from diversion, and it suggests that doing less may be just as effective as doing more—that placement in the juvenile justice process may not be the best way of dealing with *some* juvenile offenders.

Programs That Use Basic Counseling and Casework Services

Other diversion programs provide basic counseling and casework services to youths. One example is the Adolescent Diversion Project in Michigan, evaluated between 1976 and 1981. The subjects involved in this project consisted of both status and criminal offenders, but youths charged with Crime Index offenses against persons and youths already on probation were not included. The two most common offenses committed by the subjects were larceny-theft (34%) and breaking and entering (24%). The average age of the subjects was 14 years, 84% of the subjects were male, and slightly less than 30% were African American or Hispanic.[65] The longer of the two studies of the Adolescent Diversion Project began in 1976 and lasted until 1981. It examined the effectiveness of six different interventions provided by university students working for a course grade. Each student received specialized training in the type of intervention to which he or she had been assigned, and each worked with clients for 6 to 8 hours a week for 18 weeks. The students were monitored and evaluated by graduate students, by their clients, and by the professors responsible for the study. The interventions included family-focused behavioral contracting and client advocacy for youths provided by trained students; behavioral contracting and client advocacy for youths provided by trained students and a juvenile court employee; individually tailored client interventions designed by students given minimal training; relationship building led by specially trained university students; and standard court processing.[66]

A second study, which began in 1979 and lasted until 1981, investigated diversion interventions consisting of behavioral interventions and child advocacy. As in the other study, some youths were returned to the referral source for normal processing. However, the types of people who provided the services expanded to include graduate students and community volunteers. Again, each service provider was trained to provide diversion services to one client, each spent 6 to 8 hours a week with the client, and each was carefully monitored and evaluated.[67]

The two studies of the Adolescent Diversion Project indicate that a number of the interventions appear to have been effective in reducing subsequent delinquent activity, at least when compared with normal court processing. The interventions that were the most effective were those that provided behavioral contracting and child advocacy services and those that focused on relationship building. Particularly effective were those interventions led by community volunteers. The least effective involved the provision of behavioral contracting and child advocacy services by students and a juvenile court employee and standard court processing. These findings support the view of societal reaction theorists, who claim that formal intervention can have a negative effect on youths.[68]

Another type of program that has received attention in recent years is mentoring programs. Clearly, the most well-known of these programs is the Big Brothers Big Sisters program. Importantly, this program, which involves carefully chosen matches between a mentor (Big Brother or Big Sister) and a youth, has been found to be an effective intervention for many youths. A study of Big Brothers Big Sisters in the early 1990s found that children who were matched with a Big Brother or Big Sister were less likely to begin using illegal drugs or alcohol, skip school or class, or hit someone. Also, the mentored children were more confident of their performance in schoolwork and showed improved family interaction compared to children who were not matched.[69]

■ Contemporary Diversion Strategies and Programs

Today, as in the past, diversion is frequently used in juvenile justice. Also as in the past, diversion strategies are of two basic types: radical nonintervention and involvement in a diversion program. A police policy that entails the warning and release of some juvenile offenders is an example of radical nonintervention. A policy that encourages police officers to refer youths (and possibly their parents) to a diversion program is an example of the second type of strategy.

Contemporary diversion programs are operated by various types of juvenile justice and community agencies. Many communities have diversion programs operated by the local juvenile court or probation agency, others have diversion programs run by the police department, and still others have diversion programs operated by separate public or private agencies. Whether operated by juvenile justice or community agencies, diversion programs may employ a variety of interventions intended to reduce the continued involvement of youths in delinquency. The interventions include providing basic casework services to youths; truancy courts; crisis intervention to assist youths and their families; individual, family, and group counseling; conflict resolution; mentoring; participation in restitution and community service programs; and teen courts, among others. Following are three examples of more contemporary diversion programs that have received some attention in the evaluation literature.

Youth Enhancement Services (YES)

Youth Enhancement Services (YES) was intended to provide a variety of services to primarily minority youths in Daupin County, Pennsylvania (Harrisburg), who were felt to be at risk of delinquency as well as those who had contact with the police. Its major goals were to improve school performance and behavior, and reduce rates of arrest and recidivism among minority youths. Program referrals came from a variety of community sources, including the police, and services were provided to clients by a range of community agencies, including Girls Inc., the Boys Club, and the YMCA. These services consisted of some combination of client needs assessments, mentoring and adult support, peer group discussions, family support services, neighborhood and community projects, educational assistance, and job readiness training.[70]

An evaluation of YES compared outcomes for youths who were referred to the program but chose not to participate (44% of the youths referred to the program), those who participated and received less than 30 hours of service over the course of a year (24% of referrals), and those who received at least 30 hours of service (32% of the referrals). As might be expected, youths who received more than 30 hours of service had characteristics that were likely to increase the odds of program success. They were less likely than other youths to have a prior arrest, to be a self-referral, or to be referred by their family, and on average they did better in school. The evaluation revealed no significant differences among the three groups on school performance or behavior. The program did, however, appear to reduce minority youths' likelihood of contact with the police. Over a 2-year follow-up period, 50.6% of those who were referred to the program but did not participate had at least one new arrest compared to 41.3% in the low attendance group, and 25.8% in the high attendance group.[71]

Teen Courts

Another type of diversion program that has grown in popularity in recent years is teen court. A distinguishing feature of these courts is that juveniles, under the guidance of adults, play important roles in case processing and determining the types of dispositions used. Typically, these courts handle relatively young offenders with no prior arrests, and in most instances they deal primarily with cases involving theft, minor assaults, disorderly conduct, possession and use of alcohol, and vandalism.[72]

Advocates of teen court programs maintain that these programs make use of youths' desire to be accepted by their peers to encourage them to take responsibility for their behavior and engage in more socially acceptable behavior in the future. Moreover, supporters argue that not only does teen court help youths who are the defendants in cases, but also those youths who staff the court as well as the community as a whole. By involving a number of youths and adults in the teen court process, advocates believe that teen courts engender greater community involvement in the legal system, encourage community cohesion, promote law-abiding behavior, and do all of this in a cost-effective manner.[73]

At present, four basic teen court models are found around the country:

- *Adult judge.* An adult judge presides over the court and oversees court operations. Youths act as attorneys, jurors, clerks, bailiffs, and other court officers and staff. This is the most popular model around the country.
- *Youth judge.* The format is similar to the previous model, but a youth serves as the judge.
- *Tribunal.* Youth attorneys present their cases to a panel of three youth judges who determine the appropriate disposition. A jury composed of youths is not used in this model.
- *Peer jury.* This model does not use youth attorneys. Instead, the case is presented to the peer jury by a youth or an adult. In this model, the peer jury questions the defendant directly.[74]

Because the growth in teen courts is relatively recent, there is not yet an extensive body of research documenting their effectiveness. Some program evaluations have reported recidivism rates of 3 to 8% between 6 and 12 months after program completion. In addition, there is good evidence that at least some of these programs produce improved perceptions of authority, justice, and the legal system. Other studies, however, have reported recidivism rates of 24 to 32%.[75] A more recent evaluation of teen courts in Alaska, Arizona, Maryland, and Missouri, found that youths in two of the four sites displayed significantly lower rates of recidivism compared to youths who did not go through the program.[76] To date, the research indicates that some, but not all of these programs appear to have a positive impact on youths behavior. Until more comprehensive studies of these programs are completed, however, more definitive statements about the viability of teen court programs are premature.

Restorative Justice/Conflict Resolution/Mediation Programs

Restorative justice programs consist of a variety of interventions designed to mediate or resolve conflicts between juvenile offenders and their victims. Proponents of restorative justice argue that simply focusing on punishment of the offender hinders efforts to achieve justice. In contrast to present efforts that focus solely on the offender, **restorative justice** is concerned with achieving justice for both the offender and the victim. As a result, these programs are intended to provide rehabilitative services to offenders and to deliver some therapeutic benefit for victims. Moreover, they are believed to be cost-effective, and the focus on offender accountability and victim restitution and/or community service is appealing to the public.

restorative justice
A process designed to repair the harm done to the victim and the community that entails bringing the interested parties together in a non-confrontational setting where an agreeable resolution to the youth's illegal behavior can be reached.

One type of restorative justice program that is designed to divert youths from the formal juvenile justice process is *restorative justice conferencing*. In a restorative justice conference, the youth who committed the offense, the victim, and supporters of the

COMPARATIVE FOCUS

Restorative Justice Has a Long History and Has Played a Major Role in Responding to Crime in Many Cultures

According to criminologist John Braithwaite, "Restorative justice has been the dominant model of criminal justice throughout most of human history for all the world's peoples."[81] Indeed, for much of human history victims and people in communities were responsible for responding to violations of community norms and laws. Moreover, many of these traditions have continued and can be found in indigenous cultures, such as the Maori in New Zealand, Aboriginal tribes in Australia, the Inuits in Alaska, and the First Nations tribes in Canada.[82]

offender and victim meet with a trained facilitator to discuss the offense and examine the harm that it caused. Supporters have an opportunity to discuss how they have been affected by the offense, and at the conclusion of the conference, a written agreement is completed specifying how the offender can make amends to those who have been harmed by the incident. Typically, the agreement involves an apology and restitution to the victim, but it also can include other elements such as community service or requirements that youths improve their school performance or behavior at home.[77]

Although there is limited research on restorative justice conferencing with young offenders, research on programs using some form of restorative justice have produced generally favorable results.[78] One evaluation that employed an experimental design and looked at restorative justice conferencing for first-time young offenders (median age 13 years) who had committed nonserious offenses noted several positive outcomes. An analysis of interviews with victims revealed that those who participated in conferencing expressed much higher levels of satisfaction with the way their cases were handled compared to those in other diversion programs. Levels of satisfaction were similar for youths and parents involved in conferences and other types of diversion programs. Youths randomly assigned to participate in conferences compared to those assigned to other diversion programs, however, had significantly higher program completion rates than those in other diversion programs and significantly lower levels of rearrest 6 and 12 months after their initial arrest. These findings indicate that restorative justice conferencing is a promising approach for diverting young offenders from the formal justice process and can produce high levels of satisfaction among conference participants.[79]

A more recent evaluation of a restorative justice program in Maricopa County, Arizona by Nancy Rodriguez also uncovered positive program effects. Rodriguez found that youths who participated in the restorative justice program were less likely to have a new petition filed with the juvenile court than comparison youths who participated in a probation-department-designed diversion program during a 2-year follow-up period. She also found that girls and youths with minor or no criminal histories appeared to benefit the most from the restorative justice intervention.[80]

■ The Effectiveness of Diversion

Supporters of diversion strategies and programs maintain that such programs decrease the number of youths involved in the formal juvenile justice process, reduce offending among youths who receive diversionary treatment, minimize the stigma associated with formal intervention, are more cost-effective than formal processing, reduce the level of coercion employed by juvenile justice agencies, and reduce minority overrepresentation in the formal juvenile justice process.[83] As noted earlier, a number of evaluation studies of diversionary programs have found them to be superior to formal court processing in a number of respects,[84] but some evaluation studies have found that they often fall short of their goals. In fact, although some evaluation studies indicate that diversion programs can reduce recidivism or the programs are at least as effective as formal processing at reducing recidivism,[85] other studies have found that some diversion programs are associated with higher levels of subsequent offending.[86]

In addition to possible increased recidivism, other problems have been found to plague some diversion programs. Although these programs are often touted as a way to reduce the number of youths involved in the juvenile justice process, some diversion strategies may actually lead to **net-widening**.[87] Indeed, research on diversion programs has found that some programs increase the number of youths involved in the juvenile justice process by 33–49%.[88]

net-widening Increasing the availability of programs to handle youth offenders outside of the formal juvenile justice process, thereby causing youths who would otherwise have been left alone to be placed under the control of juvenile justice programs.

A rationale for the development of diversion programs is the belief that such programs are less stigmatizing than formal court processing. However, there is some evidence that involvement in some diversionary programs can be as stigmatizing as involvement in formal court programs. For example, one study that compared the self-concepts of incarcerated youths to those of youths in diversion programs found no significant differences between the two groups.[89]

The coercive nature of many diversion programs is another concern. In recommending diversion, the 1967 President's Commission on Law Enforcement and the Administration of Justice called for voluntary diversion programs. However, in some jurisdictions, potential clients are given no real choice.[90] For example, a juvenile offender might be told that unless he or she agrees to participate in the diversion program, the case will be referred for court action. Under such circumstances, most potential clients will feel compelled to participate. (Recall the choices given to youths and parents in the Sacramento 601 Diversion Program.)

Still another concern is the cost. As the cost of institutional placement as well as other formal responses has risen, diversion has been seen by many as a cost-effective alternative to formal processing. However, research that has examined the costs associated with diversion indicates that, whereas some diversion programs are less expensive than many formal programs, others are actually more costly.[91] Like other programs, diversion programs often require staff and support, which means that resources that might be devoted to other needs have to be allocated to diversion. Furthermore, if net-widening occurs, even though the per-client cost of diversion may be lower than the per-client cost of a formal program, the increased volume of cases can make the diversion program more expensive.

A final concern is the possibility that diversion programs may deny youths **due process**.[92] Diversion may be coercive and may consist of intrusive interventions, such as a requirement that a youth, and possibly the youth's family, participate in counseling or some other type of treatment. Coercion is problematic at any level of the juvenile justice process, but it is particularly problematic at the preadjudicatory stage of the juvenile justice process, where youths have not been proven guilty of an offense.

due process Refers to the actions that are necessary at each step of the juvenile justice process to ensure that youths are treated fairly and that their rights are protected.

In sum, the research on diversion has produced mixed results. A number of problems have been associated with diversion programs, but despite the problems associated with some programs, diversion appears to have some merit. As diversion programs are developed and implemented, however, problems that plague many diversion programs should be addressed, and diversion programs should be carefully evaluated after they are implemented to ensure that they are accomplishing their objectives.

Legal Issues

Due process, which is intended to ensure that an individual receives certain protections against government intervention, in some ways, conflicts with the idea of diversion. First, in many jurisdictions, eligibility for diversion may be decided by police officers who are unfamiliar with the proper criteria for diversion or how to apply them. Thus, they are ill-equipped to make diversion referrals. As a result, some youths who commit the same offenses will be diverted and others will be arrested. Second, the availability of diversion programs may increase the likelihood that a youth will become formally involved with juvenile justice agencies as a result of net-widening. Third, if coercion is used to get youths to participate in diversion programs, they are being denied their rightful due process protections. Fourth, involvement by diversion staff in the lives of family members on a "voluntary" basis may be just as intrusive as if it was done by court order. Fifth, many diversion programs make referrals to the juvenile court when "clients" are felt to be uncooperative, which is another way these programs can be coercive and widen the net of court control.

Chapter Summary

This chapter examined diversion, which encompasses responses to juvenile offenders that are intended to avoid formal court processing. The use of diversion is typically motivated by the belief that formal responses to youths who have violated the law do not always serve the best interests of the youths (or the best interests of the community) and that such responses often increase rather than decrease the likelihood of recidivism. Moreover, many people recognize that juvenile court intervention may ultimately harm a youth who becomes enmeshed in juvenile correctional programs, particularly juvenile institutions. Consequently, the immediate purpose of diversion is to minimize or avoid contact between youths and formal criminal justice agencies, such as the police or the juvenile courts.

Although diversion has a long history, it became more popular in the late 1960s and early 1970s. During this period, the rising level of juvenile crime, the willingness of many Americans to question the operation of basic American institutions, and the popularization of societal reaction theory/labeling theory (which raised serious questions about the potential detrimental effects of formal processing) served as the impetus for the development of a range of diversionary responses to juvenile offenders.

Diversion currently encompasses a wide range of responses, from radical nonintervention to the referral of youths and possibly their parents to a diversionary program intended to reduce the chance of future delinquent behavior. Despite the widespread use of diversion, evaluations of diversion programs have produced mixed results. Although studies of a number of diversion programs have produced considerable support for the concept of diversion, particularly when it is compared with formal juvenile justice processing, research has also produced evidence that some diversion programs have little effect on recidivism. Indeed, some studies indicate that diversion programs may increase the likelihood of subsequent delinquent activity. In addition, some of the supposed benefits of diversion, such as cost-efficiency and the lack of stigma associated with these programs, may be illusory. Nevertheless, some diversion programs have proven to be an effective and cost-efficient response to many juvenile offenders. As a result, diversion will continue to receive considerable attention as a juvenile justice strategy.

Key Concepts

decriminalization: The act of redefining status offenses as social problems to be dealt with by welfare agencies as opposed to criminal actions to be handled by juvenile courts.

deterrent: Refers to actions taken to influence a person's behavior such as imposing or threatening to impose a legal sanction such as probation, incarceration, a fine, etc.

diversion: The processing of juvenile offenders in ways that avoid the formal juvenile justice process.

due process: Refers to the actions that are necessary at each step of the juvenile justice process to ensure that youths are treated fairly and that their rights are protected.

experimental design: Research design in which subjects are randomly assigned to experimental and control groups.

Juvenile Awareness Project: A program in which long-term prison inmates meet with juveniles in counseling, educational, or confrontational sessions for the purpose of exposing the youths to the harsh realities of prison life. Scared Straight! is an example of this type of program.

labeling: The process by which a derogatory or otherwise negative term comes to be associated with a person. Being labeled with a term like *delinquent* or *chronic offender* may cause a person to develop negative self-perceptions and cause others to respond to the person based on the label. Furthermore, labeling can limit the opportunities available to the person and can lead him or her to develop a deviant identity, increasing the likelihood of subsequent offending.

net-widening: Increasing the availability of programs to handle youth offenders outside of the formal juvenile justice process, thereby causing youths who would otherwise have been left alone to be placed under the control of juvenile justice programs.

quasi-experimental design: Research designs that can be used to explore possible cause-and-effect relationships when true experimental designs are not possible. What distinguishes these designs from experimental designs is that they do not involve random assignment of subjects to experimental and control groups.

radical nonintervention: The strategy of leaving youth offenders alone (i.e., not placing them in the juvenile justice process) if possible.

restorative justice: A process designed to repair the harm done to the victim and the community that entails bringing the interested parties, such as the victim, offender, and a community representative, to a nonconfrontational setting where an agreeable resolution to the youth's illegal behavior can be reached.

societal reaction perspective: The point of view according to which social responses, particularly formal social responses, can contribute to subsequent delinquent behavior. This perspective favors the diversion and deinstitutionalization of juvenile offenders when possible and questions the wisdom of responding through formal means to each act of youth misbehavior (e.g., by arresting and trying the offender).

youth services bureau: An agency that provides counseling, tutoring, mentoring, advocacy, and job referrals for youths in order to keep them out of the formal juvenile justice process.

Review Questions

1. What is the definition of diversion, and what are the goals of diversion?
2. Explain the principle of radical nonintervention as it relates to youth delinquency and diversion.
3. What are youth services bureaus, and what role do they play in the diversion of youths from the formal juvenile justice process?
4. How did the social climate in the 1960s and 1970s contribute to the movement toward diversion programs?
5. Why do sociologists such as Edwin Lemert support some types of diversion programs?
6. What criminological theory supports the use of diversion?
7. What is a "scared straight" program, and do such programs work as effectively as diversion programs?
8. What is the Sacramento 601 Diversion Project, and did it work?
9. How effective is diversion as a mechanism for reducing or preventing recidivism? Support your position.
10. What are five drawbacks to using diversion for youth offenders?

Additional Readings

Braithwaite, J. (2007). Encourage restorative justice. *Criminology and Public Policy*, *6*(4), 689–696.

Bullington, B., Sprowls, J., Katkin, D., & Phillips, M. (1978). A critique of diversionary juvenile justice. *Crime and Delinquency, 24*(1), 59–71.

Cressey, D. R. & McDermott, R. A. (1974). *Diversion from the juvenile justice system.* Washington, DC: U.S. Department of Justice.

Lundman, R. J. (2001). *Prevention and control of juvenile delinquency* (3rd ed.). New York: Oxford University Press. See chapters on diversion.

Schur, E. M. (1973). *Radical nonIntervention: Rethinking the delinquency problem.* Englewood Cliffs, NJ: Prentice Hall.

Notes

1. President's Commission on Law Enforcement and Administration of Justice. (1967). *Task force report: Juvenile delinquency and youth crime.* Washington, DC: U.S. Government Printing Office.
2. Carter, R. M. & Klein, M. (Eds.). (1967). *Back on the street: The diversion of juvenile offenders.* Englewood Cliffs, NJ: Prentice Hall.
3. Schur, E. M. (1973). *Radical nonintervention: Rethinking the delinquency problem.* Englewood Cliffs, NJ: Prentice Hall.
4. Lundman, R. J. (2001). *Prevention and control of juvenile delinquency* (3rd ed.). New York: Oxford University Press.
5. Whitehead, J. T. & Lab, S. P. (1996). *Juvenile justice: An introduction* (2nd ed.). Cincinnati, OH: Anderson.
6. Krisberg, B. & Austin, J. F. (1993). *Reinventing juvenile justice.* Newbury Park, CA: Sage.

7. Platt, A. M.(1977). *The child savers: The invention of delinquency* (2nd ed.). Chicago: University of Chicago Press.

8. President's Commission on Law Enforcement and Administration of Justice, 1967, p. 2.

9. Rubin, H. T. (1985). *Juvenile justice: Policy, practice, and law* (2nd ed.). New York: Random House.

10. Roberts, A. R. (1998). The emergence and proliferation of juvenile diversion programs. In A. R. Roberts (Ed.), *Juvenile justice: Policies, programs, and services* (2nd ed.). Chicago: Nelson-Hall.

11. Roberts, A. R. (1998). Community strategies with juvenile offenders. In A. R. Roberts (Ed.), *Juvenile justice: Policies, programs, and services* (2nd ed.). (p. 117). Chicago: Nelson-Hall.

12. Roberts, 1998b.

13. Rubin, 1985.

14. Roberts, 1998a, p. 81.

15. Roberts, 1998b.

16. President's Commission on Law Enforcement and Administration of Justice, 1967.

17. Lilly, J. R., Cullen, F. T., & Ball, R. A. (2007). *Criminological theory: Context and consequences* (4th ed.). Thousand Oaks, CA: Sage.

18. Pfohl, S. J. (1994). *Images of deviance and social control: A sociological history* (2nd ed.). (p. 370). New York: McGraw-Hill.

19. Platt, 1977.

20. Pfohl, 1994.

21. Lilly, 2007.

22. Garfinkel, H. (1956). Conditions of successful degradation ceremonies. *American Journal of Sociology, 61*, 420–424.

Tannenbaum, F. (1938). *Crime and the community.* New York: Columbia University Press.

23. Frazier, C. (1976). *Theoretical approaches to deviance: An evaluation.* Columbus, OH: Merrill.

24. Becker, H. S. (1963). *Outsiders: Studies in the sociology of deviance.* New York: The Free Press.

25. Lemert, E. (1951). *Social pathology: A systematic approach to the theory of sociopathic behavior.* (p. 48). New York: McGraw-Hill.

26. Empey, L.T., Stafford, M. C., & Hay, C. H. (1999). *American delinquency: Its meaning and construction* (4th ed.). Belmont, CA: Wadsworth Publishing.

27. Lemert, E. M. (1967). The juvenile court: Quest and realities. In President's Commission on Law Enforcement and Administration of Justice, *Task force report: Juvenile delinquency and youth crime.* Washington, DC: U.S. Government Printing Office.

28. Schur, 1973.

29. Empey, Stafford, & Hay, 1999.

30. Schur, 1973, p. 169.

31. Cullen, F. T. & Gilbert, K. (1982). *Reaffirming rehabilitation.* Cincinnati, OH: Anderson.

32. Bernard, T. J. (1992). *The cycle of juvenile justice.* New York: Oxford University Press.

33. Empey, Stafford, & Hay, 1999.

34. President's Commission on Law Enforcement and Administration of Justice. (1967). *Task force report: The challenge of crime in a free society.* (p. 165). Washington, DC: U. S. Government Printing Office.

35. Bartollas, C. & Miller, S. J. (2008). *Juvenile justice in America* (5th ed.). Upper Saddle River, NJ: Pearson-Prentice Hall.

Krisberg & Austin, 1993.

36. Miller, J. G. (1998). *Last one over the wall: The Massachusetts experiment in closing reform schools* (2nd ed.). Columbus, OH: Ohio State University Press.

37. Snyder, H. N. & Sickmund, M. (2006). *Juvenile offenders and victims: 2006 national report.* Washington, DC: Office of Juvenile Justice and Delinquency Prevention.

38. Vito, G. F. & Wilson, D. G. (1985). *The American juvenile justice system.* Beverly Hills, CA: Sage.

39. Vito & Wilson, 1985.

40. Agnew, R. (2001). *Juvenile delinquency: Causes and control.* Los Angeles: Roxbury.

Gold, M. (1966). Undetected delinquent behavior. *Journal of Research in Crime and Delinquency, 3*, 27–46.

Short, J. & Nye, F. I. (1958). Extent of unrecorded delinquency: Tentative conclusions. *Journal of Criminal Law, Criminology, and Police Science, 49*, 296–302.

41. Huizinga, D. & Elliott, D. S. (1987). Juvenile offenders: Prevalence, offender incidence, and arrest rates by race. *Crime and Delinquency, 33,* 206–223.

42. Whitehead & Lab, 1996. This resource cites Lab, S. P. (1982). *The identification of juveniles for non-intervention.* Doctoral dissertation, Florida State University.

43. Finckenauer, J. O. & Gavin, P. W. (1999). *Scared straight: The panacea phenomenon revisited.* Prospect Heights, IL: Waveland Press.

Lundman, 2001.

44. Finckenauer & Gavin, 1999.

Lundman, 2001.

45. Finckenauer & Gavin, 1999.

46. Finckenauer & Gavin, 1999.

47. Lundman, 2001.

48. Finckenauer & Gavin, 1999.

Petrosino, A., Turpin-Petrosino, C., & Buehler, J. (2004). "Scared Straight" and other juvenile awareness programs for preventing juvenile delinquency. *Campbell Systematic Reviews.* Philadelphia, PA: Campbell Collaboration.

49. Petrosino, Turpin-Petrosino, & Buehler, 2004, pp. 25–26.

50. Finckenauer & Gavin, 1999.

51. Baron, R. & Feeney, F. (1976). *An exemplary project: Juvenile diversion through family counseling.* Washington, DC: U.S. Department of Justice.

52. Lundman, 2001.

53. Baron & Feeney, 1976.

54. Lundman, 2001.

55. Lundman, 2001.

56. Lundman, 2001.

57. Lundman, 2001.

58. Palmer, T., Bohnstedt, M., & Lewis, R. (1978). *The evaluation of juvenile diversion projects: Final report.* Sacramento, CA: California Youth Authority.

59. Palmer, Bohnstedt, & Lewis, 1978.

60. Lundman, 2001.

61. Lundman, 2001.

62. Lundman, 2001.

63. Dunford, F. W., Osgood, D. W., & Weichselbaum, H. F. (1981). *National evaluation of diversion projects: Final report.* Washington, DC: Office of Juvenile Justice and Delinquency Prevention.

64. Lundman, 2001.

65. Lundman, R. J. (1993). *Prevention and control of juvenile delinquency* (2nd ed.). New York: Oxford University Press.

66. Davidson, W. S., Redner, R., Amdur, R. L., & Mitchell, C. M. (1990). *Alternative treatments for troubled youth: The case of diversion from the juvenile justice system.* New York: Plenum.

67. Davidson, Redner, Amdur, & Mitchell, 1990.

68. Davidson, Redner, Amdur, & Mitchell, 1990.

Lundman, 1993.

69. Tierney, J. P., Grossman, J. B., & Resch, N. L. (1995). *Making a difference: An impact study of Big Brothers Big Sisters.* Philadelphia: Public/Private Ventures.

70. Welsh, W. N., Jenkins, P. H., & Harris, P. W. (1999). Reducing minority overrepresentation in juvenile justice: Results of community-based delinquency prevention in Harrisburg. *Journal of Research in Crime and Delinquency, 36*, 87–110.

Welsh, W. N., Harris, P. W., & Jenkins, P. H. (1996). Reducing overrepresentation of minorities in juvenile justice: Development of community-based programs in Pennsylvania. *Crime and Delinquency, 42*, 76–98.

71. Welsh, Jenkins, & Harris, 1999.

72. Butts, J. A. & Buck, J. (2000). Teen courts: A focus on research. *Juvenile Justice Bulletin.* Washington, DC: Office of Juvenile Justice and Delinquency Prevention.

73. Butts & Buck, 2000.

74. Nessel, P. A. (2000). Youth court: A national movement. *Technical Assistance Bulletin, 17.* Chicago, IL: American Bar Association.

75. Butts & Buck, 2000.

76. Butts, J. A., Buck, J., & Coggeshall, M. B. (2002). *The impact of teen court on young offenders.* Washington, DC: Urban Institute Press.

77. McGarrell, E. F. (2001). Restorative justice conferences as an early response to young offenders. *Juvenile Justice Bulletin.* Washington, DC: Office of Juvenile Justice and Delinquency Prevention.

78. Galaway, B. (1988). Crime victim and offender mediation as a social work strategy. *Social Service Review, 62*, 668–683.

Marshall, T. & Merry, S. (1990). *Crime and accountability: Victim–offender mediation in practice*. London: HMSO.

Rodriguez, N. (2007). Restorative justice at work: Examining the impact of restorative justice resolutions on juvenile recidivism. *Crime and Delinquency, 53*, 355–379.

Schneider, A. L. (1986). Restitution and recidivism rates of juvenile offenders: Results from four experimental studies. *Criminology, 24*, 533–552.

Umbreit, M. & Coates, R. (1993). Cross-site analysis of victim–offender mediation in four states. *Crime and Delinquency, 39*, 565–585.

79. McGarrell, 2001.

80. Rodriguez, 2007.

81. Braithwaite, J. (1999). Restorative justice: Assessing optimistic and pessimistic accounts. In M. Tonry (Ed.), *Crime and justice: A review of the research*. (Vol. 25). Chicago: University of Chicago Press.

82. Crawford, A. & Newburn, T. (2003). *Youth offending and restorative justice: Implementing reform in youth justice.* Portland, OR: Willan.

Weitekamp, E. G. M. (1999). The history of restorative justice. In G. Bazemore & L. Walgrave (Eds.). *Restorative justice: Repairing the harm of youth crime*. Monsey, NY: Criminal Justice Press.

83. Lundman, 2001.

Whitehead & Lab, 1996.

84. Butts & Buck, 2000.

Butts, Buck, & Coggeshall, 2002.

Lundman, 2001.

McGarrell, 2001.

Quay, H. C. & Love, C. T. (1977). The effect of a juvenile diversion program on rearrests. *Criminal Justice and Behavior, 4*, 377–396.

Regoli, R., Wilderman, E., & Pogrebin, M. (1985). Using an alternative evaluation measure for assessing juvenile diversion programs. *Children and Youth Services Review, 7*, 21–38.

Rodriguez, 2007.

85. Baron, R. & Feeney, F. (1973). Preventing delinquency through diversion: The Sacramento County 601 Diversion Project. *Federal Probation, 37*, 13–18.

Baron & Feeney, 1976.

Butts, Buck, & Coggeshall, 2002.

Davidson, Redner, Amdur, & Mitchell, 1990.

Dunford, Osgood, & Weichselbaum, 1981.

Palmer, Bohnstedt, & Lewis, 1978.

Lundman, 2001.

McGarrell, 2001.

Quay & Love, 1977.

Regoli, Wilderman, & Pogrebin, 1985.

Rodriguez, 2007.

86. Davidson, Redner, Amdur, & Mitchell, 1990.

Elliott, D. S., Knowles, B. A., & Dunford, F. W. (1978). *Diversion: A study of alternative processing practices.* Boulder, CO: Behavioral Research Institute.

Finckenauer, J. (1982). *Scared Straight! and the panacea phenomenon.* Englewood Cliffs, NJ: Prentice Hall.

Finckenauer & Gavin, 1999.

Lincoln, S. B. (1976). Juvenile referral and recidivism. In R. M. Carter & M. W. Klein (Eds.), *Back on the street: Diversion of juvenile offenders.* Englewood Cliffs, NJ: Prentice Hall.

Lundman, 2001.

87. Whitehead & Lab, 1996, p. 320.

88. Blomberg, T. G. (1979). Diversion from juvenile court: A review of the evidence. In F. L. Faust & P. J. Brantingham (Eds.), *Juvenile justice philosophy: Readings, cases and comments* (2nd ed.). St. Paul, MN: West.

Palmer, T. & Lewis, R. V. (1980). *Evaluation of juvenile diversion.* Cambridge, MA: Oelgeschlager, Gunn and Hain.

89. Paternoster, R., Waldo, G., Chiricos, T., & Anderson, L. (1979). The stigma of diversion: Labeling in the juvenile justice system. In P. L. Brantingham and T. G. Blomberg (Eds.), *Courts and diversion: Policy and operations studies.* Beverly Hills, CA: Sage.

90. Blomberg, 1979.

91. Dunford, Osgood, & Weichselbaum, 1981.

Palmer & Lewis, 1980.

92. Whitehead & Lab, 1996.

Preadjudication Processes in Juvenile Justice

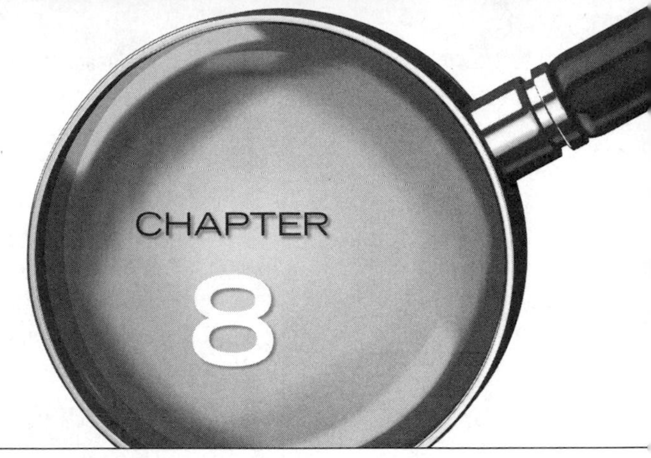

CHAPTER OBJECTIVES

After studying this chapter, you should be able to

- Describe the intake process in juvenile court
- Describe the role of discretion in the intake process
- Describe recent trends in intake decision making
- List the screening criteria commonly used in making intake decisions and explain the uses of case conferences in intake screening
- Explain the use of and need for the detention hearing or preliminary hearing and describe the due process issues surrounding the detention of juveniles
- Explain the functions that plea bargaining and sentence bargaining play in the intake and preadjudication process

CHAPTER OUTLINE

- Introduction
- The Juvenile Justice Intake Process
- Additional Juvenile Justice Preadjudication Processes
- Legal Issues
- Chapter Summary
- Key Concepts
- Review Questions
- Additional Readings
- Notes

Introduction

arrest The deprivation of a person's freedom by legal authority; the apprehension of the juvenile by the police; the physical seizing of a person by the authorities for the purpose of making that person answer for a criminal charge.[1]

This chapter examines a variety of pretrial or preadjudication processes that occur in juvenile courts. These pretrial processes include a variety of hearings, conferences, and decision-making events in which determinations about the best methods of handling juvenile court referrals are made. Like the decision to **arrest** a juvenile, decisions made at this stage of the juvenile court process are critical for a number of reasons. First, they play a major role in determining which youths penetrate further into the juvenile court process. Second, they have a direct bearing on the freedom of juvenile offenders because they often focus on youths' placement pending further court action. Third, they can lead to formal court processing and assignment of formal labels, such as "delinquent" or "undisciplined child." Finally, they can halt or set in motion a variety of responses to juvenile offenders that have the potential to exacerbate or reduce the likelihood of subsequent delinquent behavior.

The preadjudication processes that are examined in this chapter come between two important events in the juvenile justice process: the police (arrest) or social agency referral to the juvenile court and the trial or adjudication hearing in which the court decides whether or not to take formal jurisdiction of a case. One purpose of the preadjudication process is to screen out cases that may be better handled outside of court, meaning youths and their families (and even the victims of their misbehavior) will be spared any harm associated with court involvement. Conversely, if they do not effectively screen cases, more youths and their families will be subject to possibly unnecessary and expensive court intervention. Another purpose of the preadjudication process is to determine which cases are appropriate for adjudication. Thus, the decisions made at this stage of the juvenile justice process determine which youths become "official juvenile delinquents."

The Juvenile Justice Intake Process

intake The process of introducing youths who come to the police and court's attention into the juvenile justice system and determining how far they go into the system.

When the decision to arrest a youth is made or when a social agency or a public school refers a youth to the juvenile court, several things occur. Although most people, when thinking of arrest, probably picture a suspect being handcuffed, taken to the station to be booked, and placed in a lockup, in many instances when a juvenile is arrested, the youth is released into the custody of his or her parents or guardian. The police officer then completes the juvenile arrest report, which is forwarded to the **intake** unit of the juvenile court for screening. Similarly, when social agencies or schools make a referral to the juvenile court, it is the intake unit that initially receives the referral.

The purpose of the intake unit is to determine the most appropriate way to handle cases that are referred for possible juvenile court action. As noted earlier, juvenile codes are constructed so broadly that a wide range of juvenile behaviors are potentially subject

FYI

Because A Youth Is Arrested Does Not Mean He or She Is Guilty of an Offense

It should be kept in mind that being arrested does not mean that one is guilty. Guilt or innocence is determined by the juvenile court at the adjudication hearing. Thus, even though a juvenile is arrested for an offense they are not guilty of that offense until the alleged offenses have been proven at the adjudication. Moreover, youths are not considered "official" delinquents until they have been adjudicated.

MYTH VS REALITY

Formal Court Action Is Not Necessary in Every Juvenile Case

Myth—Juveniles should always receive some type of formal sanction when they engage in delinquent behavior.

Reality—Delinquency encompasses many types of offenses, many of which do not represent a substantial threat to public or individual well-being. Moreover, most youths engage in delinquency and are not caught, yet they manage to grow up to become healthy adults. Also, there is an informal juvenile justice process, as well as diversion programs, that effectively handle many cases each year. It should be remembered that, although formal court action can and does serve the best interests of children and the community in many cases, there are other instances when the effects of formal court actions are far from positive.

intake screening The process of determining whether youths should be diverted from the juvenile system or be referred for formal or informal services. Screening criteria usually include seriousness of the offense(s), family history, prior delinquency record or court contacts, legal mandates regarding diversion, attitude, age, school performance, parental cooperation, employment history, and restitution issues. Unfortunately, screening criteria may also include, despite their inappropriateness, race, gender, and socioeconomic status.

to juvenile court jurisdiction. As a result, complaints alleging that juveniles have thrown snowballs at a school bus, broken a window during a baseball game, engaged in a petty property dispute, disobeyed their parents, stayed out past curfew, been involved in a minor altercation in the neighborhood, as well as many more serious matters such as serious assaults, burglaries, and property damage cases, are regularly referred to juvenile courts for action. It is the intake unit of the court that has the responsibility of determining which of these many types of complaints warrant formal court action and which might be better handled in some other way.

Traditionally, **intake screening** was performed by probation officers employed by the juvenile court or separate state agencies responsible for performing probation and intake functions, although it is common for a member of the prosecuting attorney's office to review most complaints for **legal sufficiency**. The purpose of this review is to determine if the elements of a chargeable offense are present. Essentially, the prosecuting attorney makes a determination that there are legal grounds for the court to proceed and that the prosecuting attorney's office is prepared to represent the state's interests at the adjudication. This model is still employed in most states. After the complaint is reviewed for legal sufficiency, an intake officer (typically a probation officer performing the intake role) screens the complaint in order to make a decision about how the case should be processed. This screening often involves a consideration of the type and seriousness of the offense alleged in the complaint (e.g., harm to the victim) as well as factors such as the age of the

legal sufficiency The requirement that an alleged criminal offense and the alleged perpetrator meet certain criteria before the alleged perpetrator can be charged with a criminal act. A prosecuting attorney, district attorney, or state's attorney assesses the criminal charges and the characteristics of the alleged perpetrator to ensure the criteria are met.

FYI

A Variety of Factors Discourage Police from Making Arrests in Juvenile Cases.

There are a number of reasons police might decide not to arrest a juvenile. First, many police believe that juvenile courts are ineffective in dealing with delinquents and that all of the work required to complete a case (i.e., writing up a police report for prosecutorial review, contacting witnesses, and testifying in court) is not worth the likely outcome. Second, police are not obligated to make arrests in every case in which they come into contact with a juvenile who has engaged in illegal conduct, and although they are likely to make an arrest if a serious offense has been committed, they often make "informal adjustments" in instances involving minor delinquent activities. Third, the lack of secure detention beds for juveniles in some jurisdictions places limits on the number of youths who can be detained. A major problem confronting juvenile detention facilities, like adult jails, is chronic overcrowding. As a result, the majority of youths placed in juvenile detention are detained in overcrowded facilities.[2]

youth and his or her offense and court history (if available). In addition, information in the complaint regarding parental response to the offense, the reaction of the youth, and the behavior of the youth in the community, at school, and at home may be considered. To better understand one intake officer's perspective of his job, see Box 8-1.

BOX 8-1 Interview: Frank Weichlein, Administrator, Kalamazoo County Juvenile Home, Ninth Judicial Circuit Court, Family Division, Kalamazoo, MI

Q: What is your educational and employment background?

A: I received a BA in sociology from Western Michigan University in 1978, and an MA in public administration from WMU in 1992. I started out my work in juvenile justice by interning at the Kalamazoo County Juvenile Home and upon graduation I was hired to work there as a child guidance worker. I moved to probation after two years and did that from 1980–86. I was the intake supervisor from 1986 until 2005. I am presently the Administrator of the Juvenile Home.

Q: What were your duties and responsibilities as intake supervisor?

A: First of all, I was responsible for overseeing the entire intake process, which includes accepting and screening of all complaints and petitions that come to the court. I made or oversaw the initial decision as to whether a child or family receives diversion or gets referred to the formal docket. Second, I supervised 11 intake staff. Third, I was responsible for the development, implementation, and supervision of diversion programs run by the court. Also, with the unification of courts in Michigan, I had added responsibilities related to domestic cases and Personal Protection Orders. For example, we screened all domestic case motions and conducted case reviews on motions regarding parenting time, custody, and change of domicile. And we reviewed requests from citizens for Personal Protection Orders.

Q: In addition to your education and employment background, have you had any specialized training?

A: One thing the court has been very good about is providing staff training opportunities. I have been able to attend numerous seminars and training sessions, all of which have been helpful in my work.

Q: What part of your job in intake did you find the most rewarding?

A: Giving young people the opportunity to change behaviors in diversion without having to have a formal delinquency record gave me the most satisfaction. Also, with the added responsibilities that have resulted from court unification, I had an opportunity to be responsive to a broad range of people who come to the court for assistance.

Q: What part of your intake job did you find the least rewarding?

A: Responding to, and working under, legislation requirements that are not well thought out. The political posturing that results in legislation being passed without consulting the people that will be responsible for implementing it causes a great deal of frustration. An example of this is the victim rights legislation involving restitution. Although I'm in favor of restitution, you cannot treat juveniles like adults in that regard; but that's what the legislature has tried to do. Other questionable legislation, in my opinion, includes the registration of sex offenders, school safety legislation, changes in waiver statutes, and the development of "designation cases." While these efforts may be well-intentioned, they can produce outcomes that are detrimental to young people, and changes should be considered.

Q: What skills and abilities would you tell college students they should cultivate to work in your field?

A: Patience, resiliency, the ability to accept diversity, and flexibility are some of the main personality features anyone going into the field needs to cultivate. In addition, those who work in juvenile justice need to be able to think on their feet and have excellent oral and written communication skills. An intake worker spends a great deal of time communicating with the public, police, victims, juveniles, family members, and the court. Finally, an intake officer must be a person who can think independently, but work under supervision.

(*continues*)

Q: What particular strengths and abilities do you feel you brought to your work in intake?

A: Patience and the ability to see the "big picture," and I am flexible enough to respond to a changing work environment. I believe I have good communication skills and an ability to lead and motivate people without being coercive. It is also important to understand the fundamentals of each of my employees' positions, which is needed for effective supervision.

Q: What were the major challenges that you faced as intake supervisor?

A: Certainly my most challenging problem was dealing with serious delinquents on the one hand, while continuing to advocate for prevention and diversion programs. I genuinely believe in the importance and the effectiveness of prevention and rehabilitation, but serious delinquents take the public's focus off diversion. In addition, the development of domestic case intake processes has certainly been one of the biggest challenges we have faced.

Q: Who did you report to, and where are you in the court administration when you were Intake Supervisor?

A: I reported to the associate court administrator, who is in charge of intake and probation services. I was considered a department head and I was solidly in middle management.

Q: What unique programs were offered in intake that you believe were particularly effective?

A: Our youth diversion program is unique in that it is run by the court, not a youth services bureau outside of the court. With good caseworkers and the backup of the court, it has proven to be very effective. Our inquiry process also works very well. We have four workers with nonlegal backgrounds, but who have been specially trained, and meet with youth and parents to discuss pending delinquency charges. There are a full range of alternatives available, including formal court action, and these workers do a nice job of screening out and diverting cases. Without them, the court would be overwhelmed with cases. Our prosecuting attorney's office has been very cooperative in allowing our staff this discretion also. They trust our judgment, and we certainly listen to them in dealing with serious delinquents. The relationship we have developed with the schools in Kalamazoo County has proven effective in trying to intervene early in some families, particularly our system for processing educational neglect petitions. Also, we have worked with a number of community agencies in the development and implementation of programs based on the philosophy of Balanced and Restorative Justice.

Q: What types of offenses come in to intake on a regular basis?

A: Sadly, assaultive cases have overtaken retail fraud—shoplifting—as the number one offense that we are seeing, and many of these assaults involve domestic violence where juveniles are referred for assaulting their parents. These cases concern everyone at the court and they require the development of interventions specifically for these cases.

Q: What was your salary as intake supervisor?

A: $62,000.

Q: What are the most serious issues that you see today in the area of juvenile justice?

A: Effective programming for the serious delinquent offender is very important, while at the same time not forgetting the need for early intervention to prevent the circumstances that give rise to serious offending. Another serious issue is dwindling resources for all systems that work with children and families, particularly community mental health. We are seeing an increase in the number of severely emotionally disturbed juveniles involved with the court.

Q: Did you have much hands-on contact with juveniles in your position as intake supervisor?

A: Not as much as I would have liked, but I conducted inquiries on all school-related cases (truancy and educational neglect) as well as neighborhood nuisance offenses, mostly curfew violations, at a community-based diversion program operated out of the Boys and Girls Club. I also handled crisis walk-ins and coordinated our Teen Jury program.

Q: Is there anything else you would like to tell me about your role at the court?

A: Working in juvenile justice is vibrant and exciting, even after almost 30 years. Every day brings new challenges. You never know what is at the other end of the phone call or waiting outside the door.

FYI

The Adjudication Is the Juvenile Court Equivalent to the Trial

Adjudication is the trial, final plea, or fact-finding stage of the juvenile process, the stage at which the hearing official determines formal guilt or innocence. It is the juvenile court equivalent to the trial in adult court. Adjudication involves a "formal" hearing that is (usually) recorded or transcribed and due process protections are afforded the juvenile (i.e., the right to counsel, the right to jury trial [where available], the right to confrontation and cross-examination, the right to subpoena witnesses, and the right to testify or remain silent).

petition The request for formal court action that contains the written accusation against a juvenile, specifies the criminal or delinquent charges being made, and is properly verified as to age and venue by the complainant (the person bringing the charge).

For certain types of cases, the intake decision is routine. Indeed, in some cases formal processing is mandatory. For example, intake guidelines typically specify that certain types of serious offenses cannot be diverted and must be petitioned to court. A **petition** is a request for formal court action containing a statement of charges that is presented at the adjudication or trial stage of the juvenile justice process. In other cases, the intake officer has considerable discretion in determining how a case should be handled. The intake officer may decide that dismissing the case or warning the juvenile is the best course of action. Indeed, information contained in the police report may make it clear that the complainant does not desire formal action, or the complaint may indicate that the matter was so trivial that an expenditure of court resources is unwarranted or inappropriate. Some courts operate special programs for certain types of offenders, such as youths who commit minor shoplifting or property damage offenses, and eligible youths are routinely referred to these programs by intake officers. In other instances, intake officers may refer youths, and in some cases youths and their families, to a variety of community agencies for assistance. Unfortunately, in many communities there are a limited number of good diversion programs that court personnel can use.

Although many complaints receive rather routine screening at the intake stage, others require considerably more of the intake officer's time. Because police complaints typically contain little information other than that pertaining directly to the alleged offense, intake officers often attempt to collect additional data that they feel will assist them in making appropriate screening decisions. Consequently, it is not uncommon for an intake officer to arrange a **case conference** with the juvenile and his or her parent(s). The purpose of the case conference is to allow the intake officer an opportunity to collect additional information that might aid him or her in making a screening decision. In other instances, the intake officer might contact a victim or talk with school or social

case conference A meeting attended by an intake officer, the juvenile, and the juvenile's parents that is designed to collect additional information to aid the intake officer in screening a case.

FYI

Statutes or Court Rules Govern the Time Frames for Holding Adjudications

In most states, statutes or court rules set time limits for holding an adjudication after a petition is filed. In situations in which a youth is detained, an adjudication must be held within 10 to 180 days (depending on the state) from the date that the petition is filed. In situations in which a youth is not detained, an adjudication must be held within 30 to 180 days.[3]

FYI

The Purpose of Intake Is to Make Decisions About How to Handle Cases

Regardless of who performs the intake function—a probation officer who works for the juvenile court, a separate state agency, or a member of the district attorney's staff—the purpose of intake is to make decisions regarding how cases should be processed by the juvenile court.

agency personnel in an effort to collect information that they feel will help them in making a screening decision.

Although the probation officer or intake officer model is the traditional model employed in juvenile justice, in recent years there has been a move toward increasing the involvement of the prosecuting attorney in the intake process. In fact, in some jurisdictions (e.g., Colorado, Arizona, and Washington), intake screening is now the domain of the district or prosecuting attorney's office.[4] In other jurisdictions (e.g., Louisiana), the prosecutor has been given greater authority over intake decision making.[5]

The increasing involvement of prosecuting attorneys in the intake process reflects the increasing legalization and punitiveness of the juvenile justice process found in a number of jurisdictions. The increased use of prosecuting attorneys in juvenile justice decision making is an example of a shift toward a **punitive model of juvenile justice** intended to respond more forcefully to juvenile offenders and to circumvent the discretion employed by probation officers in the traditional intake-screening model. For example, Washington, South Dakota, and Wyoming have placed the responsibility for intake screening in the hands of the prosecutor, while Florida and Indiana require that all intake officers bring all cases to the prosecutor for review. Other states (Arizona, New Mexico, Virginia, and Texas) mandate that cases involving youths with prior records be screened by the prosecutor, and still others (California, Maryland, and Texas) require that serious offenses and weapons offenses be reviewed by the prosecutor.[6]

punitive model of juvenile justice A model of juvenile justice that is intended to act as a counter to what are believed to be traditional "soft" approaches to juvenile crime and supports more severe sanctions for illegal behavior, particularly for more serious forms of delinquency. The punitive model of juvenile justice employs the same criteria used in charging adults, and the focus is on punishment rather than rehabilitation.

In Michigan, as in a number of states, the prosecuting attorney is the only person who can petition the juvenile court in delinquency matters.[7] Furthermore, prosecutorial screening decisions are based on explicit criteria: offense seriousness, prior record, and the age of the youth. This removes traditional discretionary decision making from the probation staff, although police agencies continue to "exercise their traditional discretion regarding whether to refer or adjust incidents involving juveniles."[8]

Although some degree of discretion is present in the juvenile court intake process, limits on discretion are imposed by state statutes and juvenile court rules of procedure,

FYI

Discretion Is Involved in Intake Screening

Regardless of who makes intake decisions, discretion is still involved. Recent trends toward shifting intake decision making to the prosecuting attorney's office simply move the discretionary power from one office to another. Prosecuting attorneys still exercise considerable discretion in deciding which cases to petition. Therefore, the issue is not so much whether discretion is exercised, but who exercises it and what the outcome of the decision making will be.

which often note the types of cases that can be handled informally and those that must be handled formally. For example, the California Rules of Court, Rule 5.516 indicates that intake staff should consider nine factors in deciding whether a matter can be handled at intake. These factors include the following:

> (1) Whether there is sufficient evidence of a condition or conduct to bring the child within the jurisdiction of the court; (2) If the alleged condition or conduct is not considered serious, whether the child has previously presented significant problems in the home, school, or community; (3) Whether the matter appears to have arisen from a temporary problem within the family that has been or can be resolved; (4) Whether any agency or other resource in the community is available to offer services to the child and the child's family to prevent or eliminate the need to remove the child from the child's home; (5) The attitudes of the child, the parent or guardian, and any affected persons; (6) The age, maturity, and capabilities of the child; (7) The dependency or delinquency history, if any, of the child; (8) The recommendation, if any, of the referring party or agency; and (9) Any other circumstances that indicate that settling the matter at intake would be consistent with the welfare of the child and the protection of the public.[9]

These and similar criteria allow intake officers to employ considerable discretion in deciding how cases should be handled.

The Role of Discretion in the Intake Process

Those who make intake decisions determine who will penetrate further into the formal juvenile justice process. Although intake guidelines require intake officers to send certain cases forward for adjudication, intake decision makers often have considerable discretion in deciding how the majority of cases referred to the juvenile court should be handled. Options might include dismissing the case, having the offending youth and the parents in for a case conference to collect additional information to assist in making the intake decision, referring the youth to a diversion program (e.g., informal probation or counseling at a community agency), filing a petition (a request for formal court action that represents a legal version of the police complaint and indicates the specific charges used in the adjudication), and waiver or transfer of the case to the adult court.

Intake Legal Requirements

Regardless of who is responsible for making intake decisions for the juvenile court, certain things must happen. First, a decision must be made, based on the documentation presented (usually a police report), that there are sufficient facts to justify an identifiable criminal charge. Consequently, the intake screener, prosecuting attorney, or whoever is making this decision needs to be familiar with the elements of criminal offenses and

> ***FYI***
>
> **Many Cases Are Diverted from Juvenile Courts by Intake and Still Receive Some Type of Sanction**
>
> Of the estimated 1,697,900 delinquency cases referred to juvenile courts in 2005, approximately 56% had petitions filed. However, of the cases not referred to court, only 40% of those cases were dismissed. Sizable percentages of cases not petitioned were dealt with in a variety of ways, such as placement on informal probation (22%), or they were given some other type of disposition or sanction (38%).[10]

FYI

Intake Often Involves Screening Cases for Legal Sufficiency

A common practice in instances in which probation officers make intake-screening decisions is to have someone from the prosecuting attorney's office assess the complaint for legal sufficiency (i.e., if the elements for a chargeable offense are present).

determine not only that a crime or act of delinquency occurred, but also what the exact charge should be.

Second, a petition or complaint must be prepared and filed with the court. One of the fundamental requirements is that the petition be verified—that someone signs it who vouches as to the *accuracy* of the charges. By signing the petition and vouching for the accuracy of the charges, the "petitioner" also agrees to be responsible for going forward with the case. Essentially, this step commits the prosecuting or state's attorney to represent the state's interest in court.

Third, the petition or complaint must be presented to the court for the court to take action. The petition can be filed and a detention hearing requested, at which time the juvenile and his or her family can be advised of the charges and of their legal right to a hearing, to an attorney, to confront adverse witnesses, to make a statement in court, to have bail set (where bail is used for juveniles), and to have a formal trial. Juveniles also may present evidence at detention hearings and contest their detention.

If no detention hearing is set, then the petition can be discussed with the juvenile at an informal (off the court record) case conference or preliminary inquiry. The case conference is a meeting attended by a court official, the juvenile, and the juvenile's parents or guardian (or other responsible adult). The purpose of the case conference is to obtain additional information that can assist the intake officer in making a decision about how a case should be processed. At the case conference, the juvenile and his or her parents or guardian may be advised of alternatives, such as diversion or trial, and various rights that the juvenile has, such as a right to have an attorney represent him or her before the court, and the juvenile may ask questions about the case. In some instances, a juvenile may have an attorney represent him or her at the case conference, although this is not common.

Several states use a pretrial hearing in which a **plea** is taken. At this hearing, the youth typically admits to the charges, does not contest the charges, or denies the allegation(s) contained in the petition. The hearing typically occurs after the prosecutor and the attorney for the juvenile have met to discuss the case and have negotiated a plea, which may be the result of a plea bargain. Far more delinquency cases are settled by pleas than by trial, and the primary purpose of a pretrial hearing is to provide time for plea negotiations. If no plea can be agreed upon at the hearing, then the attorneys can exchange names of witnesses to be called at trial, set any pretrial motions for decision, and delineate for the court any legal or factual issues of importance.

plea The juvenile's formal response to criminal charges

Research on Intake Decision Making

A number of factors have been found to influence intake screening decisions, including the seriousness of the offense,[11] the youth's prior record,[12] and the youth's demeanor.[13] Youths who have committed serious offenses, those with prior records, and those who are uncooperative are more likely to be petitioned. In addition, variables such as age, socioeconomic

FYI

Decisions Made at Intake Influence Decisions Made at Later Stages
Intake decision making is important because there is evidence that decisions made at the earliest stages of the juvenile justice process influence decisions made at later stages. Moreover, as noted, a substantial body of literature exists indicating that there are disparities in the treatment of juveniles in some jurisdictions, at least some of the time. This means that decisions to treat minority youths more severely at the intake stage of the court process often will result in more severe treatment at subsequent stages such as the adjudication.[19]

status, race, and gender have been found to influence intake decision making in some jurisdictions, but the influence of these variables is not completely clear.[14] For instance, in one study of court decision making, both legal factors, such as the seriousness of the offense, and "nonlegal" factors, such as gender, race, and socioeconomic status, were found to influence intake decisions. The findings of this study indicated that the seriousness of the offense was related to the decision to file a petition, but it appeared to be a more important factor for lower-class African American males with a prior record than for other groups.[15] Another study that examined decision making at each step of the juvenile justice process found that minorities received harsher treatment at each decision-making point, including intake.[16]

Overall, the research on intake decision making suggests that, at least in some communities, lower-class minority males are more likely than other groups to receive more formal responses at the intake stage.[17] In addition, there is considerable evidence that female status offenders are likely to be treated more harshly than their male counterparts, at least in some jurisdictions.[18] Although this research indicates that bias in the intake process is not a problem in every jurisdiction, it raises serious questions about the quality of justice available to juveniles in some communities.

Additional Juvenile Justice Preadjudication Processes

In addition to intake screening, a variety of preadjudication processes can occur prior to the adjudication or trial. These include hearings such as waiver or transfer hearings (examined in detail in Chapter 9), detention hearings, preliminary hearings, arraignments, plea bargaining negotiations, and bail determinations. The remainder of this chapter describes the operation of these preadjudication processes, discusses legal issues that influence their operation, and examines relevant research on the preadjudication stage of the juvenile justice process.

detention hearing
A hearing in front of a judicial officer that is required after a juvenile has been detained or in order to get a juvenile detained. At this hearing, the juvenile has a right to be present, have his or her parents or significant adult present, be informed of the charges, have his or her legal rights explained, make a plea, have a trial date set, and request bond (where permissible).

Detention Intake and Detention Hearings

A special type of intake decision making occurs when police or other juvenile justice agents request the detention of a minor. As noted in Chapter 6, police officers, as well as other juvenile justice agents such as probation officers, at times request that youths be detained in juvenile detention units or, in some cases, adult jails. In these instances, an intake officer makes a decision about the appropriateness of the request for detention. It is important to note that it is the intake officer, acting as an officer of the juvenile court, who makes the decision to detain a youth, not the police officer.

The decision to detain a youth, however, is followed up in practice by a **detention hearing**. Each state's family or juvenile code mandates that a hearing be held when a

COMPARATIVE FOCUS

In New Zealand, Family Group Conferences Play a Major Role in Determining how Cases are Handled

In New Zealand, except in special circumstances, young people who have committed offenses and admit their involvement must be referred to a family group conference before any formal charges are made. In cases that are referred to the youth court, they must be referred to the family group conference before sentencing. The family group conference involves the youth offender, the offender's family members, and whoever they invite (e.g., teachers and friends), the youth offender's attorney (usually only in court-referred cases), the victim or victims and their representatives and supporters, the police, and sometimes, a social worker. The conference is facilitated by a youth justice coordinator.

In the family group conference, all the participants are given an opportunity to participate in the discussion of the case and to be involved in deciding the case outcome. Also, at some point in the conference, the young offender and the family are given an opportunity to discuss the case privately and to reach agreement about how they think the case should be handled. After this, the conference reconvenes, a plan for resolving the case is discussed, the agreement of all the participants is sought, and modifications to the plan are made, if necessary. In determining the outcome of the case, the conference participants are obligated to consider the offense, the circumstances surrounding the offense, and the interests of the victim. Possible outcomes consist of apologies, reparations, community service, participation in a training program, caseworker supervision, short-term residential placement, and, in some instances, a period of custody.[20]

minor is deprived of his or her liberty. These hearings are conducted by juvenile court judges, referees, or other hearing officers. At these hearings, the youth and his or her parents are advised of the juvenile's rights, an attorney may be assigned to represent the youth, and several important determinations are typically made. These determinations include assessing the facts leading to the youth's custody to determine if probable cause exists to hold the youth. In addition, a decision regarding the placement of the youth is made. The court may decide to continue holding the youth in detention, place the youth in the custody of his or her parents or guardian, or make some other placement decision. Finally, the court may authorize the filing of a petition and set dates for subsequent hearings to review the case or for an adjudication.

Trends in the Use of Detention

Although detention is used for a variety of reasons (e.g., holding youths who are awaiting trial, disposition, or transfer to another jurisdiction), there are clear indications that the use of detention for delinquency grew during the 1990s, but has declined in recent years. For example, data on national detention trends reveal an 11% increase in the use of detention for delinquency cases between 1990 and 1999. However, the number of youths detained prior to an adjudication for person, property, drug, or public order offenses declined 7.5% between 1999 and 2006.[21] In 2006, most of the youths who were detained prior to adjudication were detained for offenses against persons (31.9%), followed by property offenses (23.0%), and technical violations (22.7%).[22] The percentage of youths detained for various offenses prior to their adjudication in 2006 can be seen in Table 8-1.

A troubling reality that plagues juvenile justice in many communities is disproportionate minority contact (DMC). One area where DMC has been evident is the disproportionate

TABLE 8-1 Number and Percent of Juveniles Detained Prior to Adjudication by Type of Offense, 2006

Offense	Number of juveniles	Percent of juveniles
Person	4,232	32.0
Property	3,059	23.0
Technical violations	3,017	22.7
Public order	1,440	10.8
Drug	1,171	8.8
Status	361	2.7
Totals	**13,280**	**100.0**

Source: Data from Sickmund, M., Sladky, T. J., Kang, W., & Puzzanchera, C. (2005). *Easy access to the census of juveniles in residential placement.* Retrieved November 3, 2008 from http://ojjdp.ncjrs.gov/ojstatbb/ezacjrp/.

confinement of minority youths in correctional placements such as detention. DMC exists when "the proportion of juveniles detained or confined in secure detention facilities, secure correctional facilities, jails, and lockups who are members of minority groups ... exceeds the proportion such groups represent in the general population."[23] In most states, minorities are disproportionately represented at each stage of the juvenile justice process and they are disproportionately represented in detention and other secure placements.[24] For example, in 2006, African American youths made up 16.4% of the juvenile population younger than 18 years of age, but they accounted for 40.9% of the youths detained prior to adjudication. In contrast, white youths made up 77.6% of the juvenile population, but they accounted for only 30.4% of the youths detained prior to adjudication.[25]

In an effort to deal with this problem, federal legislation in the form of the 1988 amendments to the Juvenile Justice and Delinquency Prevention Act of 1974 were passed that requires states to address DMC in order to receive formula grants for juvenile justice. As a result, many states have begun to develop plans for reducing DMC.[26] Some of these plans may be having some positive impact on the problem of DMC, at least at some stages of the juvenile justice process. A comparison of racial disparities in the juvenile justice process for the years 1992 and 2002 found reductions in racial disparities in arrests and transfers to criminal court in 2002. However, disparities at other decision-making points, such as the detention of youths, were still found.[27]

In addition to minorities, males and older juveniles are also overrepresented in detention cases. Males make up the majority of youths placed in detention prior to their adjudication. In 2006, for example, males accounted for 81.3% of all juvenile cases involving detention prior to adjudication. Also, males from 16 to 17 years old accounted

FYI

The Reasons for Racial Disparities in Juvenile Justice Decision Making Are Not Clear

Although there is ample evidence of DMC at each stage of the juvenile justice process, the reasons for DMC are not clear.

for 62.2% of all youths younger than 18 years of age who were detained prior to their adjudication.[28]

In contrast to the increase in the use of detention for criminal or delinquent offenses, detentions for status offenses declined significantly in the late 1980s. For example, between 1986 and 1990, detentions for status offenses declined by 35%.[29] More recently, however, the number of youths being detained prior to an adjudication has varied considerably. On the date that the Census of Juveniles in Residential Placement was taken in 1997, there were 556 juveniles younger than the age of 18 years placed in detention prior to an adjudication. By 1999, the number had decreased to 384, it rose to 566 in 2003, and it decreased again to 357 in 2006.[30] The overall decline in the detention of status offenders is due in large measure to two reasons:

- New state and federal laws restrict the use of detention for runaways, truants, and incorrigible children.
- Licensing requirements for detention facilities in some states set age limits for the detention of juveniles. Because many status offenders are young, these licensing requirements result in a reduction in the number of status offenders detained.[31]

There is good evidence that the number of status offense cases being detained has been reduced significantly over time. Nevertheless, the reductions noted may not be as large as they first appear. Indeed, many juvenile court judges have been reluctant to abandon their jurisdiction over status offenders and others feel compelled to use more restrictive sanctions when status offenders fail to follow court orders to attend school or obey the commands of their parents. This has led some judges to redefine repeat status offenders as delinquents by charging them with violations of court orders and then detaining them for a delinquent offense, a process known as **"bootstrapping."**

bootstrapping The process whereby a status offender is ordered to follow certain conditions set by the court (e.g., attend school, adhere to a curfew, and obey the commands of the parents) and the youth is redefined as a delinquent if he or she violates the court's orders. The youth's new status as a delinquent then allows the court to use detention as a response to violation of the court's orders.

Juvenile Detention Regulations

All states mandate that a detention hearing be held within a time frame specified by state law, typically within 24 to 72 hours. In addition, most states specify a time frame within which the adjudication must be held when a youth is detained. As noted earlier, the time frame for adjudication ranges from 10 to 180 days.[32] The question remains as to how long a juvenile should be detained before being given his or her day in court. This question has been answered differently in different states. Although it is important to have a speedy trial, there should be enough time prior to trial to allow the attorney to prepare a meaningful defense. (A lack of time to adequately prepare is a problem that plagues many attorneys who practice in juvenile court.)[33]

Requiring a juvenile to stand trial within 10 to 14 days after being detained may sound enlightened, but if the juvenile is charged with a serious offense, 14 days may not be enough time for preparing and presenting an adequate defense at trial. Note that defense attorneys, particularly those who do public defender work, often have large caseloads.[34] In addition, a variety of tasks must be accomplished in order to construct an adequate defense. Witnesses need to be interviewed, police reports need to be read, and motions may need to be filed. Because attorneys who represent clients in juvenile court often lack adequate resources for performing these tasks,[35] forcing a trial on short notice encourages plea bargaining.

Clearly, the use of detention for youths charged with a delinquency offense has increased substantially in recent years. This increase in detention is no doubt helped by the

preventive detention Detention of an alleged offender prior to conviction so as to prevent the alleged offender from committing any crimes during the preadjudication period.

preliminary hearing See detention hearing and arraignment.

arraignment The initial "coming before" the judge or judicial officer at which time the juvenile is informed of the charges pending and of his or her constitutional rights; a plea is entered; a trial date is set, if necessary; and bail is determined, if needed.

ability of juvenile courts to use **preventive detention**. In an important U.S. Supreme Court case, *Schall v. Martin*, the court ruled that preventive detention was appropriate in juvenile cases in order to protect both the juvenile and society from offenses that the juvenile might commit prior to the adjudication. (For a discussion of this case, see Chapter 6.)

Preliminary Hearings and Arraignments

Another type of preadjudicatory hearing found in the juvenile court is the **preliminary hearing** or **arraignment**. Although the terms for these hearings and their purposes vary somewhat from state to state, they are primarily designed to take pleas from juveniles and consider the appropriate placement of the juveniles prior to adjudication. In some jurisdictions, a preliminary hearing or arraignment is used to advise the youth of his or her rights and to determine if the youth plans to contest the charges contained in the complaint or referral. If the charges are contested, then a date for the adjudication is often set. In instances when the charges are not contested, then a plea may be accepted at that time and a disposition entered by the court. In other jurisdictions, a preliminary hearing is typically used to consider the removal of the juvenile from his or her home.

Preliminary hearings or arraignments are key events in the juvenile justice process. For example, the decision by the court to deprive a youth of his or her liberty is a potentially ominous determination. As noted earlier, processing decisions made at early stages of the juvenile justice process appear to affect decisions made at later stages.[36] Likewise, a decision by a juvenile to not contest a charge or to enter a guilty plea means that the juvenile court may assume formal jurisdiction over the youth. In these circumstances, the court determines that it will take jurisdiction over the youth, which subjects the juvenile to the range of dispositions open to the court, including removal from the home and institutional placement.

The American Bar Association Juvenile Justice Center, the Juvenile Law Center, and the Youth Law Center reports:

> It is critical that counsel appear early in the life of a case. At first appearances in court, if judges ask about events surrounding alleged offenses, the circumstances of arrests, the roles of other youth involved, or clients' other contact with the juvenile justice system, and attorneys do not have answers, they lose the initial opportunity to present clients' cases in a favorable light. Judges are left to review the uncontradicted allegations in the charging petitions. Based on incomplete reviews, judges make early determinations regarding detention that may influence cases all the way until their dispositions.[37]

Given the gravity of the decisions made at preliminary hearings or arraignments, it is common for juveniles to be represented by attorneys at these hearings in many jurisdictions. However, the extent to which attorney representation is typical is not known.

FYI

The Disposition Is the Juvenile Court Equivalent to the Sentencing in Criminal Court

The disposition is the juvenile court analog of the sentencing in criminal court. At the disposition, the court makes a determination about how it will formally respond to a juvenile who has been adjudicated and comes within the jurisdiction of the court. The most common disposition is probation.

Indeed, there is evidence that attorney representation is more of an exception than the rule in some courts and that the quality of representation that is provided to juveniles in some jurisdictions is inadequate.[38] Indeed, some of the research that has examined attorney effectiveness in juvenile court has found that youths who were represented by counsel were more likely to receive harsher punishments than youths without attorneys.[39] (The issue of attorney representation in the juvenile court process is discussed in more detail in Chapter 10.)

The Use of Bail in Juvenile Justice

Given the U.S. Supreme Court's endorsement of preventive detention in juvenile cases, it is not surprising that there is considerable variability in the use of bail in juvenile justice. The wide discretion that juvenile courts have in using preventive detention stands in marked contrast to most adult courts, which make bail available to offenders unless offenders are charged with capital crimes, they have jumped bail in the past, or they have committed another offense while on bail.[40] The use of bail in juvenile justice is less common than in the adult criminal justice system. Indeed, very little attention has been given to the use of bail in juvenile cases. Typically, juvenile codes ether allow bail at the discretion of the court, deny youths the right to bail, or ignore it altogether.[41] Presently, some states, such as Georgia, Colorado, and Oklahoma, indicate that juveniles have the same right to bail as adults. Still others, such as Nebraska, Tennessee, Vermont, and Minnesota, make bail available to juveniles at the discretion of the judge. In other jurisdictions, such as Hawaii, Oregon, and Connecticut, juveniles have no right to bail.[42] Court rulings that deny juveniles bail are usually predicated on the assumption that because the actions of juvenile courts are intended to correct or help young offenders, because juvenile courts are able to release youths to their parents, and because detention hearings are common in juvenile justice, bail is unnecessary.[43]

Even in states that allow bail for juvenile offenders, it may not be widely used.[45] Indeed, the use of bail in juvenile cases raises a number of questions. One question is, who should be able to post bond for a juvenile? Parents or relatives seem like logical choices, but what about a boyfriend or girlfriend? One possible rule is that the juvenile should be released only to a person who is legally responsible for his or her care, custody, and control. However, not all parents are in a position to post bail, and in other instances parents may be unwilling to post bail.

Where does a juvenile get money for bail when the parents are poor or are angry and unwilling to post it? Would it be appropriate for bail bondsmen to provide bail money for juveniles?[46] If, for some reason, the juvenile absconds on bail, who should be responsible, the child or the parent? Because the purpose of bail has traditionally been to ensure the appearance of the alleged offender at subsequent hearings, if the parent posts bail, would this be sufficient encouragement for the juvenile to appear?

FYI

Bail Is Sometimes Used in Juvenile Justice

"Bail or a bail bond is a surety in the form of money or property that may be posted by a bonding company or others, including defendants themselves, to obtain their temporary release from custody and to ensure their subsequent appearance at trial."[44]

These questions reflect the ambiguity that surrounds the issue of bond in the juvenile courts. Despite these questions, however, many courts have attempted to develop guidelines for setting bond in juvenile cases.

Plea Bargaining in Juvenile Justice

plea bargain
An agreement between a prosecuting attorney and a juvenile, usually through the juvenile's attorney, according to which the juvenile will admit to a lesser charge in exchange for the reduction of the charge or admits to some charges in exchange for the dismissal of other charges.

Plea bargaining is another preadjudicatory process that plays a role in juvenile justice. It consists of negotiations between the prosecution and the defense counsel, often with the consent of the court, whereby the defendant agrees to plead guilty in return for the prosecuting attorney's promise to reduce the charge or to drop some of the charges pending against the defendant.[47] The idea is that each party gives up something in order to achieve a reasonably certain outcome—an outcome that could not be guaranteed if the case went to adjudication. On the one hand, the plea agreement reached guarantees the prosecution a conviction, although the prosecution gives up the opportunity to have the defendant adjudicated on a greater number of charges or on more serious charges. On the other hand, the defendant gives up his or her right to contest the charges at trial as well as other constitutional rights. Further, the defendant gives up the opportunity to avoid juvenile court jurisdiction altogether by voluntarily submitting to the court's jurisdiction through his or her plea. Indeed, as a part of the plea agreement, the defendant admits to fewer offenses or to less serious charges.

Plea bargaining and the process through which juveniles make guilty pleas have received scant attention in the research literature. Although plea bargaining appears to be rare in some juvenile courts,[48] in other jurisdictions it is a common occurrence. Today, most states attempt to exert some control over plea bargaining. Twenty-five states have established formal rules that govern the plea negotiation process, 13 states have passed legislation that spells out a judge's responsibilities when accepting a guilty plea, and 3 states attempt to regulate plea negotiations through the appellate review process. Only 10 states to date have failed to establish guidelines or legislation that regulates plea bargaining.[49]

Although plea bargaining is recognized as a part of the juvenile justice process in most states, in practice the extent and quality of regulation varies considerably across jurisdictions. As Joseph Sanborn found in a study of the plea negotiation process across the country, some jurisdictions do not require judges to remind defendants that they are giving up constitutional rights, make them aware of possible dispositional outcomes, determine if the pleas were made voluntarily, and verify that the pleas had a factual basis. Indeed, only two states, Delaware and California, require judges to discuss each of

FYI

Plea Bargaining Is Common in Many Juvenile Courts

Although plea bargaining is a common practice in many juvenile courts, most juvenile cases involve open admission to the charges. Although the percentage of cases across the nation that include plea negotiations is not known, Joseph Sanborn's research on juvenile courts in Philadelphia found that about 20% of the cases processed resulted in a negotiated plea. The majority of the negotiations focused on reducing the sentence, usually probation in lieu of an out-of-home placement, or, in cases where there was an out-of-home placement, on securing a less restrictive placement.[52]

these issues with juvenile defendants.[50] Moreover, Sanborn's research revealed that urban courts are more likely than suburban and rural courts to employ formal procedures governing the plea bargaining process. As Sanborn notes, his research raises important questions about the continuing informality of many juvenile courts at this crucial stage of the juvenile justice process. Given the increasing punitiveness of at least some juvenile courts, it is essential that defendants clearly understand the potential consequences of the decisions that they make or that others make on their behalf.[51]

Legal Issues

One legal issue worthy of attention is the advisability of having a nonjudicial officer make the detention decision. Indeed, the decision to detain a juvenile is the most important decision that the juvenile justice system can make. According to the Michigan Supreme Court in *Reist v. Bay County Circuit Judge* (396 Mich 326, 241 NW2d 55 [1976]):

> The interest of a parent and child in their mutual support and society are of basic importance in our society and their relationship occupies a basic position in this society's hierarchy of values. *Clearly any legal adjustment of their mutual rights and obligations affects a fundamental human relationship.* The rights at stake are protected and encompassed within the meaning of "liberty" as used in the due process clause (emphasis added).

The central issue in the removal decision is whether to remove a child from the family home. Without question, this is a "legal adjustment" of the family's mutual rights and obligations. However, should this fundamentally important decision—one that can trigger a variety of **due process** rights for children and their families—be left to someone who has a legal background or to someone who has a social work background? If the detention decision is made by a supervisor of the detention facility without legal advice or without the benefit of clear legal guidelines or court policies, is this not allowing the "gatekeeper" to determine who should be detained? Moreover, doesn't this raise the issue of conflict of interest that was considered in *Gault* (see Chapter 5)?

due process The constitutionally guaranteed right to be treated with fundamental fairness by the law.

Another legal issue that should be addressed at the preadjudicative stage of the juvenile justice process concerns the setting of bail. Most states have few or no guidelines for setting bail in juvenile cases, and the lack of guidelines raises serious questions about the fairness or equity of the bail determination process.

A third legal issue concerns the validity of plea bargaining and sentence bargaining in juvenile courts. There are many arguments made by proponents and opponents of plea bargaining and **sentence bargaining**, but most fail to address the fundamental reason behind the use of these "tools." The law is not just or fair in itself and the operations of the "system" are not just or fair; only people can do justice. Importantly, plea and sentence bargaining introduces flexibility to the system and allows people to see that justice is done. This flexibility is necessary because justice can only be achieved on an individual and case-by-case basis by looking at the offender as a human being, by looking at the victim as a human being, and by then reaching a humane resolution. The other reasons for bargaining—that the system could not try all the cases and that prosecutors routinely overcharge—are good reasons for bargaining, but they are not the fundamental basis for plea bargaining.

sentence bargaining Negotiations between a juvenile, usually through his or her attorney, and the court over the punishment or other consequences that the juvenile will receive in return for a guilty plea.

Chapter Summary

This chapter examined the pretrial or preadjudicative process in juvenile court. This process includes a variety of hearings, conferences, and decision-making events whose purpose is to determine the best methods of handling juvenile court referrals. Although the hearings, conferences, and negotiations that take place at the preadjudication stage are less glamorous and less formal than the adjudication and the disposition, decisions made at this stage can have a profound effect on later court hearings. For example, during the intake screening process, a variety of critical decisions are made about cases referred to juvenile courts. These decisions determine which charges against the defendant will be authorized and whether the case will go forward to the adjudication and face the prospects of a formal juvenile court response, be transferred to an adult criminal court for trial, or be dismissed. These decisions are very important because they can affect the child's well-being and that of the community.

Another type of intake decision regularly made in juvenile courts concerns the detention of juvenile suspects pending further court action. Initial decisions to detain juveniles are typically made by intake officers, but state laws require that such decisions be reviewed in a detention hearing within a short time after the juvenile is detained (typically 24 to 72 hours). Similarly, preliminary hearings and arraignments, which are used in some jurisdictions, consider some of the same issues addressed in detention hearings. These hearings are primarily designed to determine if the youth plans to contest the charges contained in the complaint or referral, to take pleas from the youth, and to consider the appropriate placement of the youth prior to the adjudication. If the charges are being contested, then a date for the adjudication is often set. In instances where the charges are not contested, then a plea may be accepted at that time and a disposition entered by the court. These hearings play an important role in the processing of juvenile cases, because a variety of important decisions are made that can influence subsequent case processing. For example, it is at these hearings that juveniles may have an attorney appointed to represent them, where decisions about the appropriateness of filing a petition are made, and where decisions about bail (where it is used in juvenile cases) are rendered. Indeed, not only do decisions made at this stage of the juvenile court process determine restrictions on the youth's freedom, but they can increase the likelihood of more severe dispositions in subsequent hearings. Moreover, like other preadjudicative decisions, those made at detention hearings, preliminary hearings, and arraignments can affect the child in the community in positive or negative ways.

Because the preadjudicative stage of the juvenile justice process has such significant consequences, juveniles' rights must be protected during this stage. However, there is considerable evidence that some courts provide insufficient protections. For example, in some jurisdictions many juveniles are not represented by attorneys, and in other jurisdictions, attorney representation is inadequate during the preadjudicative phase of juvenile court processing. The lack of adequate attorney representation, of course, raises serious concerns about the quality of justice that juveniles receive. These are issues that must be taken seriously by everyone concerned with ensuring justice for juveniles.

Key Concepts

arraignment: The initial "coming before" the judge or judicial officer at which time the juvenile is informed of the charges pending and of his or her constitutional rights; a plea is entered; a trial date is set, if necessary; and bail is determined, if needed.

arrest: The deprivation of a person's freedom by legal authority; the apprehension of the juvenile by the police; the physical seizing of a person by the authorities for the purpose of making that person answer for a criminal charge.

bootstrapping: The process whereby a status offender is ordered to follow certain conditions set by the court (e.g., attend school, adhere to a curfew, and obey the commands of the parents) and the youth is redefined as a delinquent if he or she violates the court's orders. The youth's new status as a delinquent then allows the court to use detention as a response to violations of the court's orders.

case conference: A meeting attended by an intake officer, the juvenile, and the juvenile's parents that is designed to collect additional information to aid the intake officer in screening a case.

detention hearing: A hearing in front of a judicial officer that is required after a juvenile has been detained or in order to get a juvenile detained. At this hearing, the juvenile has a right to be present, have his or her parents or significant adult present, be informed of the charges, have his or her legal rights explained, make a plea, have a trial date set, and request bond (where permissible).

due process: The constitutionally guaranteed right to be treated with fundamental fairness by the law.

intake: The process of introducing youths who come to the police and court's attention into the juvenile justice system and determining how far they go into the system.

intake screening: The process of determining whether youths should be diverted from the juvenile system or referred for formal or informal services. Screening criteria usually include seriousness of the offense(s), family history, prior delinquency record or court contacts, legal mandates regarding diversion, attitude, age, school performance, parental cooperation, employment history, and restitution issues. Unfortunately, screening criteria also may include, despite their inappropriateness, race, gender, and socioeconomic status.

legal sufficiency: The requirement that an alleged criminal offense and the alleged perpetrator meet certain criteria before the alleged perpetrator can be charged with a criminal act. A prosecuting attorney, district attorney, or state's attorney assesses the criminal charges and the characteristics of the alleged perpetrator to ensure that the criteria are met.

petition: The request for formal court action that contains the written accusation against a juvenile, specifies the criminal or delinquent charges being made, and is properly verified as to age and venue by the complainant (the person bringing the charge).

plea: The juvenile's formal response to criminal charges.

plea bargain: An agreement between a prosecuting attorney and a juvenile, usually through the juvenile's attorney, according to which the juvenile admits to a lesser charge in exchange for the reduction of the charge or admits to some charges in exchange for the dismissal of other charges.

preliminary hearing: See detention hearing and arraignment.

preventive detention: Detention of an alleged offender prior to conviction so as to prevent the alleged offender from committing any crimes during the preadjudication period.

punitive model of juvenile justice: A model of juvenile justice that is intended to act as a counter to what are believed to be traditional "soft" approaches to juvenile crime and supports more severe sanctions for illegal behavior, particularly for more serious forms of delinquency. The punitive model of juvenile justice employs the same criteria used in charging adults, and the focus is on punishment rather than rehabilitation.

sentence bargaining: Negotiations between a juvenile, usually through his or her attorney, and the court over the punishment or other consequences that the juvenile will receive in return for a guilty plea.

Review Questions

1. What purpose does an intake unit serve in a juvenile court?
2. What are the typical factors used to determine whether a juvenile is referred for formal court action?
3. What factors may influence the intake decision? Which of these factors are appropriate to use?
4. How are case conferences used in the intake process?
5. Why are some jurisdictions using prosecuting attorneys to make intake decisions for the juvenile court?
6. What is a detention (or preliminary) hearing, and how does it fit into the intake process?
7. What is an arraignment, and how does it fit into the intake process?
8. Does plea bargaining exert a positive or negative influence on the intake process?
9. What legal rights must be observed by the court before a juvenile can be detained?
10. What is bail, and how does it work? What arguments could be made for and against the use of bail in juvenile justice?
11. What impact has *Schall v. Martin* had on the preadjudication process?
12. What is a sentence bargain?

Additional Readings

Feld, B. C. (1999). *Bad kids: Race and the transformation of the juvenile court.* New York: Oxford University Press.

Pope, C. E. & Feyerherm, W. H. (1990). Minority status and juvenile justice processing: An assessment of the research literature. *Criminal Justice Abstracts, 22,* 327–335.

Puritz, P., Burrell, S., Schwartz, R., Soler, M., & Warboys, L. (1995). *A call for justice: An assessment of access to counsel and quality of representation in delinquency proceedings.* Washington, DC: American Bar Association.

Sanborn, J. B., Jr. (1992). Pleading guilty in juvenile court: Minimal ado about something very important to young defendants. *Justice Quarterly, 9,* 126–150.

Notes

1. Black, H. C. (1968). *Black's law dictionary* (4th ed.). (p. 140) St. Paul, MN: West Publishing Co.
2. Lubow, B. & Barron, D. (2000). Resources for juvenile detention reform. *OJJDP Fact Sheet.* Washington, DC: Office of Juvenile Justice and Delinquency Prevention.
3. Snyder, H. N. & Sickmund, M. (1995). *Juvenile offenders and victims: A national report.* Washington, DC: Office of Juvenile Justice and Delinquency Prevention.

4. Feld, B. C. (2000). *Cases and materials on juvenile justice administration.* St. Paul, MN: West Group.

Rubin, H. T. (1980). The emerging prosecutor dominance of the juvenile court intake process. *Crime and Delinquency, 26*, 299–318.

5. National Conference of State Legislatures. (1993). *State Legislature Summary.* Washington, DC: National Conference of State Legislatures.

6. Sanborn, J. B., Jr. & Salerno, A. W. (2005). *The juvenile justice system: Law and process.* Los Angeles: Roxbury.

7. Juvenile code. Mich. Comp. Laws 712A.2(a)(1).

8. Schneider, A. L. & Schram, D. D. (1986). The Washington state juvenile justice system reform: A review of findings. *Criminal Justice Policy Review, 1*, 211–235.

9. California Rules of Court. *Court Rules.* Retrieved November 1, 2008, from http://www.courtinfo.ca.gov/rules.

10. Puzzanchera, C. & Kang, W. *Juvenile Court Statistics Databook.* Retrieved November 1, 2008 from http://ojjdp.ncjrs.gov/ojstatbb/jcsdb/.

11. Bell, D., Jr. & Lang, K. (1985). The intake dispositions of juvenile offenders. *Journal of Research in Crime and Delinquency, 22*, 309–328.

Cohen, L. E. & Kluegel, J. R. (1978). Determinants of juvenile court dispositions: Ascriptive and achieved factors in two metropolitan courts. *American Sociological Review, 43*, 162–176.

Fenwick, C. R. (1982). Juvenile court intake decision making: The importance of family affiliation. *Journal of Criminal Justice, 10*, 443–453.

Minor, K. I., Hartmann, D. J., & Terry, S. (1997). Predictors of juvenile court actions and recidivism. *Crime and Delinquency, 43*, 328–344.

McCarthy, B. R. & Smith, B. L. (1986). The conceptualization of discrimination in the juvenile justice process: The impact of administrative factors and screening decisions on juvenile court dispositions. *Criminology, 24*, 41–64.

12. Cohen & Kluegel, 1978.

Fenwick, 1982.

McCarthy & Smith, 1986.

13. Bell & Lang, 1985.

Fenwick, 1982.

14. Bell & Lang, 1985.

Cohen & Kluegel, 1978.

McCarthy & Smith, 1986.

Thornberry, T. P. (1979). Sentencing disparities in the juvenile justice system. *Journal of Criminal Law and Criminology, 70*, 164–171.

15. Thomas, C. W. & Sieverdes, C. M. (1975). Juvenile court intake: An analysis of discretionary decision-making. *Criminology, 12*, 413–432.

Bell & Lang, 1985.

McCarthy & Smith, 1986.

Thornberry, 1979.

16. Fagan, J., Slaughter, E., & Hartstone, E. (1987). Blind justice? The impact of race on the juvenile justice process. *Crime and Delinquency, 33*, 224–258.
17. Pope, C. E. (1995). Equity within the juvenile justice system: Directions for the future. In K. K. Leonard, C. E. Pope, and W. H. Feyerherm (Eds.), *Minorities in juvenile justice*. Thousand Oaks, CA: Sage.
18. Chesney-Lind, M. & Shelden, R. G. (2004) *Girls, delinquency and juvenile justice* (3rd ed.). (pp. 198–203). Belmont, CA: Thompson/Wadsworth. This resource reviews the literature.
19. Pope, C. E. & Feyerherm, W. H. (1990). Minority status and juvenile justice processing: An assessment of the research literature. *Criminal Justice Abstracts, 22*, 327–335.
20. Morris, A. (2004). Youth justice in New Zealand. In M. Tonry and A. N. Doob (Eds.), *Youth crime and youth justice: Comparative and cross-national perspectives* (Vol. 31). Chicago: University of Chicago Press.
21. Sickmund, M., Sladky, T. J., Kang, W., & Puzzanchera, C. (2008). *Easy access to the census of juveniles in residential placement*. Retrieved on November 4, 2008, from http://ojjdp.ncjrs.gov/ojstatbb/ezacjrp/.
22. Sickmund, M., Sladky, T. J., Kang, W., & Puzzanchera, C., 2008.
23. Devine, P., Coolbaugh, K., & Jenkins, S. (1998). Disproportionate minority confinement: Lessons learned from five states. *Juvenile Justice Bulletin*. Washington, DC: Office of Juvenile Justice and Delinquency Prevention. This reference cites the 1988 amendments to the Juvenile Justice and Delinquency Prevention Act of 1974.
24. Office of Juvenile Justice and Delinquency Prevention. (1999). Minorities in the juvenile justice system. *Juvenile Justice Bulletin: 1999 National Report Series*. Washington, DC: Office of Juvenile Justice and Delinquency Prevention.
25. Puzzanchera, C., Finnegan, T., & Kang, W. (2007). *Easy access to juvenile populations: 1990–2007*. Retrieved December 1, 2008, from http://www.ojjdp.ncjrs.gov/ojstatbb/ezapop/.

    Sickmund, M., Sladky, T. J., Kang, W., & Puzzanchera, C., 2008.
26. Pope, C. E., Lovell, R., & Hsia, H. M. (2002). Disproportionate minority confinement: A review of the research literature from 1989 through 2001. *Juvenile Justice Bulletin*. Washington, DC: Office of Juvenile Justice and Delinquency Prevention.
27. Snyder, H. N. & Sickmund, M. (2006). *Juvenile offenders and victims: 2006 national report*. Washington, DC: Office of Juvenile Justice and Delinquency Prevention.
28. Puzzanchera, Finnegan, & Kang, 2007. Data was extracted from this resource.
29. Austin, J., Krisberg, B., DeComo, R., Rudenstine, S., & Del Rosario, D. (1995). *Juveniles taken into custody: Fiscal year 1993, Statistics report*. Washington, DC: Office of Juvenile Justice and Delinquency Prevention.
30. Puzzanchera, Finnegan, & Kang, 2007. Data was extracted from this resource.
31. Mich. Comp. Laws 712A.15(3).

    Mich. Court Rules 5.935(D).

    West's California Welfare and Institutions Code, sec. 632.

Illinois S. H.A., chap. 37, para. 703-1.

Juvenile Justice and Delinquency Prevention Act of 1974.

Teitelbaum, L. E. (1980). *1977 IJA-ABA Joint Commission on Juvenile Justice Standards*. Cambridge, MA: Ballinger Publishing Company.

32. Snyder & Sickmund, 1995.

33. Puritz, P., Burrell, S., Schwartz, R., Soler, M., & Warboys, L. (1995). *A call for justice: An assessment of access to counsel and quality of representation in delinquency proceedings*. Washington, DC: American Bar Association.

34. Puritz, Burrell, Schwartz, Soler, & Warboys, 1995.

35. Puritz, Burrell, Schwartz, Soler, & Warboys, 1995.

36. Bailey, W. C. & Peterson, R. D. (1981). Legal versus extra-legal determinants of juvenile court dispositions. *Juvenile and Family Court Journal, 32*, 41–59.

Clarke, S. H. & Koch, G. G. (1980). Juvenile court: Therapy or crime control, and do lawyers make a difference? *Law and Society Review, 14*, 263–308.

37. Puritz, Burrell, Schwartz, Soler, & Warboys, 1995, p. 9.

38. Feld, B. C. (1988). In re Gault revisited: A cross-state comparison of the right to counsel in juvenile court. *Crime and Delinquency, 34*, 393–424.

Feld, B. C. (1991). Justice by geography: Urban, suburban, and rural variations in juvenile justice administration. *Journal of Criminal Law and Criminology, 82*, 156–210.

39. Burruss, G. W. Jr., & Kempf-Leonard, K. (2002). The questionable advantage of defense counsel in juvenile court. *Justice Quarterly, 19*, 37–68.

Clarke & Koch, 1980.

Feld, 1988.

Guevara, L. C., Spohn, C., & Herz, D. (2004). Race, legal representation, and juvenile justice: Issues and concerns. *Crime and Delinquency, 50*, 344-371.

40. Senna, J. J. & Siegel, L. J. (1992). *Juvenile law: Cases and comments* (2nd ed.). St. Paul, MN: West Publishing Co.

41. Feld, 2000.

42. Cox, S. M., Conrad, J., & Allen, J. (2002) *Juvenile justice: A guide to practice and theory* (5th ed.). New York: McGraw-Hill.

43. Rubin, H. T. (1985). *Juvenile justice: Policy, practice, and law* (2nd ed.). New York: Random House.

Rubin, H. T. (2003). *Juvenile justice: Policies, practices, and programs*. Kingston, NJ: Civic Research Institute.

44. Champion, D. J. (2007). *The juvenile justice system: Delinquency, processing and the law* (p. 153). Upper Saddle River, NJ: Pearson-Prentice Hall.

45. Champion, 1992.

46. Simonsen, C. E. (1991). *Juvenile justice in America* (3rd ed.). New York: Macmillan.

47. Michigan Judicial Institute. (n.d.). *Handbook of legal terms*. Lansing, MI: Michigan Judicial Institute.

48. Senna & Siegel, 1992.

49. Sanborn, J. B., Jr. (1992). Pleading guilty in juvenile court: Minimal ado about something very important to young defendants. *Justice Quarterly, 9*, 126–150.

50. Sanborn, 1992.

51. Sanborn, 1992.

52. Sanborn, 1992.

The Transfer of Juveniles to Criminal Court

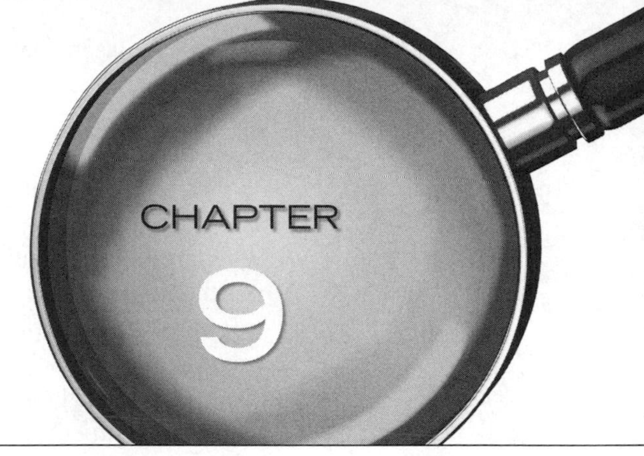

CHAPTER OBJECTIVES

After studying this chapter, you should be able to

- Describe what is meant by waiver or transfer of juvenile offenders from juvenile court to adult criminal court
- Describe the different methods by which youths are transferred or waived to the adult justice system
- Describe the general criteria that must be satisfied in order to transfer a juvenile to the adult justice system
- Present a detailed discussion of the reasons why transfer is used from the perspective of the community, the political and law-making bodies, and the juvenile and adult courts
- Describe recent trends in the use of transfer
- Present a detailed discussion of the legal and due process issues involved in transfer
- Describe how transferred juveniles are handled by the adult criminal courts
- Present arguments on the pros and cons of waiver

CHAPTER OUTLINE

- Introduction
- The Purpose of Transfer of Juveniles to Criminal Court
- Mechanisms for the Transfer of Juveniles to Criminal Court
- Trends in the Use of Transfer
- Due Process and Transfer Decisions
- Research on the Use and Effectiveness of Transfer
- Other Developments in the Use of Transfer
- Standards Governing Transfer Decisions
- Correctional Programming for Juveniles Convicted in Criminal Court
- Legal Issues
- Chapter Summary
- Key Concepts
- Review Questions
- Additional Readings
- Cases Cited
- Notes

Introduction

With the establishment of the first formally organized juvenile court in Illinois in 1899 and the spread of juvenile courts to other states, jurisdiction over juveniles charged with law violations shifted from the criminal courts to the juvenile courts. However, many people believed that in certain circumstances, juveniles who committed serious offenses might be more appropriately dealt with by criminal (or adult) courts. Consequently, most early juvenile codes contained provisions for transferring some types of juvenile cases to criminal courts, although the criteria used to make such decisions were often vague and subjective.[1]

Today, all states and the District of Columbia make it possible for some juveniles to be tried in criminal courts. Moreover, most states have multiple ways by which some juveniles can be given adult sanctions.[2] The process by which juveniles are bound over to criminal courts for trial is called **waiver**, *certification*, *remand*, *bindover*, or *transfer*. The use of transfer has increased dramatically in recent years as more policy makers have adopted a "get tough" approach in dealing with juvenile crime. However, the decision to try juveniles in criminal courts frequently appears to be a political and emotional decision that critics claim receives little support from research on the effectiveness of transfer.

waiver The process by which a juvenile is moved from juvenile court jurisdiction to adult criminal court jurisdiction for an offense committed while still a juvenile. Also known as bindover, certification, remand, and transfer.

Regardless of the reasons for the transfer of a juvenile case to a criminal court, the decision to go forward with the transfer can have profound consequences for youths who are subject to prosecution in adult courts, their family and friends, and the community. Historically, juvenile courts have attempted to shield youths from the negative consequences associated with being prosecuted in open criminal courts and being exposed to more mature and hardened offenders in the adult criminal justice process. Furthermore, juvenile courts have sought to offer youths "protection against loss of civil rights, against disqualification for public employment, and against personal status degradation and restriction of legitimate opportunities that often follow criminal conviction."[3] After a youth is transferred to a criminal court, however, these protections are lost, which can have important long-term consequences for the juvenile and his or her significant others. Moreover, if the transfer decision fails to influence the juvenile in ways that reduce the likelihood of recidivism, then community safety is also jeopardized.

The option of transferring juveniles to criminal courts raises important questions about the best response to adolescent offenders, particularly those who have committed serious offenses. In particular, to what extent does the transfer of juveniles to criminal courts benefit the youths themselves or the communities most will return to at

COMPARATIVE FOCUS

Many Countries Have a Mechanism Whereby Some Juveniles may be Prosecuted in Adult or Criminal Courts

For example, England, Scotland, Canada, the Netherlands, New Zealand, Belgium, and Japan make it possible to prosecute certain juvenile cases, such as those involving very serious crimes, in adult courts. However, the ages at which transfer can occur and the types of offenses and other criteria that must be considered prior to transfer vary from country to country. In addition, some countries, such as Denmark and Germany, do not have specialized juvenile justice systems, but youths in those countries are treated more leniently than adults in criminal courts.[4]

some point? This question invariably arises in discussions of transfer, and it needs to be carefully addressed if effective policies for dealing with serious juvenile crime are to be developed.

This chapter examines the use and consequences of transfer. Specifically, it examines the purpose of transfer, different mechanisms by which juveniles may be transferred to criminal courts for trial, the legal requirements for the transfer of juveniles to criminal courts, trends in the use of transfer, and the effectiveness of criminal court handling of juveniles. It concludes by examining a number of problems associated with the transfer of juveniles to criminal courts.

The Purpose of Transfer of Juveniles to Criminal Court

Transfer occurs when jurisdiction over a juvenile case is turned over to a criminal court. The waiver or transfer of jurisdiction from juvenile court to criminal court is predicated on the assumption that some juveniles are not appropriate for processing in juvenile court and can be more effectively dealt with by criminal courts. In essence, it is believed that although juveniles, as a class, are treatable and therefore appropriate for juvenile court handling, particular juveniles, by virtue of their actions, do not merit the protected status given to youths in the juvenile court. Instead, protection of the community from such youths demands that they be identified and transferred to the adult criminal justice system.[5] After a juvenile is transferred to a criminal court for trial, the juvenile legally becomes an adult and is subject to the same types of correctional responses as any other adult, except for the death penalty in those jurisdictions with capital punishment statutes. The death penalty for persons younger than18 years of age was ruled unconstitutional by the U. S. Supreme Court in *Roper v. Simmons* in 2005.[6]

In order to determine which juveniles are appropriate for criminal court jurisdiction, states have established various criteria. Typically, the juvenile has to meet certain age and offense criteria. For example, in Michigan, under its traditional waiver statute, a juvenile has to be at least 14 years of age (the maximum age of original juvenile court jurisdiction in Michigan is 17 years) and must be charged with a felony offense before he or she can be waived to criminal court. In addition, a juvenile court hearing must be held to determine if probable cause exists and if transfer of the case to adult court is appropriate.

FYI

Until 2005, Juveniles Could Receive the Death Penalty in the United States

The execution of people younger than 18 years of age is not unprecedented in the United States. Since colonial times, there have been at least 366 executions of juveniles in America, and 22 were carried out between 1973 and 2003.[7] However, the constitutionality of capital punishment for juveniles began to be challenged in the late 1980s, when the United States Supreme Court heard two landmark cases, *Thompson v. Oklahoma* (1988) and *Stanford v. Kentucky* (1989). The effect of these two cases was to ban the execution of youths younger than age 16 years, but permit the execution of those who committed capital offenses when they were age 16 years or older.[8] However, in an important 2005 case, *Roper v. Simmons*, the U. S. Supreme Court declared unconstitutional executions of people younger than 18 years of age.[9] Until that time, the United States was one of the few countries that permitted the execution of persons younger than 18 years of age.[10]

COMPARATIVE FOCUS

Several Countries Still Execute Youths Accused of Committing Crimes While They Were Younger than 18 Years

For example, in 2007, three countries, Saudi Arabia, Yemen, and Iran, executed persons who committed offenses when they were younger than 18 years old; in 2008, Iran executed two juveniles who were younger than 18 years of age—one juvenile was 16 years old when he was hung. Indeed, in 2008, there were as many as 140 juveniles on death row in Iran. These executions occurred even though Iran has signed the International Convention on Civil and Political Rights and the Convention on the Rights of the Child (CRC), which prohibit the execution of persons younger than 18 years of age who commit offences.[11]

■ Mechanisms for the Transfer of Juveniles to Criminal Court

judicial waiver
A type of waiver selected by a juvenile court at a hearing. It is the traditional type of waiver.

prosecutorial waiver
A type of waiver that is within the power of the local prosecutor or state attorney to choose and implement.

legislative waiver
A type of waiver that results when certain threshold criteria specified by state laws are met.

statutory exclusion
See legislative waiver.

Traditionally, the decision to transfer a case to criminal court was made in a juvenile court hearing. Today, however, there are several mechanisms (i.e., **judicial waiver, prosecutorial waiver**, and **legislative waiver** or **statutory exclusion**) by which juvenile cases are transferred to adult courts. Indeed, most jurisdictions have more than one mechanism by which the transfer or waiver of a juvenile case can be achieved. For example, in Arkansas, Colorado, the District of Columbia, Michigan, Virginia, and Wyoming, both judicial and prosecutorial waiver mechanisms are in place. In other states (Arizona, California, Florida, Georgia, Louisiana, Oklahoma, and Vermont) all three mechanisms exist.[12]

In most states (Connecticut, Massachusetts, Montana, Nebraska, New Mexico, and New York are exceptions), juveniles can be transferred to an adult criminal court by means of a separate waiver or transfer hearing in the juvenile court to determine the appropriateness of the waiver.[13] This method of waiver, which can be called *judicial waiver*, is the traditional method by which juvenile cases have been transferred to criminal courts. However, the exact process by which judicial waiver occurs and the way in which the waiver hearing in juvenile court is conducted varies from state to state. Nevertheless, these waiver hearings typically encompass four elements: (1) a determination of probable cause that the accused juvenile committed the crime(s) charged, (2) a consideration of the potential threat the youth presents to the community, (3) an evaluation of the extent to which the offender is amenable to treatment by existing juvenile justice programs, and (4) a consideration of the types of programs in the adult system that might better meet the juvenile's needs. In determining the potential threat a youth poses to the community, the court considers the seriousness of the offense and the juvenile's offense history. In evaluating the extent to which a youth is amenable to treatment, the results of previous juvenile court dispositions and information about existing juvenile justice and adult corrections programs may be examined.

A potential problem with basing waiver decisions on a youth's amenability to treatment in juvenile justice programs or adult programs is that such determinations can be

FYI

In 2005 it is estimated that 6,900 youths were transferred to adult court through a judicial waiver.[14]

FYI

Recent Changes in Transfer Laws Make It Easier to Try Some Juveniles as Adults
Since 1992, every state except Nebraska has changed its laws regarding transfer to make it easier to try juveniles as adults.[15]

highly subjective. Even if a youth has re-offended after participating in previous juvenile justice programs, how can one determine that other juvenile justice programs employing different interventions would not be successful? Furthermore, how can one determine the effectiveness of adult programs for juveniles prior to their placement? Such determinations cannot be made based on empirical evidence. Instead, they are predicated on predictions—some of which will invariably be wrong—that judges, attorneys, probation officers, and others make about individual cases.

The initiation of a waiver hearing in juvenile court usually occurs when a prosecuting attorney, believing that the transfer of a case to criminal court is the appropriate course of action, requests such a hearing. The basis of the request may be the violence of the crimes committed, but prosecuting attorneys also seek transfer for nonviolent or even minor acts of delinquency. In a few states, juveniles or their parents may request a waiver hearing.[16] One reason a juvenile may ask for a waiver is to ensure that they receive due process protections, which may not always be available in juvenile courts.

Since the 1970s, a number of states have developed additional mechanisms for transferring juvenile cases to criminal courts. One of these is referred to as *legislative waiver*. Some states have passed laws that indicate that youths who meet certain age and/or offense criteria will automatically be tried as adults, thus bypassing the juvenile justice process altogether. For example, in Georgia, Illinois, New York, and Oklahoma, juveniles who are at least 13 years old and are charged with murder are automatically tried in criminal courts. In Mississippi, juveniles who are least 13 years old and are charged with certain felonies are automatically tried in criminal courts. In other states, juveniles who have reached a certain age and have prior felony adjudications can be excluded from juvenile court jurisdiction.[17]

There is another way in which legislatures have excluded a large number of juvenile cases from juvenile court jurisdiction. When state laws establish an upper age limit for juvenile court jurisdiction of 15 or 16 years of age, they, in effect, transfer cases to criminal court jurisdiction. Indeed, data collected by the National Center for Juvenile Justice indicate that as many as 218,000 youths younger than the age of 18 years were tried in criminal court in 1996 because they were defined as adults under state laws. Unfortunately, there are no national data on the number of juveniles who are transferred to criminal courts by means of these different

FYI

Typically, Juvenile Courts Are Less Formal than Adult Courts
As noted in Chapter 5, juveniles who are processed in juvenile courts still do not enjoy all of the due process protections afforded to adults in criminal trials. Indeed, many juvenile courts still operate in an informal manner that would not be acceptable in criminal courts.

legislative waiver mechanisms.[18] However, because almost 2 million youths who are 16 or 17 years old live in states where the upper age limit of juvenile court jurisdiction is 15 or 16 years of age, a large number of youths are at risk of prosecution in criminal courts because they are defined as adults under their state laws.[19]

concurrent jurisdiction Both the juvenile and adult court share original jurisdiction. Consequently, in states with concurrent jurisdiction, prosecutors are given discretion to file cases that meet certain age and offense criteria in juvenile or criminal court.

A third method of waiver that has become possible in some jurisdictions is *prosecutorial waiver*. As of 2005, 14 states and the District of Columbia had passed legislation that makes **concurrent jurisdiction** possible in some cases. When concurrent jurisdiction exists, the prosecutor is given discretion to file certain cases directly in the adult court when a youth meets certain age and offense criteria.[20] For example, in Florida, the prosecutor has discretion to file a felony case in either the juvenile or adult court when the alleged juvenile offender has reached 16 years of age. In these instances, original jurisdiction is shared by both the juvenile and criminal courts. In Michigan, prosecutors can automatically waive juveniles at the age of 14 years if they commit specified felonies. For these particular crimes, the prosecuting attorney can choose to file a petition in juvenile court or file a complaint and warrant in adult court. Examples include the following specified juvenile violations:

- arson of a dwelling
- assault with intent to commit murder
- assault with intent to maim
- armed assault with intent to rob
- attempted murder
- first-degree murder
- second-degree murder
- kidnapping
- first-degree criminal sexual conduct
- armed robbery
- car jacking
- bank or safe robbery
- delivery or manufacture of cocaine or a narcotic drug (650 grams or more)
- possession of cocaine or a narcotic drug (650 grams or more)
- first-degree home invasion with a dangerous weapon
- assault with intent to do great bodily harm with a dangerous weapon
- escape from certain juvenile facilities[21]

These newer waiver statutes are supported by legislators who think the juvenile courts are "soft" on youth crime and are intended to circumvent the perceived leniency of the juvenile courts. The existence of different perspectives on juvenile crime is not surprising when one considers that different government entities have different roles and responsibilities, and, to some extent, they have different constituencies they must satisfy. For example, the executive branch is concerned about social control; the legislature is concerned about being perceived by the community as tough on crime; while courts have to balance community protection, fairness, rehabilitation, and due process. As a result, some see automatic waiver as a tough practical response to serious juvenile crime. Others, however, view it as a politically motivated response that wins votes and upsets the delicate balance of powers among the executive, legislative, and judicial branches of government by usurping a role traditionally played by the courts. To see one judge's perspective on dealing with juveniles, including waiving juveniles to adult court, see Box 9-1.

BOX 9-1 Interview: The Honorable Carolyn Williams, Presiding Judge, Family Division, Kalamazoo County Circuit Court

Q: How long have you been on the bench, and what was your career prior to taking the bench?

A: I have served as a juvenile court and family court judge for 18 years. Prior to my taking the bench, I was in private law practice with an emphasis on personal injury defense. This was my second experience in private practice. Early in my career, I worked for two years doing criminal defense and plaintiff's personal injury representation. In between the private practice experiences, I worked for the Department of Social Services as an administrative law judge for 10 years and took time to be a mother to my two sons, who are now grown.

Q: What is your educational background?

A: I have a BA in political science with an emphasis on international relations from George Washington University in 1964; a J. D. from George Washington University in 1968 that I earned with honors. I am also a graduate of the National Judicial College, where I have also served as a faculty member.

Q: What are the parts of your job that you find particularly rewarding?

A: I particularly enjoy those cases in which the parent and/or child have made positive changes that will enable them to be successful. To have a part in reaching people and helping them to change or, when necessary, motivating them to change is very rewarding.

Q: What are the major challenges that you see the juvenile justice system facing today?

A: I see two major challenges: out of control children, and the increasing public hostility toward all children due to this small group that is out of control. Society must recognize the causes for these children being out of control— changing family structure, substance abuse, and domestic violence. Society needs to accept its share of responsibility for this and work toward dealing with the causes. Instead, it seems that the only response by many politicians and the media is open hostility. Society seems to want to focus on punitive sanctions rather than rehabilitative solutions. In an enlightened society, I would hope that we would direct our resources toward early intervention, not toward detention.

Q: As a concept or ideal, do you believe in rehabilitation?

A: Yes, absolutely. It is what sets the juvenile/family court apart from the adult criminal system. If there is no rehabilitation, there is no hope. Interestingly, over the past 8 years I have noted an increase in the use of principles of "therapeutic jurisprudence" in both the adult and juvenile justice systems. Michigan now funds specialized courts, which focus on helping adults, juveniles, and families deal with drug and alcohol dependence. Other states have created courts, which focus on assuring compliance with mental health problems of those people involved with the criminal justice system. Communities across the country are also taking responsibility for helping to assure community safety and offender accountability using principles of balanced and restorative justice. Drug courts, therapeutic jurisprudence, [and] restorative justice are all ideas and principles that the juvenile courts have traditionally engaged in even though they may not have been identified using those words or phrases. It is truly gratifying to have been around long enough to see that those principles are now being recognized as important to the adult criminal justice system as they have always been in dealing with juveniles. As I said before, and it is as applicable to adults as it is with juveniles, any system that simply locks people away and offers them no hope, no hope of changing their behavior, or re-integrating into society, cannot in my opinion be justified or cost effective. Without rehabilitation there is no hope!

Q: Focusing on the issue of transfer or waiver, do you believe that waiver of 13 and 14 year olds, in general, is a good idea?

A: No, regardless of whatever awful act the child committed, to face a lengthy incarceration in the adult system at that age is "throwing them away" far too soon. Children at that age are impulsive. They are not adults; they don't think like adults; and, for the most part, they don't act like adults. Society needs to focus on giving children chances, not taking them away.

(continues)

Q: What is your opinion of the concept of waiver or transfer?

A: Waiver, plain and simple, is giving up on kids!

Q: What do you think about the "criminal track" alternative sentencing that exists in Michigan?

A: It, as with many tools, has benefits and drawbacks. I'm sure that for some youths, the potential of facing an adult sentence can be an incentive to work harder in the juvenile system. On the other hand, the fact that it, unlike waiver, has no lower age limit allows for very young children to face adult criminal punishment. This gives validity to the popular cry for punishment and retribution and focuses the public away from the more important focus of rehabilitation.

Q: What kinds of programming or services do you see serious delinquent juveniles needing?

A: This is not my area of expertise; but, in my experience, there are some common denominators with children. At all and any level of treatment, education must be a priority, with dollars spent on technological resources so that the children are receiving a marketable education. Second of all, in view of all the dysfunctional families we deal with, alternative families, foster care homes, and especially group homes are needed. Not only can adults in these homes model healthy relationships, but in group homes, many children feel more comfortable in not having to identify the foster parents directly in competition with their own.

Q: How do you think society should deal with serious and violent delinquents?

A: Early intervention is key. We must focus resources on parenting education, giving young people reasons to defer becoming parents, and on preventing those factors in homes that contribute to violent kids and serious delinquents—poverty, substance abuse, and domestic violence. At the other end of the spectrum, we need secure and therapeutic placements for those delinquents who must be separated from society.

Q: Do you think that separate facilities should be built to house these serious delinquents?

A: I have no objection to this concept if the goal is rehabilitation with the focus on education, substance abuse prevention, resocialization, life skills, etc. I would strongly object to any placement where a child was simply warehoused!

Q: What advice would you give to the politicians who are continually calling for stiffer penalties for juvenile offenders?

A: First, I would challenge them to promote community awareness of the underlying causes of delinquency rather than settling for the incarceration alternative. They need to get away from the political rhetoric and the "sound bite" and work at being proactive in their communities. Money needs to be spent on at-risk children at the front end, in their formative years, so that we can avoid the unsocialized, out-of-control juveniles who cause such public outcry. I would recommend a close study of the European child-care systems as models for early intervention and socialization.

Q: During your career, you have received a number of honors and awards for your work with delinquents and their families. Could you tell me about these?

A: I have received several awards, including Omega Psi Phi (Upsilon Pi Chapter) William H. Hastie Award for Significant Contributions in Law (1997); Service to Children Award of the Guidance Clinic (1993); the Kalamazoo Rotary Red Rose Citation (1997); the Glass Ceiling Award of Greater Kalamazoo Network (1997); [and] the YWCA Woman of Achievement Award (1998).

There are no comprehensive national data on the number of juvenile cases that are subject to prosecutorial waiver each year. Nevertheless, the number of such cases appears to outstrip the number that is judicially waived. In Florida, where prosecutors are given wide discretion in deciding where to file cases, nearly 5,000 juvenile cases a year were transferred to criminal courts via prosecutorial transfer during the 1990s.[22] Moreover, there are indications that in states that have both judicial and prosecutorial waiver mechanisms, prosecutorial transfer accounts for much of the growth in waiver.

For example, in 1981, Florida, which has both judicial and prosecutorial transfer mechanisms, filed two cases directly in criminal courts for every case that was judicially waived. However, by 1992, for every case that was judicially waived in Florida, more than six cases were transferred to criminal courts through prosecutorial waiver.[23]

Trends in the Use of Transfer

Since the mid-1970s, state legislatures have taken steps to expedite the transfer of juveniles to criminal courts.[24] Typically, states have made it easier to transfer youths to criminal court by taking the following actions: (1) enacting statutory exclusions, (2) lowering the minimum age for waiving youths, (3) expanding the range of offenses that can result in transfer, or (4) making judicial waiver *presumptive*[25] (i.e., where transfer for certain cases is assumed unless there are clear grounds for handling the case differently, and the burden of proof for keeping the case in the juvenile court falls on the juvenile). By 1999, 29 states had passed legislative transfer (statutory exclusion) statutes that exclude some offenses from juvenile court jurisdiction. Moreover, other states have expanded the number of offenses excluded from juvenile court jurisdiction. In 15 states, prosecutors have become more prominent players in transfer decisions because of the enactment of prosecutorial transfer (concurrent jurisdiction) laws that allow them to decide whether to file certain cases in the juvenile court or the criminal court.[26] As a result, there are several ways that juveniles can be waived to criminal court in some states.[27] Finally, some states have simplified the judicial waiver process so that it is much easier to transfer some cases to adult court for trial.

Historically, transfer to criminal court was a judicial decision, and the burden was on the prosecution to demonstrate that a youth was not amenable to treatment in the juvenile justice system and that the adult system held more promise. In some states, statutes (or court rule in Arizona) specify that certain cases are appropriate for waiver even when it is a judicial decision, and the burden is now on the defense to demonstrate that a juvenile program is more suitable for the youth and that the adult system is inappropriate.

As a result of the increased emphasis placed on "getting tough" with juvenile crime in some jurisdictions and the enactment of laws making it easier to transfer juvenile cases to adult criminal courts, there has been a substantial increase in the number of juveniles tried in criminal courts since the 1970s.[28] Indeed, between 1985 and 1994, the number of cases judicially waived to adult courts increased by 71%, from 7,200 to 12,300 cases,[29] although it declined to an estimated 6,900 cases in 2005.[30] However, although the number of judicially transferred cases has been declining recently, far more juveniles were tried in adult courts as a result of legislative and prosecutorial waiver statutes. Indeed, a study of waivers in 18 jurisdictions found that 85% of the decisions to try youths as

FYI

A Substantial Number of Children are Prosecuted in Adult Courts Each Year

As previously noted, comprehensive data on the number of persons younger than 18 years of age who are tried in adult court each year is not available.[32] Nevertheless, a study conducted by Amnesty International in the late 1990s estimated that more than 200,000 juveniles were being tried as adults each year.[33]

FYI

A Lack of Data on Youth Ethnicity Is a Problem

A lack of data on the ethnicity of youths handled by juvenile courts makes it impossible to access the differential treatment of some ethnic groups in the juvenile justice process. Moreover, it should be kept in mind that terms like "Hispanic" and "Asian" are used to refer to a wide variety of ethnically diverse groups and may mask how sub-populations within these broader categories of persons are treated.

adults were not made by judges, but resulted from statutory exclusion or prosecutorial discretion.[31] Thus, estimating the number of juveniles being tried as adults by examining only those youths who are transferred by juvenile courts through judicial waiver grossly underestimates the number of youths being tried in criminal courts.

Not only has there been a substantial change in the number of juvenile cases transferred to criminal courts, but the types of cases transferred to criminal courts, at least via judicial transfer, have also changed over time. In 1985, property offense cases were the most likely to be judicially transferred, but by 1993, the largest group of cases transferred involved offenses against people, a trend that continued through at least 2005.[34]

While the profile of cases being transferred by juvenile courts has changed over time, the gender and racial characteristics of youths who are waived has not. Presently, as in the past, youths who are transferred to adult courts are disproportionately male and they are disproportionately members of minority groups. For example, in 2005, males made up 91.5% of the cases judicially waived to adult courts. Also, African American youths accounted for 38.6% of the persons younger than 18 years of age who were waived to adult court, but they comprised only 16.4% of the population younger than 18 years of age. In addition, Native American youths accounted for 2.7% of the cases waived to adult court, but they made up only 1.3% of the population younger than the age of 18 years.[35]

Due Process and Transfer Decisions

As discussed in Chapter 5, the first juvenile court case that drew U.S. Supreme Court attention, *Kent v. United States*, addressed the issue of transfer. In this case, Morris Kent, a juvenile, was transferred to adult criminal court without a hearing or any formal notice. In its opinion, the Court set forth minimum due process standards for a valid waiver:

- Any juvenile facing waiver is entitled to an attorney.
- The juvenile is entitled to a meaningful hearing, even if it is informal and off the record.
- The juvenile is entitled to access any reports, records, and so on, used by the court in deciding waiver.
- The juvenile is entitled to know the reasons for the waiver decision.

The first requirement speaks for itself. A meaningful hearing is one in which the juvenile can present evidence, challenge evidence against him- or herself, and do this all before an impartial decision maker. This requirement is consistent with the court allowing the juvenile to challenge reports or other information that purport to be the basis for waiver. However, the Supreme Court did not require the hearings to be "on the record," and thus transcripts that indicate in detail what went on in the waiver process may not be available. Nevertheless, the court did require a "statement of reasons" for waiver, and this could form the factual basis for an appeal.

Due to the serious nature of waiver cases, it is not surprising that the Supreme Court has heard other cases that involve this issue. One of these was *Breed v. Jones* (1975).[36] In this case, a 17-year-old, accused of armed robbery, was adjudicated to be a delinquent at an adjudication hearing in juvenile court. Two weeks later, but before deciding on any sentence or disposition, the juvenile court determined at a "fitness" hearing that the juvenile was not amenable to juvenile court "care, treatment, and training," and ordered the juvenile to be prosecuted as an adult. At the subsequent adult trial, the juvenile was convicted of first-degree robbery.

In its decision, the Supreme Court held that the juvenile had been "put in jeopardy" at the original juvenile court adjudicatory hearing and that, once found within the jurisdiction of the juvenile court, he could not then be tried as an adult for the same offense. To try him as an adult would be a violation his right to be free from **double jeopardy**. *Breed v. Jones* had several ramifications. First, and most important, the Court required the state to choose waiver "up front" in the process, prior to any juvenile court adjudication, denying the state multiple opportunities to try juveniles. Second, the Supreme Court recognized the seriousness of a juvenile court adjudication and gave it the same status afforded criminal trials for the purposes of double jeopardy protection. Third, the decision emphasized the importance of the transfer or waiver hearing in the juvenile court process because these hearings determine the possible sanctions that may be imposed on juveniles.

double jeopardy
As applicable to transfer cases, a principle that requires the state authority to choose at the outset whether a juvenile is to be tried as a juvenile or as an adult. The choice must be made at the preliminary hearing or arraignment stage of the proceedings.

Research on the Use and Effectiveness of Transfer

Despite the importance of transfer and its long history of use in juvenile justice, there has been relatively little research devoted to the way transfer is used in different jurisdictions around the country and the effectiveness of transfer decisions.

The Use of Transfer

Although there is much about transfer that is not known, there is clear evidence that the use of transfer varies considerably across jurisdictions. Some jurisdictions are much more likely to transfer juveniles to adult courts than other jurisdictions. A study by Barry Feld found that juvenile court judges exercised considerable discretion in making transfer decisions and did not administer transfer statutes in an evenhanded manner. According to Feld, "Within a single jurisdiction, waiver statutes are inconsistently interpreted and applied from county to county and from court to court."[37] Another study, done by Tammy Poulos and Stan Orchowsky, examined transfers in Virginia and found that juvenile offenders in metropolitan areas were less likely to be transferred than those whose cases were processed in nonmetropolitan juvenile courts. Although the authors were not able to provide a definitive explanation for their findings, they noted the following:

> Juvenile court judges serving metropolitan jurisdictions may be less likely to send young offenders to the criminal courts for a number of reasons. Because they see so many serious offenders, their threshold for defining an offense as serious enough to warrant transfer may be higher than that of their rural counterparts. On the other hand, metropolitan judges may have at their disposal more dispositional options at the juvenile court level than their rural counterparts and thus rely less heavily on the last resort of transfer.[38]

Although one might expect that the level of serious crime in a jurisdiction has a direct influence on the number of juvenile cases waived to criminal courts, research indicates that the use of transfer is determined by a variety of factors. Likely candidates for transfer include juveniles who commit serious offenses such as murder, manslaughter, and rape; juveniles who sell drugs; older juveniles who use firearms in the commission of crimes and have prior records; and juveniles with lengthy juvenile court histories.[39] A study that examined waiver in Maricopa County, Arizona, found that previously being waived was the strongest predictor of transfer.[40]

There is evidence that the increased use of transfer in some jurisdictions is a reflection of the growing popularity of a more punitive model of juvenile justice reflected in the desire of political decision makers to appear tough on juvenile crime, rather than a direct response to higher levels of serious juvenile offending. After examining the case histories of 214 juveniles transferred to criminal court and interviewing key juvenile justice decision makers, M. A. Bortner concluded that organizational and political factors accounted for high rates of transfer. Bortner stated the following:

> Although remand represents a course of action ostensibly directed toward protection of the public, it is a policy integrally related to the juvenile system's interest in organizational maintenance.... Remand protects institutional authority by removing cases from juvenile court responsibility, thus reducing potential organizational troubles and public criticism that might damage institutional legitimacy.... In an era of fiscal uncertainty when the juvenile justice system is confronted with the necessity of reasserting its worth, maintaining its uniqueness, and redefining its mission, remanded juveniles provide a symbolic avenue for the reaffirmation of the juvenile system's commitment to public safety and rehabilitation for most juveniles. In evidencing a willingness to relinquish its jurisdiction over a small percentage of its clientele, and by portraying these juveniles as the most intractable and the greatest threat to public safety, the juvenile justice system creates an effective public gesture of retribution and punishment in the name of responsiveness to public concerns....
>
> The present analysis suggests that political and organizational factors, rather than concern for public safety, account for the increasing rate of remand.[41]

Although transfer is believed by many to be reserved for only the most serious juvenile offenders, in many jurisdictions a large percentage of the cases waived involve nonviolent offenses. Moreover, waiver cases brought against some juveniles are not very strong.[42] As noted earlier, property offense cases accounted for the greatest percentage of cases judicially waived in 1985, but by 1999, offenses against persons were the most

FYI

Efforts to "Get Tough" with Juveniles Continue

As noted in Chapter 2, there is no evidence that juvenile crime has become significantly more serious over time. Indeed, arrests of youths younger than 18 years of age for Index violent offenses have been declining since 1994. However, efforts to develop more punitive responses to juvenile crime, including the development of additional mechanisms for waiving juveniles to criminal courts, began in the mid-1970s and continue today.

common type of charge levied against youths who were subject to judicial waiver. In 2005, offenses against persons were the predominant charge in slightly over half of the cases that were judicially transferred. Nevertheless, property offenses accounted for 27.1% of the cases that were transferred, drug offenses accounted for 12.1% and public order offenses accounted for 10.3%.[43] The offense profiles of cases transferred to criminal court for selected years from 1985 through 2005 can be seen in Table 9-1.

Much less is known about the types of cases that are being waived under prosecutorial and legislative transfer laws than about judicially transferred cases. What is known is that the prosecutorial and legislative transfer statutes enacted in many states have resulted in more youths being subjected to criminal court processing.[44] In addition, there is research indicating that, at least in some jurisdictions, a large percentage of those transferred by prosecutorial transfer are nonviolent offenders. In a study of felony cases from a nationally representative sample of 300 counties, some of which were transferred by prosecutorial and legislative waiver statutes, it was discovered that only 53% of the cases transferred to and convicted in adult court involved violent offenses. In this study, 24% of criminal court convictions involved property offenses, 13% were drug offenses, and 10% involved other offenses.[45] In another study of prosecutorial waiver in two Florida counties, Donna Bishop and her colleagues found that, despite the fact that prosecutors indicated that the youths who were transferred were dangerous offenders, only 29% had committed a violent felony.[46] In fact, the majority (55%) had committed property felonies. Moreover, Bishop and her colleagues found that the tendency had been for prosecutors to transfer greater proportions of nonviolent offenders, particularly youths charged with felony drug offenses, and misdemeanants over time. After examining the characteristics of juveniles transferred in the two counties they studied, Bishop and her colleagues concluded that "few of the juveniles transferred to criminal court via prosecutorial waiver would seem to be the kinds of dangerous offenders for whom transfer is most easily justified."[47]

The Effectiveness of Transfer

A common argument supporting the transfer of juveniles to criminal courts is that it serves as a tougher response to serious juvenile offenders, and it is more likely to protect community safety than juvenile court processing. As noted earlier, however, the juveniles who

TABLE 9-1 Percent of Delinquency Cases Transferred to Criminal Court by Judicial Waiver, Selected Years, 1985–2005.

	Years			
Offense	**1985**	**1990**	**1999**	**2005**
Person	2,394 (33.3%)	2,897 (34.0%)	3,714 (42.2%)	3,480 (50.5%)
Property	3,807 (52.9%)	3,747 (43.9%)	2,823 (32.1%)	1,866 (27.1%)
Drugs	353 (4.9%)	1,218 (14.3%)	1,374 (15.6%)	831 (12.1%)
Public order	643 (8.9%)	668 (7.8%)	894 (10.1%)	708 (10.3%)
Totals	7,197	8,530	8,807	6,885

Note: Percentages rounded.

Source: Data from Sickmund, M., Sladky, A., and Kang, W. (2005). *Easy access to juvenile court statistics, 1985–2005.* Retrieved November 10, 2008, from http://ojjdp.ncjrs.gov/ojstatbb/ezajcs.

MYTH VS REALITY

Youths Who are Transferred Have Not Always Committed Serious Offenses

Myth—Transfer to criminal court is reserved for only the most serious juvenile offenders.

Reality—Older youths, youths who commit serious offenses, and those with previous juvenile court histories are more likely to be waived to adult courts.[48] There is also evidence that minority youths are disproportionately represented in transfers to adult court.[49] However, it is important to note that transfer is frequently used for youths who have committed property offenses and drug offenses, and it is sometimes used for juveniles who have committed public order offenses. In fact, transfer may have more to do with political and/or economic reasons than the usual reason given—protection of the community. Furthermore, juvenile courts, frustrated with juvenile offenders who resist rehabilitation, may opt for the waiver of these youths even though they have not committed serious offenses.

are transferred to criminal courts are not always serious or violent offenders. This section examines the extent to which transfer is an effective way to protect community safety.

Early research on transfer indicated that when juvenile offenders were transferred to criminal court, they were often treated leniently. One suggested explanation for the lenience was that the juveniles who appeared in adult courts were often young and were seen as first-time offenders.[50] In a 1978 national survey of the transfer of juveniles to criminal courts, Donna Hamparian and her associates found that only 46% of cases waived by juvenile court judges and 39% of cases transferred to criminal court by prosecutorial waiver resulted in a criminal court sanction that involved incarceration.[51] Similarly, Bortner's study, which looked at 214 juveniles waived in a western state, found that 63% of juveniles whose cases were waived received probation. Indeed, only 32% of the juveniles whose cases were waived received jail or prison sentences, 1% were given fines, and 4% had their cases dismissed.[52]

Although many juveniles transferred to adult courts have been treated leniently, others are treated more harshly than similar offenders in juvenile court. In some jurisdictions, incarceration, not leniency, appears to be the norm. For example, a study conducted by Rudman, Hartstone, Fagan, and Moore looked at 138 youths who were charged with violent offenses and were considered for transfer between 1981 and 1984 in Boston; Newark, NJ; and Phoenix. They found that 94% of these youths were convicted of violent crimes in criminal court and that 90% of the convicted youths were incarcerated. Moreover, they found that, among youths who were sentenced to a period of incarceration, those convicted and sentenced in adult courts received substantially longer sentences than those convicted and sentenced in juvenile courts.[53] Likewise, a study by Jeffrey Fagan that compared 15- and 16-year-old felony offenders in New York (where they are excluded from juvenile court jurisdiction) to similar youths in New Jersey (where they are under juvenile court jurisdiction) found that youths in New York were more likely to be incarcerated.[54]

Studies that have examined transfer indicate that there is considerable variability in how youths are treated when they are convicted in adult courts. The exact reasons for this are not completely clear, but it may be due in large part to the types of offenses juveniles are charged with. In the studies by Hamparian and colleagues and by Bortner, many of the youths transferred to criminal court were property offenders. In contrast, the youths transferred to criminal court in the study by Rudman and colleagues were convicted of violent offenses. This suggests that if juveniles are transferred to adult courts for minor property offenses, they are likely to be treated leniently, but if they are transferred to adult courts for serious offenses against people, they are likely to receive severe sanctions.

There is clear evidence that juveniles who are transferred to adult court are often treated more severely compared to juveniles who have committed similar offenses but are tried in juvenile courts. There is also evidence that juveniles are sometimes treated more severely than adults who have committed similar offenses. Indeed, in the research study noted earlier that examined juveniles convicted of felonies in criminal courts, juveniles who were convicted of murder were actually sentenced to *longer* prison terms than older people convicted of the same offense.[55] A study by Megan Kurlychek and Brian Johnson that compared the sentencing outcomes of juveniles and young adult offenders tried in adult courts in Pennsylvania found that juveniles received more severe sentences, particularly in cases involving violent crime. Moreover, they found that legal variables like offense seriousness and prior record played a smaller role in juvenile sentencing outcomes compared to those of young adults. Their research suggests that youths tried in adult court face a "juvenile penalty" which leads to more severe treatment.[56]

Contrary to what some might expect, juveniles transferred to adult courts are not always treated more severely than youths whose cases are adjudicated in juvenile courts. Indeed, in some instances, juveniles receive more severe dispositions in juvenile courts. Jeffrey Fagan found that in 1981 and 1982, youths processed in juvenile courts in New Jersey were half as likely to receive sentences mandating incarceration as youths of a similar age convicted in adult courts in New York. For example, 18% of the youths convicted of robbery in New Jersey juvenile courts were incarcerated, compared with 46% of youths convicted of robbery in criminal courts in New York. In 1986 and 1987, however, Fagan found juveniles convicted of robbery in New Jersey were more likely to be incarcerated than those handled in New York. During those 2 years, the percentage of juveniles convicted of robbery who were incarcerated in New Jersey increased to 57%, compared with 27% for New York.[58] In another study comparing youths processed in juvenile courts and those tried in criminal courts, the researchers found that, on average, criminal courts sanctioned offenders more severely for the same offense. However, the researchers also discovered that the differences were due, in part, to differences in the prior records of the offenders. What they discovered was that youths who had prior records and whose cases were adjudicated in juvenile courts were treated far more severely than youths with prior records who were tried in adult courts.[59]

MYTH VS REALITY

Sometimes Juveniles Are Treated More Harshly than Adults

Myth—Juveniles are able to use their youth as a mitigating factor to convince courts that they should not be punished for serious offenses, which results in the lenient treatment of serious juvenile offenders.

Reality—Like some adults, juveniles are sometimes treated leniently for serious offenses. However, it also should be noted that youths' age is sometimes used as an aggravating factor (i.e., a "juvenile penalty") that paints them as very dangerous offenders and results in punishments that are more severe than those sometimes given to adults. This occurs when juries or hearing officers are convinced by prosecutors that if a youth has committed a serious offense as a juvenile, he or she will likely be even more dangerous in the future.[57]

MYTH VS REALITY

Youths May or May Not Be Treated Harshly by Adult Courts

Myth—Youths who are transferred to criminal courts typically receive severe sanctions, such as long periods of incarceration.

Reality—Many youths transferred to criminal courts are treated severely and receive longer periods of incarceration than similar youths convicted in juvenile courts.[60] However, many others are treated leniently and even receive probation.[61] The differences in treatment may be due to differences in the types of offenses. Courts may be likely to treat juveniles who have committed violent offenses more severely than juveniles who have committed minor offenses. These courts appear willing, however, to treat youths severely when they have committed violent offenses.

Also, it is worth noting that the research evidence on transfer does not demonstrate that public safety is enhanced by waiver, nor does it clearly show that alternative treatments are usually exhausted before youths are transferred to criminal court.[62] Even when youths are subjected to incarceration in criminal courts, there is little evidence that the punishment has a deterrent effect. For example, studies of legislative waiver in New York and Idaho failed to uncover a deterrent effect associated with the waiver statutes of these two states.[63] Also, Jeffrey Fagan's study of youths charged with felony burglary or robbery in both juvenile and criminal courts found that, among youths charged with robbery, those processed in juvenile courts were significantly less likely to be rearrested and reincarcerated than those tried in adult courts. Furthermore, among recidivists, the length of time before they were rearrested was significantly longer for youths processed in juvenile courts than for those tried in criminal courts.[64] In a 1-year follow-up that compared case outcomes for close to 3,000 juveniles transferred to criminal court in Florida with youths retained in juvenile court, Donna Bishop and her colleagues discovered that transferred youths had a higher rearrest rate, were more likely to be rearrested for more serious offenses, and were rearrested within a shorter time frame than youths who were processed in juvenile court.[65] Although juveniles who had not been transferred had equaled transferred youths with respect to the proportion that had been arrested nearly 6 years after the initial study, transferred youths who were rearrested were rearrested in a shorter time period and reoffended more times, on average, than youths retained in juvenile court.[66] Unfortunately, only a small amount of research has compared the effects of juvenile court sanctions and the effects of criminal court sanctions. Nevertheless, the research that does exist fails to demonstrate that criminal court sanctions have more of an impact on recidivism than sanctions levied by juvenile courts.[67] Instead, the research suggests that, in general, juvenile court sanctions appear more effective than criminal court sanctions.

designation An action by a prosecutor or juvenile court that makes available adult court sentencing alternatives even though the juvenile is being tried under the jurisdiction of the juvenile court.

■ Other Developments in the Use of Transfer

Designation

In addition to the passage of prosecutorial and legislative transfer laws, states have implemented other laws and procedures related to the transfer of juveniles to adult court. Michigan has developed an alternative to prosecutorial waiver called **designation** that allows a juvenile to be tried in juvenile court in the same manner as an adult is tried in criminal court, including having a probable cause, preliminary type hearing, judgment by a jury of 12, and a formalized sentencing hearing. A juvenile can be designated in two ways. The prosecutor can designate a youth who has committed a "specified juvenile violation" or the court itself can designate a youth for any other delinquent offense. If the prosecutor designates, then the juvenile is on the criminal track. If the court designates, then the juvenile is entitled to a designation hearing to contest the referral to the criminal track. At a designation hearing, the court is required to balance the best interests of the juvenile and the public by considering a variety of factors, such as the seriousness of the offense, the culpability of the juvenile committing the offense in light of aggravating or mitigating factors, the juvenile's prior delinquency record, the juvenile's history in previous correctional programs, the adequacy of the punishments and/or programs available in the juvenile justice system, and the different disposition options available to the court. These are essentially the same factors that the juvenile court would consider in determining the appropriateness of judicial waiver.

After designation, the juvenile essentially has an adult criminal trial in juvenile court, with all of the accompanying due process rights. Upon conviction, the juvenile court has to hold a sentencing hearing to review the available alternatives. These include the following options:

- sentencing the juvenile as a juvenile
- sentencing the juvenile as an adult
- sentencing the juvenile on a delayed basis (if the disposition is not successful, the court can re-sentence the juvenile as an adult)
- sentencing the juvenile to a boot camp for a period of 90–180 days, with 120–180 days in aftercare

In order to impose a delayed adult sentence, the court must hold a hearing in which the rehabilitation of the juvenile is balanced against the juvenile's risk to public safety. The court is required to consider the following factors at the sentence review hearing:

- the extent of the juvenile's conformity to the directives in the original disposition, including participation in education, counseling, work programs, and so on
- the juvenile's willingness to accept responsibility for his or her delinquent behavior
- the juvenile's behavior in his or her current placement
- the juvenile's prior record, character, and physical and mental maturity
- the juvenile's potential for violent conduct, as shown by prior behavior
- recommendations from the caretaker institution
- other information submitted by the juvenile and/or the prosecutor

Certainly, designation gives the juvenile court another option in dealing with serious juvenile offenders. One of the concerns about the Michigan statute, however, is that there is no lower age limit for designation. In other words, a juvenile of any age can be designated. Although designation became effective January 1, 1997, it does not appear to be widely used. Moreover, to date there has been no evaluation of its effectiveness.

Reverse Waiver

At least 25 states have laws that make it possible to transfer juvenile cases from adult court to juvenile court.[68] This makes it possible for the criminal court to transfer cases to juvenile court that have been initially waived via prosecutorial transfer, legislative waiver, or, in some cases, judicial transfer. Generally, reverse waiver occurs after a hearing in criminal court where the criminal court judge considers the appropriateness of handling the case in criminal court or in the juvenile court. At this hearing, the adult court employs the same "best interest" standards used in juvenile courts to determine the appropriateness of transfer to the criminal court.[69] However, at least two related problems are likely to affect many youths seeking reverse waiver. First, youths who lack economic resources are not likely to have the legal assistance necessary to seek a reverse waiver. Second, effective legal counsel will almost always be required in these cases because a petition for reverse waiver must be filed within a specific time frame and the burden of proof needed to support a transfer back to juvenile court falls on the youth. Without effective counsel, presenting "clear and convincing evidence" that the case should be transferred to juvenile court will be difficult.[70]

Once an Adult, Always an Adult

At least 34 states have "once an adult, always an adult" requirements built into state law. Almost all of these states' laws contain provisions that indicate that the youth must be convicted of the initial offense that resulted in waiver for the "once an adult, always an adult" provision to be in effect.[71] In contrast, in some states (e.g., California, Delaware, and Mississippi), a conviction on the initial charge is not always necessary. Also, some states (e.g., Iowa, Michigan, and Texas) mandate that certain offenses (e.g., felonies) must be heard in adult court after a youth has been found guilty in adult court on a previous charge.[72]

Blended Sentencing/Determinate Sentencing

blended sentencing
Allows either the juvenile court or the criminal court to impose a disposition that involves placement in a juvenile correctional institution or an adult institution, or to serve periods in both juvenile and adult programs.

A number of states also have **blended sentencing**/determinate sentencing provisions that allow either the juvenile court or the adult court to render a disposition that involves placement in a juvenile or adult correctional institution or both. These laws have often resulted from concerns over perceived increases in serious juvenile crime and a belief that young juvenile offenders who commit serious offenses need additional supervision beyond that normally provided by the juvenile court.[73] Under blended sentencing or determinate sentencing options, depending on the state, either the juvenile or criminal court can sentence youths to longer periods in juvenile facilities than was traditionally possible under previous law, can sentence a youth to placement in an adult facility, or can sentence a youth to a juvenile facility for a period of time followed by placement in an adult facility. For example, under Texas's determinate sentencing law, youths can be sentenced for up to 40 years' incarceration by the juvenile court. They would serve their sentence in juvenile facilities operated by the Texas Youth Commission until they reach their eighteenth birthday, at which time they would be transferred to the Texas Department of Corrections to serve the remainder of their sentence.[74]

■ Standards Governing Transfer Decisions

Although there has been a recent trend toward making it easier to transfer juveniles to adult courts, the transfer of many nonviolent juvenile offenders to criminal court appears to be encouraged by a lack of clear standards for transfer decisions in some jurisdictions. In their study of prosecutorial waiver in Florida, Bishop and her colleagues found that most juvenile division chief prosecutors had not established formal policies to guide transfer decisions. Instead, they relied on informal guidelines to direct the waiver decisions made by attorneys under their supervision. Moreover, many of the prosecutors indicated that they thought the standards used by other division chiefs to limit attorney discretion were too low.[75]

MYTH VS REALITY

Waiver Laws Appear to Do Little to Deter Juvenile Crime

Myth—Threatening youths with waiver and sending more youths to adult courts will deter youths from delinquent behavior.

Reality—Studies that have examined the potential deterrent effects of state transfer laws have failed to find that they deter juvenile crime.[78]

The available evidence suggests that the increased use of transfer to criminal courts is the product of a number of factors. To some extent, the increased use is likely due to periodic fluctuations in the incidence and seriousness of juvenile crime and the difficulty that overburdened and underfunded juvenile justice agencies have in dealing with the juvenile crime problem. It is also likely due, in part, to the perception that the juvenile courts do not impose severe enough sanctions on some juvenile offenders and that they often lack reliable data information systems capable of tracking offenders involved in serious and repetitive criminality.[76] In addition, factors such as offense seriousness, age, offense history (including a history of previous waiver), a lack of clear criteria for transfer, and a desire among policy makers to appear tough on juvenile crime have also contributed to the increased use of transfer as a response to juvenile crime.[77]

Correctional Programming for Juveniles Convicted in Criminal Court

The increase in the number of adolescents being subjected to incarceration in adult correctional facilities has, in some instances, led states to develop additional strategies for handling young offenders. Moreover, in some jurisdictions, the increase in the number of juvenile offenders tried as adults has placed pressure on those who are responsible for the development of adult institutional correctional policy to implement programs that better respond to the needs of youthful prisoners, who are typically more vulnerable than inmates traditionally handled by these facilities. Around the country, four basic approaches to responding to adolescent offenders sentenced to periods of incarceration in the adult system presently exist: (1) straight adult incarceration, (2) **graduated incarceration**, (3) **segregated incarceration**, and (4) the designation of certain offenders as **youthful offenders**.[79]

graduated incarceration
An incarceration strategy according to which a convicted juvenile begins his or her sentence in a juvenile facility and then moves to an adult facility upon reaching the appropriate age.

segregated incarceration
The practice of providing a separate correctional facility designated for juvenile offenders within the adult justice system.

youthful offender
A special legal status existing in some jurisdictions that allows young offenders to receive special consideration as regards sentencing, privacy, and rehabilitation, even as first-time adult offenders.

Straight Adult Incarceration

Most states make it possible to place juveniles who are sentenced as adults in adult correctional facilities, either with young adult offenders or in the general correctional population, providing the juvenile has reached a certain age (e.g., in North Dakota and California, a youth must be at least 16 years old to be placed in an adult correctional facility). Only six states (Arizona, Hawaii, Kentucky, Montana, Tennessee, and West Virginia) prohibit the placement of juveniles in adult correctional facilities or require juveniles to be segregated from adults.[80] On June 30, 2005, counts of state prison and jail inmates indicated that there were 6,759 persons younger than 18 years of age being held in local jails and another 2,266 incarcerated in state prisons.[81]

Although state laws prohibit children from voting, smoking, drinking alcohol, buying a car or serving in the military, laws in many states also indicate that these same children are not too young to be treated as adults and serve time in adult prisons for violating criminal laws. Although many youths who are sentenced as adults have

FYI

The Number of Persons Younger than 18 Years Incarcerated in State Institutions Peaked in 1995. Since 1995, However, It Has Been Decreasing.

At mid-year 1995, there were 5,309 persons younger than the age of 18 years who were incarcerated in state prisons. However, this number had dropped to 2,266 by mid-year 2005.[82]

committed serious offenses, including murder, treating them as adults raises a number of important questions. Is sentencing a youth to life in prison, with little or no hope of parole, justified? Can placing a youth in an adult correctional facility be justified based on the youth's threat to society or because there is evidence that the youth can never be rehabilitated? Is it in the best interest of society for young people to do "adult time?" For all of the juveniles sentenced to adult prisons, what can they expect from their prison experience? These questions need to be answered in order to understand the intended and unintended consequences of placing juveniles in adult prisons.

As long ago as 1980, in the case of *United States v. Bailey,* United States Supreme Court Justice Blackmun noted:

> The atrocities and inhuman conditions of prison life in America are almost unbelievable; surely they are nothing less than shocking.... A youthful inmate can be subjected to homosexual gang rape his first night in jail, or, it has been said, even in the van on the way to jail. [83]

Many people are concerned about preventing child abuse and child pornography and protecting children from materials with sexual content, yet many of these same individuals seem to be less concerned about protecting children from sexual exploitation in prison.[84] Should children forfeit some of their fundamental rights as human beings if they are sentenced as adults?

Prison rape not only causes victims to act out their pent up rage and powerlessness on other inmates or in society when they are released, but it may increase the likelihood of suicide, the transmission of sexually transmitted diseases, psychosis, or other mental disorders.[85] Clearly, young children are at risk of physical, emotional, and mental trauma from abusive conditions in adult prisons from older and stronger inmates. There is also disturbing evidence that they may be at greater risk than adults of being beaten and abused by guards and other staff.[86] As will be discussed in Chapter 12, juveniles, when incarcerated in juvenile facilities, have a right to treatment. Under the traditional theory of *parens patriae* does the obligation of the state to provide treatment to juveniles disappear because they have been sentenced as adults? Clearly, equal protection considerations are present when children are sentenced to adult prisons, victimized, and not cared for.

Another issue that comes to the forefront in examining the implications of children sentenced as adults is the legal issue of competency to stand trial as an adult. In an adult criminal prosecution, it is an established legal principle that a defendant must be capable of meaningful participation in his or her defense. This means several things:

1. The defendant must have sufficient ability to consult with his or her attorney.
2. The defendant must have a reasonable degree of understanding of what his or her attorney is communicating to him or her.

3. The defendant must have a reasonable rational understanding of the charges against him or her.
4. The defendant must demonstrate a reasonable and rational understanding of the proceedings against him or her.[87]

What this indicates in practical terms is that the defendant must be able to communicate with his or her attorney and understand what the attorney is advising. The defendant must be able to understand his or her individual rights and be able to make decisions about exercising them or not exercising them during the various stages of the criminal process and trial. However many youths involved in the juvenile and criminal justice process suffer from developmental delays, inexperience in making life-affecting decisions and lack of life experience, which may affect their competency in a criminal trial.[88] For example, most adult defendants understand that legal rights are something that they are "entitled" to, but many youths view these "rights" as "conditional" because youths often believe that these rights can be retracted by the authorities at their discretion.[89] Many states have assumed that juveniles who are "experienced" in the juvenile justice system are likely to have a better understanding of court procedures. However, current research does not support that presumption. Some children learn from past experience, but many do not.[90]

As mentioned before, decision making is crucial to a criminal defense. A very important element of decision making is the ability to foresee the consequences of a potential decision. Therefore, "[a defendant must] be able to imagine hypothetical situations, envisioning conditions that do not now exist and that they have never experienced."[91] However, this skill or ability requires life experience that young children do not have. This ability only begins to develop in preadolescence; for youths with learning disabilities or other mental health issues, this ability would be delayed even further. Consequently, courts need to be aware of this seriously important issue of competency, provide meaningful evaluations for juveniles charged as adults, and allow legal findings of incompetency caused by developmental immaturity. Courts also should require a higher threshold for competency for juveniles being tried as adults.[92]

Graduated Incarceration/Blended Sentencing/Determinate Sentencing

Another process that has the same impact as transfer and is used in a number of states is graduated/blended/determinate sentencing. Blended sentences may be imposed by either a juvenile court or an adult court. Juvenile blended sentences are available in 15 states and allow juvenile courts to impose a combined juvenile disposition and a suspended criminal sentence for youths who meet certain age and offense criteria. Eleven of the 15 states (Alaska, Arkansas, Connecticut, Illinois, Kansas, Massachusetts, Michigan, Minnesota, Montana, Ohio, Vermont) that have this option use an *inclusive model* of blended sentences. In these states, juveniles are given a juvenile disposition as well as an adult sentence; however, the adult portion of the sentence can be suspended, provided that the youth meets certain conditions such as successfully completing the juvenile court disposition. In four states, a *contiguous model* of blended sentencing is used. In three of these states (Colorado, Rhode Island, and Texas) youths are sentenced to periods of incarceration that exceed the juvenile court's age of maximum jurisdiction. In these states, youths begin their incarceration in a juvenile facility until a specified age, usually 18 years, at which time they may be transferred to a traditional adult facility to serve out the remainder of their sentences or they may be released. In one state (New Mexico) that

FYI

In Some States, Youths May Be Designated as Youthful Offenders

One example of a youthful offender statute is Michigan's Holmes Youthful Trainee Act. This act, originally adopted in 1967, recognizes that youthful offenders, even if tried as adults, should not necessarily be treated as hardened criminals.[96] According to this statute, "youthful trainee status" is available only once to a youth charged with a felony offense (excluding offenses that result in life imprisonment or offenses that constitute a major drug offense). Originally, this act applied to youths between 17 and 20 years of age. However, amendments to the original act moved the upper age limit to 21 years and the lower age limit to 16 years, making youthful trainee status available to youths who are transferred to adult court. Upon a guilty plea, the adult criminal court does not enter a conviction, but assigns the youth to youthful trainee status. This status, which is revocable at the court's discretion, usually involves supervision and probation for up to 3 years. The focus, even if the juvenile is sent to jail as one of the conditions, is on work and education. If the youth completes the program, the criminal proceedings are dismissed, no criminal conviction is entered, and all of the proceedings are closed to public inspection, in recognition of the stigma that attaches to a person accused of serious crimes, no matter what the eventual legal outcome.

uses a type of contiguous model, the juvenile court judge has the option of imposing an adult sentence instead of a juvenile disposition.[93]

Segregated Incarceration

Some states, such as Florida and South Carolina, provide separate programs within their adult correctional systems for young adult inmates. In these states, the department of corrections operates separate facilities for adolescent adult offenders within a certain age range (18 to 21 years of age or 18 to 25 years of age). When juveniles in these states are convicted and sentenced as adults, they may become eligible for placement in these facilities because of their age.[94]

Designation of Certain Juveniles as Youthful Offenders

Another strategy used in some states is to designate certain juveniles as "youthful offenders" (other terms may also be used). Such a designation often provides special legal protections to a juvenile so designated, such as the sealing of the court record if the juvenile successfully completes his or her sentence. In addition, special programming for these youths may be provided in the juvenile or adult correctional system. Some states that presently have a version of the youthful offender designation are Colorado, Kentucky, New Mexico, and Michigan.[95]

Legal Issues

From a purely philosophical perspective, it can be asked whether it is proper to transfer juveniles to the adult criminal justice system. Moreover, a growing body of scientific evidence indicates that juveniles are developmentally different than adults, and thus, should be treated differently.[97] If a fundamental premise of juvenile justice is that juveniles should be treated differently than adults, why should any exceptions be made and waiver allowed? A counter argument can be made that waivers are necessary because the juvenile justice process has failed to do an adequate job of rehabilitating the types of youths that are the subjects of waiver. For those children who are waived at the beginning of their juvenile court history, it can be argued that the system, be it in the form of a child welfare agency or a juvenile court, failed to recognize them as problem children at an early enough age for meaningful intervention to take place.

Is the increasing trend toward legislative, prosecutorial, or presumptive waiver a statement by the legislative and executive branches that the judicial branch either is not competent to make these decisions or is consistently making poor decisions? Under either theory, the advent of nonjudicial transfer is a clear attack on the judicial branch. Any time one branch of our tripartite government invades the province of another branch, it should be cause for grave concern. Clearly, many legislators and prosecutors believe, despite the lack of supporting statistics, that the juvenile courts are too easy on serious delinquents.

The crux of the matter seems to be politics. Being hard on serious delinquents is a popular stand, and one that is hard to oppose. Because the courts cannot easily or ethically move into the political arena to protect their authority and discretion, they are vulnerable to attack. Elected officials' first responsibility is to get reelected. Politically popular stands—even those without a factual basis—garner votes. Yet, it is important to recognize that political grandstanding does not always help youths or protect the community.

Chapter Summary

Prior to the development of a separate court for juveniles at the end of the 1800s, youths accused of illegal behavior were subject to the jurisdiction of adult criminal courts. With the establishment of juvenile courts, however, jurisdiction over the criminal behavior of youths shifted to the newly organized courts, which were intended, in part, to shield youths from the harmful effects of criminal court processing and commitment to adult correctional institutions. Although jurisdiction over criminal and status offense behavior shifted to juvenile courts after they were established, state juvenile codes still made it possible to try some juveniles as adults. The transfer of a juvenile to a criminal court, which was typically accomplished by holding a juvenile court hearing, was generally thought to be appropriate if the juvenile had committed a serious offense.

Today, each state and the District of Columbia have a mechanism by which some juveniles can be sent to adult courts for trial (called bindover, waiver, transfer, remand, or certification, depending on the state). In fact, the ways in which juveniles can be transferred to adult courts have expanded in recent years. Traditionally, transfer or waiver was accomplished by holding a juvenile court hearing to consider the appropriateness of the action. Some states, in an effort to "get tough" on juvenile crime, have implemented prosecutorial waiver mechanisms that allow prosecuting attorneys the discretion to waive juveniles who meet certain age and offense criteria. Other states have established legislative waiver mechanisms that automatically transfer some juveniles to adult courts for trial if they meet certain age and offense criteria. In addition, some states have developed presumptive waiver laws, as well as blended and determinate sentencing laws, that make it easier to try certain youths in adult courts, maintain jurisdiction over juveniles for longer periods of time, and sentence them to adult correctional facilities.

The "get tough" movement within juvenile justice has led to the development of additional transfer mechanisms and has produced a substantial increase in the number of juveniles being tried as adults. Today, many serious juvenile offenders are tried in adult courts. However, a lack of clear criteria to guide prosecutorial decisions about transfer in some jurisdictions has contributed to the transfer of many juveniles who have engaged in nonserious crimes, such as property crimes.

As the number of youths tried in adult courts increases, so will the number of youths placed in adult correctional programs. Unfortunately, there is a lack of convincing evidence that adult correctional programs are equipped to meet the needs of juvenile prisoners. Indeed, there is little evidence that criminal court responses to juvenile offenders serve their best interests or increase the likelihood that they will be rehabilitated or deterred from subsequent offending. Consequently, claims about the importance of transferring youths to adult courts as a way of protecting the community, at least in the long run, do not currently rest on a sound scientific basis. Yet, despite the paucity of supporting evidence, the continued use of waiver will likely be with us for some time.

Key Concepts

blended sentencing: Allows either the juvenile court or the criminal court to impose a disposition that involves placement in a juvenile correctional institution or an adult institution, or to serve periods in both juvenile and adult programs.

concurrent jurisdiction: Both the juvenile and adult court share original jurisdiction. Consequently, in states with concurrent jurisdiction, prosecutors are given discretion to file cases that meet certain age and offense criteria in juvenile or criminal court.

designation: An action by a prosecutor or juvenile court that makes available adult court sentencing alternatives even though the juvenile is being tried under the jurisdiction of the juvenile court.

double jeopardy: As applicable to transfer cases, a principle that requires the state authority to choose at the outset whether a juvenile is to be tried as a juvenile or as an adult. The choice must be made at the preliminary hearing or arraignment stage of the proceedings.

graduated incarceration: An incarceration strategy according to which a convicted juvenile begins his or her sentence in a juvenile facility and then moves to an adult facility upon reaching the appropriate age.

judicial waiver: A type of waiver selected by a juvenile court at a hearing. It is the traditional type of waiver.

legislative waiver: A type of waiver that results when certain threshold criteria specified by state laws are met.

prosecutorial waiver: A type of waiver that is within the power of the local prosecutor or state attorney to choose and implement.

segregated incarceration: The practice of providing a separate correctional facility designated for juvenile offenders within the adult justice system.

statutory exclusion: See legislative waiver.

waiver: The process by which a juvenile is moved from juvenile court jurisdiction to adult criminal court jurisdiction for an offense committed while still a juvenile. Also known as bindover, certification, remand, and transfer.

youthful offender: A special legal status existing in some jurisdictions that allows young offenders to receive special consideration as regards sentencing, privacy, and rehabilitation, even as first-time adult offenders.

Review Questions

1. What is meant by the terms waiver, transfer, certification, remand, and bindover?
2. What is judicial waiver, and how does it compare with legislative waiver and prosecutorial waiver?
3. What are the advantages and disadvantages of prosecutorial waiver?
4. How has the double jeopardy provision of the U.S. Constitution been applied to juvenile court transfer proceedings?
5. What are the minimally required due process rights of a juvenile at a judicial waiver hearing?
6. What is designation, and how does it differ from waiver?
7. What is youthful offender status, and how does it differ from waiver?
8. What are graduated incarceration and segregated incarceration?
9. What evidence exists to indicate that waiver benefits juveniles or the community?
10. What are the thresholds that juveniles must cross in order to be eligible for waiver?
11. What are the "political" issues that influence legislative policies regarding transfer of juveniles to adult courts?
12. In what ways has the use of transfer changed over time?
13. How does the juvenile justice system benefit from transfer?
14. What does concurrent jurisdiction mean?

Additional Readings

Bonnie, R. J. & Grisso, T. (2000). Adjudicative competence and youthful offenders. In T. Grisso & R. G. Schwartz (Eds.), *Youth on trial: A developmental perspective on juvenile justice*. Chicago: University of Chicago Press.

Mole, D. & White, D. (2005). *Transfer and waiver in the juvenile justice system*. Washington, DC: Child Welfare League of America.

Snyder, H. N. & Sickmund, M. (2006). *Juvenile offenders and victims: 2006 national report*. Washington, DC: Office of Juvenile Justice and Delinquency Prevention.

Wizner, S. (1984). Discretionary waiver of juvenile court jurisdiction: An invitation to procedural arbitrariness. *Criminal Justice Ethics*, *3*, 41–50.

Cases Cited

Breed v. Jones, 421 U.S. 519 (1975).

Dusky v. United States, 362 U.S. 402 (1960).

Roper v. Simmons, 543 U.S. 551 (2005).

Stanford v. Kentucky, 492 U.S. 361 (1989).

Thompson v. Oklahoma, 487 U.S. 815 (1988).

United States v. Bailey, 444 U.S. 394, (1980). Page 444 U.S. 421–422.

Notes

1. Forst, M. L. (Ed.). (1995). *The new juvenile justice.* Chicago: Nelson-Hall.
2. Snyder, H. N. & Sickmund, M. (2006). *Juvenile offenders and victims: 2006 national report.* Washington, DC: Office of Juvenile Justice and Delinquency Prevention.
3. Bishop, D. M., Frazier, C. E., & Henretta, J. C. (1989). Prosecutorial waiver: Case study of a questionable reform. *Crime and Delinquency, 35*, 179–201.
4. Tonry, M. & Doob, A. N. (Eds.). (2004). Youth crime and youth justice: Comparative and cross-national perspectives: Vol. 31. *Crime and justice series.* Chicago: University of Chicago Press.

   Winterdyk, J. A. (Ed.). (2002). *Juvenile justice systems: International perspectives.* Toronto: Canadian Scholars' Press.

   Friday, P. C. & Ren, X. (Eds.). (2006). *Delinquency and juvenile justice systems in the non-Western World.* Monsey, New York: Criminal Justice Press.
5. Gardner, M. C., Jr. (1973). Due process and waiver of juvenile court jurisdiction. *Washington and Lee Law Review, 30*, 591–613.

   Gasper, J. & Katkin, D. (1980). A rationale for the Abolition of the juvenile court's power to waive jurisdiction. *Pepperdine Law Review, 7*, 937–951.
6. Legal Information Institute. (2005). Roper v. Simmons. In *Supreme Court collection.* Retrieved March 5, 2005, from http://supct.law.cornell.edu/supct/html/03-633.ZS.html.
7. Streib, V. L. (2003). *The juvenile death penalty today: Death sentences and executions for juvenile crimes, January 1, 1973–June 30, 2003.* Retrieved August 20, 2009, from http://www.internationaljusticeproject.org/pdfs/juvenile.pdf.
8. Hass, K. C. (1998). Too young to die? The U.S. Supreme Court and the juvenile death penalty. In A. R. Roberts (Ed.), *Juvenile justice: Policies, programs, and services* (2nd ed.). Chicago: Nelson-Hall.
9. Legal Information Institute, 2005.
10. Amnesty International. (2008). *Iran: Spare four youths from execution, immediately enforce international prohibition on death penalty for juvenile offenders.* Retrieved November 1, 2008, from http://www.amnesty.org/en/for-media/press-releases/iran-spare-four-youths-execution-immediately-enforce-international-prohi.
11. Amnesty International, 2008.
12. Snyder & Sickmund, 2006.
13. Snyder & Sickmund, 2006.
14. Griffin, P. (2008). *Different from adults: An Updated analysis of juvenile transfer and blended sentencing laws, With recommendations for reform.* Pittsburgh, PA: National Center for Juvenile Justice. Retrieved from Models for Change at http://www.modelsforchange.net/publications/181.
15. Snyder & Sickmund, 2006.
16. Snyder & Sickmund, 2006.
17. Snyder & Sickmund, 2006.
18. Snyder, H. N. & Sickmund, M. (1995). *Juvenile offenders and victims: A national report.* Washington, DC: Office of Juvenile Justice and Delinquency Prevention.

19. Sickmund, M. (2003). Juveniles in court. *Juvenile Offenders and Victims: National Report Series, Bulletin.* Washington, DC: Office of Juvenile Justice and Delinquency Prevention.

20. Griffin, 2008.

21. Snyder, H. M. & Sickmund, M. (1999). *Juvenile offenders and victims: 1999 national report.* Washington, DC: Office of Juvenile Justice and Delinquency Prevention.

Mich. Comp. Laws 712A.2(a)(1).

22. Snyder & Sickmund, 1999.

23. Snyder & Sickmund, 1995.

24. Coordinating Council on Juvenile Justice and Delinquency Prevention. (2005). *Combating violence and delinquency: The National Juvenile Justice Action Plan.* Washington, DC: Office of Juvenile Justice and Delinquency Prevention.

Krisberg, B. (2005). *Juvenile justice: Redeeming our children.* Thousand Oaks, CA: Sage.

25. Snyder, H. N., Sickmund, M., & Poe-Yamagata, E. (2000). *Juvenile transfers to criminal court in the 1990's: Lessons learned from four studies.* Washington, DC: Office of Juvenile Justice and Delinquency Prevention.

26 Puzzanchera, C. M. (2003). Delinquency cases waived to criminal court, 1990–1999. *OJJDP Fact Sheet.* Washington, DC: Office of Juvenile Justice and Delinquency Prevention.

27 Snyder & Sickmund, 1999.

28. Krisberg, B., Schwartz, I., Litsky, P., & Austin, J. (1986). The watershed of juvenile justice reform. *Crime and Delinquency, 32*, 5–38.

29. Butts, J. A. (1977). Delinquency cases waived to criminal court, 1985–1994. *OJJDP Fact Sheet.* Washington, DC: Office of Juvenile Justice and Delinquency Prevention.

Snyder & Sickmund, 1995.

30. Griffin, 2008.

Sickmund, M., Sladky, A., & Kang, W. (2005). *Easy access to juvenile court statistics: 1985–2005.* Retrieved November 10, 2008 from http://ojjdp.ncjrs.gov/ojstatbb/ezajcs/.

31. Juszkiewicz, J. (2000). *Youth crime/adult time: Is justice served?* Retrieved August 31, 2001, from http://www.buildingblocksforyouth.org/ycat/ycat.html.

32. Snyder & Sickmund, 1999.

33. Amnesty International. (1998). *Betraying the young: Human rights violations against children in the US justice system.* New York: Amnesty International.

34. Sickmund, Sladky, & Kang, 2005.

35. Puzzanchera, C., Finnegan, T., & Kang, W. (2007). *Easy access to juvenile populations: 1990–2007.* Retrieved from http://www.ojjdp.ncjrs.gov/ojstatbb/ezapop/.

Sickmund, Sladky, & Kang, 2005.

36. *Breed v. Jones*, 421 U.S. 519 (1975).

37. Feld, B. C. (1987). The juvenile court meets the principle of the offense: Legislative changes in juvenile waiver statutes. *Journal of Criminal Law and Criminology, 78*, 471–533.

38. Poulos, T. M. & Orchowsky, S. (1994). Serious juvenile offenders: Predicting the probability of transfer to criminal court. *Crime and Delinquency, 40*, 3–17.

39. Fagan, J., Forst, M., & Vivona, T. S. (1987). Racial determinants of the judicial transfer decision: Prosecuting violent youth in criminal court. *Crime and Delinquency, 33*, 259–286.

Fagan, J. & Deschenes, E. P. (1990). Determinants of judicial waiver decisions for violent juvenile offenders. *Journal of Criminal Law and Criminology, 81*, 314–347.

Nimick, E., Szymanski, L., & Snyder, H. (1986). *Juvenile court waiver: A study of juvenile court cases transferred to criminal court.* Pittsburgh, PA: National Center for Juvenile Justice.

Podkopacz, M. R. & Feld, B. C. (1996). The end of the line: An empirical study of judicial waiver. *Journal of Criminal Law and Criminology, 86*, 449–442.

Poulos & Orchowsky, 1994.

40. Lee, L. (1994). Factors determining waiver in a juvenile court. *Journal of Criminal Justice, 22*, 329–339.

41. Bortner, M. A. (1986). Traditional rhetoric, organizational realities: Remand of juveniles to adult court. *Crime and Delinquency, 32*, 53–73.

42. Juszkiewicz, 2000.

43. Sickmund, Sladky, & Kang, 2005.

44. Bishop, Frazier, & Henretta, 1989.

45. Brown, J. & Langan, P. (1998). *State court sentencing of convicted felons, 1994.* Washington, DC: Bureau of Justice Statistics.

Snyder & Sickmund, 1999.

46. Bishop, Frazier, & Henretta, 1989.

47. Bishop, Frazier, & Henretta, 1989, p. 193.

48. Snyder & Sickmund, 1999.

Snyder & Sickmund, 2006.

49. Juszkiewicz, 2000.

Mole, D. & White, D. (2005). *Transfer and waiver in the juvenile justice system.* Washington, DC: Child Welfare League of America.

Snyder & Sickmund, 2006.

Torbet, P., Griffin, P., Hurst, H., Jr., MacKenzie, L. R. (2000). *Juveniles facing criminal sanctions: Three states that changed the rules.* Washington, DC: Office of Juvenile Justice and Delinquency Prevention.

50. Snyder & Sickmund, 1995.

51. Hamparian, D. L., Estep, L., Muntean, S., Priestino, R., Swisher, R., Wallace, P. et al. (1982). Youth in adult courts: Between two worlds. Washington, DC: Office of Juvenile Justice and Delinquency Prevention.

52. Bortner, 1986.

53. Rudman, C., Hartstone, E., Fagan, J., & Moore, M. (1986). Violent youth in adult court: Process and punishment. *Crime and Delinquency, 32*, 75–96.

54. Fagan, J. (1991). *The comparative impacts of juvenile and criminal court sanctions on adolescent felony offenders.* Washington, DC: National Institute of Justice.

55. Brown & Langan, 1998.

Snyder & Sickmund, 1999.

56. Kurlychek, M. C. & Johnson, B. D. (2004). The juvenile penalty: A comparison of juvenile and young adult sentencing outcomes in criminal court. Criminology, *42*, 485–517.

57. Dobbs, A. (2004). The use of youth as an aggravating factor in death penalty cases involving minors. *Juvenile Justice Update, 10*, 1–2, 13–16. This resource examines the issue.

Kurlychek & Johnson, 2004.

58. Fagan, 1991.

59. Greenwood, P. W., Lipson, A. J., Abrahamse, A., & Zimring, F. (1983). *Youth crime and juvenile justice in California: A report to the legislature.* Santa Monica, CA: Rand Corporation.

60. Rudman, Hartstone, Fagan, & Moore, 1986.

61. Bortner, 1986.

62. Gillespie, L. K. & Norman, M. D. (1984). Does certification mean prison: Some preliminary findings from Utah. *Juvenile and Family Court Journal, 35*, 23–35. This resource provides additional information.

63. Jensen, E. L. & Metsger, L. K. (1994). A test of the deterrent effect of legislative waiver on violent juvenile crime. *Crime and Delinquency, 40*, 96–104.

Singer, S. & McDowall, D. (1988). Criminalizing delinquency: The deterrent effects of the New York Juvenile Offender Law. *Law and Society Review, 22*, 521–535.

64. Fagan, 1991.

65. Bishop, D. M., Frazier, C. E., Lanza-Kaduce, L., & Winner, L. (1996). The transfer of juveniles to criminal court: Does it make a difference? *Crime and Delinquency, 42*, 171–191.

66. Winner, L., Lanza-Kaduce, L., Bishop, D. M., & Frazier, C. E. (1977). The transfer of juveniles to criminal court: Reexamining recidivism over the long term. *Crime and Delinquency, 43*, 548–563.

67. Snyder & Sickmund, 1999.

68. Snyder & Sickmund, 2006.

69. Griffin, P., Torbet, P., & Szymanski, L. (1998). *Trying juveniles as adults in criminal court: An analysis of state transfer provisions.* Washington, DC: Office of Juvenile Justice and Delinquency Prevention.

70. Mole & White, 2005.

71. Snyder & Sickmund, 2006.

72. Griffin, P. (2003). Trying and sentencing juveniles as adults: An analysis of state transfer and blended sentencing laws. *Technical Assistance to the Juvenile Court: Special Project Bulletin.* Pittsburgh, PA: National Center for Juvenile Justice.

73. Podkopacz, M. R. & Feld, B. C. (2001). The back-door to prison: Waiver reform, "blended sentencing," and the law of unintended consequences. *The Journal of Criminal Law and Criminology, 91*, 997–1071.

74. NCJJ. (2006). *State juvenile justice profiles*. Retrieved Nov. 20, 2008 from http://www.ncjj.org/stateprofiles/profiles/TX06.asp?state=%2Fstateprofiles%2Fprofiles%2FTX06.asp&topic=Profile.

Mears, D. P. & Field, S. H. (2000). Theorizing sanctioning in a criminalized juvenile court. *Criminology, 38*, 983–1019.

75. Bishop, Frazier, & Henretta, 1989.

76. Coordinating Council on Juvenile Justice and Delinquency Prevention, 2005.

77. Bishop, Frazier, & Henretta, 1989.

Bortner, 1986.

Fagan, Forst, & Vivona, 1987.

Lee, 1994.

Nimick, Szymanski, & Snyder 1986.

Poulos & Orchowsky, 1994.

78. Jensen & Metsger, 1994.

Redding, R. E. (2008). Juvenile transfer laws: An effective deterrent to delinquency? *Juvenile Justice Bulletin.* Washington, DC: Office of Juvenile Justice and Delinquency Prevention.

Singer & McDowall, 1988.

79. Torbet, P., Gable, R., Hurst, H., IV, Montgomery, I., Szymanski, L., & Thomas, D. (1996). *State responses to serious and violent juvenile crime: Research report.* Washington, DC: Office of Juvenile Justice and Delinquency Prevention.

80. Lis, Inc. (1995). Offenders under age 18 in state adult correctional systems: A national picture. *Special Issues in Corrections.* Longmont, CO: Department of Justice, NIC Information Center.

81. Harrison, P. M. & Beck, A. J. (2006). Prison and jail inmates at midyear 2005. *Bureau of Justice Statistics Bulletin.* Washington, DC: U.S. Department of Justice.

82. Harrison & Beck, 2006.

83. *United States v. Bailey* 444 U.S. 394, (1980). Page 444 U.S. 421–422.

84. Mariner, J. (2001). *The latest trend in child sexual exploitation: Rape in adult prisons.* Retrieved November 28, 2008 from http://writ.news.findlaw.com/mariner/20010125.html.

85. Jackson, D. (2001). *When children act out in violence, they are still children: An evaluation and proposed reform of Florida's "adult crime, adult time" brand of justice.* Retrieved November 1, 2008 from http://www.kentlaw.edu/honorsscholars/2002students/Jackson.html.

86. Jackson, 2001.

Redding, R. E. (1997). Juveniles transferred to criminal court: Legal reform proposals based on social science research. *Utah Law Review, 1997*, 709–797.

87. *Dusky v. United States*, 362 U.S. 402, (1960).

88. Grisso, T. (1997). *Juvenile competency to stand trial: Questions in an era of punitive reform.* Retrieved December 1, 2008 from http://www.abanet.org/crimjust/juvjus/12-3gris.html.

89. Grisso, 1997.

90. Grisso, 1997.

91. Grisso, 1997.

92. Grisso, 1997.

93. Snyder & Sickmund, 2006.

94. Torbet, Gable, Hurst, Montgomery, Szymanski, & Thomas, 1996.

95. Torbet, Gable, Hurst, Montgomery, Szymanski, & Thomas, 1996.

96. Holmes Youthful Trainee Act. Mich. Comp. Laws 762.11 and MSA 28.853(11).

97. Bonnie, R. J. & Grisso, T. (2000). Adjudicative competence and youthful offenders. In T. Grisso & R. G. Schwartz (Eds.), *Youth on trial: A developmental perspective on juvenile justice.* Chicago: University of Chicago Press.

Scott, E. S. & Grisso, T. (1997). The evolution of adolescence: A developmental perspective on juvenile justice reform. *The Journal of Criminal Law and Criminology, 88*, 137–189.

Steinberg, L. & Schwartz, R. G. (2000). Developmental psychology goes to court. In T. Grisso & R. G. Schwartz (Eds.), *Youth on trial. A developmental perspective on juvenile justice.* Chicago: University of Chicago Press.

The Contemporary Juvenile Court

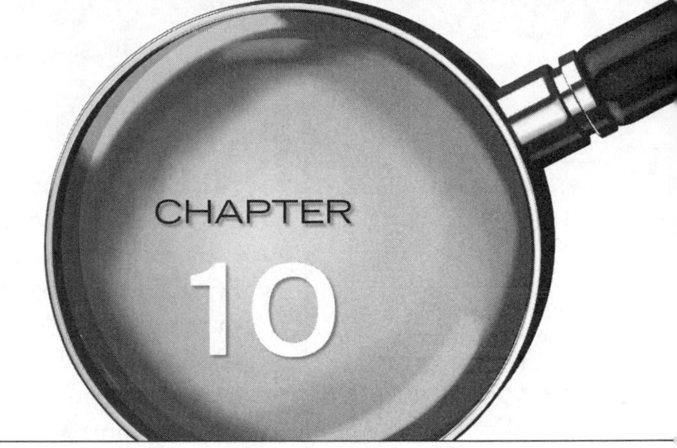

▸ ▸ CHAPTER OBJECTIVES

After studying this chapter, you should be able to

- Delineate the ways in which the juvenile court is the center and focus of the juvenile justice process
- Describe how the juvenile court, through its leadership and employees, acts as an advocate for children and families
- Describe the structure of the modern juvenile court and where the key personnel fit in that structure
- Describe the primary functions of key juvenile court personnel
- Describe the purpose and importance of the adjudication and disposition hearings conducted in the contemporary juvenile court
- Describe the due process protections that juveniles presently have when they appear in juvenile court
- Explain the role of attorneys in the adjudicative and dispositional hearings
- Describe how the juvenile court's place in the local court hierarchy can affect the resources made available to the juvenile court
- Describe the roles that legal and extralegal factors play in juvenile court dispositions

▸ ▸ CHAPTER OUTLINE

Introduction

juvenile court The court responsible for holding hearings and making decisions regarding the disposition of juveniles who have entered the juvenile justice process. This court plays many roles, including children's advocate, program leader, fund-raiser, consensus builder, parent, and protector of youths and the community.

The **juvenile court** is the heart of the juvenile justice process. It is in the juvenile court that previous decisions made by law enforcement agencies, probation officers, child welfare workers, school personnel, and prosecuting attorneys are supported or altered. It is there that additional decisions—ones that can have a lasting effect on children, their families, and the community at large—are made. Indeed, the juvenile court is the most powerful institution within juvenile justice. Not only does the juvenile court determine the outcomes of individual cases (outcomes that can result in the removal of children from their homes and, in some cases, the termination of parental rights), but also, through its legal authority, it can determine how other institutions and agencies respond to children. For example, by interpreting laws, issuing various orders, and developing policies, juvenile courts can determine which categories of youths are subjected to and which are diverted from formal court processing.

In addition, the juvenile court typically plays a central role in the child welfare system. Because the juvenile court is often seen as the institution having the greatest responsibility for responding to delinquent youths, it often is looked to for leadership in efforts to understand juvenile crime and to develop more effective responses to delinquency. In many instances, the burden of acting as an advocate within the political arena on behalf of families and children falls on the shoulders of the juvenile court. Children have little political power, especially children who are poor, and judges and other court personnel frequently testify in front of legislative committees on issues affecting children and families. Local and state bar organizations often have family law and juvenile law committees made up of judges or referees, who are in positions to influence juvenile law practice. Furthermore, many charitable organizations, such as the United Way, seek input from juvenile court personnel about children and family concerns.

Many juvenile courts also find themselves in the role of consensus builder, fostering agreement among the various community agencies, such as social service and mental health agencies, that work with children and families. Much of the difficulty in dealing with child and family problems involves the scarcity of funding for meeting the needs of children and families. In some communities, juvenile courts are more consistently and better funded and staffed than local social service agencies. As a result, these agencies regularly look to the courts for assistance in their efforts to serve clients.

best interests of the child A catch phrase that serves as a reminder that the primary focus of a juvenile court should be on the rehabilitation of the children who come before it.

Like other social institutions, the juvenile court is made up of many individuals who perform a variety of functions. Ultimately, the individuals who occupy various roles within the juvenile court strive (at least theoretically) to protect public safety, serve the **best interests of children** and their families, and ensure the smooth and efficient operation of the juvenile justice process. Yet, like other institutions, it is sometimes unable to meet its goals, and occasionally some of its goals are displaced or overridden by

FYI

Juvenile Courts Perform Multiple Roles

Today, juvenile courts and their personnel perform multiple advocacy roles for children and families. Courts and their personnel act as legal experts, political advocates, community consensus builders, and consultants on children's issues.

other goals. This chapter examines the structure, organization, and operation of the contemporary juvenile court. It also examines critical decision-making events in the juvenile court, the legal context within which juvenile court decisions are made, and the important players who influence juvenile court practice.

Case Trends and Types of Cases Processed in Juvenile Court

Most delinquency cases are referred to courts by law enforcement agencies, and these agencies have been sending an increasing number of cases to juvenile courts. For example, in 2005 juvenile courts processed approximately 1.6 million delinquency cases. This represented a 26% increase in the number of delinquency cases processed since 1990. Moreover, the number of drug law violation cases increased 179%, public order offense cases increased 102%, and person offense cases increased 67%, although the number of property offense cases declined 24% (see Figure 10-1).[1] As the previous data indicate, the increase in the number of cases being formally processed by juvenile courts since the 1990s has been caused by the increasing number of youths who are being referred to courts for drug offenses, public order offenses, and person offenses. However, it is important to note that most of the increase in cases being referred to juvenile courts occurred during the 1990s. Indeed, fewer cases were processed in juvenile courts during 2005 compared to 2000. Between 2000 and 2005, the number of person offense cases increased 10% and the number of public order offense cases increased 8%; however, the number of drug offense cases declined by about 1% and property offense cases declined by 12%.[2] As Figure 10-1 indicates, the sharp increase in the number of person, drug, and public order offense cases referred to juvenile courts slowed between 2000 and 2005 and

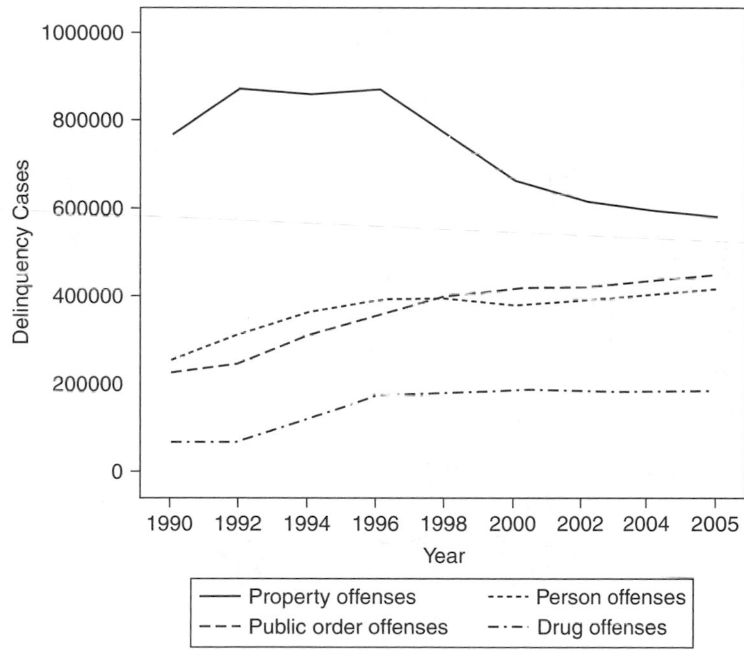

Figure 10-1 Delinquency Cases Processed in Juvenile Courts by Type of Offense, 1990–2005

Source: Data from Sickmund, M., Sladky, A., & Kang, W. (2005). *Easy access to juvenile court statistics, 1985–2005.* Retrieved November 25, 2008 from http://ojjdp.ncjrs.gov/ojstatbb/ezajcs/.

referrals for property offense cases continued to decline. This trend mirrors the trends in juvenile arrests noted in Chapter 2.

Overall, there has been a substantial change over time in the types of cases referred to the juvenile courts. For example, in 1990, property offense cases comprised a majority (58.4%) of the cases referred to juvenile courts, followed by person offense cases (19.1%), public order offense cases (17.3%), and drug offense cases (5.2%). By 2005, however, property offense cases accounted for slightly more than one-third (35.3%) of case referrals. Showing substantial increases over time were person, drug, and public order offense cases. Indeed, by 2005 public order offense cases accounted for 27.9% of case referrals, followed by person offense cases (25.3%) and drug offense cases, which had doubled from 5.2% in 1990 to 11.5% in 2005.[3]

formally process
Handle the case in a way that leads to a formal court hearing.

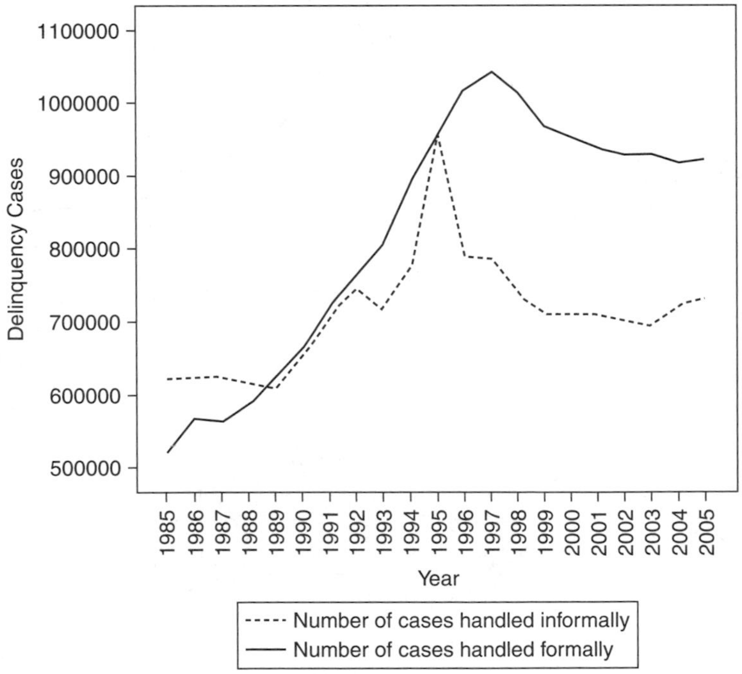

The increase in the number of person, drug offense, and public order cases being handled by juvenile courts is an important trend. Although juvenile courts are, in general, more likely to **formally process** serious offenses than nonserious offenses, a majority of person, property, drug, and public order offense cases are formally processed by juvenile courts. For example in 2005, 58% of person offense cases, 53% of property offense cases, 56% of drug offense cases, and 56% of public order offense cases were formally processed by juvenile courts. Although juvenile courts formally processed fewer cases in recent years compared to the mid-to-late 1990s, the long-term trend has been to process cases formally. As can be seen in Figure 10-2, prior to 1989, a majority of the cases referred to juvenile courts were handled informally. However, since that time, there has been an increasing trend toward formal processing of cases, which has been particularly evident since the mid-1990s. For example, in 1985, approximately 46% of the cases referred to juvenile courts were handled formally, but this percentage had increased to 56% by 2005.[4]

Figure 10-2 Number of Cases Handled Formally and Informally by Juvenile Courts, 1985–2005

Source: Data from Sickmund, M., Sladky, A., & Kang, W. (2005). *Easy access to juvenile court statistics, 1985–2005.* Retrieved November 25, 2008 from http://ojjdp.ncjrs.gov/ojstatbb/ezajcs/.

FYI

Getting Up-to-date Information on Cases Processed in Juvenile Courts is Becoming More Difficult
As this edition is being completed in spring 2009, the most recent data available on cases processed by juvenile courts is from 2005.

Another set of facts that are worthy of note is that case referrals to court increase with the age of the juvenile, and referrals are more likely for males and for many minority youths. Moreover, older youths, males, and youths who are members of particular minority groups are more likely to be formally processed by juvenile courts. For example, in 2005 the *formal case processing rate* for youths 12 years of age (the number of cases formally handled by juvenile courts for every 1,000 12-year-old youths in the population) was 10.6 per 1,000 compared to 27.3 per 1,000 for 17-year-old youths.[5] Indeed, the likelihood of formal court processing for 17-year-old youths was much higher than that for 12-year-old youths, even though a number of states do not process youths who are 17 years of age because they are considered adults. Moreover, the formal case processing rate for males was 49.7 per 1,000 compared to 14.8 per 1,000 females. For white youths, the formal case processing rate was 23.9 per 1,000 youths, compared to 67.2 per 1,000 African American youths, 29.2 per 1,000 American Indian youths, and 9.4 for youths categorized as Asian or Pacific Islander.[6]

The Philosophy Behind Case Processing in the Juvenile Court

The early juvenile courts were characterized by informality,[7] and they paid little attention to due process protections—protections recognized as the cornerstone of adult criminal court operation because they help prevent governmental abuse of power. In the juvenile courts, however, an emphasis on due process protections was thought to be unnecessary and possibly counterproductive. It was thought to be unnecessary because the juvenile courts had been created to serve the best interests of children. It was thought to be potentially counterproductive because it could hinder the efforts of the courts to respond quickly to the needs of children.

Although contemporary juvenile court operation is still marked by considerable informality in many jurisdictions, U.S. Supreme Court rulings such as *Kent v. United States*, *In re Gault*, *In re Winship*, and *Breed v. Jones*, as well as state court rulings, have led to more formal procedures in juvenile courts. For example, in many jurisdictions, it is now common practice for juveniles to be represented by counsel in juvenile court hearings, for records to be made of juvenile proceedings, for courts to carefully detail the rights of youths and parents when they appear before the court, and for courts to follow the same procedural rules used in adult criminal courts. Nevertheless, it is also true that the extent to which courts employ more formalized procedures and protect due process rights varies considerably across jurisdictions. Indeed, in some states many youths still appear before juvenile courts without representation by counsel.[8] Furthermore, the quality of representation that many youths receive is inadequate.[9]

These realities highlight the sharp differences in juvenile court operation found in the United States. These differences include not just variations in the quality of representation, but also differences in the structure of the courts, as explained in the following section.

FYI

Attorneys Do Not Always Provide Adequate Representation to Juvenile Clients

Unfortunately, youths and their parents often get what they pay for. Most attorneys that appear in juvenile courts are not retained by the parents of the juvenile to represent their child or themselves. Most are court appointed, and the pay is minimal. New attorneys, recently admitted to the bar, typically try to get on court-appointed lists as a way to get court experience and to earn some income. The lack of experience on the part of some attorneys that practice in juvenile court may be one factor that accounts for the inadequate representation that juveniles sometimes receive in juvenile courts. In addition to a lack of experience, there are several other factors that may contribute to inadequate representation of youths. These factors include high caseloads and a lack of resources for mounting a strong defense.[10] Moreover, attorney perspectives of their role in the juvenile court may influence the type of representation that is provided. Attorneys who believe that their role to is act as a surrogate guardian for the child or those who believe that their role is to assist the court in serving the best interests of the child,[11] may be less inclined to mount vigorous defenses for clients because such tactics may not be seen as serving the best interests of the child.

Of course, effective and competent representation is found in the juvenile courts. Many veteran attorneys practice in juvenile courts as appointed counsel, not so much for the money, but because they want to have a positive impact on young lives. Many attorneys take the position that if they can help juvenile offenders straighten out their lives, they may not have to provide legal representation to these same individuals after they become adults.

The Structure of the Juvenile Court

Each state, as well as the District of Columbia, has at least one court with authority over minors who engage in illegal behavior. Generally, courts that handle delinquency cases are either part of the highest court of general trial jurisdiction or part of a lower trial court where less serious criminal cases or limited-claim civil matters are heard. In some states, however, delinquency cases are heard in separately organized **family courts** by judges who specialize in juvenile and domestic relations matters.

family court A unified trial court where all cases involving families are heard, including divorce, adoption, custody, guardianship, paternity, neglect and abuse, and delinquency cases. Family court advocates claim that when all family matters are handled by one court, there is less chance for overlapping services, redundant reports, and fragmented intervention.

Although a court that hears delinquency cases is generally referred to as "the juvenile court," in most states this term is not used as the official name of the court. In fact, the structure of courts with juvenile jurisdiction varies considerably from state to state, and can even vary within a state. A court with jurisdiction over juveniles may be part of a district, superior, circuit, county, or family court. It typically has a separate division that handles juvenile cases involving criminal or status offense behaviors, and it may handle abuse and neglect cases as well as adoption, termination of parental rights, and emancipation of minors.[12] For example, the juvenile court in North Carolina is part of the district court, a lower general trial court within the state. One or more district court judges, depending on the size of the jurisdiction, volunteer to specialize in juvenile cases, or if there are no volunteers, one or more judges may be appointed by the chief district court judge to hear juvenile cases. In California and Alaska, the juvenile court is a part of the superior court, which is also a court of general trial jurisdiction, whereas in Rhode Island, the juvenile court is part of a separately organized statewide family court system. In Michigan, the juvenile court was a division of the probate court until 1998, at which time it became part of the family division of circuit court, the general trial court in that state. Colorado has a hybrid model: Denver has a separately organized municipal juvenile court, but in the remainder of the state the juvenile court is part of the district court, which is a court of general trial jurisdiction.

As noted earlier, in some states juvenile courts are part of lower trial courts. National standards recommend that states establish family courts as a division of the general jurisdiction trial court. This would allow these courts to hear a wide range of juvenile and family-related issues and theoretically would allow better coordination of cases when families are involved with multiple courts. However, there has been only limited movement in this direction to date. Some states (e.g., Rhode Island, New York, Delaware, and South Carolina) have placed jurisdiction for delinquency matters in a family court.[13] One state that has recently taken some steps in this direction is Massachusetts. In Massachusetts, there is a statewide juvenile court under the Administrative Office of the Trial Court that has 11 divisions around the state, has its own chief justice, and holds court in more than 40 locations around the state. The Massachusetts juvenile court has general jurisdiction over delinquency, children in need of services (CHINS), care and protection **petitions**, adult contributing to the delinquency of a minor cases, adoption, guardianship, termination of parental rights proceedings, and youthful offender cases.[14] Also, a number of other states (e.g., Alaska, California, Colorado, Florida, Illinois, and Wisconsin) have moved jurisdiction over delinquency cases to the highest court of general trial jurisdiction. Altogether, a variety of juvenile court models exist—indeed sometimes more than one type of court structure can be found within the same state (e.g., Colorado).

petition A pleading to initiate a matter in juvenile court. A petition sets forth the alleged grounds for the court to take jurisdiction of a case and requests court intervention.

According to H. Ted Rubin, who has studied courts extensively, "The structure of any court is significant because it affects the status of the court, in part the quality of the judges of the court, and frequently the budget and the adequacy of the staff of the court."[16] Within each state, the state supreme court is the court of highest status. At the trial level, however, the court of highest status is the **general trial court**, whatever it happens to be called (e.g., circuit court, superior court, district court, or court of common pleas). The general trial court hears felony cases and civil claims that have no maximum dollar limit.[17]

general trial court A court of high status within a state's legal system that hears felony criminal cases and civil cases with unlimited dollar amounts.

When a juvenile court is part of the general trial court, it occupies a prestigious position within the state court structure. As a result, it is in a better position to attract a larger share of state resources than lower courts, which command less prestige. It also has higher paid personnel, including judges and others who staff the court; better facilities; and more support services than it would if it were part of a lower trial court. In some cases, the additional resources allow more effective and efficient court administration.[18] Conversely, when a juvenile court is part of a lower trial court of limited jurisdiction, it may have difficulty attracting sufficient resources to support its mission.

COMPARATIVE FOCUS

The United States Is Unusual Because It Has More Than 51 Juvenile Justice Systems

There is not a single national juvenile justice process in the United States. Although there are many commonalities in juvenile justice practice across each state, the District of Columbia, and American Indian courts that handle misdemeanor offenses committed by juvenile members of recognized tribes, there are also a number of differences across these jurisdictions. In many countries juvenile laws are passed at the national level. Thus, juvenile justice processes are more uniform across the country. In the United Kingdom, for example, juvenile laws are passed by the National Parliament, unlike in the United States where each state develops its own juvenile laws.[15]

When the early juvenile codes were initially enacted, few felt there was a need for a full-time court to handle delinquency matters. Consequently, juvenile judges spent only part of their time hearing juvenile cases. These judges were generalist judges who heard all types of legal matters. Indeed, in many instances judges who had the least seniority or had the lowest status among the sitting judges were assigned to hear juvenile cases, regardless of their level of interest in juvenile law. The chance of a juvenile case being heard by a judge with little interest in juvenile law was especially high in small one-judge jurisdictions.[19]

Despite the trend toward placing juvenile courts within more prestigious general trial courts, some juvenile courts are still part of lower trial courts, perhaps because of a belief held by some that lower trial courts may be able to relate more effectively to parents and youths within the community.[20] The more likely reason, however, is that state legislatures, as well as the legal profession itself, have traditionally seen juvenile courts as less important than the courts that deal with adult crime. In addition, as mentioned earlier, the juvenile courts' clients have never been politically powerful, and the courts' prestige may have suffered as a result. Furthermore, the early juvenile courts were viewed as playing a parental role and had the responsibility not only of dealing with juvenile crime, but also of protecting children. Thus, many attorneys practicing in juvenile courts viewed themselves more as **guardians *ad litem*** (persons appointed by courts to represent children and serve their best interests) than as ardent defenders of their clients' legal rights.

guardian *ad litem* A person, usually an attorney, who is appointed by the court to represent the best interests of a child involved in legal proceedings and with social service agencies.

As a result of the *Kent* and *Gault* decisions in the 1960s and the introduction of due process into the juvenile courts, attorneys, judges, litigants, and legislators are more likely to view these courts as "real" or "normal." Even so, more intense scrutiny of the juvenile courts by the U.S. Supreme Court and the public has its drawbacks. For example, many legislators and the federal government are questioning the effectiveness of juvenile courts in dealing with serious juvenile offenders. In response, many juvenile courts are attempting to implement a more **balanced approach to juvenile justice**, one that protects the rights of juveniles and families but lets the courts address the needs of other community groups, such as victims and those who feel that juveniles should be held accountable for their actions. This balance will be discussed later in the section on adjudication hearings.

balanced approach to juvenile justice The belief that juvenile justice should give equal attention to public safety, offender accountability, the needs of victims, and the correction and treatment of juvenile offenders.

Regardless of whether a juvenile court is part of an upper or lower trial court, generalist judges can still be found on the bench. This raises the question of whether juvenile matters are being given adequate attention by these judges. Indeed, if a judge is required to hear juvenile matters but is not particularly interested in the types of cases that come before the juvenile court, these cases will likely get less attention than they deserve. Even if a generalist judge has an interest in juvenile matters, he or she may have difficulty developing expertise in juvenile law, not to mention adolescent and family issues, because of the variety of cases the judge must hear.[21]

Concerned that generalist judges are devoting insufficient attention to juvenile matters, leading juvenile court advocates have called for a separate juvenile or family court with specialist judges. These advocates maintain that specialization is required if judges are to understand the complex legal, family, and adolescent issues involved in juvenile cases. They also argue that specialist juvenile court or family court judges, unlike generalists, will be able to more effectively oversee juvenile cases that sometimes take considerable time and resources.[22]

An innovation beginning to receive attention is the idea of "one judge, one family." The thinking behind this idea is that keeping one judge with a family allows that judge to get to know the entire dynamics of that family and thus to take a "holistic" approach in handling the case. Those who support the "one judge, one family" concept argue that, even though a judge may not be a family court expert, he or she could become an expert on a particular family and its problems. Of course, each judge would be responsible for multiple families, but even then the support staff, referees, and caseworkers would have the ability to know the families individually. Instead of trying cases, judges would preside over families. Although this is an intriguing concept, there are some clear drawbacks. One drawback is the possibility that a judge could lose his or her objectivity and become unable to treat the family fairly.

Another trend has been toward the establishment of a **unified trial court**. In this approach, a single trial court capable of dealing with all matters requiring legal intervention would be established in each community. Within this single trial court would be more specialized courts dealing with family matters, delinquency, traffic, civil litigation, and adult criminal behavior. Advocates of the unified trial court maintain that tying separately organized courts together would make court administration more efficient and effective, would eliminate overlapping jurisdictions, and would allow better utilization of court personnel and other resources.[23] Some argue that a unified court encompassing several specialist courts, such as courts devoted to delinquency and family matters, would be better positioned to deal with delinquency cases than a general trial court. Although there is little evidence of a trend toward the development of separately organized juvenile courts, there has been some movement toward more specialized courts dealing with juvenile and family matters within a unified trial court.

unified trial court
Comprehensive courts encompassing several specialist courts, such as courts devoted to delinquency and family matters, as well as other types of courts. Also see family courts.

Still another type of specialized juvenile court that has been established in some jurisdictions is the **juvenile drug court**. The first juvenile drug court was implemented in 1989 in Miami, Florida. Since that time, juvenile drug courts have been started in a number of states. Drug courts are special courts that handle cases "involving substance-abusing offenders through comprehensive supervision, drug testing, treatment services and immediate sanctions and incentives."[24] It is important to note about drug courts that not only do they hear cases and deal with legal issues around substance possession and use, but they also play a critical role in the coordination of treatment for youths with substance abuse problems. Indeed, these courts follow the therapeutic jurisprudence model that is designed to: (1) provide immediate intervention, (2) use nonadversarial adjudication, (3) employ active judicial intervention, (4) utilize treatment programs, and (5) have clear rules and goals.[25] Although these courts hold youths accountable for their behaviors, the judge, attorneys, probation staff, and treatment providers also work as a team with the youth and family to encourage behavioral changes.

juvenile drug court
Specialized juvenile court that handles cases involving substance-abusing offenders through comprehensive supervision, drug testing, treatment services, and immediate sanctions and incentives.

FYI

Not All Juvenile Cases Are Heard By Juvenile Courts

As noted in the previous chapter, criminal courts hear cases of juveniles who are transferred to adult courts. Also, because there is no separate federal juvenile justice system, juveniles who are arrested by federal law enforcement officials may be handled by United States District Courts.[26] Moreover, various tribal courts handle misdemeanor cases of juvenile members of recognized American Indian tribes, and federal courts have jurisdiction over felony offenses committed by tribal members.[27]

■ Juvenile Court Personnel: The Key Players

The juvenile court, of course, is more than just a structure. It is made up of a number of individuals who work within the organizational framework of the court. These individuals perform a variety of roles—roles that regularly require them to make important decisions about the lives of youths and families. Collectively, these roles and the decisions made determine the quality and quantity of justice dispensed by the court. The following sections describe some of the key decision makers who make up the juvenile court.

The Juvenile Court Judge

The most important decision maker in the juvenile court is the judge. The judge has ultimate responsibility for the operation of the court and for the legal direction the court takes. The judge exerts influence through (1) **judicial administration** of the court and (2) **judicial leadership**.

judicial administration
The daily and long-range management of a court. The judge, as head of the court, is ultimately responsible for judicial administration, but can choose to delegate many of the particular duties to others, such as a court administrator.

judicial leadership
Guidance provided by a judge in the areas of programming, personnel, and budget administration. Also shown in his or her efforts to act as children's advocate and consensus builder within the community.

As the ultimate leader of the court organization, the juvenile or family court judge may be responsible for hiring and firing of court personnel, court policies, work rules, and the level and priority of court staffing. Usually, a court administrator is hired to handle the day-to-day operation of the court, hire and fire staff, do the budget work, and perform other administrative and policy tasks. The amount of authority delegated to the court administrator and other middle management personnel in the court, however, often depends on the philosophy of the judge. If the judge takes a "hands-on" approach, he or she will take an active interest in the day-to-day operation of the court. The judge may want to review all personnel, policy, and budget decisions before they are implemented. Furthermore, court staffing may well reflect the judge's view as to what the court's emphasis should be. For example, if the judge believes strongly in probation and the development and operation of programs for youths, then extra resources might go to hiring probation officers and developing and staffing various treatment programs, assuming that the court budget makes these activities feasible. In another court, the emphasis might be on community protection, and the judge might want money spent on a detention center and out-of-home placements. The main disadvantage of a hands-on approach is that the judge's involvement in administrative matters takes time away from his or her courtroom duties. It also can lead to frustration on the part of the court administrator and other staff, who may resent a judge's constant incursions into their professional domains.

If the judge views his or her role as setting broad administrative guidelines and policies and letting the hired managers run the court, a different atmosphere is present. A hands-off approach allows court managers to do what they do best—run the court. The advantages are obvious. People in the court are able to perform the roles their professional training prepared them for. The disadvantages include the possibility that the judge will become isolated from the daily life of the court and lose touch with the employees. Another concern is that administrators, because they are hired and not elected, may not be attuned to community priorities and concerns to the same extent as judges. Thus, the ideal is for the judge to set broad guidelines for the administrator and the other managers, but to continue to actively review the progress and outcomes of court programs.

In the judicial arena, the judge is paramount. He or she is the role model for all of the quasi-judicial officers such as referees, masters, and magistrates employed in the court. The judge decides how the courtroom will be run, what cases have priority, what outcomes are preferred, and the extent to which due process is emphasized for juveniles and families. For example, if the judge decides that status offenders are not going to be dealt with by the court, that decision not only affects the types of cases heard by the court, but it also can influence the types of staff the court will need and the types of programs operated by court personnel.

Judicial leadership is not simply relegated to the court, but is found in the community as well. Judicial leadership carries into the community, as judges commonly serve on numerous community boards and committees. How the citizens of the community perceive the judge and react to that perception will have much to do with their support for the court as a whole and for individual court programs. The judge must maintain favorable relations with business and community leaders and elected officials. Many times, the court's budget is controlled by the county board or council, and how they feel about the judge may be translated into dollars and cents. Importantly, most juvenile court judges are elected officials. Once elected, they are always "running" for re-election when in the public eye, and their ability to deal with the political pressure they experience is crucial to their success and the success of the court. For instance, they may be under considerable public pressure to take a tough stand on juvenile crime, which can sometimes conflict with their desire to help youths.

BOX 10-1 Interview: The Honorable Gerald W. Hardcastle, 8th District Judicial Circuit Court, District Judge Assigned to Family Division

Judge Hardcastle spent 18 years practicing family law; is a former prosecuting attorney in Clark County, Nevada; and has served on the family court bench for 12 years in a combined family court system that hears all neglect, delinquency, adoption, divorce, custody, and guardianship cases.

Q: Judge, how did you come to serve on the family court bench?

A: The six elected Family Division district court judges chose among themselves who will serve in each area of the family court's responsibility. My assignment for 8 years has been to juvenile court.

Q: What types of cases do you hear?

A: All child abuse and neglect cases. In Nevada, we have referees, sometimes called masters, who help judges with the caseload.

Q: How heavy is your caseload?

A: Last year, myself and three referees heard 64,000 cases! Las Vegas and Clark County contain 65% of the population of the state of Nevada, and our caseloads have grown tremendously in the recent past.

Q: What is your educational background?

A: I graduated from Weber State University as an undergrad and attended law school at the University of Utah.

Q: Do you support the family court concept as the best way to make the system better for families and children?

A: Yes and no. I certainly believe that matters involving families need to be dealt with in one place and by one court for the sake of convenience for the public and efficiency, but I believe that delinquency cases and child abuse and neglect cases need to be separated from the other family court areas. Special judicial expertise is required in handling these cases, and our present system of 3-year rotation does not allow any one judge an opportunity to acquire and cultivate that expertise.

(continues)

Q: Could you expand on your concerns about the need for specific judicial expertise in delinquency and neglect cases?

A: Certainly. Juvenile court judges have a number of roles to fulfill. They are looked at by the community to be leaders in advocating for the needs of children. It takes some time to understand these needs and the political and economic factors influencing them. Any court needs consistent policies, and if these change every 3 years with a new judge—which they tend to do—the court processes are continually disrupted. For example, if the outgoing judge advocates for foster placement of delinquents and the new judge advocates for institutionalization, the whole course of services and funding is impacted.

Q: In Clark County does the juvenile court have responsibility for delivering services to children and families?

A: No, and this is frustrating to a certain extent. Judges, as elected officials, have priorities and are expected by the public to possibly have more control over children than we do. I'm lucky, however, to have an excellent family services director in Kirby Burgess, and he and I work closely in planning for services.

Q: One of the "burning" issues in juvenile justice today is the issue of waiver or transfer of juveniles to adult courts. What are your opinions on this issue?

A: I am glad you asked. One of the most important challenges facing juvenile courts today is the need to effectively deal with the chronic serious delinquent offender. Juvenile courts need to accept this challenge and find ways to reach these kids and protect the community. Unfortunately, legislatures across the country are taking any opportunity for the juvenile courts to meet this challenge away from them with automatic transfer legislation. This is done even though there is ample evidence that the adult criminal justice system has no better solution to the problem. It doesn't work with adults, why would anyone think it will work with juveniles?

Q: Judge, could you share your thoughts on how you would change the system?

A: I believe the most effective intervention is on the local or neighborhood level. The juvenile court should use neighborhood centers for intake and mediation resolution. Only if this did not work would the formal court process be used. Along with this idea, the only way to treat families is over the long term, with a consistent continuum of services. In this area, the court needs to work closely with the Department of Family and Youth Services.

Q: What have been your most gratifying experiences while sitting as a juvenile court judge?

A: I believe that any success with families comes about when dedicated workers try any number of things to help kids and families and when the court has the flexibility to be willing to try a number of things. We have good workers in the court and the department who really care about what they do. They have a commitment to children from the director on down. My goal is to help children. My focus is to be sure that they're treated fairly. The judge must be the person who establishes the perception of fairness.

In large urban counties with several juvenile court judges, the judges typically elect one of their number to be chief judge, usually on the basis of seniority,[28] although the chief judge in some jurisdictions is chosen by the state's highest court. In a statewide juvenile court system like that found in Massachusetts, there is a chief justice who is appointed by the governor and acts as the overall administrator for the juvenile court divisions around the state. Regardless of how the chief judge is chosen, and the range of their administrative authority (e.g., county or statewide system), the chief judge has the ultimate responsibility for court administration and judicial leadership, although daily court operations may be delegated to a court administrator. In large courts, the burden of this responsibility can be extremely taxing, and therefore good administration is even more important in these courts, whether administration is directed by the judge, an administrator, or a judge and administrator team.

FYI

Most Juvenile Court Judges Are Elected

Juvenile court judges are selected in a variety of ways. In most states, they are elected in partisan or nonpartisan elections. In other states, they are appointed by the governor from a list of candidates chosen by a screening board. In Connecticut, a Judicial Selection Commission identifies and recommends qualified candidates to the Governor, who must select an individual from the list of candidates. The nominee is then sent to the General Assembly's Judiciary Committee, who must confirm the appointment after a public hearing. Finally, both chambers of the state legislature must approve the nominee.[29] In South Carolina, candidates must be screened and found qualified by the Judicial Merit Selection Commission. Then qualified candidates are elected by a vote of the legislature.[30] Also, about a dozen states have adopted the Missouri plan. Under this plan, an elected official, usually the governor, appoints a candidate from a list compiled by a commission. Once appointed to the bench, however, incumbent judges must run on their records in nonpartisan and uncontested elections.[31] Although there are many dedicated and competent judges who sit on the juvenile court bench, present methods of selecting juvenile court judges do have some drawbacks. Because successful political campaigns require substantial amounts of money, a heavily bankrolled politician can prevail over a more qualified candidate. Moreover, the appointment of candidates to the bench is often a highly political affair, and work for the party in office may be treated as more important than judicial qualifications.[32]

Juvenile court judges wield tremendous power over juveniles and families who come before the court, and they exercise wide discretion in how they respond to cases. This is most clearly seen when judges remove children from their homes. Indeed, the ability to take juveniles from their homes and detain them or impose some other out-of-home placement represents a conspicuous example of the power vested in judges. Yet, juvenile court judges exercise tremendous power in a variety of other ways—they issue orders that require youths to seek treatment, obey their parents, avoid unsavory persons or places, attend school, cooperate with probation officers, adhere to curfews, and engage in other actions judges feel are appropriate. In addition, juvenile court judges may also order parents to engage in (or refrain from) certain actions. For example, a judge may order parents to attend counseling, transport their child to court-ordered counseling, clean their home, ensure that their child attends school, cooperate with probation officers, and pay some or all of the costs of the services provided by the court. Furthermore, a juvenile court judge may hold the parents or guardian of a child in contempt of court and have them jailed for not complying with court orders.

Clearly, juvenile court judges have considerable power. Not surprisingly, some juvenile court judges abuse their power and act like tyrants or dictators when they are on the bench. Many others, however, are dedicated and able jurists. These judges are careful

FYI

Judges Are Rewarded With Substantial Salaries

In 2005, the median salary of a judge in a general jurisdiction trial court in the United States was \$112,777.[33]

to protect the rights of juveniles and families before the court, and they strive to balance the best interests of children and families against the need to protect public safety.

Juvenile Court Referees and Other Quasi-Judicial Hearing Officers

quasi-judicial hearing officer Hearing officer empowered by the court to hear cases. Hearing officers are often attorneys and hear a variety of cases, although state law typically places some limits on their authority.

Among the critical players in many contemporary juvenile courts are individuals appointed by judges as **quasi-judicial hearing officers**. These individuals, who are usually attorneys, are referred to by such titles as "referee," "commissioner," "master," "administrative law judge," and "magistrate." No matter what they are called, their primary role is to hear cases. Indeed, in some jurisdictions, the great majority of cases heard in juvenile courts are presided over by quasi-judicial hearing officers who are not judges.[34]

Although a variety of factors often enter into the selection of referees and similar quasi-judicial hearing officers, three stand out: (1) knowledge of and expertise in juvenile law, (2) judicial demeanor and interpersonal skills, and (3) ability to assist the court in handling its caseload. As noted earlier, the vast majority of judges are elected. Unfortunately, however, judicial elections are often treated as less important than those for other offices. Furthermore, voters frequently have a hard time distinguishing one candidate from another because ethical considerations prevent judges from campaigning "against" their opponents. As a result, judges are often elected because of name recognition or political connections rather than expertise in the appropriate area of law.

Referees or similar quasi-judicial hearing officers who are appointed by judges, on the other hand, are frequently chosen for their expertise. In many instances they constitute the true repository of knowledge regarding the applicable case and statutory law in the area of juvenile justice, and in some courts they have far more experience on the juvenile court bench than the sitting judges. In addition to their legal knowledge, their judicial demeanor (i.e., how they conduct themselves on the bench) and their interpersonal skills can be highly valued by juvenile court judges.

It has long been recognized in many states that judges alone cannot handle the large volume of cases that are referred to the juvenile court. There are also a number of minor judicial tasks that, from a practical and an economic point of view, could better be performed by someone else. Consequently, as the volume of work in juvenile courts has increased, the use of quasi-judicial hearing officers has expanded to keep pace. In many courts, these people preside over most of the same types of hearings as judges, and their recommendations are treated as having the same force and effect as the judges' orders. Nevertheless, most states place some limits on the authority of referees and other quasi-judicial hearing officers, such as preventing them from conducting waiver hearings or presiding over jury trials where they are available.[35]

MYTH VS REALITY

Quasi-Judicial Hearing Officers Often Have Considerable Expertise in Juvenile Justice

Myth—Juvenile court referees (or masters, commissioners, and magistrates, as they are called in some states) lack the expertise of juvenile court judges.

Reality—Quasi-judicial hearing officers may be hired by a juvenile court judge precisely because of their knowledge of juvenile and family law. In addition, some quasi-judicial hearing officers have far more experience on the juvenile court bench than some juvenile court judges.

FYI

Referees Have Substantial Authority to Act on Behalf of Judges

Referees and other quasi-judicial hearing officers are often **on call** or on weekend duty because of state laws that require a judicial review to be conducted within 24 to 48 hours following the placement of a juvenile in detention. Their role as on-call judicial officers gives referees substantial authority to determine which juveniles should be kept in detention and for what types of offenses. They also conduct preliminary hearings or arraignments, where they make important decisions about bond and the need for further court action. This screening function is crucial for the efficient processing of cases.

on call Availability for emergency service. Some jurisdictions require that a judicial officer be available to make emergency decisions about the placement of alleged juvenile offenders.

Administratively, referees and other quasi-judicial hearing officers are often looked to as the people to go to for legal advice in the court. Increasingly, they are required to be attorneys, and in many courts they are more accessible to the line staff than the judges. In some courts, quasi-judicial hearing officers act as legal advisors to the court administrators and thus have a significant influence on court policy. They are also popular in many jurisdictions because they are uniformly less expensive than judges. Their salaries are usually significantly less, and they may require less support staff. For example, a judge may need a personal secretary and a court recorder. Referees frequently end up doing their own recording and using the court's clerical pool for processing orders, reports, and other legal documents. As caseloads and docket pressures have increased in many jurisdictions, the creation of quasi-judicial hearing officer positions has become a popular way of dealing with these pressures, and they are seen as economically efficient alternatives to the creation of new judgeships.

The use of referees and other quasi-judicial hearing officers in juvenile courts does have some potential problems. In jurisdictions where quasi-judicial hearing officers hold a majority of the hearings, they are arguably "judicial substitutes rather than judicial supplements,"[36] despite claims that they are intended to assist judges, not replace them.[37] Moreover, heavy reliance on referees may send a message that juvenile court matters are not sufficiently important to merit more judges, thus diminishing the stature of the juvenile court. Still another problem is that judges, in situations where they are required to review the findings and orders of referees, may do little more than rubber-stamp them. Finally, because referees are hired by judges and serve at their pleasure, a newly hired referee would not have the same degree of power and independence that a judge in a newly created judgeship would have.[38]

The Juvenile Court Administrator

The court administrator is the manager of the court. He or she has primary responsibility for (1) personnel, (2) the budget, and (3) programming. The court administrator hires and fires employees, interviews new employees, oversees employee performance evaluations, negotiates with any collective bargaining units, and acts as a liaison with other government agencies on employment-related matters. For example, many juvenile courts are subject to the financial controls of state or local governments (county or city). Usually, these governments have personnel offices and directors who work with local employees in animal control, police, fire, mental health, and parks and recreation departments. In some jurisdictions, court employees are also local government employees and are subject to the same policies as other such employees. These situations require the

BOX 10-2 Interview: R. Scott Ryder, Tribal Court Administrator for the Nottawaseppi Huron Band of Potawatomi Indians and Associate General Counsel.

Mr. Ryder is the former Juvenile Division Director/Referee, 45th Circuit Court, Family Division, St. Joseph County, Michigan and Chief Referee for the Family Division, Ninth Judicial Circuit Court, Kalamazoo, Michigan

Age: 55

Education: BA History, Wittenberg University; JD, Indiana University

Salary: $60,500

Q: Could you tell me about your experiences as court administrator and referee?

A: I would be happy to.

Q: What were your responsibilities as an administrator and how did you like them?

A: As a court administrator I conducted hearings from time to time—I have presided over more than 20,000 hearings in my career—primarily I worked on financial issues, including preparing and overseeing the budget; personnel issues, hiring, firing, and contract negotiating; court scheduling and docket control; legal research and policy preparation; acting as a liaison between the court and other public and private agencies dealing with juveniles; and finally just managing the day-to-day activities of the court. I enjoyed the challenge and the opportunity to be involved in all facets of the court's operations.

Q: You worked for over 25 years as chief referee in a juvenile court and you still heard cases. What did you like about the referee position?

A: I liked the challenge of making decisions that can have a positive impact on children and families. I also liked the challenge of articulating community standards and the challenges that arise in the course of hearings, and I enjoyed very much the people I worked with.

Q: What were the most difficult challenges you faced as chief referee?

A: Being in a supervisory position is always a challenge, especially when you are responsible for supervising independent judicial officers, because there is not always a simple "right" decision. My approach has always been to ensure that the right procedures and policies are followed. Of course, any time one works in government there are a variety of political issues you must face. Unfortunately, political decisions that influence juvenile justice are often made without input from people who work in the field. Another challenge was working with families. There is a lot of poor parenting, and it can be difficult trying to get parents to be more responsible and positively involved in their children's lives.

Q: What was a typical presenting problem in the cases you heard in court?

A: An absence of responsible parenting. Many of the parents didn't properly supervise their children, impart appropriate values to them, or support their children in positive activities. Too many parents were so caught up in their own lives that they failed to play a positive role in the lives of their children.

Q: What qualities does it take to be a successful hearing officer?

A: You have to know family and child law, how the court works (what it can and cannot do), and how it fits into a continuum of services for youths and families. You have to know the fundamentals of appropriate parenting. You must be able to be confrontational when necessary, but you also need to be patient and understanding. I also think it is important to have a deep compassion for those involved in the court and a genuine desire to help them make their situations better.

Q: What do you feel is unique to the referee position in the court?

A: Referees can bring some unique qualities to the court that, unfortunately, are not always recognized. Because referees are not elected, they can be more independent. Also, referees can bring tremendous experience to the court that can take judges years to develop.

(continues)

Q: What are the challenges that you faced as a court administrator?

A: St. Joseph County is a small rural county in southwestern Michigan. The entire court budget was slightly under $1 million, including employees' salaries and fringe benefits. The court always had to be diligent in overseeing how the money was spent because economic resources influenced our ability to help children and families. Consequently, the court had to be creative in utilizing community resources and do a better than "good job" at determining which youths threaten community safety and should be removed from the community. It also meant the court had to be very proactive, not only in developing local programs, but also in networking with local public and private agencies to coordinate services for youths.

Q: What qualities does it take to be successful as a court administrator?

A: You have to know family and child law, how the court works, legally what it can and cannot do. You must understand the local services culture, which agency does what, who you need to contact so that resources can be directed to those who need them most. Most of all you must understand how to manage, motivate, and lead people. You must understand the strengths and weaknesses of your employees, how to best motivate them, and what incentives or discipline works with each. You must be a good listener, open always to both sides of the story, and, finally, you must be decisive. Nothing paralyzes an organization more than leaders who do not lead. I was a "middle" manager, so I had to know my position in the county hierarchy and understand that my real bosses were the citizens and taxpayers of the county. Political understanding, "moxey," can never be overlooked in importance. Finally, you have to understand the judge you work for, how he or she works, his or her philosophy, and how he or she intends to implement it.

court administrator to act as a liaison and coordinator between the court and the local government. If the court is a separate government unit, then the court administrator is primarily responsible for developing and implementing personnel policies, conducting and overseeing employee evaluations, and engaging in other employment-related functions. His or her actions, however, may be subject to the final approval of the judge.

In addition to their employment-related responsibilities, juvenile court administrators also have budget responsibilities. Many juvenile courts have large budgets, and it is the administrator's job to create a responsible budget and then make sure the court uses its monetary resources wisely. Budgetary items needing consideration may include (1) capital expenses for buildings and grounds, (2) employee salaries and fringe benefits, (3) placement costs for out-of-home care, (4) juvenile detention center costs, (5) court-appointed attorney fees, (6) witness fees, (7) security costs, (8) equipment costs and amortization, (9) training costs, and (10) mileage and travel costs. It is evident from this list that a great deal of economic tradeoff and balancing must be done to meet the needs of children, families, and the community without forgetting the employees of the court who provide services to the court's clientele.

Finally, the court administrator must exercise leadership in developing and implementing programs to serve children and families and to accomplish the mission of

FYI

Court Administrators Are Some of the Highest Paid Professionals in Juvenile Justice

The median salary for a state court administrator in 2005 was $114,205.[39]

the court. To do this effectively, the administrator must be familiar with the needs and resources of the community, the preferences of the judges, and the economic constraints of the court budget. Consequently, the court administrator must be knowledgeable in a variety of areas, including (1) community corrections, (2) diversion alternatives, (3) detention resources and secure placement options, (4) group and foster care programs, (5) substance abuse treatment, (6) mental health options, (7) domestic violence programs and shelters, and (8) institutional placement options. In many jurisdictions, the court actually operates some of these programs and must fit them with other programs to provide a **continuum of care** for children. The creation of a continuum of care requires not only staffing and managing court-operated programs, but also connecting with other state and local resources and agencies that operate programs for children and families. The administrator must be able to develop needed programs, see that they are properly staffed and managed, evaluate their effectiveness, and know when resources outside of the court must be used to aid in the treatment of juveniles and to protect the community.

continuum of care
A range of comprehensive and connected services for families and children.

The Prosecuting Attorney

The prosecuting district, state, or commonwealth attorney is the chief law enforcement officer of the county, district, or local government. All police work goes to the prosecutor for review, and in most cases only the prosecutor can issue charges of delinquency against juveniles (see Figure 10-3). Indeed, the prosecutor plays a key role in determining which cases will go to court, what the specific charges will be, which cases will be considered for waiver to adult courts, and what the disposition of each case will be.[40] Given the amount of discretion vested in the office of the prosecutor, the prosecutor can virtually control the juvenile court's delinquency docket. Moreover, by deciding what kinds of cases are to be charged, the prosecutor helps determine community standards of acceptable conduct. The decision, for example, to enforce a community curfew makes a statement to youths and their parents about when minors should be at home in the evening. By enforcing a standard for behavior, the prosecutor articulates the community standard, sets forth consequences for violation of that standard, and influences the behavior of some youths and their parents.

Prosecutors, in many states, also have legislatively mandated obligations to victims. In a very real sense, when prosecuting a crime, a prosecutor becomes the victim's attorney. Only recently have victims been recognized as more than just necessary witnesses.

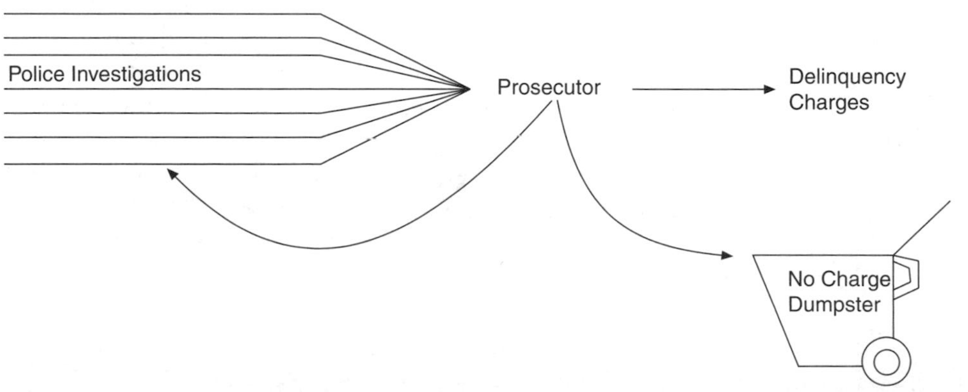

Figure 10-3 The Role of the Prosecuting Attorney

Source: Courtesy of R. Scott Ryder.

As a result, the prosecutor may have an obligation to seek restitution on behalf of a victim or the victim's estate and to take special measures to protect the victim both in the court proceedings and out of court.

Prosecutors also play a crucial role in ensuring that the court operates as efficiently as possible. Any court has a finite amount of courtroom time available. An attempt by a prosecutor to charge each juvenile and adjudicate each case would more than use up available courtroom time and would completely exhaust judges, referees, and other court personnel. By careful use of plea negotiating, the prosecutor can save the state's resources for the most important cases and keep the juvenile court docket from getting clogged up.

The Defense Attorney

It is the defense attorney's job to represent the wishes of his or her client (e.g., an accused juvenile), ensure that the client's rights are not violated, and present the client's case in the most favorable light, regardless of the client's involvement in illegal activity.[42] In order to provide effective legal representation, the juvenile defense attorney must establish a good relationship with the client and strive to understand the client's needs and wishes.[43] Theoretically, the attorney must follow the wishes of the client and vigorously defend the client's interests,[44] but many attorneys believe they also should act as an advisor. As mentioned earlier, a defense attorney typically wants to ensure that his or her client receives any help and assistance needed, but the attorney must also ensure that the court has legal grounds for exercising authority over the youth.[45]

What usually happens is that the defense attorney insists on the juvenile's due process rights at the initial detention hearing and continues to insist on them through the adjudicative or trial stage. After a conviction by trial or plea has been obtained, however, the defense attorney may become more of a guardian *ad litem*, focusing on the juvenile's best interests. A guardian *ad litem* is a person appointed by the court to represent the best interests of a child involved in legal proceedings and with social service agencies. When the attorney assumes this role, then he or she gives more consideration to the needs of the client as a child than to the expressed wishes of the client as a legal defendant.

Defense attorneys also can influence the court's docket by deciding when they will take a case to trial and when they will attempt to negotiate a plea. For example, in 1997, Michigan enacted a sexual offender registration act that requires all people who are convicted of a sexual offense to be placed on a list available to local law enforcement agencies. Before the registration law, many attorneys would plea bargain sexual offense cases out of concern for their clients' future and possible need for treatment. Now most of these cases are going to trial because of attorneys' desire to keep their clients from suffering the stigmatization that goes with appearing on the sexual offender list. This is a good example of the unintended consequences associated with new legislation. The legislature

FYI

Prosecutors Should Balance Juveniles' Interests with Those of the Community

According to standards developed by the National District Attorneys Association, prosecutors are obligated to protect the safety and welfare of the community, including crime victims. However, they also should consider the interests and needs of juveniles to the extent possible without compromising their obligations to the community.[41]

wanted communities to know who convicted sexual offenders were, for obvious public safety reasons, but the unintended results of the law appear to include an increase in sexual offense trials and a corresponding increase in court expenses.

To understand the role of defense attorneys in juvenile court, it is necessary to distinguish the ways attorneys become involved in the court. Usually, attorneys who appear in delinquency cases are (1) client retained, (2) court appointed, (3) legal aid or public defender attorneys, or (4) members of a consortium of attorneys.

Client-Retained Attorneys

These attorneys typically represent clients in criminal proceedings and are hired by the youth's parents to represent the youth in court. In many instances, however, privately retained attorneys know little about how the juvenile court operates. In fact, it may be to a client's advantage to have a court-appointed attorney, because an attorney who practices regularly in juvenile court knows the court process and personnel. On the other hand, court-appointed attorneys may have relationships with the court or prosecution that prevent them from acting as zealous advocates for their clients.[46]

Court-Appointed Attorneys

Court-appointed attorneys usually must apply to the chief judge or court administrator to be placed on a list of attorneys available for representing clients in the juvenile court. Minimum levels of training and expertise are often required for qualification, and continuing education is frequently mandated. Court-appointed attorneys are usually called by the court on a rotating basis, and parents of children represented are often required to meet with the court's finance officer to determine the family's ability to reimburse the court for attorney fees.

Legal Aid or Public Defender Attorneys

These attorneys often appear in juvenile courts because of a contract between the court and their agency to provide legal services. The extent of the services provided is determined by the contract.

Attorney Consortiums

These attorneys belong to a group of attorneys who have joined together in order to offer their services to the court for a set amount of money. Although a contract between the court and the consortium dictates the terms of the representation, these private attorneys have a specific interest in juvenile court practice.

Probation and Other Casework Staff

In addition to hearing officers and attorneys, a variety of caseworkers play important roles in the juvenile court process, including court hearings. One of the most important is the probation officer. Probation officers perform six basic roles in juvenile courts: (1) screening cases in intake, (2) conducting presentence investigations, (3) supervising and monitoring youths' adherence to their rules of probation, (4) providing assistance to youths placed on probation, (5) providing ongoing assessments of clients' needs, and (6) completing a variety of job-related administrative tasks. In other words, probation officers are the people who may visit youths' homes, locate youths who are not where they are supposed to be, confront parents when they fail to assist their children, check with school and other agency personnel to monitor youths' behavior and progress, and when necessary, request that a youth's probation be revoked. They may also conduct investigations, provide testimony and reports for court hearings, work with attorneys in developing dispositional plans, serve summons, and make arrests.

Probation officers or other caseworkers may be employed in specialized programs operated by or on behalf of courts. For example, juvenile courts or other government agencies may operate a variety of programs, such as diversion programs and foster care. Furthermore, they may also operate their own shelter care units, group homes, and detention centers. Probation officers or other caseworkers may be used to staff these programs or be given responsibility for providing casework services to youths in these programs.

Probation and other caseworkers who work for or with juvenile courts clearly perform a number of vital tasks. Indeed, those who occupy these positions can influence which youths penetrate further into the juvenile justice process. They can influence youths' involvement in a variety of juvenile justice programs, from diversion to probation to institutional placement. They can also influence how long youths are involved with the juvenile court. In short, probation officers and other caseworkers act as the eyes and ears of the court. These individuals gather and analyze different types of information and make important recommendations to the court that almost always carry weight with judges and other juvenile justice decision makers.

The Adjudication and Disposition Hearings

All of the key players in the juvenile court—judges and other quasi-judicial hearing officers, prosecuting and defense attorneys, probation officers, and other caseworkers—make decisions that can have a profound effect on youths and families. Although decisions about youths and families are made at many points in the juvenile justice process, they typically are made either in preparation for critical events or during critical events. Previous chapters examined several of these critical events, including the arrest, the juvenile court intake process, and the detention hearing. This section examines two more critical events in juvenile justice: the adjudication hearing and the disposition hearing. These hearings are significant decision-making points in the juvenile justice process because they determine which youths will fall under the formal jurisdiction of the court and how the court will handle those youths.

The Adjudication Hearing

The **adjudication hearing** is a critical event in the juvenile justice process that serves as the juvenile court equivalent to a criminal trial.[47] It is the adjudication hearing that determines whether the juvenile comes within the jurisdiction and under the formal authority of the court. If the juvenile is adjudicated, the court has the power to issue orders affecting the juvenile and his or her parents (or custodian or guardian). Without an adjudication, the court has no legal authority to intervene in the life of the juvenile and his or her family.

adjudication hearing
The juvenile court equivalent of an adult court trial. It is at the adjudication that the juvenile court attempts to ensure due process, find the facts, uphold the law, and make decisions concerning jurisdiction over the juvenile.

The adjudication hearing can take one of two forms. It can be a plea-taking hearing, in which the juvenile admits to a delinquency offense, or it can be a trial. A plea-taking hearing is by far the most common type of adjudication hearing.[48] Ideally, at plea-taking hearing, the purpose and potential outcomes of the plea and the rights that the juvenile and parents have are clearly explained to them. If the plea is the result of plea negotiating (or plea bargaining) between the prosecuting attorney and the juvenile, which is common in juvenile justice, the terms of the negotiation must be placed on the court record, including what the juvenile gets from the plea and what he or she gives up.[49] Usually, the juvenile admits to a less serious delinquency charge or fewer charges, the tradeoff

plea bargaining Negotiation of an agreement between the prosecuting and defending attorneys, often with the consent of the court, to have the accused plead guilty to a reduced charge or to fewer charges than originally brought. The prosecutor benefits by gaining a certain conviction and the defendant benefits by avoiding the possibility of a more severe sentence.

opinion testimony Testimony based on one's opinions as opposed to objective facts.

hearsay evidence Second-hand evidence, such as a witness's testimony that he or she heard someone say something.

rules of impeachment Process of demonstrating that a witness is not telling the truth or does not have the knowledge to provide specific testimony.

pretrial discovery Part of the pretrial process in which the defense and the prosecution request information from the other side in an effort to discover pertinent facts regarding a case.

FYI

Plea Bargains Are Common Events in Juvenile Justice

A **plea bargain** is a deal between the defense attorney (who represents the defendant) and the prosecuting attorney (who represents the state), according to which the defendant agrees to plead guilty to a lesser charge or to fewer charges than originally at issue. Like other deals, both sides get something and give up something. What the prosecutor gets is a conviction and the court's jurisdiction and power over the juvenile. What the prosecutor gives up is the opportunity at a trial to convict the juvenile of all of the charges or of the more serious charge. What the juvenile gets is an adjudication on a less serious charge or on fewer charges. What the juvenile gives up is the opportunity to have his or her day in court and be found not guilty.

being that the juvenile avoids the chance of a more severe sentence and the prosecuting attorney is assured of a conviction.

The juvenile's due process rights at the adjudicative phase are virtually identical to those of adults at trial. Consequently, in order to take a valid plea, a complete advice of rights should be given and an understanding waiver or "giving up" of those rights should be placed on the record. Unlike an adult proceeding, however, the juvenile's parents should also be consulted about whether they agree with the waiver of rights. By accepting a plea at the adjudication, the court not only acquires formal jurisdiction over the child, but it will likely exercise authority over the parents as well.

If the juvenile chooses to contest the delinquency charges, the only resolution of the matter is to have a contested adjudication hearing or trial. In many jurisdictions, contested adjudications proceed much like trials in criminal courts, although there can be considerable variation in the court rules and procedures that govern juvenile court adjudications. In some jurisdictions, many, if not most, of the procedural rules used in criminal trials are used in juvenile proceedings. In other states, there are no clear statutory requirements regarding rules of evidence in the areas of "**opinion testimony**, **hearsay evidence**, **rules of impeachment**, and so forth."[50] As a result, hearsay evidence from social investigations is sometimes used in juvenile court adjudications. This practice would not be acceptable in criminal courts because information about how children are doing in school, their peer associations, and their family circumstances are not relevant to their guilt or innocence.[51] In still other states, civil law rules of evidence are used in delinquency matters.[52] There are also differences between jurisdictions in the extent to which adult process such as **pretrial discovery** is afforded to juveniles. In Michigan, for

FYI

Having a Record of Juvenile Court Proceedings Is an Important Due Process Safeguard

The term *on the record*, when used to describe hearing procedures, means that an audio or audiovisual recording was made of the hearing. Having this record of the proceedings is a fundamental due process safeguard for the juvenile, because any appeal of the adjudication or disposition is based on this record. Having such a record helps prevent judges and quasi-judicial hearing officers from acting capriciously or inappropriately.

example, juveniles are entitled to limited pretrial discovery;[53] they have the right to use the defense of alibi, insanity, diminished capacity, or mental illness;[54] and motion practice is governed by virtually the same rules as in adult criminal proceedings.[55]

In regards to due process protections, there also can be differences between juvenile adjudications and criminal trials. For example, in many states juveniles have most, but not all, of the due process protections available to adults, such as the right to be proven guilty "beyond a reasonable doubt,"[56] to have the assistance of counsel, to receive written and timely notice, to cross-examine witnesses, and to remain silent.[57] However, in an important U.S. Supreme Court case, *McKeiver v. Pennsylvania*, the Court refused to guarantee a jury trial to juveniles,[58] although a number of states have granted juveniles this right. Even in these states, jury trials rarely occur, however, and they typically result in the same outcomes as uncontested cases.[59]

Given the "get tough" approach toward many juvenile offenders and the increasingly harsh punishment alternatives available to juvenile courts, it is important that juvenile defendants be afforded meaningful due process protections at the adjudication. (A complete list of the due process rights afforded juveniles in one Michigan Court is displayed in Exhibit 10-1). Undoubtedly, the most important of the protections available to juveniles at the adjudication is the right to be represented by counsel. Representation by counsel is critical because attorneys are trained in the law and have an obligation to ensure that the rights of their clients are protected. Despite the obvious importance of this due process right, many juveniles appear in juvenile court without attorney representation. For example, in a study based on juvenile court data in six states, Barry Feld found that, in cases where a petition had already been filed, only about half of the juveniles were represented by counsel.[60] Similar findings have been presented in other studies of attorney representation in juvenile courts.[61] These findings raise questions about the extent to which due process protections are in fact a reality in many juvenile courts, despite U.S. Supreme Court rulings mandating these protections.

Hearing officers in juvenile courts must be sufficiently cognizant and protective of the due process rights of the juvenile if the court is to have credibility with those it is intended to serve. If, after a conviction at trial, the juvenile or his or her family believes that the trial was unfair or that their rights were slighted by the court, cooperation at the disposition will be affected. Even though many juveniles will continue to proclaim their innocence after trial, it is crucial that the adjudication be conducted in a clear and fair manner in order to facilitate subsequent court decisions made at the disposition hearing.

The Disposition Hearing

The **disposition hearing** is "the primary feature that distinguishes the juvenile system from the adult criminal court."[63] Unlike the sentencing in criminal court, where a sanction

disposition hearing
The sentencing phase of the juvenile court process. During this phase, the court tries to establish individualized plans for juveniles that balance rehabilitation and community safety.

MYTH VS REALITY

Juveniles Have Some but Not All of the Same Rights as Adults

Myth—Juveniles have the same due process protections at trial as adults.

Reality—Juveniles do not have a right to a jury trial in all states, and even in states where this right is available, it is rarely exercised. Moreover, juvenile hearings are often characterized by practices that would not be acceptable in criminal court proceedings.

Judge/Referee ____________ APA ____________ Date ____________

❑ Minor waives Attorney ❑ Parents/Guardian agree with attorney waiver
Minor's Attorney ____________ ❑ Parents/Guardian present

❑ Proceedings on the Record:
Plea of ❑ Guilty ❑ Original Charge ❑ Lesser ____________
❑ Nolo Contendere

1. ❑ Court has stated why plea is appropriate.
2. ❑ Evidence presented to support finding of guilt.

ALL PLEAS—MINOR PERSONALLY ADVISED

❑ Nature of charge
❑ Disposition Court could impose
❑ Minor on probation
❑ Plea admits violation of probation

❑ Parents/Guardian contest verbally to minor waiving these rights. (Be sure both parents state so on record.)

MINOR ADVISED THAT BY PLEA HE GIVES UP

❑ Jury trial
❑ Trial by just judge without jury
❑ Presumption of innocence
❑ Proof beyond reasonable doubt
❑ Compulsory process
❑ Confront witness
❑ Remain silent
❑ No adverse inferences from silence
❑ Right to testify

DISPOSITION AGREEMENT

❑ Terms on record
❑ Petitioner agrees ❑ Minor agrees ❑ Minor's attorney agrees
❑ Minor's parents/guardian agree
❑ Court states if prior agreement in plea or disposition

MINOR ASKED

❑ Promised anything beyond stated disposition
❑ Threatened
❑ His choice to plead ❑ Parents agree

COURT SATISFIED

❑ Plea freely, voluntarily made
❑ Crime committed (Question minor for elements)
❑ Minor involved or took part in
❑ Plea accepted

Disposition Date ____________

Exhibit 10-1 Probate Court for the County of Kalamazoo, Michigan, Juvenile Division: Delinquency Guilty Plea Checklist

Source: Courtesy of R. Scott Ryder.

FYI

Efforts to Provide Juveniles with Quality Legal Representation Early in the Juvenile Justice Process are Made in Some Communities

One example is the Youth Advocacy Project (YAP), which provides legal representation to youths in Massachusetts juvenile courts in Boston, Roxbury, Dorchester, or West Roxbury. YAP takes a holistic approach to working with clients by examining problems that led to the youths' involvement in the juvenile justice process and by coordinating services with community agencies in order to meet youths' needs during and after court involvement.[62]

is applied to the offender, the juvenile court disposition is intended to assist the youth and protect the community. It is at the disposition hearing that formal plans designed to meet the various needs of the youth, the family, and the community are initiated. It is also at this hearing that the judge or other hearing officer attempts to balance the "best interests" of the youth and the need for community safety. Judges and other quasi-judicial hearing officers often have great latitude and discretion in making dispositional decisions.

In formulating a disposition, the court usually seeks a great deal of input and information, which is gathered and interpreted for the hearing officer by a caseworker (e.g., a probation officer) in the form of a predisposition report or social history. In preparing the predisposition report, the caseworker pulls together information from various sources in order to present a detailed social history of the youth and the family. The sources may include (1) school reports, (2) victim impact reports and restitution reports, (3) psychological evaluations, (4) substance abuse assessments, (5) financial statements and tax returns, (6) letters from the friends and family of the juvenile, (7) criminal histories of the juvenile and other family members, (8) child abuse and neglect history, and (9) the caseworker's own observations and conclusions. The purpose of this report, which basically outlines "problems" perceived by the caseworker and recommended responses to those problems, is to ensure the juvenile receives "individualized" justice.[64]

Juvenile court hearing officers often rely heavily on the information and recommendations in these reports in determining how best to respond to juvenile offenders. The amount and type of information contained in these reports is determined by state laws, court procedures,[65] and local custom. In some jurisdictions, predisposition reports are very lengthy and reflect a detailed investigation into the background and present circumstances of the juvenile and the family. In other jurisdictions, only limited information about the juvenile's background is included.[66]

In most instances, hearing officers follow the recommendations contained in the predisposition report.[67] However, this may be because probation officer recommendations are typically well within the hearing officer's beliefs about what is appropriate in particular types of cases. Indeed, as discussed in the next section, there is considerable evidence that both legal factors (such as the seriousness of the offense, the youth's prior record, and earlier decisions to detain a youth) and nonlegal factors (such as race, ethnicity, and gender) play a significant role in court dispositions.

FYI

The Predisposition Report Is an Important Document in the Juvenile Court

Although predisposition reports aid judges and quasi-judicial hearing officers in deciding on appropriate dispositions, they contain a considerable amount of opinion and hearsay evidence. Indeed, much of the information contained in a typical predisposition report would not be legally admissible during the adjudicatory hearing. Consequently, the hearing officer must not see any of this information prior to the trial or plea. Furthermore, the caseworker should not begin his or her investigation and preparation of the report until after jurisdiction is obtained at the adjudication. Initiating a sentencing investigation, which clearly implies a presumption of guilt before any guilt has been formally determined, undermines the presumption of innocence and gives the juvenile reason to doubt the court's objectivity. Unfortunately, predisposition investigations are sometimes started prior to adjudication, and hearing officers sometimes view the information gathered prior to adjudication.

Given the importance of the disposition hearing, it is critical that the juvenile receive competent representation by counsel. Moreover, it is important that the juvenile and his or her counsel have access to copies of all documents that are considered by the court in formulating the disposition—a practice supported by the U.S. Supreme Court in *Kent v. United States*[68] (see Chapter 5). Also, all documents considered by the court should be marked as exhibits and entered as evidence, **on the record**, at the dispositional hearing. Despite the obvious importance of attorney representation, in some jurisdictions many, if not most, juveniles are not represented by counsel at the disposition hearing.[69] Moreover, site visits by researchers from the American Bar Association Juvenile Justice Center, the Juvenile Law Center, and the Youth Law Center revealed that when attorneys did represent youths at dispositions, they often felt ill-prepared because of high caseloads and the lack of resources and support staff.[70]

on the record Recorded in some fashion. Making audio or audiovisual recordings of critical hearings, including adjudication and disposition hearings, can facilitate the appeal of a verdict and, in general, provides due process protection.

Although the predisposition report plays an important role in formulating the disposition, the hearing officer is not bound by the recommendations of the caseworker. Furthermore, state statutes, local resources, and funding level place some limits on the dispositional authority of juvenile courts. Nevertheless, hearing officers often have rather broad discretion in determining the disposition,[73] although courts are usually required to use the **least restrictive available alternative** that meets the child's needs and ensures public safety. Dispositional alternatives available to courts may include the following options:

least restrictive available alternative An available placement that restricts a juvenile offender's freedom least while ensuring the safety of the juvenile and the community.

- probation in the juvenile's own home (the most common disposition employed by courts)[74]
- placement in a relative's home on probation, or placement in a foster home on probation
- probation with restitution to the victims or probation with community service
- commitment to the state for placement in a state facility
- detention for a specified time period, then release on probation
- placement in a private institutional setting funded by the court and/or the state or a state agency
- placement on intensive probation or house arrest
- placement in a boot camp program
- in-state or out-of-state placement in a private correctional facility
- some combination of the previous alternatives

COMPARATIVE FOCUS

Most Countries Provide Attorneys to Represent Children Facing Criminal Penalties

Most countries, even many that do not provide attorneys for the average citizen, provide attorneys to represent children who face criminal penalties.[71] Moreover, according to the United Nations' Convention on the Rights of the Child, Article 37, "Every child deprived of his or her liberty shall have the right to prompt access to legal and other appropriate assistance, as well as the right to challenge the legality of the deprivation of his or her liberty before a court or other competent, independent, impartial authority and to a prompt decision on any such action."[72]

Research on Factors Influencing Dispositional Decision Making

Research on juvenile court dispositions suggests that a variety of factors may play a role in dispositional decision making, including *legal factors* (e.g., prior record, severity of offense, and prior juvenile justice processing decisions) and *nonlegal factors* (e.g., race, gender, social class, family structure, and age). The research results consistently show that prior record and severity of offense are strongly related to severity of disposition.[75] In addition, there is considerable evidence that previous juvenile justice decisions, such as the decision to detain a youth, influences juvenile court dispositions.[76]

Research on the effect of nonlegal variables on dispositions has produced mixed findings. Although the effects of age on dispositions has rarely been examined, an early study by Robert Terry found that older youths were more likely to receive severe dispositions.[77] As regards the effects of race, social class, and gender, some studies indicate that minority youths, lower-class youths, and females are more likely to receive harsher

COMPARATIVE FOCUS

Comparative Focus on Indian Justice

One of the constant challenges for the contemporary juvenile court is to continually search for programs and ideas that will efficiently and effectively use tax payers' dollars and that will work in rehabilitating juveniles in the court system. To that end, juvenile courts would be well advised to look at the emerging Indian tribal courts for philosophy as well as programming.

Tribal courts have played an increasingly important role in the movement toward tribal sovereignty that began in earnest in 1934 with the Indian Reorganization Act, which is also known as the Wheeler-Howard Act. The movement was given additional impetus in 1979 when Congress passed the Indian Child Welfare Act (ICWA).[81] Among its other provisions to preserve Indian families and to protect Indian children from non-Indian adoptions was recognition of the authority of tribes and tribal courts to intervene in state child welfare cases and even have the cases removed to tribal courts for adjudication and disposition. Not only did ICWA recognize the existence of tribal courts, it also acknowledged their primacy and competency in handling matters involving Indian children and Indian families. Since 1979, tribal courts have continued to develop and refine how they work with children and families, often combining both traditional and contemporary practices.

In understanding how tribal courts work, it is imperative that Indian perspectives on the role of courts in their society and the underlying philosophy about courts and the human condition be explored and explained. "To become human again"[82] is the phrase used by Judge/Magistrate Mike Jackson, the Keeper of the Circle of the Kake people in Alaska, to describe what it means for tribal members who have committed crimes in the community to rejoin the community as full members once again. In this phrase, he has captured the essence of Indian philosophy for courts in Indian country. But to fully understand what this phrase really means, certain legal concepts and commonly held beliefs have to be discussed.

Western legal philosophy, as embodied in English common law, assumes that: (1) the best way to arrive at the *truth* in any dispute is to use the adversarial system; (2) out of two zealous advocates opposing each other in court, the actual facts will emerge; and (3) blame or responsibility must be assigned to someone, although, in fairness, blame can be apportioned or shared by more than one person or party. These concepts were brought to the Americas by the European colonists and were, in large part, imposed upon the existing native population. The people who colonized the Americas never gave much thought to the fact that the native people had their own system of dispute resolution and justice that worked very well for them. This traditional system is worth looking at because of the valuable lessons and principles that it contains.

(continues)

Indian legal and justice concepts are closely intertwined with their religious and spiritual beliefs. Although each tribe has its own individual belief system, there are certain common themes:

- Ancestors are respected, remembered, and present in community life. What was thought and taught by them is followed as part of daily life.
- Every living object has a spirit that needs to be recognized and respected. Because human beings also have spirits, they are connected in this way with all other living things. Before tapping the maple trees for sap in the spring to make maple syrup, prayers were said and thanks were given by the Nottawaseppi Huron Band of Potawatomi Indians to the maple trees for what the tribe was taking from them. To many Indians, life is seen as a "seamless continuity" in which human existence and spiritual existence were connected and coexist.
- Observations made by Indians of the living creatures that they share the earth with demonstrate that these creatures have an important place in the overall natural system and that their innate intelligence, how they adjust to the conditions of their existence, not only shows their essential spirit, but provides valuable examples for humans to follow.[83]
- Indians focus on the connectedness of all things. The tribe is a unit that shares common ancestors, customs, and traditions that connect them with each other, with their ancestors, and with the world around them.
- The peace and harmony of the community is very important and central to the focus of Indian justice concepts.
- Ethical behavior for human beings goes far beyond human society and requires respect for and sensitivity toward all living things.

With these concepts in mind, the primary focus for Indians in dealing with criminal offenders is to reconcile them with the community. This is accomplished by the community or community representatives meeting with the offender and victim. The purpose of the meeting is to personally reconcile the offender and victim as well as arriving at a consensus of the community as to how this reconciliation can best be accomplished. The natural antagonism between offender and victim is not emphasized as much as the need for the community to be restored to peace and harmony. Focus is on the relationship, not on potential individual consequences. When the offender is reconciled with the community, then and only then has that person "become human again."

What lessons can the juvenile justice system learn from the Indians? First, rather than focusing on blame and consequences, focus should be put on community peace and reconciliation. Second, both the offender and the victim are, in reality, members of the same community and may very well end up living in that community after their involvement in the juvenile justice process. This makes the community's interest in harmony paramount! Third, systems of justice should focus on how people are connected rather than what separates them. The system should focus on restoring the sense of community rather that estranging people from the community by jailing them or otherwise removing them. Only in the very worst cases is separation felt to be appropriate, because, historically, separation from the tribe was tantamount to a death sentence.

There may be important lessons that other citizens can learn from the approach to justice practiced by Indian peoples. Perhaps, it is time to take a long look at that system and its traditions and consider how it might help us respond more effectively to youth offending, and what it could do to reduce the present prison and detention populations and costs.

dispositions.[78] However, other studies have failed to find that these variables influence dispositional decisions.[79] At least one study found that males are more likely to be treated more severely than females for criminal offenses but that females receive harsher dispositions for status offenses.[80]

Overall, the research on disposition decision making reveals mixed results. Although some studies fail to find evidence of bias in the dispositional phase of the juvenile court process, other studies suggest that, in some jurisdictions, nonlegal variables such as gender, race, and social class influence dispositions, raising serious questions about bias in dispositional decision making.

Chapter Summary

As noted at the beginning of this chapter, the juvenile court is the heart of the juvenile justice process. It is a place where previous decisions made by personnel in other agencies, including law enforcement officers and prosecuting attorneys, are supported or altered and where a variety of decisions are made that can have an indelible effect on children, their families, crime victims, and the community at large. Furthermore, not only does the juvenile court determine the outcome of individual cases, but through its legal authority it can determine how other institutions and agencies respond to children.

Although the juvenile court occupies a central position in juvenile justice, its position as a legal institution is far from secure. Indeed, there are various trends underway that will impact the juvenile court in the near future. For example, the number of cases being processed by juvenile courts continues to grow, and this growth will tax the personnel, financial, and other resources of the juvenile court for some time to come.

Another trend is toward reformation of juvenile court organizational structures. Presently, a variety of structures are found around the country. Unfortunately, some do not allow the juvenile court the status or the resources that it deserves. Moreover, some jurisdictions still employ generalist judges who hear a variety of cases and whose commitment to juvenile and family matters is suspect. Although some jurisdictions have moved toward family courts or unified court structures that may lend more prestige to juvenile and family law matters, the extent to which these changes will continue is unknown.

Of course, juvenile courts, like other juvenile justice agencies, are made up of people who perform a variety of roles. Key personnel include the judges, a variety of quasi-judicial hearing officers (referees, masters, or commissioners), court administrators, prosecuting and defense attorneys, and probation officers and other caseworkers. Clearly, the most important figure in the juvenile court (and juvenile justice in general) is the juvenile court judge. Indeed, the juvenile court judge makes decisions that not only influence the lives of youths and their parents, but also affect the operation and mission of the juvenile court as well as other juvenile justice agencies. Of course, the juvenile court judge and other hearing officers do not make these decisions without assistance. They are surrounded by a supporting cast of professionals who exercise considerable discretion and who themselves make a number of important decisions, including decisions that determine which youths move to the adjudicatory and dispositional stages of the juvenile justice process.

The adjudication and disposition hearings determine which youths come within the formal jurisdiction of the court. Moreover, it is in these hearings that attorneys, probation officers, and others provide information that assists the court in determining guilt and innocence and in making decisions about how to respond to the adjudicated delinquent. Yet, despite the significance of these hearings, there is considerable evidence that youths are not always treated fairly. In some jurisdictions, youths are not afforded representation by counsel; when youths are represented by counsel, that representation is sometimes inadequate; adjudicatory and dispositional decisions are, at times, based on opinion and hearsay evidence; and gender, race, and social class bias sometimes impact judicial decision making.

Clearly, serious problems confront juvenile courts in the United States, and in some jurisdictions these problems undoubtedly affect the quality of justice that juveniles receive. However, it should not be forgotten that most juvenile court personnel, from judges to caseworkers, are highly motivated and skilled individuals who are committed to serving the best interests of children and protecting community safety—and who are frequently able to achieve both of these sometimes conflicting goals.

Key Concepts

adjudication hearing: The juvenile court equivalent of an adult court trial. It is at the adjudication that the juvenile court attempts to ensure due process, find the facts, uphold the law, and make decisions concerning jurisdiction over the juvenile.

balanced approach to juvenile justice: The belief that juvenile justice should give equal attention to public safety, offender accountability, the needs of victims, and the correction and treatment of juvenile offenders.

best interests of the child: A catch phrase that serves as a reminder that the primary focus of a juvenile court should be on the rehabilitation of the children who come before it.

continuum of care: A range of comprehensive and connected services for families and children.

disposition hearing: The sentencing phase of the juvenile court process. During this phase, the court tries to establish individualized plans for juveniles that balance rehabilitation and community safety.

family court: A unified trial court where all cases involving families are heard, including divorce, adoption, custody, guardianship, paternity, neglect and abuse, and delinquency cases. Family court advocates claim that when all family matters are handled by one court, there is less chance for overlapping services, redundant reports, and fragmented intervention.

formally process: Handle the case in a way that leads to a formal court hearing.

guardian *ad litem*: A person, usually an attorney, who is appointed by the court to represent the best interests of a child involved in legal proceedings and with social service agencies.

general trial court: A court of high status within a state's legal system that hears felony criminal cases and civil cases with unlimited dollar amounts.

hearsay evidence: Second-hand evidence, such as a witness's testimony that he or she heard someone say something.

judicial administration: The daily and long-range management of a court. The judge, as head of the court, is ultimately responsible for judicial administration, but can choose to delegate many of the particular duties to others, such as a court administrator.

judicial leadership: Guidance provided by a judge in the areas of programming, personnel, and budget administration. Also shown in his or her efforts to act as children's advocate and consensus builder within the community.

juvenile court: The court responsible for holding hearings and making decisions regarding the disposition of juveniles who have entered the juvenile justice process. This court plays many roles, including children's advocate, program leader, fund-raiser, consensus builder, parent, and protector of youths and the community.

juvenile drug court: Specialized juvenile court that handles cases involving substance-abusing offenders through comprehensive supervision, drug testing, treatment services, and immediate sanctions and incentives.

least restrictive available alternative: An available placement that restricts a juvenile offender's freedom least while ensuring the safety of the juvenile and the community.

on call: Availability for emergency service. Some jurisdictions require that a judicial officer be available to make emergency decisions about the placement of alleged juvenile offenders.

on the record: Recorded in some fashion. Making audio or audiovisual recordings of critical hearings, including adjudication and disposition hearings, can facilitate the appeal of a verdict and, in general, provides due process protection.

opinion testimony: Testimony based on one's opinions as opposed to objective facts.

petition: A pleading to initiate a matter in juvenile court. A petition sets forth the alleged grounds for the court to take jurisdiction of a case and requests court intervention.

plea bargaining: Negotiation of an agreement between the prosecuting and defending attorneys, often with the consent of the court, to have the accused plead guilty to a reduced charge or to fewer charges than originally brought. The prosecutor benefits by gaining a certain conviction and the defendant benefits by avoiding the possibility of a more severe sentence.

pretrial discovery: Part of the pretrial process in which the defense and the prosecution request information from the other side in an effort to discover pertinent facts regarding a case.

quasi-judicial hearing officer: Hearing officer empowered by the court to hear cases. Hearing officers are often attorneys and hear a variety of cases, although state law typically places some limits on their authority.

rules of impeachment: Process of demonstrating that a witness is not telling the truth or does not have the knowledge to provide specific testimony.

unified trial court: Comprehensive courts encompassing several specialist courts, such as courts devoted to delinquency and family matters, as well as other types of courts. Also see family courts.

Review Questions

1. What is the basis for claiming the juvenile court is the center of the juvenile justice process?
2. What responsibility does the juvenile court judge have as an advocate for children?
3. How may being part of a lower-level trial court influence juvenile court operations? How may being part of an upper-level trial court influence juvenile court operations?
4. Who are the key personnel in the juvenile court, and what functions do they perform?
5. What is an adjudication hearing?
6. What due process protections exist at the adjudication hearing?
7. What is the disposition hearing, and what type of information is usually presented to the court at this hearing?
8. What is a family court, and how does it differ from a juvenile court?

9. What is the role of the prosecuting attorney in the juvenile court?
10. What is the role of the defense attorney in the juvenile court?
11. What does it mean to say that a judge is a "hands-on administrator"? Is this a good thing to be?
12. What function does the court administrator have in the juvenile court?
13. How many different roles or functions does a caseworker (such as a probation officer) have in the juvenile court?
14. What are the multiple factors that impact the disposition decision? How do courts weigh them?

Additional Readings

Bernard, T. J. (1992). *The cycle of juvenile justice.* New York: Oxford University Press.

Feld, B. C. (1993). *Justice for children: The right to counsel and the juvenile courts.* Boston: Northeastern University Press.

Puritz, P., Burrell, S., Schwartz, R., Soler, M., & Warboys, L. (1995). *A call for justice: An assessment of access to counsel and quality of representation in delinquency proceedings.* Washington, DC: American Bar Association.

Rubin, H. T. (1985). *Behind the black robes: Juvenile court judges and the court.* Beverly Hills, CA: Sage. This resource contains an insightful view of juvenile court judges.

Senna, J. J. & Siegel, L. J. (1992). *Juvenile law: Cases and comments* (2nd ed.). St. Paul, MN: West Publishing Co.

Notes

1. Sickmund, M., Sladky, A., & Kang, W. (2005). *Easy access to juvenile court statistics: 1985–2005.* Retrieved November 25, 2008 from http://ojjdp.ncjrs.gov/ojstatbb/ezajcs/.
2. Sickmund, Sladky, & Kang, 2005.
3. Sickmund, Sladky, & Kang, 2005.
4. Sickmund, Sladky, & Kang, 2005.
5. Sickmund, Sladky, & Kang, 2005. This resource was used in the calculations.

   Puzzanchera, C., Finnegan, T., & Kang, W. (2007). *Easy access to juvenile populations: 1990–2007.* Retrieved November 26, 2008 from http://www.ojjdp.ncjrs.gov/ojstatbb/ezapop/. This resource was used in the calculations.
6. Sickmund, Sladky, & Kang, 2005. This resource was used in the calculations.

   Puzzanchera, Finnegan, & Kang, 2007. This resource was used in the calculations.
7. Bernard, T. J. (1992). *The cycle of juvenile justice.* New York: Oxford University Press.

   Schwartz, I. M. (1989). *(In)justice for juveniles: Rethinking the best interests of the child.* Lexington, MA: Lexington Books.
8. Feld, B. C. (1988). In re Gault revisited: A cross-state comparison of the right to counsel in juvenile court. *Crime and Delinquency, 34,* 393–424.

   Feld, B. C. (1993). *Justice for children: The right to counsel and the juvenile courts.* Boston: Northeastern University Press.

Jones, J. B. (2004). Access to council. *Juvenile Justice Bulletin.* Washington, DC: Office of Juvenile Justice and Delinquency Prevention.

9. Feld, 1988.

Feld, B. C. (1989). The right to counsel in juvenile court: An empirical study of when lawyers appear and the difference they make. *Journal of Criminal Law and Criminology, 79,* 1185–1346.

Puritz, P., Burrell, S., Schwartz, R., Soler, M., & Warboys, L. (1995). *A call for justice: An assessment of access to counsel and quality of representation in delinquency proceedings.* Washington, DC: American Bar Association.

10. Puritz, Burrell, Schwartz, Soler, & Warboys, 1995.

11. Sanborn, J. B., Jr. & Salerno, A. W. (2005). *The juvenile justice system: Law and process.* Los Angeles: Roxbury.

12. Snyder, H. N. & Sickmund, M. (1995). *Juvenile offenders and victims: A national report.* Washington, DC: Office of Juvenile Justice and Delinquency Prevention.

13. NCJJ. (2006). *State juvenile justice profiles.* Retrieved November 28, 2008 from http://www.ncjj.org/stateprofiles/asp/using.asp.

Rubin, H. T. (1998). The juvenile court landscape. In A. R. Roberts (Ed.), *Juvenile justice: Policies, programs, and services* (2nd ed.). Chicago: Nelson-Hall.

14. Massachusetts Court System. (n.d.). *Juvenile court department.* Retrieved November 28, 2008 from http://www.mass.gov/courts/courtsandjudges/courts/juvenilecourt/index.html.

15. Rubin, H. T. (2003). *Juvenile justice: Policies, practices, and programs.* Kingston, NJ: Civic Research Institute.

16. Rubin, H. T. (1985). *Juvenile justice: Policy, practice, and law* (2nd ed.). (p. 350). New York: Random House.

17. Rubin, 1998.

18. Siegel, L. J. & Senna, J. J. (1997). *Juvenile delinquency: Theory, practice, and law* (6th ed.). St Paul, MN: West Group.

19. Rubin, 1985.

20. Siegel & Senna, 1997.

21. Rubin, H. T. (1989). The juvenile court landscape. In A. R. Roberts (Ed.), *Juvenile justice: Policies, programs, and services.* Chicago: Dorsey Press.

22. Rubin, 1998.

23. Rubin, 1989.

24. National Association of Drug Court Professionals. (n.d.). *Facts on drug courts.* Retrieved August 23, 2004, from http://www.nadcp.org/whatis/.

25. Hora, P. F., Schma, W. G., & Rosenthal, J. T. A. (1999). Therapeutic jurisprudence and the drug treatment court movement: Revolutionizing the criminal justice system's response to drug abuse and crime in America. *Notre Dame Law Review, 74,* 439–537.

26. Snyder, H. N. & Sickmund, M. (2006). *Juvenile offenders and victims: 2006 national report.* Washington, DC: Office of Juvenile Justice and Delinquency Prevention.

27. Rubin, 2003.

28. Rubin, 1985.

29. State of Connecticut Judicial Branch. (n.d.). *Online Media Center*. Retrieved November 28, 2008 from http://www.jud.state.ct.us/external/media/faq.htm#become_judge.

30. South Carolina Judicial Department. *How judges are elected in South Carolina*. Retrieved November 28, 2008 from http://www.sccourts.org/judges/howJudges-Elected.cfm.

31. Siegel & Senna, 1997.

32. Rubin, H. T. (1985). *Behind the black robes: Juvenile court judges and the court*. Beverly Hills, CA: Sage. This resource contains an insightful view of juvenile court judges.

33. National Center for State Courts. (2005). *Survey of judicial salaries* (Vol. 30). Williamsburg, VA: National Center for State Courts.

34. Rubin, 1998.

35. Rubin, 1998.

36. Rubin, 1989, p. 114.

37. Drowns, R. W. & Hess, K. M. (1995). *Juvenile justice* (2nd ed.). St. Paul, MN: West.

38. Rubin, 1989.

39. National Center for State Courts, 2005.

40. Cox, S. M., Conrad, J., & Allen, J. (2002). *Juvenile justice: A guide to practice and theory* (5th ed.). New York: McGraw-Hill.

41. National District Attorneys Association. (1991). *National prosecution standards* (2nd ed.). Alexandria, VA: National District Attorneys Association.

42. Bailey, F. L. (1971). *The defense never rests*. New York: Signet.

43. Puritz, Burrell, Schwartz, Soler, & Warboys, 1995.

44. Mnookin, R. H. & Weisberg, D. K. (1989). *Child, family, and state: Problems and materials on children and the law* (2nd ed.). Boston: Little, Brown & Co.

45. Puritz, Burrell, Schwartz, Soler, & Warboys, 1995.

46. Puritz, Burrell, Schwartz, Soler, & Warboys, 1995.

47. Puritz, Burrell, Schwartz, Soler, & Warboys, 1995.

48. Bernard, 1992.

Fox, S. J. (1977). *The law of juvenile courts in a nutshell* (2nd ed.). St. Paul, MN: West.

49. Puritz, Burrell, Schwartz, Soler, & Warboys, 1995.

50. Senna, J. J. & Siegel, L. J. (1992). *Juvenile law: Cases and comments* (2nd ed.). (p. 130). St. Paul, MN: West Publishing Co.

51. Krisberg, B. & Austin, J. F. (1993). *Reinventing juvenile justice*. Newbury Park, CA: Sage.

52. Senna & Siegel, 1992.

53. Mich. Court Rules 5.922(A).

54. Mich. Court Rules 5.922(B).

55. Mich. Court Rules 5.922(C).

56. *In re Winship*, 397 U.S. 385 (1970).

57. *In re Gault*, 387 U.S. 1 (1967).

58. *McKeiver v. Pennsylvania*, 403 U.S. 528 (1971).

59. Mahoney, A. R. (1985). Jury trial for juveniles: Right or ritual? *Justice Quarterly, 2*, 553–565.

60. B. Feld, 1988.

61. Brooks, K. & Kamine, D. (Eds.). (2003). *Justice cut short: An assessment of access to counsel and quality of representation in delinquency proceedings in Ohio.* Washington, DC: American Bar Association Juvenile Justice Center.

Feld, 1989.

Jones, 2004.

Puritz, Burrell, Schwartz, Soler, & Warboys, 1995.

LaVera, D. (2003). *President says new reports show "conveyor belt justice" hurting children and undermining public safety*. Press release. American Bar Association.

Miller-Wilson, L. S. (2003). *Pennsylvania: An assessment of access to counsel and quality of representation in delinquency proceedings.* Washington, DC: American Bar Association Juvenile Justice Center.

62. Jones, 2004.

Youth Advocacy Project. (2005). Retrieved December 4, 2008 from http://www.youthadvocacyproject.org/index.htm.

63. Puritz, Burrell, Schwartz, Soler, & Warboys, 1995.

64. Mnookin & Weisberg, 1989.

65. Siegel & Senna, 1997.

66. Siegel & Senna, 1997.

67. Jacobs, M. D. (1990). *Screwing the system and making it work: Juvenile justice in the no-fault society*. Chicago: University of Chicago Press.

Siegel, L. J., Welsh, B. C., & Senna, J. J. (2003). *Juvenile delinquency: Theory, practice, and law* (8th ed.). Belmont, CA: Wadsworth.

68. *Kent v. United States*, 383 U.S. 541 (1966).

69. Puritz, Burrell, Schwartz, Soler, & Warboys, 1995.

Walter, J. D. & Ostrander, S. A. (1982). An observational study of a juvenile court. *Juvenile and Family Court Journal, 33*, 53–69.

70. Puritz, Burrell, Schwartz, Soler, & Warboys, 1995.

71. Zalkind, P. & Simon, R. J. (2004). *Global perspectives on social issues: Juvenile justice systems.* Lanham, MD: Lexington Books.

72. Office of the United Nations High Commissioner for Human Rights, 1989. *Convention on the Rights of the Child*. Retrieved December 4, 2008 from http://www2.ohchr.org/english/law/crc.htm#art37.

73. Senna & Siegel, 1997.

74. Snyder & Sickmund, 2006.

75. Carter, T. J. (1979). Juvenile court dispositions: A comparison of status and non-status offenders. *Criminology, 17*, 341–359.

Clarke, S. H. & Koch, G. G. (1980). Juvenile court: Therapy or crime control, and do lawyers make a difference? *Law and Society Review, 14*, 263–308.

Cohen, L. E. & Kluegel, J. R. (1978). Determinants of juvenile court dispositions: Ascriptive and achieved factors in two metropolitan courts. *American Sociological Review, 43*, 162–176.

Dannefer, D. & Schutt, R. K. (1982). Race and juvenile justice processing in court and police agencies. *American Journal of Sociology, 87*, 1113–1132.

McCarthy, B. R. & Smith, B. L. (1986). The conceptualization of discrimination in the juvenile justice process: The impact of administrative factors and screening decisions on juvenile court dispositions. *Criminology, 24*, 41–64.

Staples, W. G. (1987). Law and social control in juvenile justice dispositions. *Journal of Research in Crime and Delinquency, 24*, 7–22.

Thomas, C. W. & Cage, R. J. (1977). The effect of social characteristics on juvenile court dispositions. *Sociological Quarterly, 18*, 237–252.

76. Bailey, W. C. & Peterson, R. D. (1981). Legal versus extra-legal determinants of juvenile court dispositions. *Juvenile and Family Court Journal, 32*, 41–59.

Clarke & Koch, 1980.

Frazier, C. E. & Bishop, D. M. (1985). The pretrial detention of juveniles and its impact on case dispositions. *Journal of Criminal Law and Criminology, 76*, 1132–1152.

McCarthy, B. R. (1987). Preventive detention and pretrial custody in the juvenile court. *Journal of Criminal Justice, 15*, 185–200.

77. Terry, R. M. (1967). Discrimination in the handling of juvenile offenders by social-control agencies. *Journal of Research in Crime and Delinquency, 4*, 218–230.

78. Bishop, D. M. & Frazier, C. E. (1996). Race effects in juvenile justice decision-making: Findings of a statewide analysis. *Journal of Criminal Law and Criminology, 86*, 392–414.

Carter, 1979.

Chesney-Lind, M. & Shelden, R. G. (2004) *Girls, delinquency and juvenile justice* (3rd ed.). Belmont, CA: Thompson/Wadsworth.

Cohen & Kluegel, 1978.

Feld, B. C. (1999). *Bad kids: Race and the transformation of the juvenile court.* New York: Oxford University Press.

Guevara, L. C., Spohn, C., & Herz, D. (2004). Race, legal representation, and juvenile justice: Issues and concerns. *Crime and Delinquency, 50*, 344–371.

Lieber, M. J. & Mack, K. Y. (2003). The individual and joint effects of race, gender, and family status on juvenile justice decision-making. *Journal of Research in Crime and Delinquency, 40*, 34–70.

Scarpitti, F. R. & Stephenson, R. M. (1971). Juvenile court dispositions: Factors in the decision-making process. *Crime and Delinquency, 17*, 142–151.

Terry, 1967.

Thornberry, T. P. (1973). Race, socioeconomic status and sentencing in the juvenile justice system. *Journal of Criminal Law and Criminology, 64*, 90–98.

79. Carter, 1979.

Clarke & Koch, 1980.

Horwitz, A. & Wasserman, M. (1980). Formal rationality, substantive justice, and discrimination. *Law and Human Behavior, 4*, 103–115.

Tracy, P. E. (2002). *Decision making and juvenile justice: An analysis of bias in case processing.* Westport, CT: Praeger.

80. Scarpitti & Stephenson, 1971.

81. *Indian Child Welfare Act*, 25 U.S.C. 1901 *et.seq.*

82. Jackson, Mike. (April 10, 2008). Speech presented at the 2008 Federal Association Indian Law Conference, Albuquerque, New Mexico.

83. Deloria, V., Jr. (2004). Native wisdom: A new respect for old ways. *National Museum of the American Indian, Special Commemorative Issue.*

Community-Based Correctional Programs for Juvenile Offenders

CHAPTER OBJECTIVES

After studying this chapter, you should be able to

- Define community-based corrections
- Describe common objectives of community-based correctional programs
- Describe different kinds of at-home community-based programs
- Describe different types of out-of-home or residential community-based programs
- Describe standard probation and how it varies between and within jurisdictions
- Describe the basic roles performed by juvenile probation officers
- Evaluate the effectiveness of standard probation
- Describe recent trends in juvenile probation
- Describe problems that confront community-based correctional programs
- Evaluate the effectiveness of community-based correctional programs
- Explain why aftercare is a key component of juvenile corrections

CHAPTER OUTLINE

- Introduction
- At-Home Community-Based Programs for Juvenile Offenders
- Out-of-Home (Residential) Community-Based Placements for Juvenile Offenders
- The Effectiveness of Community-Based Corrections for Juvenile Offenders
- Linking Institutional and Community-Based Corrections: Aftercare Programs
- Legal Issues
- Chapter Summary
- Key Concepts
- Review Questions
- Additional Readings
- Notes

Introduction

nonformal processing
Includes various ways of processing cases that do not involve an adjudication and the initiation of a formal juvenile court record for the youth.

formal processing
Case processing that results in an adjudication and initiation of a formal juvenile court record.

community-based corrections
Noninstitutional correctional programs and interventions.

When youths are adjudicated, a number of dispositional options may be exercised by juvenile courts, although the options available in many jurisdictions are fairly narrow. Regardless of the number of options available to courts, however, they fall into three categories. One category of options is **nonformal processing**, which includes dismissal of the case, used in a small percentage of cases; referral to a community diversion program (note that some states limit the types of crimes that can be referred to diversion programs); and placement on nonformal probation. The two categories of **formal processing** options are placement in a community-based correctional program and placement in an institutional correctional program. This chapter examines the operation and effectiveness of placement in community-based correctional programs.

The term **community-based corrections** refers to diverse types of supervision, treatment, reintegration, control, and support programs for youths involved in the juvenile justice process. These programs are based on the belief that the most effective way to encourage law-abiding behavior on the part of youths is to help them assume legitimate roles within the community.[1] Examples of community-based programs are diversion, pretrial release, probation, foster care, group home placement, and parole. Some of the programs, which are referred to as *at-home community-based programs*, are designed to provide services to youths in their own homes. Other community-based programs provide services to youths who are removed from their homes, at least for short periods of time; these are referred to as *out-of-home community-based programs*.

Like other correctional interventions, community-based programs are intended to accomplish a variety of objectives, although the objectives actually pursued by different programs vary to some extent. These objectives include controlling and sanctioning youths, allowing youths to maintain existing ties with the community, helping them restore ties and develop new and positive ones with the community (reintegration), helping the individual avoid the negative consequences of institutional placement, providing a more cost-effective response to those who violate the law, and reducing the likelihood of recidivism.[2]

At-Home Community-Based Programs for Juvenile Offenders

Probation

probation
The conditional release of an adjudicated delinquent into the community under the supervision of the court.

Probation is the most frequently used correctional response to youths who are adjudicated in juvenile courts. For example, in 2005, almost 6 out of every 10 delinquency cases (59.8%) adjudicated in juvenile courts were given probation.[3] **Probation** is the conditional release of an adjudicated youth into the community under the supervision of the court. The conditions under which a youth is released constitute the rules of probation. Typical rules of probation require the probationer to obey all laws; follow home rules; be home each day by a certain time; meet with the probation officer when requested; and, if the youth is of school age, attend school each day and obey the rules of the school. However, a court may make other rules of probation if it thinks they are necessary to assist the youth and protect the community. The court, for example, might demand that the youth avoid certain people or places, attend counseling, and make restitution to the victim. The role of the probation officer is to monitor the youth and/or assist the youth in his or her efforts to adhere to the rules of probation.

Simply defining probation, however, does not reflect the complexities of probation nor the disparities that exist in the practice of probation. In fact, considerable variability in probation practice can occur in the same state and even in the same probation department. Some juvenile courts have more than one type of probation program for youths under their supervision. These different programs—sometimes called standard, moderate, and intensive probation—are intended to reflect different levels of contact between probation officers and their clients. Other courts or probation departments do not make a formal distinction between types of probation programs. Whether probation is called standard, moderate, or even intensive probation in some cases, the level of contact between a probation officer and a client can range from regular to infrequent. There are also jurisdictions that, because of large caseloads, are able to provide only "file drawer probation," which is characterized by a lack of meaningful and regular contact between probation officers and probationers. Moreover, there is considerable variability in the training, skills, motivation, and workload of individual probation officers. Some probation officers are well trained, highly skilled, motivated, and work with small caseloads. Others have received relatively little training, lack motivation, and have large caseloads, which of course prevents them from having regular contact with each of their clients.[4] In order to determine the number of probation officers available for working with (or for) juvenile courts, some states use population formulas. In other states, the determination of the number of available probation officers is made locally and may be based on the level of community resources and various political factors.[5]

Not only do probation officers differ in terms of their ability and willingness to supervise or work with clients, but communities also vary in the number and quality of services available for use by probation officers and probationers. Some communities have a range of high-quality programs appropriate for probationers, such as alternative education programs, educational support services, job training and placement programs, substance abuse treatment programs, counseling, and structured recreational programs. Unfortunately, many other communities lack such programs or have only a few high-quality programs that are accessible to court-involved youths. The availability of high-quality programs in a community is important to probation officers, because they often serve as *service brokers* who attempt to link clients with appropriate community services and thereby reduce the likelihood of further law-violating behavior. Juvenile courts and other governmental units responsible for funding or administering probation departments also vary in their support of probation services. Some probation departments receive considerably more resource support than others. Desired resources

FYI

The Level of Probation Supervision Varies Across Jurisdictions and Clients

Many probation officers develop their own levels of probation supervision for their caseload or their own system for prioritizing their cases, based on their assessment of need. For example, at any one time, some cases on a probation officer's caseload will be in crisis (e.g., because of a family fight, the suspension of a youth from school, or because the youth lost his or her job) and will require considerable expenditure of time and energy on the part of the officer. In contrast, other cases will be relatively trouble free and close to discharge from probation. As a result, the probation officer will have considerably more contact with some probationers than with others.

MYTH VS REALITY

Probation Can Mean Regular Supervision and Provision of Services

Myth—Most juvenile probation officers have large caseloads that prevent them from effectively monitoring their clients, thus endangering public safety.

Reality—Although some juvenile probation officers have caseloads of more than 200 clients, the average juvenile probation caseload is about 41 active cases (11 more than the optimal caseload of 30 suggested by probation officers themselves).[6] Moreover, most youths placed on probation are nonviolent offenders. Only 26% of the adjudicated cases that were placed on probation in 2005 involved person offenses.[7]

and supports include probation and support staff, staff training, pay and benefits, and effective supervision and administration of the department.

Although the actual practice of probation varies, probation officers usually perform six important functions in the juvenile justice process: (1) they do intake screening; (2) they conduct presentence investigations; (3) they supervise offenders and monitor the extent to which youths adhere to their probation orders; (4) they provide assistance to youths placed on probation; (5) they provide ongoing assessments of clients' needs; and (6) they complete a variety of job-related administrative tasks. In some jurisdictions, all of these functions are the responsibility of the same individual, whereas in other jurisdictions, probation officers specialize in one or more.

As noted in Chapter 8, *intake screening* involves making decisions about the most appropriate ways of processing cases referred to the juvenile court. It is at the intake stage that decisions to dismiss, petition, or handle a case in some nonformal way are made. Moreover, those who perform intake screening may employ considerable discretion in making intake decisions. In addition to this, the **presentence investigation** plays a critical role in juvenile court dispositions because hearing officers rely on these investigations to decide upon the most appropriate way of handling formal cases.

presentence investigation An investigation that is typically completed by a probation officer prior to the disposition; it is intended to assist the hearing officer in deciding the best response to an adjudicated youth.

The supervision and assistance roles of probation officers involve monitoring probationers to ensure that they are complying with the rules of probation and providing various types of help to them. The supervision role requires checking to ensure that probationers are at school, work, home, or other places at designated times or monitoring youths assigned to electronic or officer-monitored house arrest programs. It may also require collecting urine samples from youths to enable screening for substance use. The assistance role may encompass service brokerage and the provision of direct services to clients. When acting as a service broker, the probation officer attempts to link youths and possibly families with community agencies. The officer might also provide direct

FYI

Probation Officers Perform a Number of Functions

By the mid-1990s, there were an estimated 18,000 juvenile probation officers in the United States. Approximately 85% of these individuals were line staff involved in providing basic intake, investigation, and supervision services; the remaining 15% were involved in the administration of probation offices or the supervision of probation staff.[8]

services, such as basic counseling, job search assistance, basic family counseling, and crisis intervention services, as well as act as an advocate for youths in dealings with parents or other relatives, teachers, school administrators, employers, and social service personnel. In cases where probation officers have an opportunity to have regular contacts with clients, they are in a position to better understand problems that may have contributed to the youths' involvement with the court. They can also conduct investigations into the youths' and families' needs. Such investigations may uncover problems that were passed over or have arisen since the presentence investigation. By becoming aware of current problems, the probation officer will be in a better position to recommend or develop, possibly with the court's assistance, strategies for helping the youths successfully complete probation and avoid subsequent delinquent activity.

Because probation officers work for and with governmental agencies with significant legal responsibilities, probation officers also engage in a variety of job-related administrative tasks, such as filing orders and paperwork related to the processing of cases and preparing reports of casework activity. Furthermore, probation officers may be employed in specialized programs operated by or on behalf of courts. For example, juvenile courts or other government agencies may have a variety of programs, such as diversion and foster care programs, and also operate their own shelter-care units, group homes, and detention centers. Probation officers may be used to staff these programs or may be responsible for providing basic casework services to youths in these programs.

In performing their various duties, probation officers may experience conflict between their supervision and assistance roles (sometimes referred to as their "law enforcement" and "social work" roles).[9] As social control agents, probation officers have a responsibility to protect the community and to make sure that their clients follow the rules of probation. This is their *law enforcement role*. On the other hand, they may be expected to work closely with other agencies and with parents in assisting their clients. Moreover, they are expected to work closely with the youths on their caseload in order to help them make appropriate choices and avoid further law-violating behavior. This is their *social work role*. Their law enforcement and social work roles are not always compatible,[10] and individual officers may deal with the role conflict by emphasizing one role over another or by attempting to balance the roles,[11] often emphasizing different roles with different probationers. However, large caseloads cause many probation officers, regardless of their own beliefs about their roles, to operate in a crisis intervention mode. When clients' problems come to their attention, they attempt to react to these crises as best they can, but they may have very little time left over to spend in providing proactive assistance to clients.

FYI

There Is Considerable Variability in Probation Practice

Although probation involves doing intake screening, conducting presentence investigations, supervising cases, providing assistance to clients, conducting ongoing evaluations of clients' needs, and completing routine job-related administrative tasks, the amount of time that probation officers spend on the different tasks varies considerably. In some jurisdictions, probation officers spend a majority of their time dealing directly with clients or making client-related contacts designed to assist youths on probation. In other jurisdictions, much of their time is spent preparing court reports and engaging in other administrative tasks.

The Organization of Juvenile Probation

The organization and administration of juvenile probation varies from state to state. The most common model is for probation to be part of a state executive agency. This model is used in 12 states (Alaska, Delaware, Florida, Kentucky, Maine, Maryland, New Hampshire, New Mexico, North Carolina, Rhode Island, South Carolina, and Vermont). Also, state executive agencies are involved in some way in the administration of probation in 11 other states (Georgia, Louisiana, Michigan, Minnesota, Mississippi, North Dakota, Oklahoma, Oregon, Tennessee, Virginia, and Wyoming). Within those states where a state executive agency has all or some responsibility for administering probation, the most common type of state executive agency used to administer probation is a juvenile corrections agency, such as a state Department of Juvenile Justice. In other states, the executive agency responsible for probation, at least in part, is a state child protection agency, human services agency, or an adult corrections agency.[12]

In addition to the state executive agency model of juvenile probation administration, other states employ a local judicial model in which juvenile probation is placed under local juvenile courts. This is the model employed in 9 states (Arizona, Arkansas, Colorado, Idaho, Illinois, Indiana, Kansas, Pennsylvania, and Texas). Also, in 15 other states (Alabama, California, Georgia, Louisiana, Michigan, Minnesota, Mississippi, Nevada, Ohio, Oklahoma, Tennessee, Virginia, Washington, Wisconsin, and Wyoming) probation is administered by local juvenile courts in some parts of the state, such as urban areas.[13]

In other states, a state-level judicial model is used to administer probation. In this model, a state-level judicial agency, such as a State Court Administrator's Office or Administrative Office of the Courts, oversees the operation of probation services. Ten states use the state-level judicial model (Connecticut, Hawaii, Iowa, Massachusetts, Montana, Nebraska, New Jersey, South Dakota, Utah, and West Virginia). The District of Columbia also uses a judicial model in which the Family Court of the District of Columbia Superior Court oversees the operation of probation.[14]

The least common model of probation administration is the local executive model. Only one state, New York, places juvenile probation entirely under a local executive agency. In New York, 58 county probation departments are responsible for providing probation services. However, in five other states (California, Nevada, Ohio, Washington, and Wisconsin) local executive agencies are responsible for probation services in some areas. Also, in one state, Oregon, responsibility for juvenile probation is shared by a state agency (the Oregon Youth Authority) and local executive agencies (county juvenile departments).[15]

Research on Standard Juvenile Probation Effectiveness

Research on the effectiveness of juvenile probation is complicated by the wide variability in probation practices across and within jurisdictions, by the lack of research on probation in general, and by the failure of studies of probation to clearly document the types of interventions used by probation officers and the quality of services provided to youths. The lack of research on probation effectiveness is somewhat surprising considering that juvenile probation has existed since the mid-1800s. Nevertheless, some studies indicate that probation is effective at both reducing and prolonging recidivism[16] and represents a cost-effective alternative to incarceration,[17] particularly for moderately delinquent youths.[18] Other studies, however, suggest that probation does not appear to be particularly effective at reducing recidivism, particularly for those involved in serious delinquency,[19] thus questioning its ability to protect the public from chronic or serious offenders. Other research indicates that community-based programs have often failed to provide needed services to clients and have failed to reduce correctional costs.[20]

MYTH VS REALITY

Probation Is Often as Effective as Institutional Placement

Myth—Youths who are placed on probation usually engage in additional crimes while on probation or when they are discharged.

Reality—Research indicates that youths on probation in many communities avoid offending while on probation as well as subsequent to court discharge, thus indicating that probation is at least as effective at reducing recidivism as institutional placement.[25] Moreover, there is growing evidence that some types of probation programs are effective with serious or violent juvenile offenders.[26]

Given the variation in the level of supervision and quality of services provided to youths on probation, it is not surprising that the limited research on probation effectiveness has produced mixed findings. However, more recent research that has explored program effectiveness indicates that some juvenile justice programs, including programs for serious, violent, and chronic juvenile offenders, are effective. For example, a **meta-analysis** of more than 500 programs conducted by Mark Lipsey found that the typical juvenile justice program reduced recidivism. Indeed, the recidivism rate in more than 200 court supervision programs ranged from 21 to 34%.[21] Moreover, an earlier meta-analysis of 200 experimental or quasi-experimental studies of programs for serious offenders found that the typical program for these youths reduced recidivism by 12% and the most effective programs reduced recidivism by 40%.[22] In addition, an examination of juvenile probation in selected jurisdictions conducted by Howard Snyder and Melissa Sickmund found that only 15% of juvenile probationers were adjudicated for new offenses while they were on probation.[23] Not only is there increasing evidence that many court operated programs for offenders are effective at reducing recidivism, there is also evidence that community-based interventions are more cost effective than institutional programs.[24]

meta-analysis A data analysis technique that makes it possible to examine treatment effects across different studies.

house arrest See home confinement.

Recent Trends in Juvenile Probation

A recent trend in probation involves the development of intensive probation (supervision) programs, which in some jurisdictions can also involve **house arrest** (home confinement). **Intensive supervision programs** are intended to ensure that there is regular contact between probationers and probation officers. They also serve as an **intermediate sanction** for youths that is more restrictive than standard probation but less restrictive than incarceration. However, as in the case of standard probation programs, the frequency of contact between probation officers and probationers varies considerably across intensive supervision programs.[27]

intensive supervision program A program intended to ensure that there is regular contact between probationers and probation officers. Although the contact is regular, its frequency varies considerably across programs.

Although there is wide variability in what intensive supervision means in different jurisdictions, there is some indication that programs that provide frequent supervision of offenders as well as services are as effective as incarceration at reducing recidivism.[30] Programs showing particular promise are those that help youths develop interpersonal skills; use behavioral and cognitive-behavioral interventions; employ social learning techniques such as modeling and role playing; and address risk factors in the domains of the family, school, peer group, and community.[31] Moreover, there is evidence suggesting that intensive programs are more cost-effective than incarceration, providing they actually divert a sizable number of youths from institutions.[32] However, other research indicates that some intensive supervision programs may *not* be as cost-effective as other interventions that keep youths in the community, such as standard probation and programs using

intermediate sanction A diverse range of responses (e.g., intensive supervision, day treatment, and house arrest) to offenders that are more restrictive than standard probation but less restrictive than institutional placement.

MYTH VS REALITY

Probation Is Effective with Some Youths

Myth—Probation is not an effective response to more serious juvenile offenders, cannot be used for youths who otherwise would be placed in institutions, and endangers community safety.

Reality—Although standard probation programs do not appear to reduce recidivism among chronic or serious offenders, probation can be successfully used as an alternative to institutionalization for nonserious offenders who account for a sizeable proportion of those youths placed in institutional settings.[28] In addition, research that has looked at some intensive supervision programs found they were just as effective in reducing recidivism as institutional placement. In one study, William Barton and Jeffrey Butts examined a population of Michigan youths who were randomly assigned to participate in an intensive supervision program or placed in state institutions. Those assigned to the intensive supervision program were part of small caseloads, were closely monitored, and were given a variety of support services, including individual counseling, social skills training, parent counseling, and involvement in youth groups and recreational activities. After being in the program for a 12-month period, the youths were placed on standard probation. In their evaluation, Barton and Butts found that, controlling for time in the community, the intensive supervision program was as effective at reducing recidivism as institutionalization. Moreover, they found that the intensive supervision program was considerably less expensive.[29]

home confinement The state of being confined to home. A home confinement program requires the participants to spend certain hours at home. Supervision in these programs is done by probation officers who make random checks on clients or monitor compliance by electronic monitoring. Also called home arrest or house arrest.

electronic monitoring The ability to determine an individual's movement by having him or her wear a tamper-resistant electronic device that allows communication between the device and a computer housed at the monitoring agency.

cognitive-behavioral treatment.[33] Moreover, failure rates in these programs can be high, particularly among young property offenders.[34] This is not surprising because youths placed on intensive supervision programs are often more serious offenders, and enhanced monitoring of these youths increases the probability that violations of program rules or further law violations will be uncovered. Also, as we have already noted, there is considerable variation in the quality of programs offered to youths involved in juvenile justice.

Intensive supervision sometimes employs **home confinement** or *house arrest*, which began to be used with juvenile offenders in the 1970s[35] and has grown in popularity since that time. Today, home confinement or house arrest programs use two mechanisms to monitor youths: frequent probation officer contacts and **electronic monitoring** (which requires the offender to wear a tamper-resistant electronic device that automatically notifies the probation department if the juvenile leaves his or her home or another designated location). Although home confinement programs have grown in popularity around the country, there has been limited research on their effectiveness. Nevertheless, the research that has been done is somewhat encouraging. For example, in one national evaluation of seven home confinement programs that included officer monitoring, the researchers found that the programs were able to prevent both recidivism and violations of program rules. Moreover, when participants did run into problems, it was usually for violating program rules, not for committing new criminal offenses.[36]

Electronic monitoring of juvenile offenders began in the 1980s and has become more prevalent since that time.[37] However, because electronic monitoring technology is new, diverse, and being rapidly developed, relatively little research has focused on its ability to reduce recidivism and the types of clients that are likely to fare well on electronic monitoring programs. Nevertheless, there is some evidence that recidivism is no higher in electronic monitoring programs than in other community-based programs, costs can be lower, and they may help reduce institutional crowding. However, evidence suggests that serious felony offenders—those with substance abuse problems, repeat offenders, and those with long sentences—are least likely to be successful in these programs.[38]

BOX 11-1 Interview: James McIntyre, Juvenile Court Counselor, North Carolina Department of Juvenile Justice and Delinquency Prevention

Education: BS, Criminal Justice, Western Carolina University

Q: How long have you been a juvenile court counselor?

A: 15 years

Q: How did you get interested in juvenile justice?

A: Since I was about seven or eight, I knew I wanted to work in the justice field. At first I wanted to work in law enforcement, but while I was in college I worked as a volunteer with child-serving groups and really liked it. Also, after I graduated, I worked at a children's home. My job title was "resident counselor." I was part of an interdisciplinary team that provided crisis intervention, basic counseling, and supervision to emotionally disturbed kids who were placed out of their homes. In many instances these children were the victims of emotional, physical, and sexual abuse and neglect. That position provided me with an excellent experiential base for my present job.

Q: Does your department have specialized units?

A: Units no, there are only three of us in this county. But we do have specialized tasks and responsibilities.

Q: What is your role?

A: I am responsible for delinquency intake as well as maintaining a moderately sized supervision caseload of 15–18. I also serve on numerous boards and committees such as the Juvenile Crime Prevention Council and Communities-In-Schools. One of my coworkers is responsible for undisciplined intake while also maintaining a slightly larger supervision caseload of 22–27. My other coworker is responsible for a supervision caseload of 30–35. Like me, my two coworkers serve on several boards.

Q: Tell me about your caseload and how you spend your time.

A: Most of my time is focused on delinquency intake. I receive a report from law enforcement, which I then review and enter into our department's database. I then create a petition. Soon thereafter I contact the victim and then meet with the juvenile and his or her family. That interview concerns the allegations facing the juvenile as well as how things are at home, at school, and in the community. I then make a decision as to whether the juvenile should be approved for court or if a lesser community-based intervention can be utilized to address the situation. If the juvenile is an adjudicated delinquent I then prepare a report on the child and family, which includes an evaluation of risk and need factors. I also recommend to the court what measures should be taken to protect the community as well as how best to assist the juvenile and family in correcting and moving away from the problems that contributed to the juvenile's current situation. From that point, the case is supervised by me or one of my coworkers until the child is no longer under the jurisdiction of the court (usually one year).

Q: From your perspective, why do children get involved with the court?

A: There is no easy answer to that. For most children, part of the equation is certainly the breakdown of the family unit, which many times is multigenerational in nature. This breakdown often manifests itself through a lack of appropriate parenting skills. Let's face it, if these parents were not exposed to appropriate skills, such as consistent discipline and structure by their own parents, how can they accomplish that for their own children? Many of the children I see come from single-parent homes where the fathers are absent. This can have a negative impact on areas such as self-esteem and is often a root cause of many behavioral problems, such as poor anger management and substance abuse. A significant number of these families face serious economic problems as well.

Q: How dangerous is your work?

A: I don't perceive much of a risk/threat from the children or their families. However, many of these families live in dangerous neighborhoods, so it's important that we keep our eyes and ears open when we visit the area.

(continues)

Q: What do you like about your work?

A: I like working with children and families. I know that sounds ambiguous; I like the challenge and I don't take the responsibility lightly. I really like trying to serve the child's best interests, along with doing what is in the best interest of the community. I also enjoy collaborating with other agencies. We all work together in order to maximize our opportunities to successfully serve the community and its children and families. I also like the flexibility in my schedule and setting my daily agenda—except for court hearings, you don't have a choice about that.

Q: What do you not like?

A: Being told that there are no "funding streams" available to finance services for a juvenile and his or her family. Money should be the last consideration, but is often the first and highest hurdle we must overcome.

Researchers have documented some of the reasons why some jurisdictions find electronic monitoring attractive: It is a way to ease the problem of detention overcrowding; it allows youths to participate in counseling, educational, and vocational programs without endangering public safety; it allows youths to live in a natural environment, albeit with supervision; and it allows court workers to better assess the ability of youths to live in the community under standard probation after they leave the program.[39]

Other At-Home Community-Based Interventions for Juveniles

Aside from probation, juvenile courts in some jurisdictions may employ a variety of other types of community-based interventions, such as restitution or community service programs, wilderness probation programs, and day treatment programs. In practice, probation is often combined with one of these other community-based interventions. For example, it is common for courts to place a youth on probation and order him or her to make restitution to the victim.

Restitution

restitution program
A program that requires offenders to compensate victims for damages to property or for physical injuries.

Restitution programs require offenders to compensate victims for damage to property or for physical injuries.[40] The primary goal of restitution programs is to hold youths accountable for their actions. In addition, restitution programs may also seek to provide services or treatment to juveniles and/or provide services or reparation to victims.[41] There are several types of restitution. *Monetary restitution* involves making a cash payment to the victim for harm done. *Victim service restitution* requires that the offender provide some service to the victim. The last type, *community service restitution*, requires the offender to provide some assistance to a community organization.[42] Each of these types has its own rationale. Monetary restitution is intended to help the juvenile offender realize the economic costs incurred by the victim as a result of the offender's behavior. In-kind or in-person victim restitution is intended, among other things, to help humanize the victim. Community service restitution is intended to help the juvenile offender repay the community for the harm it suffered as a result of the youth's actions.

Like research on the effectiveness of other juvenile justice programs, research on the effectiveness of juvenile restitution programs shows mixed results. Data collected as part of a national evaluation of juvenile restitution programs found extremely high rates of compliance with restitution orders in many programs. One national study published in the early 1980s found that 95% of youths participating in various federally sponsored restitution programs successfully carried out their restitution orders.[43] A more recent national survey published in the late 1990s reported completion rates as high as 88%.[44] However, compliance rates appear to be considerably higher in well-managed and more structured programs than in informal and less structured programs, in which offenders are simply ordered to make restitution but there is little follow-up. One evaluation of structured and unstructured programs conducted in Wisconsin found that only 45% of those youths involved in unstructured programs completed their restitution, compared with 91% of the youths in formal programs.[45]

Other research indicates that restitution programs also can collect substantial amounts of monetary restitution for victims and can produce substantial amounts of community service and victim service hours. One national survey of restitution programs in 1991 found that the programs had collected $44.5 million in 1990. This same survey also found that restitution program participants had performed approximately 17.1 million hours of community service and 44,000 hours of direct victim service during 1990.[46]

Some studies show that restitution programs can have an impact on recidivism when they are used either prior to or after adjudication. For example, a study of Utah restitution programs found that, when used at both the preadjudicatory stage and as part of formal probation, youths involved in restitution relapsed less than those who received other nonformal responses or youths who were placed on probation alone.[47] A study of four restitution programs located in Boise, Idaho; Washington, D.C.; Clayton County, Georgia; and Oklahoma City, Oklahoma, found that the programs in Washington and Clayton County produced significantly lower rates of recidivism than traditional probation programs in those two jurisdictions. In the two remaining programs, the differences were not significant.[48] Still another study of six experimental restitution programs found that juveniles who were randomly assigned to a restitution program were less likely to relapse than youths placed on probation or given short terms of incarceration.[49] In a national survey of restitution programs conducted in 1991, the overall recidivism rate was 28.8%, however, it was 23.8% among youth who successfully completed the program.[50]

Despite the growing popularity of juvenile restitution programs,[52] as well as the ability of some programs to effectively reduce recidivism, there are problems, or potential problems, with these programs. Some programs (those that are informal or poorly managed) have low compliance rates,[53] some programs experience high recidivism rates,[54] and some hearing officers make unrealistic orders that cannot be complied with by juveniles. Moreover, concern over who should be responsible for restitution is not uncommon. Is restitution the youth's responsibility or the responsibility of the parents or guardian? The existence of such problems and concerns indicates that not all restitution programs

COMPARATIVE FOCUS

Restitution Is a Common Response of Youth Courts in Other Countries Such as England

As a result of reforms in English law in the late 1990s, reparation orders have become more common responses to youths who plea to or who are convicted of less serious types of offenses. These orders require youths to make reparations to the victim (if the victim agrees) or to the community at large. Moreover, courts are required to state the reasons for not entering such orders when they have the opportunity to do so. In entering a reparation order, the court must ensure that it is proportionate to the seriousness of the offense and it may not exceed twenty-four hours.[51]

net-widening Increasing the availability of programs to handle youth offenders, thereby causing youths who would otherwise have been left alone to be placed into programs.

wilderness probation program A program designed to reform the behavior and attitudes of juvenile offenders by involving them in wilderness activities, including solo camping. The philosophy behind such programs is partly based on the philosophy of the Outward Bound organization.

are holding youths responsible for their actions or resulting in satisfactory outcomes from the point of view of the youths involved, their parents, or the victims. Indeed, such problems and concerns can lead to negative perceptions of juvenile justice.

Two additional problems are that restitution programs can contribute to **net-widening**,[55] and restitution requirements can be subject to discretionary abuse.[56] There is always the potential that restitution programs will subject youths to juvenile court control who would otherwise not be subjected to such control. Moreover, after youths fall under the control of a court, they are at risk for additional coercive treatment. Although controlling some youths is appropriate and can even help them and protect the community, in other cases court control is clearly counterproductive: It may result in harm to youths and increased danger to the community. Finally, youths who commit similar offenses may receive substantially different restitution orders, although some jurisdictions have established restitution guidelines in an attempt to remedy this problem.[57]

Wilderness Probation (Outdoor Adventure) Programs

Wilderness probation programs, or outdoor adventure programs for juvenile offenders, are based, in part, on ideas derived from the Outward Bound organization. A basic assumption underlying these programs is that learning is best accomplished by acting within an environment where there are consequences for one's actions.[58] Consequently, these programs involve youths in a physically and psychologically challenging outdoor experience intended to help participants develop confidence in themselves, learn to accept responsibility for themselves and others, and develop trust in program staff.[59] The participants engage in such activities as camping, backpacking, rock climbing, canoeing, and sailing, and, in many programs, they complete a solo wilderness experience that lasts one or more nights.

Evaluations of wilderness probation programs have indicated that some programs may produce positive effects, such as an increase in self-esteem during and after the program, and a decrease in criminal activity after the program.[60] In a meta-analysis of wilderness programs, Sandra Wilson and Mark Lipsey found that youths in wilderness programs engaged in less post-program delinquency than comparison youths. Moreover, they found that short-term programs lasting up to 6 weeks appeared to be the most effective, along with those with distinct therapeutic components.[61] However, research also indicates that the positive effects associated with these programs may diminish over time[62] or may be no greater than the effects produced by probation programs in which regular and meaningful contact occurs between probation officers and probationers.[63] Moreover, as Wilson and Lipsey note, the evaluations done on these programs have primarily focused on white males already in the system, so the effects of these programs for other populations of youths is unknown.[64]

FYI

Serious Problems Have Been Found in Some Wilderness Programs

Although there appear to be a number of well run wilderness programs in the United States, there is clear evidence that some programs have been guilty of abusing and neglecting youths in their care. Moreover, there have been cases where youths have died as a result of their treatment in some wilderness programs.[65]

Day Treatment Programs

day treatment program
A program that provides treatment or services to youths during the day and allows them to return home at night.

Day treatment programs for juvenile offenders are operated in a number of jurisdictions around the United States. These programs, which usually target serious offenders who would otherwise be candidates for institutionalization, provide treatment or services to youths during the day and allow the youths to return home at night. Because they are viewed as alternatives to incarceration, they are thought to be cost-effective, and because they are often highly structured programs, they are capable of protecting community safety as well. The range of services or treatments may include academic remediation, individual and group counseling, job skills training, job placement services, and social skills training.

An example of a day treatment program was Project New Pride, which was developed in Denver, Colorado, and has been replicated in a number of other cities around the country. This program was intended to provide intensive services to youths who had committed serious offenses. Program services included diagnostic assessment; remedial education; special education for youths with learning disabilities; cultural, physical, and health education; job preparation; job placement; volunteer support; intensive supervision (during the initial 6 months of the program); and follow-up with clients after the initial intensive supervision phase of the intervention.[66]

One state that has begun to rely more heavily on community-based programs, such as day treatment, is Missouri. In these programs, youths spend from 8 a.m. to 3 p.m. Monday through Friday receiving academic instruction and counseling followed by some combination of tutoring and individual or family counseling. Also, like day treatment programs in other states, Missouri's programs are frequently used as "step down" programs for youths who are returning from residential placement.[67]

Unfortunately, the few studies done of day treatment programs such as New Pride have not produced clear evidence of their effectiveness. There is some evidence that some programs are no more effective at reducing recidivism than standard probation. Nevertheless, there is also evidence that other programs are as effective or more effective than institutional placement and they are far less expensive.[68] For example research in Missouri revealed that only 11% of the youths released from residential care into community based programs over a two year period were rearrested or returned to state custody within a year. Moreover, they reported significantly lower correctional costs compared to surrounding states.[69]

MYTH VS REALITY

Community-Based Interventions Can Be Effective With Some Serious Juvenile Offenders

Myth—Community-based interventions are not appropriate for serious juvenile offenders.

Reality—Some community-based interventions have been proven to be effective with some serious and chronic offenders. For example, Multisystemic Therapy (MST) has been found to be a highly effective treatment for many youths engaged in serious and chronic delinquency. MST is a theory- and research-based comprehensive treatment approach that provides intensive family-focused services and is designed to address individual, family, peer, school, and neighborhood factors that encourage serious antisocial behavior. Moreover, MST is designed to foster collaborative relationships with families and other social institutions and to mobilize these institutions in an effort to change behaviors that foster delinquency. Importantly, controlled studies have found that MST is a cost-effective treatment approach that is associated with decreased criminal activity and lower levels of incarceration among program participants.[71]

Juvenile justice practitioners in Missouri cite several keys to the operation of successful programs in that state. These keys include focusing on community safety and carefully screening youths for placement in community-based programs, hiring high-quality staff, and building a supportive constituency for community-based programs.[70] Another key component in many day treatment programs is involvement of the child's parents, guardian, or custodian. This involvement may range from something as simple as aiding in overnight monitoring to undergoing intensive family counseling. Parental involvement is key to the success of these programs because any lack of involvement or indifference shown by the parents gives a clear message to the child that cooperation is not important. At present, there is a need to conduct sound research on these programs in order to evaluate their effectiveness.

Out-of-Home (Residential) Community-Based Placements for Juvenile Offenders

community-based placement An out-of-home placement for a youth offender, often lasting from one day to several months, although occasionally of longer duration.

foster home A family-like setting to which youths are temporarily assigned. In foster care programs, foster parents are licensed to provide care for one or more youths (usually no more than three) and are paid a daily rate (per diem) for the cost of care.

placing out An effort to respond to the growing number of problem children in the cities during the 1800s. It consisted of the placement of children on farms and ranches in the Midwest and West.

Although many community-based programs are designed to provide services to youths in their own homes, other programs provide services to youths who are removed from their homes and placed in some type of residential setting. Typically, these **community-based placements** are intended to provide services to youths who are placed out of their homes for a short period of time—from several days to several months. A few programs are longer term and may last for many months or even years.

Foster Homes

Foster homes have an extensive history in the United States. For example, in colonial villages, children older than the age of 7 years who were not appropriately supervised by their parents or who engaged in "the sin of idleness" were placed with other parents.[72] Also, as we noted in Chapter 4, **placing out** was used as a response to many destitute and problem children throughout much of the 1800s and early 1900s. Indeed, the major response of child welfare organizations, whether private or public, has been to remove children from parents when those parents have been deemed unfit, and foster care has often been the result.[73] Today, foster care is still a popular means for dealing with children whose parents abuse them or fail to care for them, children who come from chaotic families, and children who exhibit behavioral problems. In 2007, some 783,000 children spent time in foster care.[74]

Unlike institutional or some other types of community-based programs, foster care is intended to resemble, as much as possible, life in a family-like setting. Foster parents are licensed to provide care for one or more youths (usually one to three) and are paid a daily rate (*per diem*) for the cost of care. Foster placements are often used by courts when a youth's home life has been particularly chaotic or harmful. In these cases, foster homes are used to temporarily separate the conflicting parties—the youth and the parents or guardian—in an effort to resolve those problems that resulted in the youth's removal. However, foster homes are also used instead of more restrictive institutional placements, such as detention or assignment to a state training school, for some nonviolent delinquent offenders. Programs that use foster homes in these two ways can be classified as *halfway-in programs*. Programs that use foster homes as transitional placements for youths who are returning from an institutional placement can be classified as *halfway-out programs*.

Although foster homes are widely used by juvenile courts in some jurisdictions, there are few sound evaluations of these programs. Clearly, many foster homes provide

BOX 11-2 Interview: Connie Loviska, Foster Parent

Q: How many years have you been a foster parent?

A: I have been a foster parent for 34 years. My husband and I started with the Kalamazoo County Juvenile Court in 1969.

Q: Who are the members of your family?

A: We have five sons, now all grown, and we adopted a girl who was a foster child. My husband and I were married for 43 years when he died suddenly in December 2002, but I continue to be a foster mother.

Q: What impact has being a foster parent had on your children?

A: A very positive impact. My children are very compassionate people who are very supportive of individuals who are disadvantaged or who are having day-to-day problems. They want to help everyone get a fair chance in life; but, at the same time, they are not tolerant of people who play the "victim" all the time. They all have a well-developed social conscience and sense of community service.

Q: What led you to become a foster parent?

A: Originally, we had relatives who needed help, and we would take care of their children off and on. One of my son's friends was having some difficulty, and the juvenile court got involved. We contacted the court about this child and ended up being licensed by the court as foster parents.

Q: What do you see as the main day-to-day challenges of being a foster parent?

A: The most difficult thing is making the foster children realize that they must obey the rules of the home. In many instances, they come from dysfunctional families that have no boundaries or inconsistent rules. We have to show them that normal families have rules, that normal people all have rules to follow. Closely related to this, of course, is the challenge of holding the foster child accountable for rule violations. Often, they don't understand about accountability or responsibility. Last, it is often difficult to cooperate with the foster child's natural family and not come across as judgmental of them. As soon as you become judgmental of the child's family—even if the judgment is accurate—you have erected a barrier and given the child ammunition to use against you.

Q: What are the rewards of being a foster parent?

A: Children that succeed! By success I mean becoming a normal functioning adult, not dependent on welfare, not involved in the criminal justice system. It's the satisfaction of knowing that you helped break the cycle of abuse and dysfunction that brought the child into your home in the first place.

Q: What types of children have you fostered, and do you specialize?

A: We have neglected, emotionally fragile, and delinquent teenagers in our home. We specialize in delinquent teens.

Q: What do you do to positively influence your foster children?

A: We teach them about families and proper family relationships, interaction, and behavior by including them in our family. We work together, we play together, we travel together. Through these family activities, we teach.

Q: Have you ever been victimized by a foster child?

A: Yes, but rarely over the long haul. I have been physically attacked, threatened, and have had property damaged. It's the property damage that's really upsetting. I can understand a flash of anger from a child I'm confronting who might physically lash out at me, but "coolly" calculated property damage hurts all the kids in the home, not just me.

Q: How do you see foster care in the overall continuum of care for juveniles?

A: Of course I believe that foster care is very important for kids in the system. The problem is that it's not used soon enough. In recent years, the courts are leaving kids in dysfunctional homes too long, which makes my job as a foster parent much more difficult. The kids' attitudes become entrenched; and the longer they go being irresponsible, the harder it is for me to teach them to be responsible!

(continues)

Q: What special training or expertise do you have?

A: No training can replace 34 years of experience. However, I regularly attend conferences to learn new techniques in dealing with difficult children. I go to eight or nine trainings each year through our licensing agency and the state foster care association. I also keep up with the available literature and with TV shows that help.

Q: How much money do you earn from being a foster parent?

A: Foster parenting is not a money-making proposition. The per diem can never reimburse you for all of the little things that are involved in raising children in a family. All of those expenses, from varsity jackets to music, cannot be predetermined or reimbursed. I don't do it for the money, nor did my husband and I do it for the money. We valued what would be called the "nonmonetary" rewards that make foster parenting so worthwhile.

Q: What advice would you give to a person who might want to consider becoming a foster parent?

A: Be certain that you have the support of your entire family and friends. Foster parenting is a lot of work, and it's easy to become isolated. Make it a family decision. When everyone is part of the decision, they aren't so ready to draw lines between "us" and "them," which always leads to conflict.

Q: In your opinion, what are the major challenges facing foster parents in the future?

A: Allegations of improper conduct. It's great that kids today have been educated about abusive behavior and dangerous situations, but this does not mean that every allegation is true or that their opinion about what is abusive or appropriate is accurate. Every foster parent lives in fear of a foster child making false allegations in an attempt to manipulate a situation. Be sure to know why you are a foster parent. As time passes, re-evaluate. You may want to change age groups or types of problems in the children you have, or you may need to take a break. It's okay to need time for yourself.

Q: Is there anything else you would like to say?

A: I intend to stay with it as long as I can. I still love the children and the challenges they bring. The lasting rewards are all of the wonderful relationships that still continue.

a supportive and caring environment for youths and help them lead law-abiding lives. However, the limited research that exists indicates that traditional foster care is not generally effective and may even be counterproductive as a strategy for reducing delinquency.[75] These findings parallel other efforts to examine foster care outcomes for nondelinquent populations that demonstrate that many foster care children fare poorly on indicators of well-being such as neurological and cognitive development, and they tend to exhibit behavioral and emotional problems compared to other youths.[76] Although some foster home placements undoubtedly help youths, the short length of time youths often spend in foster homes, the use of multiple foster care placements for youths, the poor or inadequate treatment given some youths in foster care,[77] and the problems that children may bring with them may account for less than favorable results in too many other cases.

The effectiveness of foster care depends in large part on the abilities of the individual foster parents, the characteristics of the youths, and the ability of the foster parents to establish meaningful individual relationships with their foster children. The operation of foster care programs, unfortunately, can act as a barrier to effective or humane care. According to child advocate Ben Wolf, associate legal director at the Roger Baldwin Foundation of the American Civil Liberties Union, foster parents are often mistreated by child welfare agencies. For example, they are often told they will be reimbursed for expenses but remain unpaid, they are often denied opportunities for respite or vacations, they are often denied the training and support needed to work with troubled youths, they may be given little or no information about a child's health or other needs, and they may receive very few backup services.[78]

Although there is good evidence that typical foster care placements may not adequately serve children's needs, there is evidence that treatment foster care (TFC) can be an effective response to some youths, including serious and chronic juvenile offenders. One such program was developed by the Oregon Social Learning Center. The program provides specialized training for foster parents and they receive support from a program case manager. The program is designed to reinforce youths' positive behaviors, provide close supervision, assist youths in following rules and provide appropriate consequences for rule violations, support the development of youths' academic skills and work habits, and improve family communication skills. Importantly, evaluations of TFC indicate that it can be more effective than traditional foster or group care programs, as well as many other community-based programs, at reducing delinquency, and it is more cost effective.[79] Moreover, another specialized approach to foster care, Multidimensional Treatment Foster Care (MTFC), has also been found to be associated with positive outcomes. In the MTFC model, foster parents are provided specialized training and they work as part of a treatment team. The foster parents are responsible for implementing a structured and individualized program for each youth in the home (one or two youths are placed in MTFC homes). The program is designed to build on youths' strengths and involves establishing clear rules, expectations, and limits. MTFC parents are contacted daily, Monday through Friday, by telephone and data are collected on the youths' behavior during the past 24 hours. Additional components of the program include weekly supervision and support meetings for MTFC parents; skill-focused individual treatment for youths; weekly family therapy for biological parents or guardians; frequent contact between youths and their biological parents/guardians that includes home visits, close monitoring of youths' performance in school, coordination with probation or parole workers, and coordination with psychiatric/mental health services; and medication management when needed.

Importantly, youths in MTFC report significantly fewer contacts with the juvenile justice process than youths in standard forms of group care.[80] Thus, specialized forms of foster care may represent viable community-based programs for some (including serious and chronic) juvenile offenders.

Group Homes and Halfway Houses

Similar to foster homes, **group homes** and halfway houses are open, nonsecure community-based placements[81] that may be used as halfway-in programs (i.e., as a last alternative before institutional placement) or halfway-out programs (i.e., for youths leaving institutional settings). Consequently, group homes and halfway houses are another type of intermediate sanction found in many communities. However, group homes and halfway houses are somewhat larger and are often more impersonal and less family-like than foster homes.[82] Typically, these facilities are operated by staff who work shifts, as opposed to "parents" who live in the foster home. In fact, the purpose of many group homes and halfway houses is to avoid requiring youths to accept "substitute parents," partly because many youths who need placement are in the process of developing emotional independence from parental figures.[83] Nonetheless, staff in these facilities serve as important role models, and the facilities themselves are intended to be less impersonal than more closed institutions and are often designed to individualize treatment as much as possible.

group home An open, nonsecure community-based setting, larger and often less "family like" than a foster home. Group homes provide a place for youths to reside temporarily without having to accept "substitute parents," which can be beneficial if a youth is in the process of developing emotional independence from parental figures.

They are also less expensive than more restrictive institutional placements, and because they are often located in urban or suburban neighborhoods, residents have an opportunity to take advantage of a variety of community services. For example, youths who reside in group homes and halfway houses may go to school in the community or work in the community.[84] They may also participate in recreational and cultural activities in the community and use various types of treatment and educational services. Group home and halfway house treatment programs typically arrange for individual and group counseling to be provided by group home staff or outside counselors. In addition, residents may receive academic instruction and participate in social and vocational skills-building courses within the facility or in the community.

Although group homes are used extensively in juvenile justice, there is relatively little sound research that has examined the effectiveness of these programs. One well-known study of a group home program, the Silverlake Experiment, did find that youths placed in this program for approximately 6 months were neither more nor less likely to engage in subsequent delinquency than similar youths who were randomly assigned to a more restrictive institutional setting.[85] However, evaluations that have compared MTFC with group home treatment indicate that MTFC appears to be more effective than many group home programs.[86]

There is no doubt that group home programs often face a number of obstacles that act as barriers to the effective treatment of youths. These barriers include poor pay and benefits for staff, a lack of staff training on how to work with troubled youths, a paucity of support services for staff, and local resistance to allowing group homes into the community. Moreover, short-term group home placements do not allow sufficient time to affect youths in a positive way, and large group homes are not able to offer the types of individualized treatment that many youths need.

Another problem that sometimes plagues group homes, halfway houses, and other types of residential placement settings is the victimization of residents. Although the extent of this problem is not clear, there have been instances when youths residing in such settings have been victimized by other residents or even staff. Robert Mutchnick and Margaret Fawcett, in a study of nine group homes in southwestern Pennsylvania, found that some of the homes were more likely to have high levels of resident victimization than others. Group homes with high levels of resident victimization were characterized by extremely open or closed environments, a lack of program structure, staff who displayed little concern for security, and residents who expressed attitudes of mistrust and negativism. In contrast, homes with low levels of resident victimization were characterized by open but structured environments, close staff supervision, consistent enforcement of policies and rules, staff who demonstrated concern for residents, and residents who exhibited relatively positive attitudes.[87]

Shelter-Care Placements

Shelter-care settings are like group homes, although they often hold youths for shorter periods of time. A *shelter-care program* is a type of nonsecure program that typically provides a temporary residence for youths prior to returning them home or determining the suitability of some other placement. Often shelter care is used for youths who have run away from home, who have been abused by their parents, who have engaged in minor types of delinquency, or who need temporary emergency placement. Shelter care typically involves short-term crisis intervention services, basic counseling, and basic educational programming.

Family Group Homes

Family group homes are similar to both foster homes and group homes—they are run by foster parents but house more youths than a regular foster home. Although family group homes are designed to provide youths with parent substitutes and a more family-like environment than might be found in a traditional group home, and although they have sometimes been advocated as a way of reducing community opposition to larger institutional facilities, little is known about the effectiveness of family group home programs. Moreover, these programs can suffer from the same problems that confront traditional foster care and group home programs.

The Effectiveness of Community-Based Corrections for Juvenile Offenders

As indicated earlier, the research that has been done on community-based corrections programs has produced mixed results. Studies of some community-based programs indicate they are as effective as more restrictive institutional placements at reducing recidivism. There is also evidence that community-based programs are far less costly than institutional programs, in some cases only half as costly.[88] It should be noted, however, that evaluations of other programs indicate that they have little effect on subsequent offending. Such findings are not surprising, however, given the diversity in both the types of programs and the quality of programs that are in operation around the country. Despite the mixed research results on the effectiveness of community-based programs, the efforts of some states to deinstitutionalize their juvenile offender populations and move toward a more community-based system have met with considerable success.[89]

A good example of a state that has moved toward a more community-based corrections system is Massachusetts. Under the leadership of a new commissioner of youth corrections, Dr. Jerome Miller, Massachusetts began closing its state training schools in the 1970s. At first, Miller attempted to implement reforms in the Massachusetts institutions where he observed frequent abuse of residents. When these reforms were resisted, he began closing the training schools and developed or used community-based programs for youths who were released from the institutions. Despite the abuse and inhumane conditions within Massachusetts' training schools, Miller continued to face strong opposition to his efforts to deinstitutionalize the juvenile offender population. Moreover, many politicians and others allied with the existing system of institutions argued that

MYTH VS REALITY

Many Youths in Institutions Could Be Better Served in Community-Based Programs

Myth—Most youths housed in juvenile correctional institutions pose an immediate threat to public safety.

Reality—As Jerome Miller notes in his book *Last One Over the Wall: The Massachusetts Experiment in Closing Reform Schools*, many youths in juvenile correctional facilities could be more effectively served in community-based programs without endangering community safety. Miller points out that, in Massachusetts, only about 25% of the youths committed to state correctional facilities had committed offenses against persons. Further, many of these supposedly violent offenses did not actually involve physical violence or the threat of physical violence. This shows that many of the youths who are placed in correctional facilities, including those confined for "violent offenses," do not pose a grave threat to public safety.[93]

deinstitutionalization would lead to an increase in juvenile crime as "hardened" young offenders were released into the community. However, despite the protestations of those opposed to deinstitutionalization, a juvenile crime wave did not sweep the state.[90] In fact, evaluations of deinstitutionalization in Massachusetts conducted by Harvard's Center for Criminal Justice and by the National Council on Crime and Delinquency found that the community-based programs were effective alternatives to institutions and did not threaten public safety.[91] In communities that had a variety of quality programs for offenders, recidivism decreased rather than increased, contrary to what skeptics had argued would happen.[92] The Massachusetts experiment initiated by Miller demonstrated that a network of high-quality community-based programs can serve as a viable alternative to a system of juvenile correctional institutions.

Linking Institutional and Community-Based Corrections: Aftercare Programs

aftercare The provision of services intended to assist youths in successfully making the transition from juvenile institutions to life in the community.

parole The conditional release of a youth from a correctional facility (one form of aftercare).

Aftercare involves the provision of services intended to help youths successfully make the transition from juvenile institutions to life back in the community. The services are the same as those provided by the community-based programs described earlier. Indeed, aftercare may encompass foster care, shelter care, or group home placement; home placement; or efforts to help youths live on their own (independent living). **Parole**, which consists of the conditional release of a youth from a correctional facility, is one form of aftercare. Youths on parole are required to adhere to certain conditions, such as avoidance of illegal behaviors, attendance at school, and attendance at meetings with their parole officers upon request. If youths violate these conditions, they may be returned to an institution. Parole services are typically provided by a parole officer who is a state employee. However, a variety of aftercare services are provided to youths, and these services may be provided by court probation staff or even by staff hired by an institution to provide follow-up services after a youth returns to the community.

Unfortunately, the quality of many aftercare programs is questionable, and many youths fail to receive any services after institutional release. Parole supervision may actually involve very little contact between parole officers and parolees. Moreover, large caseloads carried by aftercare workers may preclude the provision of meaningful services. In other instances, the supervision that occurs may be primarily focused on surveillance and control, and may provide little in the way of treatment or service provision. Indeed, research on parole has not found it to be particularly effective. In one study of juvenile parolees in California, it was discovered that there were no differences in the likelihood of a subsequent arrest among youths randomly assigned to parole supervision or to discharge without supervision. However, those who were given parole supervision were more likely to be arrested for a serious offense.[94] In another California study, youths who were placed on intensive parole supervision after a short institutional stay were as likely to have their parole suspended for illegal activity as youths placed in California Youth Authority facilities.[95] (Keep in mind that efforts to deinstitutionalize correctional populations have not led to increases of recidivism when a range of quality community-based programs are available.)[96]

More recently, the Office of Juvenile Justice and Delinquency Prevention (OJJDP) has supported an effort to develop an effective approach to aftercare called the Intensive Aftercare Program (IAP). Developed by David Altschuler and Troy Armstrong, the IAP is intended to be a theory-driven comprehensive approach to juvenile aftercare that overcomes the institutional boundaries that frequently prevent institutions and community-based programs from developing and implementing sound collaborative plans for youths' return to the community. A key component of the IAP model is the concept of *overarching case management* that consists of five elements:

- risk assessment designed to identify and focus services on high-risk youths
- individualized case planning that includes both institutional and community-based staff, as well as family and community members, and addresses youths' needs within their families, peers, schools, and other socializing institutions
- a mix of both intensive surveillance and services
- the use of a graduated system of incentives for positive behavior and sanctions for violations of program rules
- the development of links with community resources that can support program success.[97]

Presently, the IAP model is being implemented in demonstration sites, and while evaluation results indicate that the IAP model has received considerable support and has resulted in improved relationships between institutional and aftercare staff, there is little evidence to indicate that it has a significant impact on recidivism.[98] Although the IAP model is based on sound theory and understanding of the problems traditionally associated with aftercare, more development and research is needed before strong conclusions about its effectiveness can be reached.

Legal Issues

Community service restitution programs carry with them a possibility of civil liability. If a juvenile is injured doing community service, the supervising court, the probation officer, or the community agency could be held legally responsible for any supervisory negligence. They could also be held legally responsible if the juvenile, during the course of doing the community service, injures someone or damages property in a negligent or reckless manner.

The monitoring of delinquent youths in a community-based corrections program raises the issue of the legal authority of probation and parole officers, foster parents, and group home personnel to apprehend, take into custody, and generally control youths under their supervision. In many states, probation officers are considered to be government officials with the power to arrest, apprehend, detain, or place a child.[99] Probation officers are usually "sworn" employees of the court or government agency. Foster parents, unless they are sworn probation officers, which would be unusual, are not empowered as probation officers and would have to rely on a liaison probation officer or law enforcement officer to help them control youths placed in their homes. Group homes can be privately run or can run under court auspices. If private, then group home personnel would not be considered government officials. If the group home was run by the court, the employees could be considered probation officers.

Juvenile courts, by their very nature, deal primarily with juveniles. A juvenile court has a legitimate expectation that, for each juvenile it deals with, there will be a responsible adult or adults, such as parents, guardians, or custodians, in the home. Just as a juvenile on probation is subject to orders of the court, in many jurisdictions, adults who are responsible for the care of a juvenile can also be subject to court orders. As mentioned earlier, in many jurisdictions, adults can be ordered to make monetary restitution for their children's acts, and this restitution would necessarily be court ordered. The payment of restitution by minors raises the issue of the means or method a juvenile court can use to enforce its orders over an adult. The most commonly used means is to cite the adult for contempt of court for failure to obey a court order. Two kinds of contempt are generally recognized, civil and criminal:

1. Criminal contempt usually results from an act done in the presence of the court during a hearing or while the court is in session, such as swearing at the judge. This kind of contempt does not require a hearing but may be summarily determined by the judge, and sanctions, including jail, may be levied.
2. **Civil contempt** is usually associated with an individual's failure to obey a court order that has been legally served on the individual. In a case of restitution, the responsible parent would be given an opportunity in a court hearing to "show cause" as to why he or she is not in contempt and could usually "purge" him- or herself of contempt by obeying the order—even after a hearing that resulted in the parent being found in contempt of court. Usual consequences for contempt include monetary fines, court costs, and jail for limited time periods. One of the harsh realities of the adult criminal justice system is that, because of chronic jail overcrowding, persons held for contempt are often the first to be released if there is a shortage of beds.

civil contempt Willful failure of an individual to obey a court order that has been legally served on the individual. Consequences may include monetary fines, court costs, and jail for limited time periods.

Chapter Summary

This chapter examined the operation and effectiveness of community-based corrections programs for juvenile offenders. These types of corrections programs consist of (1) at-home programs, such as probation, restitution, community service, day treatment, house arrest, and aftercare or parole programs, which are typically designed for youths who reside in their own homes, and (2) out-of-home programs, such as foster care, group homes, shelter care, and family group home programs, which provide services to youths who are removed from their own homes and placed in residential settings.

Like other corrections programs, community-based programs attempt to accomplish a number of objectives. Although individual programs vary in the extent to which they emphasize these objectives, they typically include sanctioning and controlling youths, helping them maintain existing ties within the community, helping them restore ties that have been severed as a result of their delinquent behavior, assisting youths' efforts to develop new and positive ties to others in the community, providing a correctional response that avoids the negative consequences of institutional placement, providing a more cost-effective response to juvenile crime, and reducing the likelihood of recidivism.

Since the development of the juvenile court in the late 1800s, the most popular correctional response to juvenile offenders has been probation. Despite its long history, however, little is known about the effectiveness of probation. Moreover, what is known suggests that, although some forms of traditional probation are effective with nonserious juvenile offenders, it is not always effective with chronic or serious offenders. As a result, many jurisdictions have developed intensive supervision programs that are intended to increase levels of offender supervision and service. In addition, a number of jurisdictions have developed a variety of other interventions (e.g., restitution and community service, day treatment, and house arrest) that are often used in combination with probation.

Although research on these interventions has produced mixed results, there is evidence that some programs that employ these interventions help youths avoid subsequent offending behavior. Moreover, there is sound evidence that some community-based interventions, even when used with chronic and serious juvenile offenders, can be effective in reducing recidivism and avoiding the negative aspects of incarceration.

Similar to studies of at-home community-based programs, those that have examined residential community-based programs have also produced mixed results. Nevertheless, at least some studies indicate that residential community-based programs are as effective as or more effective than institutionalization at reducing recidivism. In addition, research indicates that residential community-based corrections are considerably less expensive.

Because of the paucity of research on individual programs, it is difficult to make definitive statements about the effectiveness of many community-based corrections programs. Nevertheless, there is good evidence that some programs not only provide high-quality services and successfully address a variety of client needs, but they also do an effective job of protecting public safety. There is, unfortunately, good evidence that other programs fail to do any of these things. Nevertheless, community-based programs clearly play a critical role in the juvenile corrections system, and they are likely to play an even more important role in the future.

Key Concepts

aftercare: The provision of services intended to assist youths in successfully making the transition from juvenile institutions to life in the community.

civil contempt: Willful failure of an individual to obey a court order that has been legally served on the individual. Consequences may include monetary fines, court costs, and jail for limited time periods.

community-based corrections: Noninstitutional correctional programs and interventions.

community-based placement: An out-of-home placement for a youth offender, often lasting from one day to several months, although occasionally of longer duration.

day treatment program: A program that provides treatment or services to youths during the day and allows them to return home at night.

electronic monitoring: The ability to determine an individual's movement by having him or her wear a tamper-resistant electronic device that allows communication between the device and a computer housed at the monitoring agency.

formal processing: Case processing that results in an adjudication and initiation of a formal juvenile court record.

foster home: A family-like setting to which youths are temporarily assigned. In foster care programs, foster parents are licensed to provide care for one or more youths (usually no more than three) and are paid a daily rate (per diem) for the cost of care.

group home: An open, nonsecure community-based setting, larger and often less "family like" than a foster home. Group homes provide a place for youths to reside temporarily without having to accept "substitute parents," which can be beneficial if a youth is in the process of developing emotional independence from parental figures.

home confinement: The state of being confined to home. A home confinement program requires the participants to spend certain hours at home. Supervision in these programs is done by probation officers who make random checks on clients or monitor compliance by electronic monitoring. Also called home arrest or house arrest.

house arrest: See home confinement.

intensive supervision program: A program intended to ensure that there is regular contact between probationers and probation officers. Although the contact is regular, its frequency varies considerably across programs.

intermediate sanction: A diverse range of responses (e.g., intensive supervision, day treatment, and house arrest) to offenders that are more restrictive than standard probation but less restrictive than institutional placement.

meta-analysis: A data analysis technique that makes it possible to examine treatment effects across different studies.

net-widening: Increasing the availability of programs to handle youth offenders, thereby causing youths who would otherwise have been left alone to be placed into programs.

nonformal processing: Includes various ways of processing cases that do not involve an adjudication and the initiation of a formal juvenile court record for the youth.

parole: The conditional release of a youth from a correctional facility (one form of aftercare).

placing out: An effort to respond to the growing number of problem children in the cities during the 1800s. It consisted of the placement of children on farms and ranches in the Midwest and West.

presentence investigation: An investigation that is typically completed by a probation officer prior to the disposition; it is intended to assist the hearing officer in deciding the best response to an adjudicated youth.

probation: The conditional release of an adjudicated delinquent into the community under the supervision of the court.

restitution program: A program that requires offenders to compensate victims for damages to property or for physical injuries.

wilderness probation program: A program designed to reform the behavior and attitudes of juvenile offenders by involving them in wilderness activities, including solo camping. The philosophy behind such programs is partly based on the philosophy of the Outward Bound organization.

Review Questions

1. What is the definition of community-based juvenile corrections?
2. What types of at-home and out-of-home (residential) community-based corrections programs are discussed in the text? Describe each one.
3. What are the objectives of community-based corrections programs?
4. What is probation?
5. What are some typical rules of probation?
6. What factors influence the operation of probation between and within jurisdictions?
7. What are the basic roles performed by juvenile court probation officers? What do these roles entail?
8. According to the research, how effective is standard probation?
9. What are some recent trends in juvenile probation?
10. How effective are community-based corrections programs at reducing recidivism?

11. What types of legal authority are possessed by juvenile probation officers?
12. What legal issues are raised by juvenile court-ordered restitution?

Additional Readings

Empey, L. T. & Erickson, M. L. (1972). *The Provo experiment: Evaluating community control of delinquency.* Lexington, MA: Lexington Books.

Empey, L. T. & Lubeck, S. G. (1971). *The Silverlake experiment: Testing delinquency theory and community intervention.* Chicago: Aldine Pub. Co.

Gendreau, P. (1994). Intensive rehabilitation supervision: The next generation in community corrections? *Federal Probation, 58,* 72–78.

Howell, J. C. (2009). *Preventing and reducing juvenile delinquency: A comprehensive framework* (2nd. ed.). Los Angeles: Sage.

Jacobs, M. D. (1990). *Screwing the system and making it work: Juvenile justice in the no-fault society.* Chicago: University of Chicago Press.

Jones, M. A. & Krisberg, B. (1994). *Images and reality: Juvenile crime, youth violence, and public policy.* San Francisco: National Council on Crime and Delinquency.

Miller, J. G. (1991). *Last one over the wall: The Massachusetts experiment in closing reform schools.* Columbus, OH: Ohio State University Press.

Sherman, L. W., Gottfredson, D., MacKenzie, D., Eck, J., Reuter, P., & Bushway, S. (1997). *Preventing crime: What works, what doesn't, what's promising* (Report to the U. S. Congress). Washington, DC: National Institute of Justice.

Notes

1. Ohlin, L. E., Miller, A. D., & Coates, R. B. (1977). *Juvenile correctional reform in Massachusetts: A preliminary report of the Center for Criminal Justice of the Harvard Law School.* Washington, DC: National Institute for Juvenile Justice and Delinquency Prevention.
2. McCarthy, B. R. & McCarthy, B. J. (1991). *Community-based corrections* (2nd ed.). Pacific Grove, CA: Brooks/Cole.
3. Sickmund, M., Sladky, A., & Kang, W. (2005). *Easy access to juvenile court statistics: 1985–2005.* Retrieved December 6, 2008 from http://ojjdp.ncjrs.gov/ojstatbb/ezajcs/.
4. Jacobs, M. D. (1990). *Screwing the system and making it work: Juvenile justice in the no-fault society.* (pp. 125–129). Chicago: University of Chicago Press.

   Rubin, H. T. (1998). The juvenile court landscape. In A. R. Roberts (Ed.), *Juvenile justice: Policies, programs, and services* (2nd ed.). Chicago: Nelson-Hall.
5. Mich. Comp. Laws Ann. 712A.9.

   Mich. Stat. Ann. 27.3178(598.9).
6. Torbet, P. M. (1996). Juvenile probation: The workhorse of the juvenile justice system. *Juvenile Justice Bulletin.* Washington, DC: Office of Juvenile Justice and Delinquency Prevention.
7. Sickmund, Sladky, & Kang, 2005.
8. Torbet, 1996.
9. Jacobs, 1990.

Reese, W. A., Curtis, R. L., Jr., & Richard, A., Jr. (1989). Juvenile justice as people-modulating: A case study of progressive delinquent dispositions. *Journal of Research in Crime and Delinquency, 26*, 329–357.

10. Ohlin, L. E., Piven, H., & Pappenfort, D. M. (1956). Major dilemmas of the social worker in probation and parole. *Crime and Delinquency, 2*, 211–225.

Whitehead, J. T. & Lab, S. P. (1990). *Juvenile justice: An introduction*. Cincinnati, OH: Anderson.

11. Whitehead & Lab, 1990.

12. Griffin, P. & King, M. (2006). National overviews. *State Juvenile Justice Profiles.* Pittsburgh, PA: National Center for Juvenile Justice.

13. Griffin & King, 2006.

14. Griffin & King, 2006.

15. Griffin & King, 2006.

16. Empey, L. T. & Erickson, M. L. (1972). *The Provo experiment: Evaluating community control of delinquency.* Lexington, MA: Lexington Books.

Murray, C. A. & Cox, L. A. (1979). *Beyond probation: Juvenile corrections and the chronic delinquent.* Beverly Hills, CA: Sage.

National Institute of Mental Health. (1973). *Community-based corrections.* Washington, DC: U.S. Government Printing Office.

Reiss, A. J., Jr. (1951). Delinquency as the failure of personal and social controls. *American Sociological Review, 16*, 196–207.

Scarpitti, F. R. & Stephenson, R. M. (1968). A study of probation effectiveness. *Journal of Criminal Law, Criminology, and Police Science, 59*, 361–369.

Wooldredge, J. D. (1988). Differentiating the effects of juvenile court sentences on eliminating recidivism. *Journal of Research in Crime and Delinquency, 25*, 264–300.

17. Empey & Erickson, 1972.

18. Scarpitti & Stephenson, 1968.

19. Lab, S. P. & Whitehead, J. T. (1988). An analysis of juvenile correctional treatment. *Crime and Delinquency, 34*, 60–83.

Lipton, D. S., Martinson, R., & Wilks, J. (1975). *The effectiveness of correctional treatment: A survey of treatment evaluation studies.* New York: Praeger.

Robison, J. & Smith, G. (1971). The effectiveness of correctional programs. *Crime and Delinquency, 17*, 67–80.

20. Greenberg, D. F. (1975). Problems in community corrections. *Issues in Criminology, 10*, 1–33.

Hylton, J. H. (1982). Rhetoric and reality: A critical appraisal of community correctional programs. *Crime and Delinquency, 28*, 341–373.

21. Lipsey, M. W. (2007). *A standardized program evaluation protocol for programs serving juvenile probationers.* Nashville, TN: Vanderbilt University, Center for Evaluation Research and Methodology.

22. Lipsey, M. W. & Wilson, D. B. (1998). Effective interventions for serious juvenile offenders: A synthesis of research. In R. Loeber & D. B. Farrington (Eds.), *Serious and violent juvenile offenders: Risk factors and successful interventions.* Thousand Oaks, CA: Sage.

23. Snyder, H. N. & Sickmund, M. (2006). *Juvenile offenders and victims: 2006 national report*. Washington, DC: Office of Juvenile Justice and Delinquency Prevention.

24. Aos, S., Phipps, P., Bamoski, R., & Lieb, R. (2001). *The comparative costs and benefits of programs to reduce crime*. Olympia, Washington: Washington State Institute for Public Policy.

25. Empey & Erickson, 1972.

Scarpitti & Stephenson, 1968.

Lipsey, 2007.

Snyder & Sickmund, 2006.

Wooldredge, 1988.

26. Lipsey & Wilson, 1998.

27. Armstrong, T. L. (1988). National survey of juvenile intensive probation supervision (Parts 1 and 2). *Criminal Justice Abstracts, 20*, 342–348, 497–523.

28. Snyder, H. N. & Sickmund, M. (1995). *Juvenile offenders and victims: A national report.* Washington, DC: Office of Juvenile Justice and Delinquency Prevention.

29. Barton, W. H. & Butts, J. A. (1990). Viable options: Intensive supervision programs for juvenile delinquents. *Crime and Delinquency, 36*, 238–256.

30. Barton & Butts, 1990.

Lipsey & Wilson, 1998.

Wiebush, R. G. (1993). Juvenile intensive supervision: The impact on felony offenders diverted from institutional placement. *Crime and Delinquency, 39*, 68–89.

Walker, E. N. (1989). Community Intensive Treatment for Youth program: A specialized community-based program for high risk youth in Alabama. *Law and Psychology Review, 13*, 175–199.

31. Lipsey, M. W. (1999). Can intervention rehabilitate serious delinquents? *Annals of the American Academy of Political and Social Science, 564*, 421–166.

Lipsey & Wilson, 1998.

32. Barton & Butts, 1990.

33. Robertson, A. A., Grimes, P. W., & Rogers, K. E. (2001). A short-run cost-benefit analysis of community-based interventions for juvenile offenders. *Crime and Delinquency, 47*, 265–284.

34. Ryan, J. E. (1997). Who gets revoked? A comparison of intensive supervision successes and failures in Vermont. *Crime and Delinquency, 43*, 104–118.

35. Ball, R. A., Huff, C. R., & Lilly, J. R. (1988). *House arrest and correctional policy: Doing time at home.* Newbury Park, CA: Sage.

36. Young, T. M. & Pappenfort, D. M. (1977). *Secure detention of juveniles and alternatives to its use: National evaluation program, Phase one.* Washington, DC: U.S. Government Printing Office.

37. Ford, D. & Schmidt, A. K. (1985). Electronically monitored home confinement. NIJ Reports. Washington, DC: National Institute of Justice.

Renzema, M. & Skelton, D. T. (1990) Use of electronic monitoring in the United States: 1989 update. *Research in Brief.* Washington, DC: National Institute of Justice.

38. Roy, S. (1997). Five years of electronic monitoring of adults and juveniles in Lake County, Indiana: A comparative study on factors related to failure. *Journal of Crime and Justice*, *20*, 141–160.
39. Vaughn, J. B. (1989). A survey of juvenile electronic monitoring and home confinement programs. *Juvenile and Family Court Journal*, *40*, 1–36.
40. Durham, A. M., III. (1994). *Crisis and reform: Current issues in American punishment.* Boston: Jones and Bartlett Publishers.
41. Schneider, A. L. & Warner, J. S. (1989). *National trends in juvenile restitution programming.* Washington, DC: Office of Juvenile Justice and Delinquency Prevention.
42. Siegel, L. J., Welsh, B. C., & Senna, J. J. (2003). *Juvenile delinquency: Theory, practice, and law* (8th ed.). Belmont, CA: Wadsworth.
43. Schneider, P. R., Schneider, A. L., Griffith, W. R., & Wilson, M. J. (1982). *Two-year report on the national evaluation of the juvenile restitution initiative: An overview of program performance.* Eugene, OR: Institute of Policy Analysis.
44. Schneider, P.R. & Finkelstein, M. C. (Eds.). (1998). *RESTTA national directory of restitution and community service programs.* Bethesda, MD: Pacific Institute for Research and Evaluation. Retrieved from http://www.ojjdp.ncjrs.org/pubs/restta/index.html.
45. Schneider, A. L. & Schneider, P. R. (1984). A comparison of programmatic and "Ad Hoc" restitution in juvenile courts. *Justice Quarterly*, *1*, 529–547.
46. Schneider & Finkelstein, 1998.
47. Butts, J. A. & Snyder, H. N. (1992). Restitution and juvenile recidivism. *OJJDP Update on Statistics*. Washington, DC: Office of Juvenile Justice and Delinquency Prevention.
48. Schneider, A. L. (1986). Restitution and recidivism rates of juvenile offenders: Results from four experimental studies. *Criminology*, *24*, 533–552.
49. Ervin, L. & Schneider, A. (1990). Explaining the effects of restitution on offenders: Results from a national experiment in juvenile courts. In B. Galaway & J. Hudson (Eds.), *Criminal justice, restitution, and reconciliation.* Monsey, NY: Criminal Justice Press.
50. Schneider & Finkelstein, 1998.
51. Bottoms, A. & Dignan, J. (2004). Youth justice in Great Britain. In M. Tonry and A. N. Doob (Eds.), *Youth crime and youth justice: Comparative and cross-national perspectives* (Vol. 31). Chicago: University of Chicago Press.
52. Schneider & Warner, 1989.
53. Schneider, A. L. & Schneider, P. R., 1984.
54. Bazemore, G. & Schneider, P. R. (1985). Research on restitution: A guide to rational decision-making. In *The Guide to Juvenile Restitution: A Training Manual for Restitution Program Managers.* Washington, DC: Office of Juvenile Justice and Delinquency Prevention.
55. Austin, J. & Krisberg, B. (1982). The unmet promise of alternatives to incarceration. *Crime and Delinquency*, *28*, 374–409.
56. Siegel, Welsh, & Senns, 2003.
57. Siegel, Welsh, & Senns, 2003.

58. Golins, G. L. (1980). *Utilizing adventure education to rehabilitate juvenile delinquents.* Las Cruces, NM: Educational Resources Information Center, Clearinghouse on Rural Education and Small Schools.

59. Golins, 1980.

Minor, K. I. & Elrod, P. (1994). The effects of a probation intervention on juvenile offenders' self-concepts, loci of control, and perceptions of juvenile justice. *Youth and Society, 25*, 490–511.

60. Callahan, R. (1985). Wilderness probation: A decade later. *Juvenile and Family Court Journal, 36*, 31–35.

Kelly, F. J. & Baer, D. J. (1971). Physical challenge as a treatment for delinquency. *Crime and Delinquency, 12*, 437–445.

Winterdyk, J. & Roesch, R. O. (1982). A wilderness experiential program as an alternative for probationers: An evaluation. *Canadian Journal of Criminology, 24*, 39–49.

61. Wilson, S. J. & Lipsey, M. W. (2000). Wilderness challenge programs for delinquent youth: A meta-analysis of outcome evaluations. *Evaluation and Program Planning, 23*, 1–12.

62. Winterdyk, J. & Griffiths, C. (1984). Wilderness experience programs: Reforming delinquents or beating around the bush? *Juvenile and Family Court Journal, 35*, 35–44.

Winterdyk & Roesch, 1982.

63. Elrod, H. P. & Minor, K. I. (1992). Second wave evaluation of a multi-faceted intervention for juvenile court probationers. *International Journal of Offender Therapy and Comparative Criminology, 36*, 247–262.

64. Wilson & Lipsey, 2000.

65. Community Alliance for the Ethical treatment of Youth (CAFETY). Retrieved January 6, 2009 from http://cafety.org/index.php?option=com_content&task=view&id=629&Itemid=8.

Coalition Against Institutionalized Child Abuse (CAICA). Available from http://www.caica.org/index.htm. This resource provides a list of deaths of children in residential facilities.

66. Office of Juvenile Justice and Delinquency Prevention. (1979). *Project New Pride: Replication.* Washington, DC: Office of Juvenile Justice and Delinquency Prevention.

67. Mendel, R. A. (2001). *Less cost, more safety: Guiding lights for reform in juvenile justice.* Washington, DC: American Youth Policy Forum.

68. Mendel, 2001.

Palmer, T. (1992). *The re-emergence of correctional intervention.* (Newbury Park, CA: Sage.

69. Mendel, 2001.

70. Mendel, 2001.

71. Henggeler, S. W., Mihalic, S. F., Rone, L., Thomas, C., & Timmons-Mitchell, J. (1998). Multisystemic therapy. In D. S. Elliott (Ed.), *Blueprints for violence prevention* (Book 6). Boulder, CO: Center for the Study and Prevention of Violence, Institute of Behavioral Science, University of Colorado.

72. Golden, R. (Ed.). (1996). *Disposable children: America's child welfare system.* Belmont, CA: Wadsworth Publishing.

73. Golden, 1996.

74. U. S. Department of Health and Human Services, Administration for Children and Families. (2008). *Trends in foster care and adoption—FY 2002–FY 2007.* Retrieved January 7, 2009 from http://www.acf.hhs.gov/programs/cb/stats_research/afcars/trends.htm.

75. Elrod, H. P. & Brown, M. P. (1991). *Individual, court and offense-related variables as predictors of subsequent offense activity among juvenile court probationers: A preliminary analysis.* Paper presented at the American Society of Criminology annual meeting, San Francisco.

McCord, J., McCord, W., & Thurber, E. (1968). The effects of foster-home placement in the prevention of adult antisocial behavior. In J. R. Stratton & R. M. Terry (Eds.), *Prevention of delinquency: Problems and programs.* New York: Macmillan.

76. Vandivere, S., Chalk, R., & Moore, K. A. (2003). Children in foster homes: How are they faring? (Child Trends Research Brief Publication No. 2003-23). Washington, DC: Child Trends.

77. Rubin, H. T. (1985). *Juvenile justice: Policy, practice, and law* (2nd ed.). New York: Random House.

78. Wolf, B. (1997). Advocate's narrative: Ben Wolf. In R. Golden (Ed.). *Disposable children: America's child welfare system.* Belmont, CA: Wadsworth Publishing.

79. Chamberlain, P. (1990). Comparative evaluation of specialized foster care for seriously delinquent youths: A first step. *Community Alternatives: International Journal of Family Care, 2,* 21–36.

Chamberlain, P. & Reid, J. B. (1998). Comparison of two community alternatives to incarceration for chronic juvenile offenders. *Journal of Consulting and Clinical Psychology, 66,* 624–633.

Chamberlain, P. (1998). Treatment foster care. *Juvenile Justice Bulletin.* Washington, DC: Office of Juvenile Justice and Delinquency Prevention.

80. Chamberlain & Reid, 1998.

Chamberlain, P., Leve, L. D., & DeGarmo, D. S. (2007). Multidimensional treatment foster care for girls in the juvenile justice system: 2-year follow-up of a randomized clinical trial. *Journal of Consulting and Clinical Psychology, 75,* 187–193.

Eddy, J. M., Whaley, R. B., & Chamberlain, P. (2004). The prevention of violent behavior by chronic and serious male juvenile offenders: A 2-year follow-up of a randomized clinical trial. *Journal of Emotional and Behavioral Disorders, 12,* 2–8.

Leve, L. D., Chamberlain, P., & Reid, J. B. (2005). Intervention outcomes for girls referred from juvenile justice: Effects on delinquency. *Journal of Consulting and Clinical Psychology, 73,* 1181–1185.

81. Finckenauer, J. O. (1984). *Juvenile delinquency and corrections: The gap between theory and practice.* Orlando, FL: Academic Press.

82. Finckenauer, 1984.

83. Simonsen, C. E. (1991). *Juvenile justice in America* (3rd ed.). New York: Macmillan.

84. Simonsen, 1991.

85. Empey, L. T. & Lubeck, S. G. (1971). *The Silverlake experiment: Testing delinquency theory and community intervention*. Chicago: Aldine Pub. Co.

86. Chamberlain & Reid, 1998.

Chamberlain, Leve, & DeGarmo, 2007.

Eddy, Whaley, & Chamberlain, 2004.

Leve, Chamberlain, & Reid, 2005.

87. Mutchnick, R. J. & Fawcett, M. (1991). Group home environments and victimization of resident juveniles. *International Journal of Offender Therapy and Comparative Criminology*, *35*, 126–142.

88. Greenwood, P. W., Model, K. E., Chiesa, J., & Rydell, C. P. (1996). *Diverting children from a life of crime: Measuring costs and benefits* (Rev. ed.). Santa Monica, CA: Rand.

Steele, P. A., Austin, J., & Krisberg, B. (1989). *Unlocking juvenile corrections: Evaluating the Massachusetts Department of Youth Services—Final report*. San Francisco: National Council on Crime and Delinquency.

Lipsey, M. W. (1992). Juvenile delinquency treatment: A meta-analytic inquiry into the variability of effects. In T. D. Cook, H. Cooper, D. S. Cordray, H. Hartmann, L. V. Hedges, R. J. Light, T. A. Louis, and F. Mosteller (Eds.), *Meta-analysis for explanation: A casebook*. New York: Russell Sage Foundation.

Rutherford, A. & Benger, O. (1975). *Community-based alternatives to juvenile incarceration: Final report* (National Evaluation Program, phase 1 assessment) (pp. 27–31). Washington, DC: U.S. Government Printing Office.

Yoshikawa, H. (1994). Prevention as cumulative protection: Effects of early family support and education on chronic delinquency and its risks. *Psychological Bulletin*, *115*, 1–27.

Yoshikawa, H. (1995). Long-term effects of early childhood programs on social outcomes and delinquency. *The Future of Children*, *5*, 51–75.

Zavlek, S. (2005). Planning community-based facilities for violent juvenile offenders as part of a system of graduated sanctions. *Juvenile Justice Bulletin*. Washington, DC: Office of Juvenile Justice and Delinquency Prevention.

89. National Council on Crime and Delinquency. (n.d.). *Juvenile justice: Tough enough?* San Francisco: National Council on Crime and Delinquency.

Lundman, R. J. (2001). *Prevention and control of juvenile delinquency* (3rd ed.). New York: Oxford University Press.

90. Miller, J. G. (1991). *Last one over the wall: The Massachusetts experiment in closing reform schools*. Columbus, OH: Ohio State University Press.

91. Miller, 1991, pp. 220–222. This reference cites Ohlin, L. Coates, R., & Miller, A. (1975). *Report on Massachusetts Department of Youth Services, 1975*. Cambridge, MA: Center for Criminal Justice, Harvard University Law School.

Miller, 1991, pp. 223–225. This reference cites Steele, Austin, & Krisberg, 1989.

92. Miller, 1991, pp. 220–222. This reference cites Ohlin et al., 1975.

93. Miller, 1991, pp. 193–198.

94. Jackson, P. (1983). *The paradox of control: Parole supervision of youthful offenders.* New York: Praeger.

95. Lerman, P. (1975). *Community treatment and social control: A critical analysis of juvenile correctional policy.* Chicago: The University of Chicago Press.

96. Miller, 1991.

97. Altschuler, D. M. & Armstrong, T. L. (1994). *Intensive aftercare for high-risk juveniles: A community care model summary*. Washington, DC: Office of Juvenile Justice and Delinquency Prevention.

Gies, S. V. (2003). Aftercare services. *Juvenile Justice Bulletin.* Washington, DC: Office of Juvenile Justice and Delinquency Prevention.

98. Wiebush, R. G., Wagner, D., McNulty, B., Wang, Y., & Le, T. N. (2005). *Implementation and outcome evaluation of the intensive aftercare program: Final report.* Washington, DC: Office of Juvenile Justice and Delinquency Prevention.

99. Mich. Court Rules 5.903(11).

Institutional Corrections Programs for Juvenile Offenders

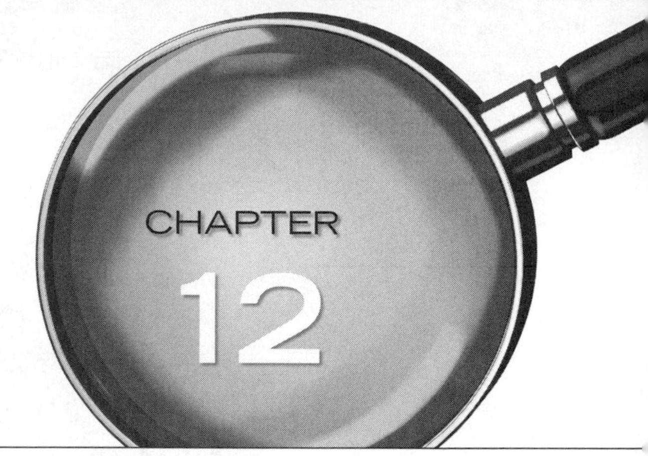

CHAPTER OBJECTIVES

After studying this chapter, you should be able to

- Describe the different types of juvenile corrections facilities found in the United States
- Describe the basic functions of juvenile detention centers, how these institutions operate, and recent trends in the use of detention
- Articulate the different rationales that are used to support detention
- Describe the two essential components of a juvenile detention program
- Define a token economy and explain how it is used in juvenile institutions
- Describe the problems associated with the detention of juveniles
- Describe the problems confronted by correctional personnel and juveniles when youths are placed in jail
- Describe the characteristics of juvenile training schools
- Describe the types of educational, vocational, and treatment programs found in state training schools
- Describe the problems that confront today's training schools
- Describe the basic features of diagnostic and reception centers, outdoor programs, boot camps, and private institutional settings
- Describe recent trends in juvenile incarceration
- Describe the research that has looked at the effectiveness of juvenile institutions
- Describe the challenges that presently face juvenile corrections institutions
- Discuss important legal issues related to juvenile incarceration

CHAPTER OUTLINE

- Introduction
- Overview of Institutional Corrections for Juveniles
- Types of Juvenile Corrections Institutions
- Trends in Juvenile Incarceration
- The Effectiveness of Institutional Corrections Programs for Juveniles
- Challenges Facing Juvenile Corrections Institutions
- Legal Issues

- Chapter Summary
- Key Concepts
- Review Questions
- Additional Readings
- Notes

Introduction

Institutional or residential placement is the most restrictive type of placement used by juvenile courts. Indeed, a distinguishing characteristic of institutional placement is that it restricts youths' access to the community to a greater degree than does assignment to a community-based program. Institutional and residential placement settings include a variety of residential facilities, such as juvenile detention facilities, jails, diagnostic and reception centers, forestry camps, ranches, boot camps, and state training schools, among others. Although many of these facilities house offenders who have committed a variety of offenses, others house specialized offender populations, such as sex offenders and youths with mental health and substance abuse problems.

Although institutions for juvenile offenders have existed since the early 1800s and have played an important role in juvenile corrections, they have become increasingly popular as tools for responding to juvenile crime in recent years. The growing popularity of institutional placement appears to be, in part, a result of increased violence among some juvenile offenders, but it also appears to reflect the belief by many policy makers that the best approach to juvenile crime is to "get tough" with juvenile offenders. Presently, institutional placement is used for many youths charged with violent juvenile offenses. However, today, as in the past, this type of placement is used not only for youths charged with and adjudicated for serious offenses, but also for those charged with and adjudicated for minor offenses. Indeed, in many instances it is used for youths who could be better and more cheaply served by community-based programs.

Despite the lengthy history of juvenile institutions and their increasing popularity, there is little evidence that the incarceration of youths is an effective correctional response to juvenile crime. As we noted in Chapter 4, early institutions for juveniles faced a number of problems, including overcrowding, mixing minor or status offenders with serious offenders, poor programming, inhumane treatment of residents, inadequate or nonexistent aftercare, and high recidivism rates. Although the state of institutional juvenile corrections has generally improved since the 1800s, none of these problems has been completely eliminated. As a result, critics charge that institutions have often failed to achieve their stated goals of rehabilitating youths or protecting the public.

This chapter examines contemporary institutional placements for juvenile offenders. It begins with a general overview of juvenile corrections institutions, then examines types of juvenile corrections institutions found in the United States, with special attention paid to the two most common: detention centers and state training schools. Throughout, the chapter focuses on the purpose and operation of these institutions as well as their problems. The chapter concludes with an examination of trends in juvenile incarceration and the effectiveness of juvenile corrections institutions.

Overview of Institutional Corrections for Juveniles

As noted previously, juvenile corrections institutions represent the most restrictive option available to juvenile courts. These institutions, however, vary in the extent to which they focus on custody and control. For example, some employ a variety of security hardware and structures, such as perimeter fencing or walls, and surveillance and detection devices, such as motion detectors, sound monitors, and security cameras.[1] These juvenile institutions are classified as **secure facilities** (or **closed facilities**) and their purpose is to closely monitor the movements of residents within the facility and restrict their access to the community. In many ways these institutions resemble secure adult institutions. Most public and private detention centers, reception and diagnostic centers, and state training schools are secure or closed facilities.[2]

Other juvenile institutions rely much less on security devices. These facilities have no perimeter fencing, and some do not lock entrances or exits at night.[3] These facilities are classified as **open institutions**,[4] and they rely more heavily on staff than on security hardware, structures, and devices. Most private facilities as well as most public shelters, ranches, forestry camps, and farms are considered open institutions.[5] Juvenile corrections institutions also differ in a number of ways besides the degree of reliance on security hardware, structures, and devices. For example, some institutions are *privately operated* and some are *publicly operated*. In fact, a majority of juvenile correctional facilities are privately operated, although the majority of youths are held in public institutions. When the 2006 Census of Juveniles in Residential Placement (CJRP) was taken, approximately 69% of the youths in juvenile facilities were residing in public or tribal facilities and about 31% were residing in private facilities.[6]

As one might expect, there are often important differences between **private facilities** and **public facilities**. For example, private facilities tend to be smaller than those that are publically operated.[7] Another difference is the types of youths that are placed in public and private institutions. Although some private facilities specialize in treating violent or seriously delinquent offenders, as a general rule, public facilities contain a much higher percentage of youths who have committed serious delinquent offenses than do private facilities. Conversely, private facilities typically house a much higher percentage of youths who are in custody for committing status offenses and for other nondelinquent reasons.[8] The differences in the types of youths admitted to public and private facilities is likely due to several factors, including the ability of private facilities to choose who they will serve, the existence of specialized programming (e.g., substance abuse or mental health treatment) in many private facilities, as well as economic considerations that require private facilities to accept youths who have insurance or the ability to pay for placement.

secure facility A facility that closely constrains and monitors residents' activities and physically restricts their access to the community typically through the use of various hardware devices, such as locked metal doors, security cameras, motion and sound detectors, fences with climb resistant mesh, and razor wire to limit residents' movement within the institution and to prevent escape.

open institution A facility that is not locked down, has no walls or fencing, and relies on staff to control inmates.

closed Facility See secure facility.

private facility A facility that is privately owned and operated. Private facilities can generally choose whom to accept, allowing them to focus on specific types of juveniles, such as sexual offenders or less serious offenders.

public facility A government-run facility. Public facilities have less ability than private facilities to screen out difficult and violent offenders and usually house more of both.

FYI

The Census of Juveniles in Residential Placement (CJRP) Provides Valuable Data on Youths in Institutional Placements

The Census of Juveniles in Residential Placement (CJRP) is conducted every 2 years and collects data on the types of facilities in which youths are held, the characteristics of youths placed in facilities, and the types of services youths are provided.

FYI

Private Facilities Often House Different Populations than State Facilities

In 2006, status offenders accounted for 12% of the residents of private facilities, but only 2% of the residents of state facilities. Moreover, while 31% of the residents of private facilities had committed offenses against persons, 43% of the residents in state facilities had committed offenses against persons.[9] Indeed, a substantial percentage of youths admitted to private facilities are "voluntarily" admitted by their parents or by school officials.[10]

Historically, publicly operated facilities were more likely to hold greater proportions of nonwhite youths, males, and older youths.[11] More recently, however, the demographic characteristics of youths in public and private facilities appear more similar, although private facilities hold a greater proportion of youths younger than 15 years of age and state facilities hold a greater proportion of youths who are 18 to 20 years of age.[12] Another way that public institutions differ from private institutions is that public institutions are more likely to be overcrowded. According to the 2004 Juvenile Residential Facilities Census (JRC), 11% reported that they were operating above their bed capacity compared to only 1% of private institutions. However, a larger percentage of private institutions (33%) indicated that they were operating at 100% capacity compared to public institutions (16%).[13]

Juvenile correctional institutions vary in other ways as well. For example, the majority of both private and public institutions are small, housing 20 residents or less. However, 3% of juvenile institutions have a legal capacity of 200 or more residents.[15]and some institutions house 800 or more youths.[16] Also, some institutions are coed institutions, whereas others are single-gender institutions.

Facilities also differ in the average length of time that residents stay. Typically, youths committed to state facilities by juvenile courts stay longer than those placed in private facilities or local detention facilities.[17] In addition, some private and public facilities are designed for short-term placement, but others are intended for long-term stays. Detention centers, jails, and diagnostic and reception centers tend to be short-term facilities,

MYTH VS REALITY

Public Versus Private Facilities

Myth—Private facilities provide more humane care for juveniles than do public facilities.

Reality—There are many well-run private facilities that provide excellent care for delinquent and problem youths. However, just because a juvenile corrections facility is privately operated does not automatically make it better than many publicly operated facilities. Indeed, some privately operated facilities have treated their charges as badly as the worst public facilities. Research by David Shichor and Clemens Bartollas that looked at public and private facilities used by a large probation department in southern California found that private facilities did not always provide promised services to clients, that many private facility staff members did not have appropriate qualifications, and that private facilities did not always separate violent and older youths from younger and less serious offenders.[14] Keep in mind, however, that there are many well-run publicly operated programs around the United States.

FYI

White Youths and Males Tend to Stay Longer in Institutional Placements

In 2003, the median number of days that youths spent in juvenile facilities was 68. The median number of days in placement for males was 71 days compared to 48 days for females. The median number of days in placement for white youths was 72 days compared to 64 days for minority youths.[19]

and ranches, forestry camps, farms, and training schools are generally considered long-term facilities. However, there is considerable variation in the average length of time youths spend in institutions classified as long term.[18]

Another way that juvenile institutions differ is in the types of programming and the quality of institutional life they provide. Almost all juvenile correctional institutions offer basic educational and counseling programs for their residents. Moreover, more than half of all institutions provide family counseling and employment counseling, hold peer group meetings, or use behavioral contracts or a point system. However, there is wide variability in the extent to which programming that goes beyond basic education and counseling is offered.[20] Although a majority of all institutions provide some substance abuse screening and treatment services, and about half of all institutions provide mental health screening of youths,[21] the quality of these services is unknown in many cases.

As noted in Chapter 4, children have often been subjected to abuse and inhumane treatment in juvenile correctional institutions. Certainly, the overall quality of institutional life in juvenile correctional facilities has vastly improved since youths were first placed in the houses of refuge and early reform and training schools. Many juvenile institutions are administered by competent, caring, and professional administrators who oversee skilled and caring staff involved in the delivery of a variety of quality services to residents. Yet, in other instances, many of the problems that have historically plagued juvenile correctional institutions are still evident. Some of the problems facing juvenile correctional institutions are considered later in this chapter.

Types of Juvenile Correctional Institutions

Juvenile detention centers, jails, diagnostic and reception centers, forestry camps, ranches, boot camps, and state training schools are among the main types of institutions used by the juvenile courts. These types of facilities differ from each other, of course, but there are often considerable differences between facilities of the same type. This section provides a general overview of juvenile institutions employed in juvenile justice. For purposes of organization, it divides juvenile institutions into two types: **preadjudicatory institutions**, which are primarily intended to house youths awaiting adjudication, and **postadjudicatory institutions**, which are more often used for youths after adjudication.

preadjudicatory institution A temporary institution for juveniles who are awaiting trial or awaiting sentencing.

postadjudicatory institution An institution to which a juvenile offender has been sentenced.

Preadjudicatory Institutional Placements

Detention Centers

Detention centers (sometimes called juvenile halls or juvenile homes) are common in the United States. Indeed, many large counties and even many small ones operate their own detention facilities for juvenile offenders. In addition, many states operate regional detention centers that may be used to house youths taken into custody around the state. Detention centers, whether local or regional, house youths in physically restrictive settings "pending juvenile court action, or following adjudication pending disposition, placement, or transfer" to some other setting.[22] Although they are designed primarily to house youths awaiting adjudication, disposition, or placement in another correctional setting, such as a state training school, they are also used for postdispositional placement in some jurisdictions.[23]

Detention centers constitute the most popular type of correctional institution employed in juvenile justice and account for the great majority of admissions to juvenile correctional facilities each year. Typically, youths spend only a short time in detention. In 2006, approximately 27% of youths placed in detention centers were there for less than one week and 72% were housed in detention for one month or less. However, almost 3% of the youths placed in detention remained there for more than 6 months.[24]

Typically, youths are detained in one of two ways. First, they can be detained after a legal arrest by a police officer for an offense that is serious enough to warrant detention. Incarceration in a detention center is seldom automatic, however. During court hours, an arrested juvenile is usually brought to the court with the necessary paperwork, the juvenile is charged by the prosecutor or state attorney, and a detention hearing is held. After hours, an "on call" person designated by the judge typically makes the detention decision.

The second way juveniles are detained is by the issuance of a court order, sometimes called a warrant, complaint, pick up order, or apprehension and detention order (A & D order). This court order allows the arrest (apprehension) and incarceration (detention) of a juvenile, with a hearing to follow within a specified period of time, usually 24 hours. The legal grounds for these orders vary, but the following common grounds are included:

- The juvenile engaged in serious delinquent behavior that, if engaged in by an adult, would justify that adult's arrest and incarceration.
- The juvenile failed to appear in court after having been properly summoned by the court for a hearing.
- The juvenile ran away from a setting where he or she was ordered to be placed by the court (this may also be considered a probation violation).

Children found to be in situations that threaten their safety (e.g., neglected or abused children) can also be subject to an A & D order and be picked up by the police and placed in shelter care or foster care. These youths should not be placed in a detention center. Nevertheless, this happens in some instances.

Status offenders (see Chapter 13) should not be detained for running away unless they are under the court's jurisdiction and court ordered into a placement. Indeed, the

FYI

There May be a Trend Toward Increased Use of Detention

After declining 23% between 2001 and 2003, the number of youths residing in detention centers increased by almost 4% between 2003 and 2006 according to the Census of Juveniles in Residential Placement.[25]

MYTH VS REALITY

Youths May Be Detained for Minor Offenses

Myth—Only youths who have committed the most serious offenses are placed in detention centers.

Reality—Many youths who are detained have not committed serious crimes. According to the 2006 Census of Juveniles in Residential Placement, slightly less than 30% of the youths residing in detention centers were placed there because they had committed crimes against persons. Almost as many, about 27%, were detained for technical violations, such as violations of probation, parole, or a court order, and slightly more than 10% were detained for public order offenses. Approximately 3% were detained for status offenses. Other offenses for which youths were detained included property offenses (22.4%) and drug offenses (8.1%).[26]

Juvenile Justice and Delinquency Prevention Act of 1974, as amended, prohibits the detention of status offenders, although federal regulations have interpreted the act to permit detention for a period not to exceed 24 hours when status or nonoffenders first come into contact with the police. However, some courts have skirted the act by holding hearings and ordering youths to stay at home or attend school. If youths violate these orders, then they are detained for violation of a legitimate court order, not the offense of running away or failing to attend school.

The key point to remember is that taking a child into custody may or may not be a judicial decision, but keeping a juvenile in secure detention must be a judicial decision.

The detention of juveniles is supported on a number of legal grounds:

- The juvenile, if released, may commit other offenses, injuring other persons or property.
- The juvenile, if released, may be harmed or injured by others.
- The juvenile, if released, is not likely to appear at future court hearings.
- Unless the juvenile is detained, an uncooperative parent may remove him or her from the state, making the juvenile unavailable for trial.
- There is no responsible adult willing or able to provide supervision or care for the juvenile.[27]

In addition to the statutorily authorized grounds for detention, there are a number of reasons that detention is favored as a juvenile justice response in some localities. According to H. Ted Rubin, who has extensive experience as a juvenile justice practitioner and researcher, these include the following reasons:

- Detention is used as a form of punishment.
- Detention is used to fill up detention beds and demonstrate a need for detention.

FYI

The Juvenile Court, or Those Authorized By the Court, Make Detention Decisions

It should be noted that police officers and others must request that juveniles be detained. This is because the decision to detain a youth is typically made by the juvenile court or those authorized by the juvenile court to make detention decisions. In practice, however, such requests are rarely denied.

- Detention represents a convenient response to juvenile offenders that can be exercised by juvenile justice agents.
- Detention placates law enforcement officials who demand tough responses to juvenile offenders.
- Detention is used as an emergency mental health placement because such placements are often in short supply.[28]

The Operation of Detention After a juvenile is brought to a detention facility, a detention intake or admittance procedure is followed. Although procedures vary across detention facilities, detention intake procedures tend to have some common features. One of the first things done by the staff, if this has not already been done by the arresting officer, is to attempt to make contact with the minor's parents or guardian. At the same time, a variety of other steps are taken to ensure the safety of the juvenile and detention personnel and to acclimate the juvenile to the facility:

- The youth is searched. A strip search may be used to make sure the youth is not in possession of any contraband, drugs, or weapons.
- The youth's personal possessions are collected and stored until the youth is released.
- The youth is screened to determine if he or she is under the influence of drugs or alcohol, if any medical intervention is required, and if the youth has a medical condition that may affect other residents.
- Basic information about the youth, such as gender, age, address, complete name, date of birth, parents' names, school, and the reasons for detention, is collected.
- The rules that residents are expected to follow while in the facility are explained. The youth may be given a handbook that explains facility rules and may be required to take a test to demonstrate understanding of the rules prior to placement with the other residents.

detention program The total set of organized activities in which youths are involved while in a detention facility.

token economy A type of behavior modification program in which juveniles, in return for appropriate behaviors, earn points that can be redeemed for privileges, snack foods, and so on.

In a detention facility, the **detention program** "is the sum of all activities in which the child engages while in the detention home."[30] Two important components of a detention program are (1) the rules that youths are expected to follow while in detention and (2) the interventions, such as the educational curriculum and the behavior management system, used by detention personnel to control resident behavior.

Many detention facilities try to manage resident behavior through use of a **token economy**. In a token economy, youths earn tokens (or points) for exhibiting appropriate behavior. For example, youths may earn points for using appropriate language and

FYI

Detention Admission Procedures Have Sometimes Been Abusive

Given the need to protect the safety of residents and staff, the admission steps listed previously seem reasonable. However, there have been many instances of abuse associated with intake admissions, including dehumanizing searches intended to threaten and intimidate residents, such as pelvic examinations and vaginal searches of females, isolation of youths, and physical abuse.[29] Unfortunately, the extent to which such practices still occur is unknown. For information about the abuse of children in institutional settings, see the Coalition Against Institutionalized Child Abuse at http://www.caica.org/index.htm.

gestures, helping clean the facility, completing homework assignments, and following other program rules. As they accumulate points, they are able to purchase or earn privileges, such as more recreational time, more telephone time, and a later bedtime. Supporters of token economy programs argue that they help reinforce positive behavior among youthful detainees. These programs are based on operant conditioning principles that assume that behaviors that are reinforced are more likely to be exhibited in the future.

When token economies are implemented by people who clearly understand their purpose and operation, they can help shape youths' behavior in the desired way. When these programs are not implemented properly, however, they can increase the potential for conflict among residents and also between residents and staff. A major problem in many facilities with token economy programs is inconsistency in the operation of the programs. Such inconsistency can cause residents to perceive some groups or individuals as being given differential or preferential treatment, which in turn can increase the level of hostility among residents and between residents and staff.

Another concern is that token economy programs can disguise punitive programming with a veil of scientific language. For example, many programs contain a large number of rules that can result in residents' losing points but relatively few ways for residents to earn points. As a result, the overall focus of these programs becomes punishment, not reward, and such punitive programs are inappropriate, according to critics, because many youths in detention have not been adjudicated.[31]

A variety of other interventions are found within detention facilities. For example, many detention centers, to be in compliance with state compulsory education laws, operate basic educational programs staffed by certified teachers who work for local school systems. These programs typically provide basic and individualized instruction for residents, although some facilities offer more specialized programs including vocational training or college-level instruction.[32] Unfortunately, the motivation and training of teachers in these programs and the resources devoted to them vary considerably.

It is also common for detention facilities to provide some basic health care services, such as health appraisals, instruction on how to access various health services, and routine screening for health problems.[33] Most detention centers have mental health professionals available to youths who require such services, and some type of basic counseling is offered as well. However, only 30% of detention centers report that all youth are screened for mental health problems and 8% of detention facilities report that no youths are evaluated.[34] Moreover, information on the extent to which counseling is a part of detention center programming and the quality of existing counseling programs is not available.

Many detention centers also operate some type of recreational program for residents. Although almost all centers meet American Correctional Association standards, which require youths to be allowed at least one hour of recreation that involves large muscle activity (exercise) per day and at least one hour of leisure time per day, some operate more structured recreation programs (though the number of these is not known).

Problems Confronting Juvenile Detention Today, juvenile detention faces a number of problems. One major problem is the absence of strict standards governing the use of juvenile detention and a lack of agreement over the purpose of detention in many jurisdictions.[35] Indeed, wide discretion to detain juveniles is the rule in many communities.[36] As a result, almost any youth charged with a criminal offense in some jurisdictions can be detained.[37] This has contributed to another significant problem in many jurisdictions: institutional overcrowding.

Although the intent of the juvenile court is to provide for the care, protection, and rehabilitation of youths, abuse is a potential problem in any system intended to exercise restrictive control over youths. One type of abuse occurs when youths are detained without sufficient evidence that detention is necessary. Abuse also occurs when detention becomes a form of punishment. This is particularly troublesome prior to adjudication, because youths are presumed to be innocent prior to fact finding by the court. However, punishment after the adjudication can also be abusive when the express intention of the court is to "teach the youth a lesson," and even in cases where the intent is to treat or rehabilitate a youth. Importantly, efforts to treat or rehabilitate youths often involve some element of punishment. Furthermore, when juveniles are detained in order to ensure subsequent court appearances or to protect victims from harm, those making decisions to detain youths are making predictions about their future behavior. However, accurate predictions about juveniles' future behavior are difficult to make and are often based on criteria that have little relationship to their likelihood of appearing in court or engaging in subsequent delinquent acts.[38]

This is not to say, however, that using detention as a short-term solution cannot have some benefits. Most county or local detention facilities do not have the programming to keep juveniles for an extended period of time, but short-term detention that supports the authority of the court, probation, and programs such as house arrest can have a beneficial impact on the lives of some youths.

The effectiveness of detention, unfortunately, is continually being undermined by the chronic overcrowding of local detention facilities. Many facilities have licensed limits on their capacities, and exceeding those limits can result in loss of licensing, to say nothing about increased legal liability, increased stress on the staff and the residents, and the increased financial costs associated with detention. Indeed, crowding influences all aspects of institutional life. It is related to increased levels of anger and hostility between residents, and between residents and staff, that contribute to disciplinary infractions, escapes, and violence. It also negatively impacts residents' perceptions of safety and leads to a variety of resident responses, such as joining gangs for protection or acting "crazy" so that they will be placed in isolation where they feel less threatened. Crowding also leads to a variety of staff responses, such as greater focus on control and confinement of youths in locked rooms for longer periods of time, and it makes adequate classification and placement of youths within the facility more difficult. In short, crowding leads to a variety of responses on the part of residents and staff that exacerbate many of the problems that exist in institutional settings.[39]

Unlike the adult justice system, which makes bail available to most offenders, juvenile justice relies heavily on preventive detention. Most states allow judges to deny bail to adult suspects only if they have committed capital crimes (e.g., murder), if they have jumped bail in the past, or if they have committed another offense while on bail.[41] However, in the

FYI

Overcrowding in Detention Is a Significant Problem in Some Jurisdictions

In 2006, 14% of publically operated juvenile detention facilities were operating over their standard bed capacity.[40]

1984 U.S. Supreme Court case, *Schall v. Martin*, a majority of the justices ruled that the preventive detention of a juvenile is proper if there is reason to believe that the juvenile would commit additional crimes while awaiting adjudication, although the Court also ruled that juveniles have a right to a detention hearing and a statement indicating the reasons why they are being detained. In handing down its decision, the majority justified its position by noting that such detention protects both the juvenile and society from the "hazards" related to pretrial crime. Furthermore, the Court argued that preventive detention was appropriate because juveniles, unlike adults, are always in some form of custody[42] (see the discussion of *Schall v. Martin* in Chapter 6). The Court also pointed out that many states allowed preventive detention of juveniles, which indicates that the concept did not offend any fundamental due process right. So long as juvenile courts afford juveniles timely hearings to challenge detention decisions, the constitutional right to be free from "excessive bail," found in the Eighth Amendment, takes a back seat to society's interest in the supervision and control of children.

Unfortunately, one of the results of *Schall v. Martin* has been to serve as a legal support for the overuse of detention. The excessive use of detention occurs in many jurisdictions, despite national organizations and commissions, such as the National Advisory Commission on Criminal Justice Standards and Goals, the National Advisory Committee for Juvenile Justice and Delinquency Prevention, the Institute of Judicial Administration, and the American Bar Association, that recommend that detention be used only as a last resort for youths who have allegedly committed serious delinquent acts.[43] Indeed, in many jurisdictions, the great majority of youths who are detained pending a hearing are released after their hearing, often on probation. Many, then, question why it is necessary to detain these youths, when later it seems evident they present no danger to the community.

Although a number of grounds appear to be used to justify the detention of juveniles, research on detention decisions suggests that a variety of factors appear to influence them, although the factors vary across jurisdictions. Early studies suggested that the availability of detention facilities and a lack of alternative placements in a community influenced the level of detention use.[45] Still other research revealed that the willingness of parents or guardians to accept youths back into their homes after being taken into custody by the police influenced detention decisions.[46] Some research has found that race is related to detention decisions, particularly when minority youths live in disadvantaged neighborhoods.[47] According to the Census of Juveniles in Residential Placement, minority youths are disproportionately represented in detention facilities. For example, the 2006 census found that African-American youths accounted for 41% of the youths in detention and

COMPARATIVE FOCUS

Preventive Detention Is also Used in Japan

The structure of the juvenile justice process in Japan is similar in many respects to that in the United States. In Japan, as in the United States, preventive detention of youths is possible when the family court judge believes it is in the best interests of the child. In most instances, the period of detention lasts no longer than four weeks, although it is possible to keep a youth in detention for up to eight weeks. Also, when the youth is in detention, assessments of the youth are conducted in order to develop a treatment plan for the child.[44]

Hispanic youths accounted for 23.7%,[48] although these groups made up only 16.5% and 18.5% respectively of the juvenile population between 10 and 17 years of age.[49]

Although the incarceration of females, particularly minority females, increased during the 1990s and early 2000s, there has been some reduction in the number of females placed in detention since 2001.[50] Nevertheless, the detention of girls deserves careful scrutiny. Historically, girls have been much more likely than males to be detained for status offenses, such as running away and incorrigibility.[51] Moreover, many girls who end up in detention have been victims of physical, sexual, or emotional abuse and fail to receive needed treatment while detained.[52] Since the late 1990s, however, males have comprised the majority of youths placed in detention for status offenses. Also, males account for the great majority of youths detained for person, property, drug, public order offenses and for technical violations such as violations of probation, parole and court orders.[53]

Another problem uncovered by research is that youths who are detained are more likely to be treated more severely at the next stage of the juvenile justice process than similar youths who are not detained.[54] The explanation for the difference might be that youths who are detained are less able to assist in the preparation of their defense than youths released pending a hearing. Also, hearing officers at subsequent hearings may feel that youths who have been detained represent more of a threat to community safety, making the officers more likely to impose severe sanctions on these youths than those who are not detained. Judges who are elected (and hearing officers appointed by and responsible to those judges) are not oblivious to public sentiments regarding juvenile crime. Although judges and referees may protect juveniles' due process rights at trial, detention often proves to be an easy and attractive remedy that earns them community support. Being tough on delinquent juveniles is always a popular political stand.

Not only are there problems associated with decisions to detain youths, but there are also problems associated with the daily operation of many detention facilities. One serious problem is child neglect and abuse, which has occurred despite the fact that part of the mission of detention facilities is to provide youths with a safe environment. As noted in Chapter 4, the history of juvenile institutions has been characterized by punishment and inhumane treatment touted as rehabilitation. Many youths in detention have been physically and sexually assaulted by other residents and staff, many have been subjected to intimidation and other forms of psychological abuse, and many have been exposed to a variety of degrading and dehumanizing conditions. Unfortunately, abuse and inhumane treatment have not been eradicated from all juvenile detention facilities.

Today, there is evidence that isolation, mechanical restraints (e.g., handcuffs, straitjackets, anklets, etc.), and physical restraints (i.e., tackling and body holds) are still overused in some facilities. For example, according to a large-scale study of confinement conditions in juvenile correctional facilities published in 1994, there is some evidence that isolation and "time out" is overused in many facilities. This same study found that almost three-quarters (72%) of detention facilities reported the use of mechanical restraints, usually handcuffs and/or anklets. Other problems found in some facilities include high numbers of juvenile injuries (which typically result from attacks by other residents), occasional escapes and escape attempts, and occasional self-destructive behaviors among youths.[55] Unfortunately, there is a lack of ongoing research on the conditions of confinement found in juvenile correctional institutions around the country.

Despite the problems that plague some juvenile detention facilities, there are well-run institutions managed and staffed by well-trained and caring employees who work

MYTH VS REALITY

Juvenile Correctional Facilities, Including Detention Centers, Are Not Always Safe Environments for Children

Myth—Juvenile correctional institutions are safe environments for youths.

Reality—Many juvenile correctional facilities strive to ensure that youths are safe while they are in placement. However, others are not safe environments for children and have experienced ongoing problems with ensuring youths' safety. According to a survey of states by the Associated Press, there were more than 13,000 claims of abuse in juvenile correctional institutions, including detention centers, between 2004 and 2007. Sadly, many critics believe that the actual level of abuse may be much higher.[56]

diligently to ensure that children under their supervision live in a safe, supportive, and secure environment. Such facilities share a number of program characteristics, according to David Roush, a former detention director and now the director of research and professional development at the National Juvenile Detention Association at Michigan State University. They tend to have an adequate number of qualified staff, a concern with safety, a range of positive activities, strong leadership, a focus on staff education about detention, and a strong evaluation component.[57]

Jails

In some jurisdictions, juveniles may also be placed in adult jails. Like detention centers, jails represent a type of preadjudicatory institutional placement. The jailing of juveniles often occurs in rural jurisdictions, where there are few juvenile detention facilities. According to the Bureau of Justice Statistics, more than 6,000 youths younger than the age of 18 years were detained in adult jails on a typical day in 2006.[58] Indeed, at least 6,000 youths younger than the age of 18 years resided in adult jails each year that the survey was administered between 1995 and 2006. However, the number of youths has been declining in recent years from a high of more than 9,000 in 1999.[59]

The federal government has taken an active role in attempting to have juveniles removed from adult jails since 1980, as a result of an amendment to the 1974 Juvenile Justice and Delinquency Prevention Act. In testifying before a U.S. House of Representatives subcommittee on the reauthorization of the act in 1980, the deputy attorney general of the United States, Charles B. Renfrew, noted the following:

> The jailing of children is harmful to them in several ways. The most widely-known harm is that of physical and sexual abuse by adults in the same facility. Even short term, pretrial or relocation detention exposes juveniles to assault, exploitation and injury.
>
> Sometimes, in an attempt to protect a child, local officials will isolate a child from contact with others. Because juveniles are highly vulnerable to emotional pressure, isolation of the type provided in adult facilities can have a long-term negative impact on an individual child's mental health.
>
> Having been built for adults who have committed criminal acts, jails do not provide an environment suitable for the care and maintenance of delinquent juveniles or status offenders. In addition, being treated like a prisoner reinforces a child's negative self-image. Even after release, a juvenile may be labeled as a criminal in his community as a result of his jailing, a stigma which can continue for a long period.

The impact of jail on children is reflected in another grim statistic—the suicide rate for juveniles incarcerated in adult jails during 1978 was approximately seven times the rate among children in secure juvenile detention facilities.[60]

Under the Juvenile Justice and Delinquency Prevention Act (JJDPA), states that received federal funds under the act were required to separate juveniles from adults. Furthermore, as a result of a 1980 amendment to the act, they were required to remove juveniles from adult jails within five years or lose federal funds.[61] Additional legislation, however, indicated that 75% compliance by 1985 was acceptable.[62]

The Juvenile Justice and Delinquency Prevention Act has encouraged the removal of children from adult jails, at least in some states. For example, in 1991 the state of Indiana abolished the jailing of juveniles in that state. In writing about the jail removal efforts in Indiana, Barry Krisberg and James Austin, two respected juvenile justice researchers, noted the following:

For years, the state of Indiana placed hundreds of children in dangerous situations in adult jails. Yet a bipartisan coalition led by the Department of Corrections convinced state lawmakers that better alternatives existed. The jail removal movement in Indiana demonstrated again that the goals of child protection and public safety were not antithetical. Moreover, many Indiana communities have learned that sensible alternatives to jailing, such as temporary shelter care or home detention, can be created without requiring large new expenditures of public funds. Also, elected officials in Indiana and many other states learned that they could support humane policies for children without suffering the alleged political liability of appearing to be soft on crime. Successes in the jail removal area illustrate the power of legislative action to discourage harmful practices. These stories also suggest that the long standing inferior treatment of children can be quickly remedied.[64]

Other states, such as Idaho, Ohio, and North Dakota, have also made significant headway in this area.[65]

Despite pressure from the federal government on the states, however, jail removal has not been completely successful. Although progress has been made in some jurisdictions, juveniles are still being incarcerated in adult jails. In fact, there is some pressure

FYI

Juveniles May Be Placed in Adult Jails Under Certain Conditions

The Juvenile Justice and Delinquency Prevention Act does not completely ban the jailing of juveniles. Subsequent court rulings and legislation permit juveniles to be incarcerated in adult facilities under certain circumstances—for example, when a juvenile is being tried as an adult or has been convicted of a felony. Also, a juvenile can be held in an adult jail or lockup for up to 6 hours while other arrangements are being made, providing there is "sight and sound separation" of adults and juveniles. Also, in rural areas a juvenile may be held in an adult jail or lockup for up to 24 hours if the following conditions are met:

- The minor is awaiting an initial court appearance.
- An alternative placement is not available.
- There is sight and sound separation of juveniles and adults.[63]

from "get tough on crime" advocates to place more juveniles in adult facilities.[66] As a result of such pressure, a failure of states to develop juvenile detention facilities, and an unwillingness to develop appropriate alternatives to the jailing of juveniles in some jurisdictions, a number of juveniles continue to be housed in jails each year. Moreover, "get tough" measures that have resulted in more youths being tried and sentenced as adults mean that a number of persons younger than 18 years are being jailed as adults. For example, at midyear 2006, of the 6,104 persons younger than 18 years that were being held on a typical day, 4,836 or 79% were being held as adults.[67] Although the optimistic assessment of Barry Krisberg and James Austin quoted earlier regarding the removal of juveniles from adult jails was clearly justified in the early 1990s and still holds true in some states, the jailing of juveniles continues to occur in some jurisdictions.

Problems Associated with Placing Juveniles in Adult Jails Incarcerating juveniles in adult jails creates a variety of problems for those juveniles who are incarcerated as well as those who are responsible for their safety. Indeed, many of today's jails face a number of serious problems: a deteriorated or poorly designed physical plant, overcrowding, a lack of services and constructive activities for inmates, gang conflicts, and violence. It is worth noting that many jails have difficulty meeting the needs of adult inmates, let alone the needs of juveniles.

One obvious problem facing juveniles in jails is the risk of physical and sexual assault. Although juveniles are supposed to be separated from adult prisoners, separation does not always occur. Moreover, there are often inmate groups in jails, such as trustees, who have access to juveniles even when they are separated from the regular adult population in the jail.[68] Furthermore, many jails do not do a good job of isolating juveniles who are at risk of victimization by older and more aggressive juveniles. Consider the case of Eric, a 13-year-old who was jailed for stealing a dirt bike. He was incarcerated "next to a 17-year-old who was being held for assault and who had a history of violent and sexual offenses." In 1986, a class action suit was filed by private attorneys in Portland, Maine, with help from the Youth Law Center in San Francisco, alleging that staff in the jail left the cell doors open, allowing the 17-year-old to assault Eric and force him to have oral intercourse.[69] More recently, a 2009 article in Dallas Morning News reported that four juveniles may have been abused in the Dallas County Jail.[70] Juvenile girls may be particularly vulnerable to sexual assaults by jail personnel. As Meda Chesney Lind notes, they may be "doubly disadvantaged" because they are females and juveniles.[71]

Physical and sexual assault are but two of the numerous problems associated with placing juveniles in adult facilities. Even when juveniles do not come into direct contact with adults, they may be subjected to psychological abuse by adult inmates. In a 1967 report, the President's Commission on Law Enforcement and Administration of Justice noted that jail is characterized by "enforced idleness, no supervision, and rejection. It is a demoralizing experience for a youngster at a time when his belief in himself is shattered or distorted. Repeated jailing of youth has no salutary effect on the more sophisticated youngster; on the contrary, it reinforces his delinquency status with his peers and his self-identification as a criminal. Enforced idleness in a jail gives the sophisticated juvenile ample time and reason for striking back at society."[72]

The idleness and rejection experienced by youths in jail, coupled with inadequate supervision due to limited staff who are often poorly trained to deal with juveniles, often leave youths in a fragile psychological condition. Ironically, steps taken to protect youths, such as keeping them segregated from adults, often exacerbate the effects of jailing because

FYI

Jailing Juveniles Increases Their Risk for Self-Destructive Behaviors

Jail inmates younger than 18 years of age have the highest rates of suicide. Moreover, juveniles in adult jails are 19 times more likely to commit suicide than youths of the same age in the general population.[74]

separation often amounts to solitary confinement. Under such conditions, depression is not uncommon, and the probability of youths engaging in self-destructive behaviors, such as suicide and self-mutilation, is increased. As noted earlier, suicide or attempted suicide is a well-known problem associated with placing juveniles in adult jails. Indeed, research clearly demonstrates that the rate of suicide among children held in adult jails is much greater than that among youths in juvenile detention centers or among youths in the general population of the United States.[73]

Although juveniles of any age or gender can experience problems in jail, females may be particularly vulnerable. This is because female juveniles often have a history of sexual and physical abuse, which makes them susceptible to depression and self-destructive behaviors.[75] This type of history is often related to running away from home and other status offenses—behaviors that may result in incarceration in jails. There is considerable evidence that girls are often jailed for status offenses, such as running away, or because they are neglected or dependent youth.[76] In addition, they are often younger than male juveniles who are jailed. Typically, jails are even less adequately prepared to meet the needs of female juveniles than male juveniles, and they often end up becoming victimized a second time—by the juvenile justice system supposedly there for their protection.

Not only does the jailing of children create serious problems for children, but it also creates problems for jail administrators and staff. First, the requirement to separate juveniles from adults causes serious difficulty for jail administrators, who already have the burden of segregating a variety of offender groups. For example, jails need to separate males and females and isolate inmates who have medical problems and inmates who are disruptive. They may also separate younger adults from older adults and separate offenders awaiting trial from those who are serving sentences. Another problem for administrators is that jails are often crowded and frequently lack adequate personnel and support services, such as mental health workers, to assist those in need of such services. A third problem is that the jailing of juveniles creates additional liability for jails and county or city governments. In short, when juveniles are admitted to jails, they simply increase the headaches experienced by jail staff, who have a diverse population to work with already.

The truth of the matter is that jails are often not equipped to deal with the needs of many adult offenders, much less children. Most have neither the physical facilities nor an adequate number of correctional and other staff to adequately care for juvenile offenders. Furthermore, correctional staff are rarely trained to deal with the special needs and problems presented by young adult or juvenile inmates, nor do jails typically have ancillary resources at their disposal to help them meet the needs of these inmates. The problems associated with jailing juveniles raise serious questions about the appropriateness of this practice. Despite these problems, however, the practice continues in a number of jurisdictions.

Postadjudicatory Institutional Placement

State Training Schools

State training schools, which are public facilities, often serve as the last resort for youths designated as serious or repeat offenders. Yet not all youths who reside in such facilities are there because they are violent. Indeed, in 2006, only 48% of the juveniles committed to secure state correctional institutions were held because they had committed a violent crime against a person.[77]

state training school
A restrictive facility designed to deal with repeat juvenile offenders.

Some training schools actually resemble adult prisons. These institutions have a *closed environment*. They are physically secure; they can have various combinations of physical security devices such as high walls and fences, razor wire, motion detectors, and locked cell blocks; and they are often self-sufficient (i.e., the laundry, infirmary, school, and maintenance facility are run by institutional staff). They may also have solitary confinement or isolation rooms for those who consistently misbehave, they may employ mechanical restraints, and they place considerable emphasis on physical security.

Other training schools, however, use a cottage system. In these **cottage institutions**, youths are housed in cottages, which are usually small and have a more "homelike" atmosphere. Each cottage is run by cottage "parents," or staff who work on shifts. These institutions have a more *open environment*. Rather than relying on fences and walls and other hardware for security, they rely on staff to maintain security within the institution.

cottage institution
Facility in which juvenile offenders live in a "homelike" setting. Such institutions are typically more open and smaller than other types of juvenile correctional institutions, are run by "cottage parents," and have high staff-to-juvenile ratios.

Training School Operation Programming in state training schools usually consists of a combination of (1) academic education, (2) vocational education, (3) recreation, (4) basic counseling, and (5) behavior management. In addition, a variety of specialized treatment and education programs exist in some facilities. For example, more than half of all training schools offer family counseling, health and nutrition counseling, substance abuse education, and sex offender treatment. Religious counseling and programs are also prevalent in juvenile training schools and are more common than any other type of volunteer program found in juvenile correctional facilities. It should be noted that training schools are more likely to offer programs that do not require counseling staff,[79] because institutional budgets often preclude hiring highly paid professional staff.

Many of the youths in training schools have fewer years of education than youths of the same age in the general population,[80] and many others are well below their grade level in academic performance. Indeed, recent research indicates that only about 23% of juveniles in correctional facilities read at their nominal grade level.[81] Consequently, the educational programs at these institutions focus on basic skills, and most are designed to help students obtain a general equivalency diploma (GED). The majority of youths in training schools have access to special education classes, and almost all have access to literacy or remedial reading programs. These programs are important because as many

FYI

A Majority of Youths Placed in Correctional Institutions Are Placed in Secure Facilities
While juvenile institutions vary in their use of security hardware and structures (e.g., walls and perimeter fences), the use of physical security devices is common. In 2006, 78% of juveniles who were committed to juvenile correctional facilities were housed in locked facilities. As might be expected, the great majority (88%) of long-term secure institutions were classified as locked facilities.[78]

as 40% of youths in residential facilities may have some type of learning disability.[82] Some facilities also offer college programs for residents.[83]

Today, most youths confined in state training schools have access to vocational education programs. Like the academic educational programs offered in these facilities, the vocational programs offered are also intended to give youths basic skills. The most common types of vocational courses offered are carpentry and various building trades, food services, and auto mechanics, although printing, cosmetology, secretarial training, retail trades, and forestry/agriculture are found in at least 10% of training schools.[84] One problem with many of these programs is **sex typing**—a problem that may be more difficult to avoid in single-sex institutions.[85] Another problem is that they do not necessarily prepare youths for employment after they leave the institution. After youths leave institutions, they may have difficulty joining unions or they may still lack the necessary education needed for many jobs.[86] Although some institutions provide job training, few offer job placement programs that attempt to match youths with jobs offering a chance of advancement.

sex typing Where males are taught traditionally male skills such as carpentry and females are taught traditionally female skills such as cosmetology.

Most training schools operate recreational programs for residents. Indeed, courts have ruled that the failure of institutions to allow children regular exercise and recreation violates juveniles' constitutional rights.[87] The American Correctional Association (ACA), in its *Standards for Juvenile Training Schools*, recommends that juveniles be given 2 hours of recreation each school day and an additional hour on nonschool days.[88] Most training schools have an outdoor playing field, an outdoor basketball court, an outdoor volleyball court, indoor table games, ping-pong tables, and television sets. Many also have an indoor gym. Although program administrators typically view recreational programming as an important responsibility, it is worth noting that a national study of juvenile correctional facilities in the mid-1990s found that 19% of juvenile training schools failed to meet the ACA recommended requirements for recreation.[89]

Another program component consists of counseling or treatment. According to correctional administrators, approximately half or more of the residents in their facilities are experienced with at least one of the following problems: family difficulties, drug or alcohol abuse, peer problems, parental abuse, disruptive behavior, and depression.[90] Consequently, most institutions offer basic individual counseling. This type of counseling is provided by a range of program staff and is intended to help youths deal with adjustment problems within the institution and general problems they are experiencing in their lives. Some institutions use more structured therapeutic counseling, such as psychotherapy or reality therapy. This type of counseling is conducted by trained

FYI

Many Youths Involved in the Juvenile Justice Process Suffer from Some Form of Mental Disorder

Research on mental disorders among youths involved in the juvenile justice process indicates that youths involved in juvenile justice have higher rates of mental disorders than youths in the general population. One recent study of mental disorders among youths in juvenile facilities revealed that 18.9% had some type of anxiety disorder (generalized anxiety, panic, posttraumatic stress, etc.), 9.1% had some form of mood disorder (major depressive, manic episode, etc.), 31.8% suffered from some form of disruptive disorder (conduct disorder and oppositional defiant disorder), and 49.3% had a substance use disorder.[91]

therapists and is also designed to help youths adjust to the institution as well as resolve a variety of personal problems.

In addition to individual counseling, many institutions use some type of group counseling, such as guided group interaction (GGI), positive peer culture (PPC), or milieu therapy. **Group counseling** is less costly than individual counseling, because counselors work with more than one youth at a time. The goal of group counseling is to develop a group that helps the members identify personal issues and supports individual change. Group counseling can be a powerful intervention when it is led by a skilled therapist. Unfortunately, there is little information on the quality of the different types of counseling programs found in juvenile correctional facilities and the qualification and training of staff who provide counseling to residents.

group counseling
A type of therapy designed to use group members to help individuals identify personal issues and support individual change.

In many institutions, treatment and behavior management programs include a behavior modification component. Behavior modification is based on the idea that behaviors are learned and that current and future behaviors can be shaped by a system of rewards and punishments. Like detention centers and other types of correctional programs, training schools commonly employ a *token economy program*, in which youths earn or lose points based on their behavior. Indeed, release may be based on the earning of a specified minimum number of points, indicating the youth has achieved the treatment goals set by the institution.

Problems Associated with Juvenile Training Schools One serious problem facing a majority of training schools is overcrowding. In crowded facilities, youths have less privacy, the probability of interpersonal conflict is increased, and "double bunking" (sleeping more than one resident to a room) increases the chances of physical or sexual assault. Despite the drawbacks of overcrowding, according to the 2004 Juvenile Residential Facility Census, 3 in 10 juvenile correctional facilities reported that the number of residents in the institution met or exceeded institutional capacity. Moreover, 13% of public training schools reported that the number of residents exceeded their standard bed capacity. [92] Consequently, many juveniles in training schools are housed in rooms that do not meet ACA standards for living space and are housed in units that exceed the ACA recommended standard for residents per unit (25 or less) or do not conform to local fire codes.[93]

Contributing to institutional overcrowding is the inappropriate placement of many nonserious offenders in state training schools. In 2006, less than half (47%) of all youths committed to state operated long-term secure facilities had committed a person offense. Slightly more than 28% had committed property offenses, approximately 10% were held for public order offenses, and about 7% were incarcerated for drug offenses. In addition, slightly more than 6% were incarcerated for technical violations, and almost 1% were held for status offenses.[94] Certainly, juvenile offenders that represent a clear threat to others need to be detained. However, one must question the extensive use of expensive and often ineffective state institutional placement for nonviolent offenders.

Another problem facing training schools is escapes. Data collected for 1991 indicated that, during a typical month, juvenile training schools reported 454 successful escapes and 478 unsuccessful escape attempts which would translate into roughly "5,500 escapes and more than 5,700 unsuccessful attempts per year."[95] In a more recent survey of 3,690 juvenile facilities conducted in 2000, it was found that 34% of these facilities reported at least one unauthorized absence over a one month reference period with most of these escapes coming from small non-secure facilities.[96] Although recent data on escapes from juvenile institutions is not available, there is evidence that this continues to be a problem in some jurisdictions.[97] Escapes and escape attempts can result in injury to staff and residents, can result in injury to escapees on the run, and can become a public relations nightmare for correctional administrators, who frequently face some community resistance to institutional programs.

Still another problem faced by facilities is too few staff and treatment resources. Because of inadequate funding, many facilities have lower-than-recommended security staff ratios.[98] Many also have fewer treatment staff, particularly professional treatment staff such as trained counselors, and other programming staff than they would like. Scarce resources may mean that staff members do not receive adequate training. Unfortunately, when budgets are austere, treatment programming and staff training typically suffer, and more emphasis is placed on security than treatment.

In addition to the previously identified problems, many juvenile correctional facilities are plagued by violence and the abuse of residents. For example, in testimony before the National Prison Rape Elimination Commission, Barry Krisberg noted the following:

> The California DJJ is plagued with high levels of violence and fear. Fights, assaults on staff, and riots are common occurrences. Incidents of violence, gang and racial conflicts, and staff fears have led to reliance on extended periods of lockdown, with many youth spending an average of 21 hours per day in their cells. There is virtually daily use of chemical and mechanical restraints, and many correctional staff wear equipment such as security vests and helmets that are more typical of maximum security prisons than juvenile correctional institutions. A video tape showing several DJJ staff beating up two youth was aired on national television news shows and the Internet. In the past, some of the facilities have employed guard dogs to maintain order. Suicide attempts are frequent events, and four youths took their own lives in the last two years. Sexual assaults are part of these horrific conditions.[100]

Overcrowding, inappropriate placement of nonviolent offenders in institutional settings, violence, abuse, inadequate treatment staff and programming, and poor staff training combine to adversely affect the correctional environment in many training schools. At best, mere warehousing of youths is the result. At worst, residents are exploited by other residents and, in some cases, by staff. In either case, such institutions fail to rehabilitate youths, protect them, or respond to them in ways that ensure community safety. These problems represent only a sample of the difficulties that presently confront publicly operated juvenile training schools. Additional problems for juvenile institutions in general, including training schools, are examined later in this chapter.

Other Institutions Used for Postadjudicatory Placement

Besides publicly operated juvenile training schools, there are a variety of other institutions and programs in which juveniles are placed following adjudication. These include

FYI

Many Juvenile Facilities Fail to Assess Youths' Mental Health Status

According to the 2002 Juvenile Residential Facility Census, only 64% of long-term secure facilities evaluated youths for mental health needs and only 68% evaluated all youths for suicide risk.[99]

diagnostic and reception centers; ranches, forestry camps, farms, and outdoor programs; boot camps; and private institutions. This section describes these types of facilities.

diagnostic and reception center A facility that holds juveniles temporarily while they are evaluated to determine the best ultimate placement for them.

Diagnostic and Reception Centers The goal of **diagnostic and reception centers** is to assess youths who have been committed to state care by juvenile courts and determine the best treatment plans and placements for them. After a diagnostic and reception center completes an assessment, which takes an average of 5 weeks,[101] the youth is placed in a training school or another correctional facility to complete the period of commitment. Although these centers are used in only a few states, where they do exist they play an important role in determining the placements and the treatment goals for youths under state care. In states without diagnostic and reception centers, many of the functions performed by them are performed at each individual institution.

The focus of diagnostic and reception centers is on determining each youth's short-term treatment needs. Typically, this is accomplished by conducting a general assessment that may encompass psychiatric and psychological evaluations, educational assessment, a social history, and behavioral observations as well as a review of information provided to the center by the juvenile court and other social agencies. For example, a psychiatrist may conduct a psychiatric evaluation and then treat the youth if he or she is determined to need psychiatric treatment. A clinical psychologist may administer a variety of psychological tests, such as IQ, personality, and other diagnostic tests. A social worker may compile a social history that examines the youth's family and social background. Educational staff may screen the youth for learning disabilities and other learning and academic problems. Unit supervisors may assess the youth's institutional adjustment and peer relationships. All the information acquired is compiled and summarized in order to determine the treatment needs and the best institutional placement for the youth.[102]

Ranches, Forestry Camps, Farms, and Outdoor Programs These long-term facilities and programs are for persons who do not require secure custody.[103] Although the facilities and program settings are not secure and residents have more access to the community than in some other types of institutions, their physical location serves as a barrier between them and the community. Perhaps because of their location, they rely less on formal security measures, such as institutional head counts and classification systems that separate groups of offenders, than training schools, detention centers, and diagnostic and reception centers. In addition, they typically have lower security-staff-to-resident ratios than other institutional settings.[104]

Ranches, forestry camps, farms, and outdoor programs are not new. Indeed, the basic philosophy behind these facilities and programs dates back to the placing out movement, active in the 1800s (see Chapter 4).[105] At that time, many reformers believed that children could be reformed if they were removed from the corrupting influence of the city and reared in wholesome rural environments where they could learn the value of honest work. Today, these programs focus on teaching youths responsibility and self-sufficiency and building their self-esteem. Like earlier programs, they are premised on the idea that "by placing a child in a totally foreign environment, old attitudes may be altered, and new, acceptable behavior patterns may be created."[106]

Boot Camps Although military boot camps have existed for some time, their first modern correctional equivalent was founded in 1973 by a former marine, George Cadwalader. This boot camp, called the Penikese Island School, was based on the belief that serious young offenders could be influenced in a positive way by exposing them to rigorous military-style training.[107] It was not until the 1980s, however, that boot camps began to become popular for dealing with young adult offenders, and it was not until the 1990s that their popularity began to spread to juvenile justice.[108]

juvenile boot camp
A camp for juvenile offenders that uses military-style training and discipline to control and modify their behavior.

A number of **juvenile boot camps** have been developed around the United States. They exhibit considerable variability, but share one essential feature: military-style training and discipline.[109] Consequently, drill instructors and other staff with military backgrounds play a critical role in these institutions, although teachers and case managers are also employed.[110] In addition, substance abuse treatment specialists and other treatment providers are beginning to play a more prominent role in many of these institutions.

Typical elements of a juvenile boot camp include a platoon structure (with platoons made up of 10 to 13 youths who move through the program together), military-style uniforms for residents and drill instructors, military customs and jargon, a strenuous daily routine that begins at 5:30 or 6:00 a.m. and ends with lights out at 9 or 10 p.m., immediate responses to minor rule infractions (e.g., push-ups) and progressive sanctions (including removal from the program for serious rule violations), and a public graduation ceremony.[111]

The idea of placing juvenile offenders into boot camps has become popular with some individuals, particularly those with military backgrounds. And of course the discipline they impose, the belief that they reduce institutional populations in long-term facilities, and lower cost appeals to many policy makers. However, evaluations of boot camp programs have indicated that they face a number of difficulties. Low salaries, high staff turnover and burnout, and inadequate aftercare have plagued many of these programs.[112] In addition, there is no evidence suggesting that these programs are any more effective than traditional institutional placements at reducing recidivism, nor have they been found to have a significant impact on institutional populations or correctional costs.[113]

Private Institutions Privately operated juvenile correctional institutions, which have a long if not always distinguished history, continue to play a significant role in juvenile corrections. As noted in Chapter 4, the first correctional institutions designed especially for children in this country, the houses of refuge, were privately operated. Moreover, a number of juvenile justice interventions, such as placing out and juvenile probation, were first developed as private sector endeavors.

The large number of private institutions designed to deal with problem children make up what long-time juvenile justice researcher and practitioner Ira Schwartz refers to as a "hidden system" of control.[115] This system of control comprises detention facilities, shelter homes, group homes, training schools, drug and alcohol programs, ranches, farms, forestry programs, outdoor experiential programs, and mental health facilities, among others. The quality of care given children in these institutions, however, is as varied as the types of institutions that exist. A number of them provide outstanding care to their clients—better care, indeed, than many publicly operated facilities, owing to their lower staff–resident ratios, their greater flexibility, and the greater numbers of

FYI

Privately Operated Corrections Facilities Play a Significant Role in Juvenile Justice
In 2006, private facilities housed 30.7% of all juvenile offenders and fewer than 9% of all youths placed in long-term secure facilities.[114]

FYI

A Variety of Institutions Exist for Controlling Youths

A substantial number of the admissions to private facilities are so-called **voluntary admissions**, although they result from referrals by parents or school officials, and the juveniles in question hardly have a choice. Such admissions represent the growing "medicalization of defiance."[122] The result is that behaviors like "running away from home, habitual truancy, and being in conflict with one's parents are now being classified as emotional and psychiatric disorders"[123] that are dealt with by a growing medical industry that charges insurance companies and parents as much as $200 to $600 a day for inpatient psychiatric and chemical dependency treatment.[124]

voluntary admission The placement of a juvenile into an institution by his or her parents or guardian without benefit of a court order or a review of due process. The cost of such a placement is borne, not by the public, but by the parents or guardian.

qualified professional staff they can afford to hire. In addition, many privately operated facilities are more cost effective than many publicly operated institutions,[116] and evidence indicates some may be more successful at reducing recidivism than many traditional training schools.[117] On the other hand, there is considerable evidence that many of the admissions to privately operated chemical dependency and mental health facilities are medically unnecessary.[118] There is additional evidence that the quality of care provided to children in many private facilities is woefully inadequate, despite the exorbitant prices charged. Some programs appear to be more interested in profit than quality of care or treatment.[119] Consider the following examples cited by Ira Schwartz in his book *(In)Justice for Juveniles: Rethinking the Best Interests of the Child*: "'One private psychiatric hospital [in Los Angeles County] served its new adolescent patients only a cheese sandwich and milk for each meal until they earned enough points to eat other items.' Another private hospital in the same community 'conducted routine strip searches, where female adolescents had to squat, en masse, and were, on occasion, subjected to involuntary pelvic exams for contraband.'"[120] Such practices ended only after a threat of legal action by local advocacy groups.[121]

Trends in Juvenile Incarceration

Incarceration has become an increasingly popular response to delinquency. According to the annual Children in Custody Census and the more recent Census of Juveniles in Residential Placement, which provide estimates of the number of children placed in juvenile facilities on a specific day during a particular year, the number of youths

MYTH VS REALITY

Overall, There Has Been a Trend Toward Increased Incarceration of Youths

Myth—The increase in juvenile incarceration is a product of the increasingly serious juvenile crime wave.

Reality—The overall juvenile crime rate has been relatively stable over time. Moreover, the violent juvenile crime rate did not show an increase until the late 1980s, well after the documented trend toward increased incarceration began, and it has been decreasing since 1994. It may well be that the increase in juvenile incarceration primarily reflects a change in many juvenile justice decision makers' beliefs about the best way to deal with delinquency.

FYI

Disproportionate Minority Confinement Is a Significant Problem

The 1992 amendments to the Juvenile Justice and Delinquency Prevention Act (JJDPA) require that states determine whether the proportion of juveniles in confinement exceeds the proportion of juveniles in the population. States are also required to determine whether there is disproportionate minority confinement (DMC) and, if so, to take steps to remedy the problem.[131]

placed in public and private juvenile facilities increased from approximately 74,270 in 1975 to approximately 134,011 in 1999, an increase of more than 80%.[125] Although this number had decreased to 92,854 in 2006,[126] this still represents an increase of 25% over 1975 levels.

Aside from the increase in the number of youths incarcerated, several other interesting trends in juvenile incarceration have been noted in recent years. One important trend has been the increasing privatization of juvenile corrections. Both the number of admissions and the rate of admissions to private juvenile facilities grew substantially between the late 1970s and the late 1980s. A similar trend was also evident in the 1990s. The number of youths committed to private facilities increased by 79% during the 1990s while commitments to public facilities increased by 41%.[127] This trend appears to have abated in recent years, however. Overall, little change in the percentage of youths placed in private institutions was noted between 2000 and 2006.[128]

Another significant trend since the late 1970s has been the increasing number of minority youths placed in institutional settings. As a result, by 2006, almost two-thirds (65.1%) of youths incarcerated in juvenile correctional facilities were minority youths, although they made up only 22.5% of the juvenile population between 10 and 17 years of age.[129] Moreover, disproportionate minority confinement was evident in almost every state.[130]

The Effectiveness of Institutional Correctional Programs for Juveniles

Despite the long history of juvenile correctional institutions and their increasing popularity, little research on their effectiveness has been done. Moreover, what is known is not encouraging. Although some institutional programs for juveniles appear to be effective, the bulk of the evidence indicates that many juvenile institutions have little impact on recidivism. (Not surprising, perhaps, considering the quality of life characteristic of many juvenile institutions and the woeful lack of good treatment programs in many facilities.) One review of rearrest rates in states that rely heavily on juvenile institutions found that the percentage of youths rearrested ranged from 51% to more than 70%.[132]

MYTH VS REALITY

Some Institutional Placements Are Effective

Myth—Institutional programs for juvenile offenders are not effective.

Reality—Although some institutional programs are not effective, there is clear evidence that some programs are associated with substantial reductions in recidivism.

COMPARATIVE FOCUS

The U.S. Has a Much Higher Juvenile Incarceration Rate than Canada

The juvenile incarceration rate in Canada, excluding Ontario where data were not available, was 8.2 per 10,000 youths in 2003/2004.[140] In comparison, the 2003 juvenile incarceration rate in the U. S., excluding youths placed in detention, was 18.8 per 10,000 youths, and if youths in detention are included, the rate climbs to 28.9 per 10,000.[141]

Another study, which followed up with almost 450 youths released from state training schools, discovered that the recidivism rate increased each year after discharge, reaching 54% in the fourth year.[133] An examination of the re-arrest rates of youths discharged from correctional facilities in New York state found that 51% were picked up by the police within a year of release.[134] Still more recent data collected by the Virginia Department of Juvenile Justice found that 55% of youths released from placements in Virginia, Florida, and New York were rearrested within 12 months.[135] Furthermore, a number of literature reviews have found that institutional placements and correctional treatments in general have little positive effect on recidivism.[136]

In contrast, other studies and literature reviews have presented a more favorable picture of the effectiveness of some institutional programs. For example, one large-scale review of the literature on residential programs found that such programs do produce a modest reduction in recidivism.[137] Moreover, a more recent review of institutional programs for serious juvenile offenders conducted by Mark Lipsey and David Wilson indicated that although some institutional programs have no appreciable effect on recidivism, others were associated with substantial reductions in subsequent offending.[138]

Overall, then, the research that has examined the effectiveness of institutional programs for delinquents has produced mixed results. There are quality institutional programs that treat youths humanely and that are operated by caring and professional staff. There is some evidence that small secure treatment facilities that target violent or chronic offenders and use **therapeutic**, **cognitive-behavioral**, and **skill-building interventions** are effective at reducing recidivism.[139] On the other hand, many institutional programs are characterized by a lack of genuine caring and concern for youths and poorly designed and implemented programs. Moreover, many institutions, particularly large state institutions, are little more than warehouses for youths and have little positive effect on the behavior of former residents. In fact, they may increase the likelihood that youths will engage in subsequent delinquency.

therapeutic interventions Interventions intended to have some positive rehabilitative or habilitative effect on the individual.

cognitive-behavioral interventions Interventions based on the premise that many offenders have not acquired the necessary cognitive (thinking) skills necessary for effective social interaction. Consequently, these approaches attempt to help youths identify problem thinking patterns, take responsibility for their thoughts and actions, and develop the skills necessary to engage in effective social interaction.

skill-building interventions Interventions designed to help youths develop various social and interpersonal skills needed for effective social functioning. These skills may include anger management, conflict resolution, effective communication, dealing with authority, planning and decision making, and avoiding risk.

Challenges Facing Juvenile Correctional Institutions

A number of challenges presently face juvenile correctional institutions. One challenge is a lack of comprehensive information on these facilities. Although the Office of Juvenile Justice and Delinquency Prevention, with assistance from other organizations and individuals such as the National Center for Juvenile Justice, the National Council on Crime and Delinquency, and university researchers, collects and disseminates valuable information about juvenile institutions, there is still much that is not known about contemporary correctional institutions for youths. Indeed, relatively little is known about the institutional environments found in many facilities, the quality of programs and services provided to youths, and the effectiveness of those programs and services. Until the early 1990s, much of what was known came from studies of individual institutions or studies that looked at several institutions within one state. Unfortunately, regular, nationally representative studies of juvenile facilities were uncommon. What the existing research made clear, however, was that the quality of care in residential and institutional settings varied considerably. Indeed, this research, along with law suits filed against various institutions, showed that, at least in some facilities, the exploitation and victimization of residents, both male and female, occurred all too frequently.

To some extent, the lack of knowledge about juvenile corrections was alleviated in 1994 by the publication of *Conditions of Confinement: Juvenile Detention and Corrections Facilities*. Mandated by Congress as part of the amendments to the Juvenile Justice and Delinquency Prevention Act, this study represented an important contemporary milestone in our understanding of juvenile institutions nationwide. Perhaps not surprisingly, it found that there are still a number of widespread problems confronting juvenile institutions, including the following issues:

- Crowding was a pervasive problem in juvenile facilities. As a result, a significant number of juveniles were housed in facilities where they had inadequate living space.
- A substantial number of juvenile facilities had lax security measures, which led to escapes and injuries.
- Suicidal behavior was a problem in juvenile institutions, and many institutions lacked adequate measures for preventing suicide.
- Many institutions failed to conduct health appraisals and health screenings in a timely manner. In addition, institutions often lacked good data on health care services given to residents.[142]

Other studies point to a variety of other difficulties confronting juvenile institutions. As noted earlier, one of these is the disproportionate confinement of minorities in institutional settings. For example, in the U. S. in 2006, the custody rate per 10,000 youths who were adjudicated and committed to a juvenile correctional facility was 1.17 for whites, 6.61 for blacks, 2.85 for Hispanics, 5.14 for American Indians, and .55 for Asians and Pacific Islanders.[143] In many facilities, minorities account for a large proportion of the institutional population. Indeed, there is considerable variation across states in their custody rates.

Possibly not surprisingly, racial tension is another problem found in many settings. Racial tension is often exacerbated when institutions have sizable gang populations because gangs are often divided along racial lines. Racial and gang rivalries can feed on one another, creating the potential for institutional conflict. Importantly, a majority of administrators of state-operated secure facilities indicate that the youths being admitted to their facilities are more likely to be members of gangs than was true in the past.[144] The result is an increased probability of conflict within institutions.

inmate code
An unwritten set of norms and values that stand in opposition to the expressed norms and values of the institution.

Another challenge facing juvenile institutions is the development of an organizational climate and culture conducive to positive behavioral change. Too many institutions still foster an inmate code that operates to subvert institutional treatment goals. The **inmate code** consists of a set of norms and values that stand in opposition to the expressed norms and values of the institution. For example, in a study of institutionalized juveniles, Clemens Bartollas, Stuart Miller, and Simon Dinitz found an inmate code according to which exploiting others, maintaining distance from staff, not giving in to others, and not "ratting" on peers were proper ways to behave.[145] Other research has also confirmed the existence of inmate subcultures that influence youth behaviors within institutions.[146]

One characteristic of the organizational climate and inmate code of a number of institutions is support for the exploitation of others. Such exploitation ranges from the theft of goods from weaker residents by older or more physically imposing inmates to rape and other types of physical and sexual assault by residents or staff.[147] In their 1976 study of a male training school in Ohio, Bartollas, Miller, and Dinitz found that 90% of the 150 residents were being exploited by others or were exploiting another resident. Their research found a clear hierarchy of roles within the institution and a system of exploitation based on hierarchical position. This exploitation involved taking food, clothing, cigarettes, and sex from those in lower positions.[148] In a follow-up to this study, these researchers found a similar institutional culture, although there was less sexual victimization (when residents engaged in sexual relations, these relations were more likely to be consensual).[149]

A clear challenge facing each juvenile institution is to develop a positive social organization characterized by a positive and supportive set of relationships that exist between those who work and live within the institution. As noted earlier, one type of adaptation can result in subcultures that promote violence and exploitation. However, other types of adaptations to institutional life are also possible. For example, institutionalized youths are deprived of normal heterosexual relationships, which can lead to a variety of adaptations. In some cases, a hierarchy of roles is created, and youths at the bottom of the hierarchy, who are generally less able to defend themselves, become the sexual objects of those higher up the hierarchy. In other cases, the development of sexual roles is based on the need to replace normal relationships that are denied youths who are institutionalized. Incarcerated youths may develop kinship role systems, such as *make-believe families*, in which youths take on various family roles (e.g., mother, father, husband, wife, sister, and brother). Members of the make-believe family give each other mutual advice and assistance, look out for one another's interests, and participate in institutional activities together. In addition, some youths may develop homosexual alliances. In a study of three coed and four all-female training schools, Alice Propper found that 17% of institutionalized girls reported at least one homosexual experience, ranging from kissing to intimate sexual contact. Also, about half of the girls reported taking a make-believe family role. Propper found that make-believe families were as common in coed institutions as in all-female facilities.[150] Other research has produced similar findings.[151]

secondary adjustment
A response that a juvenile makes to the unique characteristics of an institutional environment that may not be perceived as reasonable or appropriate in another context. Examples include violence, homosexual relationships, and make-believe families.

Exploitation and violence toward others, homosexual relationships, and make-believe families are examples of **secondary adjustments** that youths make to the organizational climate of the institution. These secondary adjustments are responses to the unique characteristics of the institutional environment and may not be seen by the juvenile as reasonable or appropriate in another context. For example, a youth may engage in a homosexual relationship with another youth within an institutional setting, yet not view himself as a homosexual nor consider the relationship conceivable in another social context.

social organization
Refers to the types and quality of relationships that exist between people who come together in a particular setting, such as those who live, work, or participate in a correctional institution or some other organizational or social setting.

The type of organizational culture that develops within an institution and the types of adaptations that residents make to that culture are not simply an out-growth of placing "delinquents" in institutions. To a large degree, the **social organization** of an institution is a product of the management style and philosophy employed in the institution, the available resources, and the training and correctional philosophy of the staff.[152] For example, research indicates that inmates in custody-oriented institutions report higher levels of violence than those in treatment-oriented institutions.[153] Moreover, an institutional focus on custody has been found to encourage the development of an inmate code that runs counter to institutional goals.[154]

Of course, the social organization of an institution affects not only the youths who reside there, but also the correctional staff. Consequently, another challenge facing juvenile corrections is to develop a social organization that meets staff needs. Working in a juvenile facility can be a very exciting and rewarding experience, but it can also be a frustrating and stressful experience, particularly when staff members are concerned about being assaulted by residents or feel that the residents lack needed services.[155]

Another problem that can contribute to staff stress is role conflict—conflict between the staff's custody and treatment responsibilities. Facility staff members are often expected to enforce rules and to be confidants and helpers to youths under their care. However, these roles are not simultaneously compatible, and staff may emphasize one of these roles over the other, emphasize both of these roles, or emphasize neither. Unfortunately, role conflict can lead to confusion among both residents and staff about which roles are appropriate. In addition, other factors, such as low pay, poor training, a lack of public understanding, and indifference from management can produce stress. The result can be staff burnout and high staff turnover, which are problems that confront many institutions. Consequently, it is important that staff be given the training, supervision, and support they need in carrying out their roles within the institution.

The role of institutional corrections, however, does not end when a youth is released from a corrections facility. Indeed, a major challenge facing contemporary corrections is the development of effective **aftercare** (often called parole) for released juveniles. As noted in the previous chapter, aftercare involves the provision of services intended to help youths successfully make the transition from juvenile institutions to life back in the community. Ideally, aftercare provides a planned transition from the institution to the community. Unfortunately, as Richard Snarr notes, "In many jurisdictions, aftercare (parole) is an after-thought.... Many times little or no thought and planning are given to the transition when a youth is released from an institution or residential setting and faced with living back in the community."[156] The failure to provide adequate aftercare for youths released from institutions is widely recognized as a major shortcoming in juvenile corrections, and the development of effective aftercare programs represents a significant challenge.[157]

aftercare Supervision given to children for a limited time after they are released from a correctional facility but are still under the control of the facility or the juvenile court.

Good aftercare programs do exist. For example, Kalamazoo County (Michigan) Juvenile Court has developed a day treatment program to provide aftercare for juveniles returning to the community from institutional placements. The program includes a strict phone and/or tether monitoring program, family counseling, and educational and vocational components. The program is partly based on the realization that, no matter to what extent the out-of-home placement changes the juvenile, if the home environment does not change, the chance of a successful return home will be diminished. Intensive family counseling helps all family members understand their roles in reintegrating the juvenile back into the family and into the community.

A final and serious challenge facing juvenile correctional institutions is the development of more programs that respond to children in humane ways and that protect the community. Ultimately, of course, this is the goal of most correctional programs—a goal that is, at present, unmet in far too many instances. Yet, there is some reason for optimism. There are model programs around the country that can serve as exemplars for those interested in programs that really do work. And, as discussed in Chapter 15, there are a variety of other reasons to believe that a more reasonable and effective response to juvenile offenders can be achieved.

Legal Issues

In some instances, the institutionalization of a child may be necessary in order to rehabilitate the child, deliver appropriate punishment, or protect the community. How institutional placement is accomplished and what happens to the juvenile while in the institution poses a number of important legal questions. Moreover, decisions to incarcerate youths and the conditions of their confinement influence the extent to which the juvenile justice process helps youths and protects the community.

Clearly, the decision to remove a child from his or her home is a serious one and affects not only the child, but also the family and others. Given the seriousness of this decision, it is important to consider a number of issues related to the incarceration of youths:

- What rights do juveniles have at the time that treatment decisions are made?
- Do juveniles have a right to treatment?
- Do juveniles have the right to refuse treatment?
- What types of conditions should juveniles be subjected to, and what types of treatment should juveniles receive in institutions?
- How long should juveniles remain in institutions?
- Are adult facilities appropriate for the placement of juveniles?
- How does the Civil Rights for Institutionalized Persons Act apply to juveniles in custody?
- What are the liabilities that correctional institutions and institutional personnel are subject to?

What Rights Do Juveniles Have at the Time that Treatment Decisions Are Made?

The power to remove children from their homes is an ominous one and requires that children have some protection when removal is a possibility. The importance of such protection is clearly articulated in the *Gault* ruling, which stated that juveniles have a right to counsel when facing the prospect of institutionalization. But what if the parents are the ones initiating commitment on grounds other than delinquency? One case that examined this issue was *Heryford v. Parker*,[158] in which it was determined that before a child could be committed to a training school for the "feeble minded," he or she was entitled to representation by counsel. Moreover, this ruling held that representation was necessary even though the parents had initiated the commitment. Consequently, court rulings have indicated that juveniles have a right to counsel when facing institutionalization even when their parents have initiated commitment proceedings, but these rulings tell us nothing about the quality of assistance that children should receive.

Do Juveniles Have the Right to Treatment?

On the most fundamental level, the placement of a juvenile in an institutional setting against his or her will raises a question about the purpose of incarceration. In order to pass constitutional muster, there must be a legal justification for every deprivation of liberty, regardless of age. Historically, the placement of children in institutions by the juvenile courts rests on the *parens patriae* obligation of the state to provide care for children in need. If the state assumes responsibility for the child's welfare and then fails to protect that welfare, the child's due process rights are violated.

equal protection The constitutional right of all persons to have equal access to the government and courts and to be treated equally by the law and courts.

Another persuasive argument for a juvenile's right to treatment is found in the **equal protection** safeguard. Juveniles are often "sentenced" to institutional placements for a much longer time than an adult convicted of a similar offense. The only way that this disparity can be justified is if the term of incarceration of the juvenile is for treatment, not punishment. Of course, the obligation "to treat" carries with it the presumption that treatment is available. Certainly, a juvenile should be able to make sure of the availability and suitability of treatment before being committed to that treatment. Consequently, juvenile courts may also be required to consider additional factors regarding placement, such as the location of the institution. Is it in the state or out of the state, and does the distance and the available transportation make it convenient or inconvenient for family members to be involved in treatment? If the placement is out of the state or far away, are equivalent closer facilities available? Will the placement cause the child or the family undue hardship?[159]

Do Juveniles Have the Right to Refuse Treatment?

"Treatment" is a nebulous concept, and the line between treatment and punishment is often nonexistent. In reality, treatment usually involves some level of discomfort or punishment regardless of the intentions of the treatment providers. Moreover, there are many instances in which treatment cannot be reasonably justified because of the abusive conditions that youths experience in some institutional settings. Consequently, juveniles have legitimate concerns about the actual type of treatment they are asked to receive. If asked about their treatment needs, many youths might choose a different type of treatment. Still others might opt for confinement so that they could "do their time" and be done with it.

Certainly, juveniles have the "right" not to be placed into programs that degrade or dehumanize them or that are designed to embarrass or humiliate them. The American Bar Association has promulgated standards that would allow children to refuse services except school attendance, services designed to prevent harm to their health, and services that are court ordered as conditions for noninstitutional placement (e.g., probation). The common denominator of these services is that they are "services" that society would reasonably expect to be accepted.[160] However, standards like the American Bar Association's standards invariably result in conflicts between those who feel that treatment providers and others have an obligation to provide for children's needs (i.e., that children have a right to treatment) and those who are concerned about the coercive nature of service provision.

What Types of Conditions Should Juveniles Be Subjected to and What Types of Treatment Should Juveniles Receive in Institutions?

Because children have the right to be treated, the type of treatment they receive is of utmost importance. The Eighth Amendment forbids treatment that is so foul, so inhumane as to violate basic societal concepts of decency. For example, long-term solitary confinement, inadequate hygiene facilities, dirty or unsanitary rooms or cells, and failure to allow for proper clothing have all been determined by appellate courts to be in violation of the Eighth Amendment. Unfortunately, there is ample evidence that many juvenile institutions have dealt with children in abusive and neglectful ways.

In addition, because the state has a higher degree of responsibility for children than for adults, some maintain that juveniles could suffer cruel and unusual punishment if they are denied rehabilitative opportunities such as therapy, school, and job training programs.[161] Without these, any placement becomes punishment, which raises equal protection and due process issues as well.

How Long Should Juveniles Remain in Institutions?

Determining when a juvenile has been rehabilitated is not an exact science. Rehabilitation as an outcome is as individualized as the person who is being rehabilitated. Determining that a juvenile needs to be in placement, in many instances, is not very difficult. Determining when a youth is rehabilitated is often more difficult. To keep a child in an institution longer than necessary to protect the community and effect rehabilitation would be an unjustifiable deprivation of liberty[162] and a waste of public tax dollars. This issue takes on new importance in view of the increasing privatization of treatment institutions. If profit is the goal of a placement, there will be pressure on the institution to keep the child longer than necessary in order to achieve financial objectives. Most, if not all, juvenile court statutes and applicable court rules have provisions for periodic reviews to check on the welfare and progress of juveniles removed from the home. These review requirements provide "windows of opportunity" for return home and act as safeguards against "extra" or unnecessary time in placement.[163]

Are Adult Facilities Appropriate for the Placement of Juveniles?

As noted earlier in this chapter, the jailing of juveniles presents a variety of problems for correctional personnel as well as juveniles. In recognition of the potentially harmful effects of jailing juveniles, the Juvenile Justice and Delinquency Prevention Act was amended in 1980 to abolish the confinement of juveniles in adult jails and lockups in states that received funds under the act.[164] The power of the purse can be an excellent persuader, but it has no impact on states that choose to opt out of participation. Moreover, as noted earlier in this chapter, there are circumstances under which juveniles can be placed in adult jails. As a result, juveniles continue to be placed in jails in some jurisdictions. Moreover, youths younger than 18 years are also committed to state prisons each year. For example, there were an estimated 2,639 persons younger than 18 years of age residing in state prisons at midyear 2007. Moreover, this number increased between 2006 and 2007.[165] This occurs despite the fact that good evidence exists that such commitments neither help youths nor reduce the likelihood that these youths will recidivate upon release. See Chapter 9 for a thorough discussion of issues related to the transfer of juveniles to adult court.

Civil Rights of Institutionalized Persons Act (CRIPA): Implications for Juvenile Correctional Facilities

The federal government has recognized the dangers inherent in the confinement and detention of youths. The Civil Rights of Institutionalized Persons Act (CRIPA) can help eliminate unlawful and dangerous conditions of confinement. Under its provisions, the Civil Rights Division of the United States Department of Justice has the power to investigate and bring legal actions against state or local governments for violating the civil rights of persons, including youths, institutionalized in publicly operated facilities.

Although enacted by Congress in 1980, CRIPA remains underutilized as a means to protect youth in confinement. In order to initiate an investigation by the Department of Justice, there has to be a complaint or some action requesting that the Civil Rights Division do something. Parents, advocates, and juveniles should be aware of the very significant role they can have in bringing attention to harmful or unlawful conditions of confinement for juveniles.

In addition to receiving information that would trigger an investigation, the Department of Justice must also determine that the facility is a "public institution." Public institutions are those that:

- are owned, operated, or managed by any state or political subdivision;
- are a private facility under contract to a state or local government;
- house juveniles awaiting trial;
- house juveniles for care or treatment;
- house juveniles for any other purpose than solely for education.

After these thresholds are crossed, the Department of Justice can conduct an investigation. If civil rights violations are found, the jurisdiction controlling the facility gets a "findings letter" which sets forth the alleged violations and the underlying factual basis. This letter also sets forth suggested steps to correct the violations. The Department of Justice must wait 49 days after issuing the findings letter to file suit, thus giving the parties ample opportunity to negotiate a settlement of the matter. Most CRIPA actions to date have resulted in settlements.

Because almost half of the juveniles incarcerated have multiple needs, including mental health disabilities, CRIPA can be used to ensure not only civil rights, but also the right to treatment for youths in custody. For additional discussions about remedies for youths incarcerated, see the discussion about the use of ombudsman programs found in Chapter 15.

What Liabilities Are Correctional Institutions and Institutional Personnel Subject To?

Should correctional personnel be liable for harm to juveniles in their care? Generally, states are protected from liability based on the doctrine of sovereign immunity. However, in a Kansas case, *C.J.W. by and through L.W. v. State*, the state was found liable under the state's tort claims act.[166] At issue in this case was whether the state had an obligation to ensure the safety of minors by making reasonable efforts to convey known information about an aggressive and sexually deviant juvenile inmate to protect other youths with whom this youth came into contact. The state was found liable for injuries suffered by a 12-year-old inmate who was attacked and sexually abused by an older youth on two separate occasions.[167]

Chapter Summary

A variety of types of institutions and programs are used for the placement of juvenile offenders, including detention centers, jails, training schools, ranches, farms, forestry camps, outdoor programs, chemical dependency units, and mental health facilities. Even institutions and programs of the same type can have substantial differences. For example, contemporary correctional facilities can vary significantly in their operation, in the types of programs they offer, in their social organization, and in their effectiveness. Moreover, some of these facilities are categorized as open institutions, whereas others are closed. Some institutions, such as detention centers, reception and diagnostic centers, and training schools, house a mixed population of youths, whereas other facilities focus on specialized offender groups, such as youths who are chemically dependent or who have mental health problems. Some are publicly operated, whereas others are privately operated (and these arguably constitute a growing "hidden system" of juvenile control).

The use of institutional placement has grown significantly since the mid-1970s, despite the lack of scientific evidence that juvenile crime increased during much of this time. Indeed, its increased use appears to be driven by a punitive "get tough" approach to delinquents. In any case, it is a significant trend in juvenile justice. However, it has not affected all groups of children equally. As has always been the case, poor children, minorities, and males are disproportionately represented in institutional populations. Particularly striking is the overrepresentation of minorities in publicly operated juvenile institutions.

There is little scientific evidence that indicates incarceration in general has a significant impact on the level of juvenile crime or a positive effect on juvenile offenders. A possible reason is that many juvenile correctional institutions are ill equipped to respond to the needs of juvenile offenders. They typically lack the monetary resources, facilities, treatment interventions, aftercare programs, and political support needed to effectively respond to children's needs. The worst institutions expose youths to inhumane treatment, victimizing those they are supposedly there to help and protect. These institutions protect neither youths nor the community.

Clearly, there are many challenges facing juvenile institutions, such as the provision of adequate living space, the development of better security measures, the discovery of better ways to prevent self-destructive behaviors, and the provision of better health screening. The inappropriate placement of juveniles in institutional environments, including jails and adult prisons; the overrepresentation of minorities in institutional settings; negative social organization; a lack of sound aftercare programs; and meager evaluation of existing programs are some of the problems that currently exist. In spite of these problems, however, there are institutional programs that appear to make a difference in the lives of youths and thus offer the promise of protecting the community as well. And there is no question that institutions capable of handling juvenile offenders are needed. What must be determined is which youths need to be placed in institutions and what types of institutions these should be.

Key Concepts

aftercare: Supervision given to children for a limited time after they are released from a correctional facility but are still under the control of the facility or the juvenile court.

closed facility: See secure facility.

cognitive-behavioral interventions: Interventions based on the premise that many offenders have not acquired the necessary cognitive (thinking) skills necessary for effective social interaction. Consequently, these approaches attempt to help youths identify problem thinking patterns, take responsibility for their thoughts and actions, and develop the skills necessary to engage in effective social interaction.

cottage institution: Facility in which juvenile offenders live in a "homelike" setting. Such institutions are typically more open and smaller than other types of juvenile correctional institutions, are run by "cottage parents," and have high staff-to-juvenile ratios.

detention program: The total set of organized activities in which youths are involved while in a detention facility.

diagnostic and reception center: A facility that holds juveniles temporarily while they are evaluated to determine the best ultimate placement for them.

equal protection: The constitutional right of all persons to have equal access to the government and courts and to be treated equally by the law and courts.

group counseling: A type of therapy designed to use group members to help individuals identify personal issues and support individual change.

inmate code: An unwritten set of norms and values that stand in opposition to the expressed norms and values of the institution.

juvenile boot camp: A camp for juvenile offenders that uses military-style training and discipline to control and modify their behavior.

open institution: A facility that is not locked down, has no walls or fencing, and relies on staff to control inmates.

postadjudicatory institution: An institution to which a juvenile offender has been sentenced.

preadjudicatory institution: A temporary institution for juveniles who are awaiting trial or awaiting sentencing.

private facility: A facility that is privately owned and operated. Private facilities can generally choose whom to accept, allowing them to focus on specific types of juveniles, such as sexual offenders or less serious offenders.

public facility: A government-run facility. Public facilities have less ability than private facilities to screen out difficult and violent offenders and usually house more of both.

secondary adjustment: A response that a juvenile makes to the unique characteristics of an institutional environment that may not be perceived as reasonable or appropriate in another context. Examples include violence, homosexual relationships, and make believe families.

secure facility: A facility that closely constrains and monitors residents' activities and physically restricts their access to the community typically through the use of various hardware devices, such as locked metal doors, security cameras, motion and sound detectors, fences with climb resistant mesh, and razor wire to limit residents' movement within the institution and to prevent escape.

sex typing: Where males are taught traditionally male skills such as carpentry and females are taught traditionally female skills such as cosmetology.

skill-building interventions: Interventions designed to help youths develop various social and interpersonal skills needed for effective social functioning. These skills may include anger management, conflict resolution, effective communication, dealing with authority, planning and decision making, and avoiding risk.

social organization: Refers to the types and quality of relationships that exist between people who come together in a particular setting, such as those who live, work, or participate in a correctional institution or some other organizational or social setting.

state training school: A restrictive facility designed to deal with repeat juvenile offenders.

therapeutic interventions: Interventions intended to have some positive rehabilitative or habilitative effect on the individual.

token economy: A type of behavior modification program in which juveniles, in return for appropriate behaviors, earn points that can be redeemed for privileges, snack foods, and so on.

voluntary admission: The placement of a juvenile into an institution by his or her parents or guardian without benefit of a court order or a review of due process. The cost of such a placement is borne, not by the public, but by the parents or guardian.

Review Questions

1. What are the distinguishing characteristics of institutional or residential placements?
2. What are the four most common problems of institutional placements?
3. Describe the fundamental differences between secure facilities and open facilities.
4. How do private institutional placements differ from public institutional placements?
5. What is the most common preadjudicatory institutional placement?
6. In what ways do juveniles commonly access detention?
7. What is a token economy?
8. What were the major issues and decisions of the U.S. Supreme Court in the case of *Schall v. Martin*?
9. What are the arguments against using adult jails to house juveniles?
10. How are training schools commonly used?
11. How are diagnostic and reception centers used?
12. What are the distinguishing characteristics of juvenile boot camps?
13. Describe the recent trends involving institutional and residential placements for juveniles.
14. Is aftercare necessary? Provide a sound argument for the development of aftercare programs.
15. What are secondary adjustments?
16. What legal rights do juveniles have while incarcerated or institutionalized?

Additional Readings

Bartollas, C. M., Miller, S. J., & Dinitz, S. (1976). Exploitation matrix in a juvenile institution. *International Journal of Criminology and Penology*, 4, 257–270.

Giallombardo, R. (1974). *The social world of imprisoned girls: A comparative study of institutions for juvenile delinquents.* New York: Wiley.

Lipsey, M. W. & Wilson, D. B. (1998). Effective interventions for serious juvenile offenders: A synthesis of research. In R. Loeber & D. B. Farrington (Eds.), *Serious and violent juvenile offenders: Risk factors and successful interventions.* Thousand Oaks, CA: Sage.

Miller, J. G. (1991). *Last one over the wall: The Massachusetts experiment in closing reform schools.* Columbus, OH: Ohio State University Press.

Parent, D. G., Lieter, V., Kennedy, S., Livens, L., Wentworth, D., & Wilcox, S. (1994). *Conditions of confinement: Juvenile detention and corrections facilities.* Washington, DC: Office of Juvenile Justice and Delinquency Prevention.

Polsky, H. W. (1963). *Cottage six: The social system of delinquent boys in residential treatment.* New York: Russell Sage Foundation.

Rothman, D. (1971). *The discovery of the asylum: Social order and disorder in the New Republic.* Boston: Little, Brown & Co.

Schwartz, I. M. (1989). *(In)justice for juveniles: Rethinking the best interests of the child.* Lexington, MA: Lexington Books.

Wooden, K. (1976). *Weeping in the playtime of others: America's incarcerated children.* New York: McGraw-Hill.

Notes

1. Elrod, P., Kinkade, P. T., Smith, A., Brown, F., Matthews, A. (1992). *Toward an understanding of state operated juvenile institutions: A preliminary descriptive analysis of a survey of chief correctional administrators.* Paper presented at the American Society of Criminology annual meeting, New Orleans.
2. Thornberry, T. P., Tolnay, S. E., Flanagan, T. J., and Glynn, P. (1991). *Children in custody 1987: A comparison of public and private juvenile custody facilities.* Washington, DC: Office of Juvenile Justice and Delinquency Prevention.
3. Elrod, Kinkade, Smith, Brown, Matthews, 1992.
4. Thornberry, Tolnay, Flanagan, & Glynn, 1991.
5. Thornberry, Tolnay, Flanagan, & Glynn, 1991.
6. Sickmund, M., Sladky, T. J., Kang, W., & Puzzanchera, C. (2008). *Easy access to the census of juveniles in residential placement.* Retrieved January 20, 2009, from http://ojjdp.ncjrs.gov/ojstatbb/ezacjrp/.
7. Livsey, S., Sickmund, M. & Sladky, A., et al. (2009)."Juvenile residential facility census, 2004: Selected findings." Juvenile Offenders and Victims: National Report Series. Washington, DC: Office of Juvenile Justice and Delinquency Prevention.
8. Snyder, H. N. & Sickmund, M. (2006). *Juvenile offenders and victims: 2006 national report.* Washington, DC: Office of Juvenile Justice and Delinquency Prevention.
9. Sickmund, M., Sladky, Kang, & Puzzanchera, 2008.
10. Snyder, H. N. & Sickmund, M. (1995). *Juvenile offenders and victims: A national report.* Washington, DC: Office of Juvenile Justice and Delinquency Prevention.
11. Sickmund, M. (2004). Juveniles in corrections. *Juvenile Offenders and Victims: National Report Series, Bulletin.* Washington, DC: Office of Juvenile Justice and Delinquency Prevention.
12. Sickmund, Sladky, Kang, & Puzzanchera, 2008.
13. Livsey, Sickmund, & Sladky, 2009.
14. Shichor, D. & Bartollas, C. (1990). Private and public juvenile placements: Is there a difference? *Crime and Delinquency, 36,* 286–299.
15. Livsey, Sickmund, & Sladky, 2009.
16. Elrod, Kinkade, Smith, Brown, Matthews, 1992.
17. Sickmund, Sladky, Kang, & Puzzanchera, 2008.
18. Sickmund, Sladky, Kang, & Puzzanchera, 2008.
    Thornberry, Tolnay, Flanagan, & Glynn, 1991.
19. Snyder & Sickmund, 2006.
20. Thornberry, Tolnay, Flanagan, & Glynn, 1991.
21. Snyder & Sickmund, 2006.
22. Parent, D. G., Lieter, V., Kennedy, S., Livens, L., Wentworth, D., & Wilcox, S. (1994). *Conditions of confinement: Juvenile detention and corrections facilities.* (p. 1). Washington, DC: Office of Juvenile Justice and Delinquency Prevention, 1994).
23. Snyder & Sickmund, 1995.
24. Sickmund, Sladky, Kang, & Puzzanchera, 2008.

25. Sickmund, Sladky, Kang, & Puzzanchera, 2008.

26. Sickmund, Sladky, Kang, & Puzzanchera, 2008.

27. Rubin, H. T. (1985). *Juvenile justice: Policy, practice, and law* (2nd ed.). New York: Random House.

28. Rubin, 1985.

29. Chesney-Lind, M. & Shelden, R. G. (2004) *Girls, delinquency and juvenile justice* (3rd ed.). Belmont, CA: Thompson/Wadsworth.

Parent, Lieter, Kennedy, Livens, Wentworth, & Wilcox, 1994.

Puritz, P. & Scali, M. A. (1998). *Beyond the walls: Improving conditions of confinement for youth in custody.* Washington, DC: Office of Juvenile Justice and Delinquency Prevention.

30. Griffiths, K. A. (1964). Program is the essence of juvenile detention. *Federal Probation*, *28*, 31–34.

31. Rubin, 1985.

32. Livsey, Sickmund, & Sladky, 2009.

Parent, Lieter, Kennedy, Livens, Wentworth, & Wilcox, 1994.

33. Livsey, Sickmund, & Sladky, 2009.

Parent, Lieter, Kennedy, Livens, Wentworth, & Wilcox, 1994.

34. Parent, Lieter, Kennedy, Livens, Wentworth, & Wilcox, 1994.

Snyder & Sickmund, 2006.

35. Burrell, S., DeMuro, P., Dunlap, E., Sanniti, C., & Warboys, L. (1998). *Crowding in juvenile detention centers: A problem-solving manual.* Washington, DC: Office of Juvenile Justice and Delinquency Prevention.

Pappenfort, D. M. & Young, T. M. (1980). *Use of secure detention for juveniles and alternatives to its use.* Washington, DC: Department of Justice.

36. Frazier, C. E. (1989). Preadjudicatory detention. In A. R. Roberts (Ed.), *Juvenile justice: Policies, programs, and services.* Chicago: Dorsey Press.

37. Frazier, C. E. & Bishop, D. M. (1985). The pretrial detention of juveniles and its impact on case dispositions. *Journal of Criminal Law and Criminology*, *76*, 1132–1152.

38. Frazier, 1989.

39. Burrell, DeMuro, Dunlap, Sanniti, & Warboys, 1988. This resource reviews the literature.

40. Livsey, Sickmund, & Sladky, 2009.

41. Senna, J. J. & Siegel, L. J. (1992). *Juvenile law: Cases and comments* (2nd ed.). St. Paul, MN: West Publishing Co.

42. *Schall v. Martin*, 467 U.S. 253 (1984).

43. Burrell, DeMuro, Dunlap, Sanniti, & Warboys, 1988

Schwartz, I. M., Fishman, G., Hatfield, R. R., Krisberg, B. A., & Eisikovits, Z. (1987). Juvenile detention: The hidden closets revisited. *Justice Quarterly*, *4*, 219–235.

44. Elrod, P. & Yokoyama, M. (2006). Juvenile justice in Japan. In P. C. Friday & X. Ren (Eds.), *Delinquency and juvenile justice systems in the non-Western World.* Monsey, NY: Criminal Justice Press.

45. Kramer, J. H. & Steffensmeier, D. J. (1978). Differential detention/jailing of juveniles: A comparison of detention and non-detention courts. *Pepperdine Law Review*, *5*, 795–807.

Lerman, P. (1977). Discussion of "differential selection of juveniles for detention." *Journal of Research in Crime and Delinquency*, *14*, 166–172.

Pappenfort & Young, 1980.

Poulin, J. E., Levitt, J. L., Young, T. M., & Pappenfort, D. M. (1977). *Juveniles in detention centers and jails: An analysis of state variations during the mid 1970's* Washington, DC: Office of Juvenile Justice and Delinquency Prevention.

46. Rubin, H. T. (1980). *Juveniles in justice: A book of readings.* Santa Monica, CA: Goodyear Pub. Co.

47. Maupin, J. & Bond-Maupin, L. J. (1999). Detention decision-making in a predominately Hispanic region: Rural and non-rural differences. *Juvenile and Family Court Journal*, *50*, 11–23.

48. Sickmund, Sladky, Kang, & Puzzanchera, 2008.

49. Puzzanchera, C., Finnegan, T., & Kang, W. (2007). *Easy access to juvenile populations: 1990-2007*. Retrieved January 22, 2009, from http://www.ojjdp.ncjrs.gov/ojstatbb/ezapop/.

50. Sickmund, Sladky, Kang, & Puzzanchera, 2008. For reviews of this research, see Chesney-Lind & Shelden, 2004.

McCarthy, B. R. (1987). Preventive detention and pretrial custody in the juvenile court. *Journal of Criminal Justice*, *15*, 185–200.

51. Chesney-Lind & Shelden, 2004.

52. Chesney-Lind & Shelden, 2004.

Hoyt, S. & Scherer, D. G. (1998). Female juvenile delinquency: Misunderstood by the juvenile justice system, neglected by social science. *Law and Human Behavior*, *22*, 81–107.

53. Sickmund, Sladky, Kang, & Puzzanchera, 2008.

54. Frazier & Bishop, 1985.

Henretta, J. C., Frazier, C. E., & Bishop, D. M. (1986). The effect of prior case outcomes on juvenile justice decision making. *Social Forces*, *65*, 542–562.

55. Parent, Lieter, Kennedy, Livens, Wentworth, & Wilcox, 1994.

56. Mohr, H. (2008, March 2). AP: 13K claims of abuse in juvenile detention since '04. *USA Today*. Retrieved January 23, 2008, from http://www.usatoday.com/news/nation/2008-03-02-juveniledetention_N.htm.

57. Roush, D. W. (1995). Juvenile detention programming. In R. A. Weisheit & R. G. Culbertson (Eds.), *Juvenile delinquency: A justice perspective* (3rd ed.). Prospect Heights, IL: Waveland.

58. Sabol, W. J., Minton, T. D., & Harrison, P. M. (2007). Prison and jail inmates at midyear 2006. *Bureau of Justice Statistics Bulletin.* Washington, DC: U.S. Department of Justice.

59. Sabol, Minton, & Harrison, 2007.

Sickmund, 2004.

Snyder & Sickmund, 2006.

60. Schwartz, I. M. (1989). *(In)justice for juveniles: Rethinking the best interests of the child* (pp. 66–67). Lexington, MA: Lexington Books. This reference contains a quotation from the deputy attorney general of the United States, Charles B. Renfrew.

61. Senna & Siegel, 1992.

62. Rubin, H. T. (2003). *Juvenile justice: Policies, practices, and programs.* Kingston, NJ: Civic Research Institute.

63. Snyder & Sickmund, 2006.

64. Krisberg, B. & Austin, J. F. (1993). *Reinventing juvenile justice* (p. 177). Newbury Park, CA: Sage.

65. Rubin, 2003.

66. Schwartz, 1989.

67. Sabol, Minton, & Harrison, 2007.

68. Lotz, R., Regoli, R. M., & Poole, E. D. (1985). *Juvenile delinquency and juvenile justice.* New York: Random House.

69. Schwartz, 1989.

70. Jennings, D. (2009, January 23). Jail teacher faces more sexual abuse charges. *The Dallas Morning News.* Retrieved January 23, 2008, from http://www.dallasnews.com/sharedcontent/dws/news/localnews/stories/DN-santos_23met.ART.State.Edition1.4efed73.html.

71. Chesney-Lind, M. (1988). Girls in jail. *Crime and Delinquency, 34,* 150–168.

72. President's Commission on Law Enforcement and Administration of Justice. (1967). *Task Force Report: Corrections* (p. 122). Washington, DC: U.S. Government Printing Office.

73. Campaign for Youth Justice. (2007). *Jailing juveniles: The dangers of incarcerating youth in adult jails in America.* Retrieved January 23, 2009, from http://www.campaign4youthjustice.org/Downloads/NationalReportsArticles/CFYJ-Jailing_Juveniles_Report_2007-11-15.pdf.

Community Research Associates. *Juvenile suicides in adult jails: Findings from a national survey of juveniles in secure detention facilities.* Washington, DC: U.S. Department of Justice.

Flaherty, M. (1983). The national incidence of juvenile suicide in adult jails and juvenile detention centers. *Suicide and Life-Threatening Behavior, 13*(2), 85–94.

Memory, J. M. (1989). Juvenile suicides in secure detention facilities: Correction of published rates. *Death Studies, 13,* 455–463.

74. Campaign for Youth Justice, 2007.

75. Browne, A. & Finkelhor, D. (1986). Impact of child sexual abuse: A review of research. *Psychological Bulletin, 99,* 66–77.

76. Chesney-Lind & Shelden, 2004.

77. Sickmund, Sladky, Kang, & Puzzanchera, 2008.

78. Sickmund, Sladky, Kang, & Puzzanchera, 2008.

79. Parent, Lieter, Kennedy, Livens, Wentworth, & Wilcox, 1994.

80. Snyder & Sickmund, 1995.

81. Parent, Lieter, Kennedy, Livens, Wentworth, & Wilcox, 1994.

82. Gemignani, R. J. (1994). Juvenile correctional education: A time for change. *OJJDP Update on Research.* Washington, DC: Office of Juvenile Justice and Delinquency Prevention.

83. Parent, Lieter, Kennedy, Livens, Wentworth, & Wilcox, 1994.

84. Parent, Lieter, Kennedy, Livens, Wentworth, & Wilcox, 1994.

85. Siegel, L. J., Welsh, B. C., & Senna, J. J. (2003). *Juvenile delinquency: Theory, practice, and law* (8th ed.). Belmont, CA: Wadsworth.

86. Bartollas, C. & Miller, S. J. (2005). *Juvenile justice in America* (4th ed.). Upper Saddle River, NJ: Pearson-Prentice Hall.

87. Soler, M. I., Shotton, A. C., Bell, J. R., Jameson, E. J., Shauffer, C. B., & Warboys, L. M. (1990). *Representing the child client.* New York: Matthew Bender.

88. American Correctional Association. (1991) *Standards for juvenile training schools* (3rd ed.). Laurel, MD: American Correctional Association.

89. Parent, Lieter, Kennedy, Livens, Wentworth, & Wilcox, 1994.

90. Snyder & Sickmund, 1995, p. 169. This resource cites Leiter, V. (1993). *Special analysis of data from the OJJDP conditions of confinement study.* Boston, MA: Abt Associates.

91. Wasserman, G. A., Ko, S. J., & McReynolds, L. S. (2004). Assessing the mental health status of youth in juvenile justice settings. *Juvenile Justice Bulletin.* Washington, DC: Office of Juvenile Justice and Delinquency Prevention.

92. Livsey, Sickmund, & Sladky, 2009.

93. Sickmund, M. (2002). Juvenile residential facility census, 2000: Selected findings. *Juvenile Offenders and Victims: National Report Series, Bulletin.* Washington, DC: Office of Juvenile Justice and Delinquency Prevention.

Snyder & Sickmund, 1995.

94. Sickmund, Sladky, Kang, & Puzzanchera, 2008.

95. Snyder & Sickmund, 1995, p. 172.

96. Dobrin, A. & Gallagher, C. A. (2004). Escapes from juvenile justice residential facilities: An examination of the independent and additive effects of security components. *Journal for Juvenile Justice Services, 19,* 47–57.

97. Chisamera, D. (2008, February 17). Two teens help 17-year-old murder suspect escape juvenile hall. *eFluxMedia.* Retrieved January 24, 2009, from http://www.efluxmedia.com/news_Two_Teens_Help_17_Year_Old_Murder_Suspect_Escape_Juvenile_Hall_14108.html.

98. Snyder & Sickmund, 1995.

99. Snyder & Sickmund, 2006.

100. Krisberg, B. (2006, June 1). *Stopping sexual assaults in juvenile corrections facilities: A case study of the California Division of Juvenile Justice.* Testimony before the National Prison Rape Elimination Commission, Boston.

101. Parent, Lieter, Kennedy, Livens, Wentworth, & Wilcox, 1994.

102. Bartollas & Miller, 2005.

103. Austin, J., Krisberg, B., DeComo, R., Rudenstine, S., & Del Rosario, D. (1995). *Juveniles taken into custody: Fiscal year 1993, Statistics report.* Washington, DC: Office of Juvenile Justice and Delinquency Prevention.

104. Parent, Lieter, Kennedy, Livens, Wentworth, & Wilcox, 1994.

105. Rogers, J. W. & Mays, G. L. (1987). *Juvenile delinquency and juvenile justice.* New York: John Wiley.

106. Rogers & Mays, 1987, p. 435.

107. Binder, A., Geis, G., & Dickson, D. B. (1997). *Juvenile delinquency: Historical, cultural, and legal perspectives* (2nd ed.). Cincinnati, OH: Anderson.

108. Bourque, B. B., Cronin, R. C., Pearson, F. R., Felker, D. B., Han, M., & Hill, S. M. (1996). Boot camps for juvenile offenders: An implementation evaluation of three demonstration programs. *National Institute of Justice Research Report.* Washington, DC: National Institute of Justice.

109. Binder, Geis, & Dickson, 1997.

110. Bourque, Cronin, Pearson, Felker, Han, & Hill, 1996.

111. Bourque, Cronin, Pearson, Felker, Han, & Hill, 1996.

112. Bourque, Cronin, Pearson, Felker, Han, & Hill, 1996.

113. Austin, J., Camp-Blair, D., Camp, A., Castellano, T., Adams-Fuller, T., Jones, M., et al. (2000). *Multi-site evaluation of boot camp programs, final report.* Washington, DC: The Institute on Crime, Justice, and Corrections at The George Washington University.

MacKenzie, D. L., Gover, A. R., Armstrong, G. S., & Mitchell, O. (2001). A national study comparing the environments of boot camps with traditional facilities for juvenile offenders. *National Institute of Justice Research in Brief.* Washington, DC: National Institute of Justice.

114. Sickmund, et al., n.d.

115. Schwartz, 1989.

116. Durham, A. M. (1989). Origins of interest in the privatization of punishment: The nineteenth and twentieth century American experience. *Criminology, 27,* 107–139.

117. Office of Juvenile Justice and Delinquency Prevention. (Producer). (1996). *Effective programs for serious, violent and chronic juvenile offenders teleconference* [Video]. Available from http://ojjdp.ncjrs.org/jjjournal/jjjournal1297/pubs.html.

118. Schwartz, I. M., Jackson-Beeck, M., & Anderson, R. (1984). The "hidden" system of juvenile control. *Crime and Delinquency, 30,* 371–385.

119. Schwartz, 1989.

Shichor & Bartollas, 1990.

120. Schwartz, 1989, p. 139. This reference cites B. D. Lurie.

121. Schwartz, 1989.

122. Schwartz, 1989, p. 131. This reference cites Patricia Guttridge.

123. Schwartz, 1989, p. 131.

124. Schwartz, 1989.

125. It is likely that this figure overestimates the percentage increase in the juvenile population of correctional facilities due to improvements in data collection. However, there is no doubt that the incarceration of youths for illegal behaviors has become more popular.

Flanagan, T. J. & Maguire, K. (Eds.). (1990). *Sourcebook of criminal justice statistics, 1989* (p. 559, Table 6.6). Washington, DC: U.S. Government Printing Office.

Moone, J. (1997). Juveniles in private facilities, 1991–1995. *Fact Sheet #64.* Washington, DC: Office of Juvenile Justice and Delinquency Prevention.

Moone, J. (1997). States at a glance: Juveniles in public facilities, 1995. *Fact Sheet #69.* Washington, DC: Office of Juvenile Justice and Delinquency Prevention.

Sickmund, 2004.

126. Sickmund, Sladky, Kang, & Puzzanchera, 2008.

127. Krisberg, B., DeComo, R., & Herrera, N. C. (1992). *National juvenile custody trends*, 1978–1989. Washington, DC: Office of Juvenile Justice and Delinquency Prevention.

Sickmund, 2004.

128. Sickmund, Sladky, Kang, & Puzzanchera, 2008.

129. Sickmund, Sladky, Kang, & Puzzanchera, 2008.

130. Sickmund, M., Sladky, T. J., & Kang, W. (n.d.). *Census of juveniles in residential placement databook.* Retrieved January 21, 2005, from http://OJJDP.ncjrs.org/ojstatbb/cjrp/asp/Offense_Race.asp.

Sickmund, 2004.

131. Pope, C. E., Lovell, R., & Hsia, H. M. (2002). Disproportionate minority confinement: A review of the research literature from 1989 through 2001. *Juvenile Justice Bulletin.* Washington, DC: Office of Juvenile Justice and Delinquency Prevention.

132. Krisberg, DeComo, & Herrera, 1992.

133. Miller, S. (1971). *Post-institutional adjustment of 443 consecutive TICO releases* (Unpublished doctoral dissertation). Ohio State University, Columbus, Ohio.

134. Goldman, I. J. (1974). *Arrest and reinstitutionalization after release from state schools and other facilities of the New York State Division for Youth: Three studies of youths released January 1971 through March 1973.* Albany, NY: New York State Division for Youth.

135. Snyder & Sickmund, 2006. This reference cites Virginia Department of Juvenile Justice. (2005). Juvenile recidivism in Virginia. *DJJ Research Quarterly.* Richmond, VA: Virginia Department of Juvenile Justice.

136. Whitehead, J. T. & Lab, S. P. (1989). A meta-analysis of juvenile correctional treatment. *Journal of Research in Crime and Delinquency*, *26*, 276–295.

Martinson, R. (1974). What works? Questions and answers about prison reform. *The Public Interest*, *35*, 22–54.

Wright, W. E. & Dixon, M. C. (1977). Community prevention and treatment of juvenile delinquency. *Journal of Research in Crime and Delinquency*, *14*, 35–67.

137. Garrett, C. J. (1985). Effects of residential treatment on adjudicated delinquents: A meta-analysis. *Journal of Research in Crime and Delinquency*, *22*, 287–308.

138. Lipsey, M. W. & Wilson, D. B. (1998). Effective interventions for serious juvenile offenders: A synthesis of research. In R. Loeber & D. B. Farrington (Eds.), *Serious and violent juvenile offenders: Risk factors and successful interventions*. Thousand Oaks, CA: Sage.

Lundman, R. J. (2001). *Prevention and control of juvenile delinquency* (3rd ed.). (Chapter 10). New York: Oxford University Press.

139. Krisberg, B. (1992). *Juvenile justice: Improving the quality of care.* San Francisco: National Council on Crime and Delinquency.

Lipsey & Wilson, 1998.

Palmer, T. (1996) Programmatic and nonprogrammatic aspects of successful intervention. In A. T. Harland (Ed.), *Choosing correctional options that work: Defining the demand and evaluating the supply*. Thousand Oaks, CA: Sage.

140. Statistics Canada. (2005). *Youth correctional services: Key indicators.* Retrieved September 5, 2009, from http://www.canadiancrc.com/Newspaper_Articles/STATSCAN_Youth_Crime_01DEC05.aspx.

141. Puzzanchera, Finnegan, & Kang, 2007.

Sickmund, Sladky, Kang, & Puzzanchera, 2008.

142. Parent, Lieter, Kennedy, Livens, Wentworth, & Wilcox, 1994.

143. Calculations based on data from Puzzanchera, Finnegan, & Kang, 2007.

Sickmund, Sladky, Kang, & Puzzanchera, 2008.

144. Elrod, Kinkade, Smith, Brown, Matthews, 1992.

145. Bartollas, C., Miller, S. J., & Dinitz, S. (1976). *Juvenile victimization: The institutional paradox.* New York: John Wiley & Sons, Inc.

146. Sieverdes, C. M. & Bartollas, C. (1986). Security level and adjustment patterns in juvenile institutions. *Journal of Criminal Justice, 14*, 135–145.

147. Bartollas, C. M., Miller, S. J., & Dinitz, S. (1976). Exploitation matrix in a juvenile institution. *International Journal of Criminology and Penology, 4*, 257–270.

Miller, J. G. (1991). *Last one over the wall: The Massachusetts experiment in closing reform schools.* Columbus, OH: Ohio State University Press.

148. Bartollas, Miller, & Dinitz, 1976b.

149. Bartollas & Miller, 2005.

150. Propper, A. M. (1982). Make-believe families and homosexuality among imprisoned girls. *Criminology, 20*, 127–138.

151. Giallombardo, R. (1974). *The social world of imprisoned girls: A comparative study of institutions for juvenile delinquents.* New York: Wiley.

152. Miller, 1991.

Poole, E. D. & Regoli, R. M. (1983). Violence in juvenile institutions: A comparative study. *Criminology, 21*, 213–232.

Sieverdes & Bartollas, 1986.

153. Poole & Regoli, 1983.

154. Sieverdes & Bartollas, 1986.

155. Dembo, R. & Dertke, M. (1986). Work environment correlates of staff stress in a youth detention facility. *Criminal Justice and Behavior, 13*, 328–344.

156. Snarr, R. (1996). *Introduction to corrections* (3rd ed.). Dubuque, IA: Wm. C. Brown.

157. Altschuler, D. M., Armstrong, T. L., & MacKenzie, D. L. (1999). Reintegration, supervised release, and intensive aftercare. *Juvenile Justice Bulletin.* Washington, DC: Office of Juvenile Justice and Delinquency Prevention.

158. *Heryford v. Parker*, 396 F. 2d 393 (10th Cir. 1968).

159. Mich. Court Rules 5.943(E)(2).

160. Thornton, W. E. & Voight, L. (1992).
Delinquency and justice (3rd ed.) (p. 432). New York: McGraw-Hill.

161. *Holt v. Sarver*, 309 F. Supp. 362 (E.D. Ark 1970).

162. Thornton & Voight, 1992.

163. Mich. Comp. Laws Ann. 712A.18(1)(e).
Mich. Comp. Laws Ann. 712A.18(c)(3).
Mich. Comp. Laws Ann. 712A.18(d).
Mich. Court Rules 5.944(c)(1)–(4).

164. Thornton & Voight, 1992.

165. Sabol, W. J. & Couture, H. (2008). Prison inmates at midyear 2007. *Bureau of Justice Statistics Bulletin.* Washington, DC, Department of Justice.

166. *C.J.W. by and through L.W. v. State*, 853 P. 2d 4 (Kan. 1993).

167. Del Carmen, R. V., Parker, M., & Reddington, F. P. (1998). *Briefs of leading cases in juvenile justice.* Cincinnati, OH: Anderson Publishing Company.

The Status Offender in Juvenile Justice

CHAPTER OBJECTIVES

After studying this chapter, you should be able to

- Articulate reasons supporting state intervention with status offenders
- Articulate reasons against state intervention with status offenders
- Describe how the concept of *parens patriae* has encouraged the involvement of juvenile courts in status offense cases
- Give examples of status offenses and indicate how these offenses differ from delinquent or criminal offenses
- Discuss research on the differences and similarities between status offenders and youths involved in criminal behavior
- Discuss the role that the Juvenile Justice and Delinquency Prevention Act of 1974 and subsequent amendments to that act have played in the deinstitutionalization of status offenders
- Describe the mechanisms used by different jurisdictions to incarcerate status offenders and discuss whether the incarceration of status offenders is good public policy
- Discuss different approaches that might be used with status offenders, including the strengths and weaknesses of each approach

CHAPTER OUTLINE

Introduction

Since the development of the juvenile court, incorrigible youths, runaways, and truants (i.e., **status offenders**) have often been treated the same as youths who have engaged in criminal behaviors. Historically, the ***parens patriae*** philosophy of the juvenile courts has supported juvenile court intervention in a broad range of youthful behaviors. Indeed, many parents and other adults have demanded that juvenile courts exert their authority over problem youths even when they have avoided criminal behavior. From these persons' perspective, juvenile courts have an obligation to support parental and adult authority over children. Other people feel that juvenile court involvement is necessary in order to protect status offenders who place themselves at risk of harm. Still others believe that, if left unchecked, minor forms of child misbehavior will escalate into more serious forms of law violation. From their point of view, juvenile court involvement is necessary to protect the community from potential delinquents. Because juvenile courts are also political institutions that are sensitive to community demands for assistance, they often feel obliged to intervene in the lives of troubled children and their families.

status offender A youth who comes within the jurisdiction of a juvenile court for a status offense rather than a delinquent or criminal offense. Alternative terms include *children in need of supervision (CINS or CHINS), undisciplined children, incorrigible youths, and persons in need of supervision (PINS).*[89]

parens patriae The doctrine that the state has a legal obligation to provide for the care, protection, treatment, and rehabilitation of children if the children's parents are unable to do so.

status offense "Any offense committed by a juvenile that would not be a crime if committed by an adult, according to the statutes or ordinances of the jurisdiction in which the offense was committed, and which is specifically applicable to juveniles because of their status as a juvenile."[90] Status offenses consist of two types: (1) acts and behaviors only possible for juveniles, such as incorrigibility, running away, and truancy, and (2) acts and behaviors that are legal for adults but are illegal for juveniles, such as smoking cigarettes, drinking alcohol, and sexual intimacy.

Despite the compelling nature of the arguments favoring juvenile court intervention in status offense cases, there are good reasons to question the ability of juvenile courts to effectively respond to these cases. Many status offense cases are the products of ongoing family problems, abuse, neglect, and ineffective parenting, and these cases present a number of complicated issues to juvenile courts, which often lack the resources or expertise to intervene in these matters in a helpful way. Indeed, it is possible to question the ability of formal legal institutions to resolve the interpersonal, social, and psychological problems that often underlie status offense behaviors. At best, juvenile courts can play a role, along with other social agencies, in the handling of youths who defy their parents, run away from home, or skip school. To further complicate matters, the resources needed to deal with many of the problems that underlie status offense behaviors may not exist, may not be present in the community, may be difficult to access, or may be difficult to coordinate, which leads to incomplete, ineffective, and sometimes counterproductive responses. As a result, some juvenile courts have recognized their limitations in dealing with these cases and avoid them even when legal intervention is possible.

Should juvenile courts intervene in the lives of status offenders? If so, what actions represent appropriate and realistic responses? These are critical questions that continue to be debated within communities and juvenile courts around the country. This chapter reviews a variety of arguments for and against juvenile court intervention in the lives of status offenders, examines the legal basis for such intervention, and presents a historical overview of court responses to the status offender. It also examines trends in the processing of status offenders, legal issues related to court processing, and research on the similarities and differences between status offenders and criminal offenders. It concludes with a discussion of effective responses to status offense cases.

FYI

Status Offenses Can Only Be Committed by Minors

A **status offense** is "any offense committed by a juvenile that would not be a crime if committed by an adult, according to the statutes or ordinances of the jurisdiction in which the offense was committed, and which is specifically applicable to juveniles because of their status as a juvenile."[1]

Arguments For and Against Juvenile Court Involvement in the Lives of Status Offenders

As noted earlier, the juvenile court's role in dealing with status offenders is a matter of considerable controversy. Indeed, one of the most challenging issues that has confronted juvenile justice personnel in recent years is whether the juvenile court should "maintain, restrict, or abandon jurisdiction over status offenders."[2] Today there are two opposing camps whose positions are based on different assumptions regarding the degree of similarity between delinquents' behavior and needs and status offenders' behavior and needs.[3] Those who favor continued court intervention in the lives of status offenders present a number of arguments in support of their cause. The following arguments have been used to support continued court intervention in the lives of status offenders:

- Many status offenders have "special needs,"[4] and discontinuing court intervention will deny them services that they require and would not otherwise receive.[5] If the court does not intervene, no one else will, and thus very real problems faced by children, their families, and others in the community will be ignored.
- The treatment needs of status offenders are often more pronounced and require more professional and financial resources than those experienced by many delinquent offenders. Juvenile courts can assist status offenders by seeing that the necessary range of services is provided.
- Status offenses often escalate into more serious offenses[6] and are "predictive of a criminal career."[7] Behaviors such as failing to go to school, failing to obey one's parents, and running away from home can easily lead to criminal behaviors such as burglary and robbery if the court does not intervene.
- A failure of the court to intervene could result in youths' injuring themselves because many status offense behaviors put children at risk for victimization, injury, and even death. Running away and incorrigibility may place youths in situations where they turn to drugs, alcohol, prostitution, and other high-risk behaviors.
- Court oversight of community services is often necessary to ensure that youths receive appropriate treatment and that adequate services are provided to children within the community.[8] Court orders are frequently necessary in order to "encourage" social service agencies to provide needed services to families and youths and to provide services in a timely manner.
- Because voluntary agencies do not afford youths the full range of due process protections, youths may be coerced into receiving treatment. Court oversight of the treatment process makes coercive treatment less likely.[9]
- If juvenile courts relinquish jurisdiction over status offenders, compulsory school laws would be undermined because there would be no agency with the authority to enforce these laws.[10] Moreover, everyone in the community, children and adults alike, will bear the costs of youths who fail to get an adequate education. Juvenile court jurisdiction over truancy cases is necessary in order to uphold compulsory school laws.
- Children have a responsibility to obey their parents, and juvenile courts have a responsibility to support lawful parental authority.[11] Parents would have no effective way to deal with an unruly or troubled child without support from the local juvenile court.

- If juvenile courts surrendered their authority over status offense cases, they would, in effect, be abandoning their historical mission—protecting and serving the best interests of children.[12] They would also take a step toward becoming mini-adult courts.

The opponents of maintaining juvenile court jurisdiction over status offenders claim that the previous arguments are based on a number of questionable assumptions.[13] They contend that juvenile court jurisdiction is inappropriate, frequently ineffective, and often harmful, and they recommend strict limitations on or outright abandonment of juvenile court jurisdiction over status offenders (abandonment is often called **decriminalization**). The following arguments have been used to support the decriminalization of status offenders:

decriminalization
The process of removing proscriptions from legal codes (in the context of juvenile justice, the process of removing status offenses from the juvenile code).

cruel and unusual punishment
Punishment handed down by a court that is too harsh for the offense committed. The Eighth Amendment of the U.S. Constitution explicitly prohibits the imposition of cruel and unusual punishment.

equal protection under the law
The constitutional requirement that all laws be applied to all persons fairly and without regard to race, creed, color, national origin, sex, or age. The Fourteenth Amendment of the U.S. Constitution prohibits denial of equal protection of the laws to anyone.

- Family or juvenile codes dealing with status offenses are often vague, expose youths to **cruel and unusual punishment**, and fail to adequately provide youths with **equal protection under the law**, thus raising serious questions about their constitutionality.[14] Furthermore, juvenile courts have often failed to protect the constitutional rights of status offenders.[15]
- Jurisdiction over status offenses is based on faulty assumptions that these behaviors will escalate into more serious delinquent behaviors.[16]
- There is a lack of evidence that juvenile court intervention benefits status offenders.[17] Instead of relying on the juvenile courts, which often lack the necessary resources and expertise to intervene effectively, we should recognize that other social agencies are better equipped to meet the needs of these youths.
- The lack of juvenile court success in dealing with status offense cases often results in employment of the courts' ultimate constraint—institutionalization.[18] Use of the court's full coercive powers is inappropriate in cases in which youths have not violated criminal laws.
- Youths learn disrespect for the law when they confront the reality that they are held to higher standards of behavior than adults, who are not required to go to school, stay at home, or agree with their parents.[19]
- Processing of status offense cases leads to negative labeling, which encourages additional acting out behaviors.[20]
- Most status offense behaviors are common but transitory behaviors that revolve around intra-family problems that are not suited to formal court intervention.[21] Moreover, courts often weaken parental responsibility by being too willing to intervene in family relationships.[22]
- Court intervention often removes responsibility from schools and other social agencies to develop and implement out-of-court solutions to family and community problems.[23] Indeed, because the juvenile courts deal with these problems and often promise solutions, other agencies find it easy to avoid responsibility for these cases.[24]
- Status offense laws are unequally applied against poor children and females. Children of wealthy parents are able to avoid juvenile court because their parents can afford to purchase private services for them, including inpatient mental health and substance abuse treatment. However, poor children are subject to court jurisdiction because these services are not similarly available. Also, females are more likely to be processed by courts and detained as runaways.[25] As a result, minority youths and females are at greater risk of being treated like delinquent offenders and subjected to punitive court action.

COMPARATIVE FOCUS

Some Countries Do Not Formally Handle Status Offenses

In Sweden, minors are persons who are younger than the age of 15 years. However, while youths can engage in behaviors that are prohibited for minors, no legal sanctions can be levied against the youths. For example, if a minor purchases alcohol, sneaks in to see an adult movie, or has intercourse with an adult, the adults involved would be subject to sanctions, but not the minors. When youths are involved in activities that threaten their safety, they are referred to the social welfare committee in the child's hometown, not to the court.[29]

- Courts place an undue burden on children by holding them responsible for their behavior, but fail to hold parents responsible for what they contribute to the problem.[26] In many instances, parents, guardians, school personnel, and others play a significant role in the development of those behaviors that are subject to court intervention.

As the preceding material makes clear, there are a variety of arguments for and against juvenile court intrusion into the lives of status offenders. Regardless of the pros and cons, in most jurisdictions, juvenile courts do have legal authority over status offense cases. Moreover, some courts deal with a large number of status offenders each year.[27] For example, juvenile courts formally handled 150,600 status offenses cases in 2005. Moreover, this represented a 29% increase in the number of status offense cases handled by juvenile courts since 1995.[28]

The juvenile court's authority over juvenile offenders is part of a long tradition of juvenile court intervention in a broad range of situations involving youths. The following section examines the legal basis for juvenile court jurisdiction over status offense cases.

The Legal Basis for Juvenile Court Jurisdiction over Status Offenders

The legal basis for juvenile court jurisdiction over status offenders can be traced back to the development of the legal concept of *parens patriae* and its elaboration in the legislation that established juvenile courts in the United States. As noted in Chapter 5, the first juvenile courts were designed not only to respond to the criminal behaviors of the young, but also, in keeping with their *parens patriae* focus, to deal with troublesome children—children viewed as likely to become criminals without appropriate intervention.[30]

FYI

Juvenile Codes Cover a Wide Range of Behaviors

The intent of the early reformers who worked for the establishment of juvenile courts was to avoid formal legal distinctions between delinquent, dependent, and neglected youths.[32] Consequently, statutory definitions of delinquency included acts that would be criminal if committed by an adult; behaviors that were in violation of local town, city, and county ordinances; and violations of vaguely worded laws prohibiting, for example, "vicious or immoral behavior," "incorrigibility," "profane or indecent behavior," "growing up in idleness," and "living with any vicious or disreputable person."[33]

Indeed, early juvenile courts were vested with the authority to deal with a wide range of youth behaviors and conditions, including criminal law violations, nuisance behaviors, and neglect by parents.[31]

Underlying the concept of *parens patriae* and status offense statutes is the belief that children occupy a unique position in our society. As noted in Chapter 5, juveniles are recognized by the law as having certain constitutional rights that are similar, although not identical, to those afforded adults in like circumstances. For example, the U.S. Supreme Court mandated, in the *Gault* (1967) decision, that juvenile offenders be given most of the due process protections possessed by adults—the right to assistance from counsel, the right to reasonable notice, the right to confront witnesses, and freedom from self-incrimination.[34] However, in *McKeiver v. Pennsylvania* (1971), the Court ruled that juveniles do *not* have a right to a jury trial. In part, this decision was based on the desire of the Court to maintain the informal nature of the juvenile courts and to avoid turning juvenile court adjudications into an adversarial process. In addition, the Court ruled that it was more important to allow the juvenile courts to continue to serve children's best interests—a goal not advanced by requiring these courts to have jury trials.[35] The *McKeiver* ruling not only refused to grant jury trials to juveniles, but also reinforced the idea that children enjoy a special status in society that is inconsistent with granting them all of the due process protections afforded adults.

State statutes also recognize the special status of children. For example, children (minors) cannot enter into binding contracts, possess real or personal property, marry, or seek employment except under special circumstances. From a legal standpoint, childhood is defined purely in terms of age. Yet, underlying any age-related definition of childhood is the idea that persons younger than a certain age lack the life experience and maturity to act in their own best interests. Indeed, due to their age and limited experience, they could easily be victimized by unscrupulous adults. Consequently, the law recognizes the need to protect children in most legal or business relationships.

According to the concept of *parens patriae*, the state, in certain circumstances, has a duty and an obligation to protect children from adults, including their parents. Consequently, the state has established a variety of laws that spell out the obligations that parents, teachers, and others have toward children. These laws also permit state intervention in the lives of children and families when adults do not fulfill their obligations to youths. In return, the state has imposed special legal obligations on children. For example, state legal codes often require children to

- obey the reasonable and lawful rules of parents, guardians, or custodians;
- attend school regularly and obey school rules;
- not run away or absent themselves from home without permission;
- obey curfews.

Although it can be argued that the legal obligations imposed on children are reasonable and within the state's legitimate interest because they ensure that families are supported and children are disciplined, any violation of the laws and ordinances that spell out these legal obligations is considered a status offense. These offenses consist of two types: (1) acts and behaviors only possible for juveniles, such as incorrigibility, running away, and truancy, and (2) acts and behaviors that are legal for adults but are illegal for juveniles, such as smoking cigarettes, drinking alcohol, and sexual intimacy.[36] In short, a status offense is an offense that can be committed only by children because of their

FYI

Different Terms Are Used to Refer to Status Offenders Around the Country

Although *status offender* is the most common term used to refer to those youths who come within the jurisdiction of juvenile courts for status offenses, a variety of other names have been developed to identify these youths. In the District of Columbia, they are referred to as children in need of supervision (CINS or **CHINS**); in North Carolina, they are termed undisciplined children; in Michigan, they are termed incorrigible youths; in New York State, they are persons in need of supervision (**PINS**).[38] These designations have been created in an effort to recognize the unique status of children and to differentiate status offenders from youths who engage in criminal behavior.

CHINS Children in need of supervision (also abbreviated as CINS). This term is used to refer to status offenders in some states.

PINS Persons in need of supervision. This term is used to refer to status offenders in some states.

"status" as minors. Moreover, these offenses are not inconsequential and can result in a child coming under the jurisdiction of a juvenile court.

Although many states have developed special designations for status offenders, such as children in need of supervision (CHINS or CINS) or persons in need of supervision (PINS), these labels often present a misleading picture of the "typical" status offender. For example, many children designated as ungovernable youths are, in fact, neglected or abused children who are forced onto the streets because of physical or sexual abuse or unlivable home conditions.[37] Using the power of the juvenile court to force this group of so-called status offenders to return home and obey their parents would hardly serve their best interests. In reality, many children who become status offenders come from neglectful or dysfunctional homes in which the parents or responsible adult have not cultivated the appropriate adult–child relationship and fundamental respect has broken down. Many children who have experienced domestic violence at a young age become incorrigible when they reach adolescence. In these cases, the children have lost respect for both parents and resort to making their own rules for behavior. They refuse to accept their father's authority, because he is often the abuser, and they know they cannot count on him. They initially look to their mother for protection, but she cannot protect herself, let alone them, so they soon learn not to rely on her either. At this point, incorrigibility is almost inevitable without outside intervention.

Juvenile Court Responses to Status Offenders: An Historical Perspective

Concerns about problem children are hardly new. Throughout recorded history, young people have engaged in behaviors that have been troublesome to adults. For example, in the Bible we find the following passage:

> If a man has a stubborn and rebellious son, who will not listen to them even when they chastise him, then his father and mother shall lay hands on him and bring him before the sheiks of his town at the local gateway, telling the sheiks of his town, "This son of ours is a stubborn and rebellious fellow who will not obey our orders; he is a spendthrift and a drunkard." Whereupon, all his fellow citizens shall stone him to death. So shall you eradicate evil from you, and all Israel shall hear and fear.[39]

The legislation that established the juvenile courts codified a similar attitude to that of the previous Biblical passage. Indeed, concern over the behavior of youths—particularly

MYTH VS REALITY

Each Generation of Adults Has Been Concerned with Youths' Behavior
Myth—Concern over children's refusal to bow to parental authority is a recent problem.
Reality—Adult concern about children's unwillingness to adhere to adult wishes goes back to antiquity.

poor children, who were seen as making up a population of potential criminals—played a prominent role in the development of such legislation.[40] The laws spelled out a variety of legal responses to a number of troublesome behaviors, such as incorrigibility, truancy, idleness, the use of profane language, living with disreputable persons, and immoral behavior. Moreover, these laws were used as an ancillary means by which parents and other adults might exert proper control over the young so that they would develop into moral, disciplined, and productive adults.[41] Moreover, early laws vested the authority to enforce these statutes in the juvenile court, which, operating on the basis of *parens patriae*, was supposed to "act on behalf of the child and to provide care and protection equivalent to that of a parent."[42] The result was a court that had the authority to exert control over children—control viewed as critical by the early child savers because adult courts had no authority to deal with behaviors that were not criminal.[43]

In their efforts to control problem and wayward children, juvenile courts have historically employed a variety of strategies, including warning youths, ordering them on probation, placing them out, and institutionalizing them in a variety of residential placements. These strategies were thought to be justified by the courts' mission: to support families, control children, help youths lead productive and moral lives, and protect the community.

Despite a lengthy history of responding to status offenses, many juvenile courts have come to recognize that their efforts to control and prevent such behaviors have not always been effective. For one thing, the resources and therapeutic techniques available to juvenile courts are often inadequate for responding to children's needs, and today more of these resources are being devoted to dealing with violent offenders. Moreover, many courts have come to the conclusion that efforts to control status offenders have often produced more harm than good. Unfortunately, status offenders have frequently been treated exactly like youths who have committed criminal offenses. In essence, these young offenders have been victimized by the juvenile courts, despite the courts' legal obligation to look out for their best interests.

By the early 1970s, some changes in the juvenile courts' response to status offenders were beginning to be seen in some jurisdictions. In 1972, Massachusetts, under the leadership of Jerome Miller, the commissioner of the Department of Youth Services, began to **deinstitutionalize** the juvenile offender population and moved toward a community-based system of corrections. Not only were status offenders released from institutions, but so were most youths who were committed to state care for criminal offenses.[44] In addition, the Juvenile Justice and Delinquency Prevention Act, which called for the nationwide deinstitutionalization of status offenders, was introduced in Congress that same year and was signed into law in 1974.[45]

deinstitutionalization
The removal of offenders from institutional settings. In this chapter, the term is used to refer to the removal of status offenders from detention centers and other secure institutional settings.

According to the Juvenile Justice and Delinquency Prevention Act, within two years of the passage of the act, states were required to refrain from placing in a detention or

FYI

Most States Continue to Allow Juvenile Courts to Process Status Offense Cases
Although most states continue to process status offense cases, two states, Delaware (1976) and Maine (1977), have decriminalized status offenses. As a result, juvenile courts in those states no longer process these types of cases.[46]

correctional facility any youth who had committed an offense that would not be a crime if committed by an adult. Subsequent amendments to the act and clarification of the deinstitutionalization mandate confirmed Congress's intent to remove status offenders from juvenile corrections facilities and adult jails. The amendments required states to make sufficient progress in their deinstitutionalization efforts or risk losing their eligibility to receive federal formula grant funds. In addition, West Virginia and Tennessee abolished the institutionalization of status offenders as a result of lawsuits brought in those states, whereas other states, such as Alabama, New Mexico, New York, Delaware, Louisiana, and Washington, passed legislation supporting deinstitutionalization.[47]

Although the Juvenile Justice and Delinquency Prevention Act has led to a reduction in the detention and institutionalization of status offenders, a number of tactics are used at the local level to circumvent the intent of the act. These tactics include referring status offenders to a secure mental health facility, creating or waiting for delinquency jurisdiction, developing "semi-secure" facilities that house only status offenders, and using the court's power of contempt to detain a youth for failure to obey a lawful court order rather than for the status offense. This latter method was used with approval by the family court in South Carolina in the case of *In re Darlene C.* In its ruling on this case, the Supreme Court of South Carolina stated the following:

> The issue is whether a juvenile who commits criminal contempt by running away in violation of a court order may be given a disposition reserved for delinquents who have committed offenses which would be crimes if committed by an adult. We conclude that, under the most egregious of circumstances as we have here, family courts may exercise their contempt power in such a manner that a status offender will be incarcerated in a secure facility.[48]

The "most egregious circumstances" included the fact that the juvenile was 16 years of age, was female, and was a chronic runaway and school truant. The South Carolina Supreme Court supported the use of the contempt powers of the family court as a mechanism for circumventing federal restrictions regarding the treatment of status offenders. Indeed, the use of contempt powers to control status offenders allows most jurisdictions to incarcerate status offenders by elevating them to delinquency status.[49] However, in a few states, this approach to status offenders has been rejected. In the case of *W.M. v. State of Indiana*, the Indiana Court of Appeals stated that contempt of court "should not be used to accomplish the goal that the juvenile court in the instant case attempted. It is meant to allow the juvenile courts a means of keeping order in their courtrooms.... In light of the statutory prohibition of incarcerating juveniles whose acts are not crimes for adults, [contempt] cannot be used to incarcerate them when the underlying act is a status offense."[50]

FYI

Many Status Offenders Are Treated Like Youths Who Have Committed Criminal Offenses

The passage of the Juvenile Justice and Delinquency Prevention Act of 1974, as well as changes in law in some states, has reduced the institutionalization of status offenders in the United States. Nevertheless, efforts to deinstitutionalize status offenders have not been completely successful. The 1980 amendments to the Juvenile Justice and Delinquency Prevention Act made it possible for courts to detain juveniles initially referred to courts as status offenders. For example, the amendments allow juvenile court judges to detain status offenders for a limited period of time if they have violated a court order. The amendments also gave states additional time to comply with the law.[51] On the day that the Census of Juveniles in Residential Placement was conducted in 2006, there were 4,717 status offenders residing in various types of residential placements. The majority (62%) of these were residing in group home placements, but 17% were in detention, 5% were in long-term secure facilities, and the rest were in a variety of other placement settings.[52]

These cases jointly indicate the continuing conflict within juvenile justice over the best ways to respond to status offenders. Many courts are unwilling to abandon their authority over status offenders and wish to retain their power to institutionalize these youths when they see fit. Nevertheless, some jurisdictions have recognized a need to curb the ability of courts to incarcerate youths who have not committed criminal offenses.

Although the deinstitutionalization of status offenders has not been completely successful, considerable progress has been made in efforts to remove these youths from juvenile and adult correctional facilities. However successful the movement to deinstitutionalize status offenders has been, though, it has not necessarily prepared courts or communities to deal more effectively with the status offender population. Significant numbers of status offenders continue to come to the attention of school personnel, law enforcement officers, and juvenile courts each year. The questions still remain as to who should respond to the problems presented by this population and which responses are most effective.

Status Offenders and Delinquents: Is There a Difference?

Arguments that status offense behaviors will escalate to become more serious offenses, that such offenses leave youths vulnerable to victimization, and that status offenders have special needs that can be best met by juvenile court intervention are often used to support continued court involvement in status offense cases. As the North Carolina Supreme Court ruled in the case of *In the Matter of Walker*, "In seeking solutions which provide in each case for the protection, treatment, rehabilitation, and correction of the child, it is compellingly relevant to the achievement of the state's objective that distinctions be made between undisciplined children, on the one hand, and delinquent children on the other. *The one may need protection while the other needs correction*"[53] (emphasis added).

This ruling clearly reflects the state's desire to assist children, and recognizes that status offenders and those involved in criminal behaviors may have different needs. If there are differences, they could be used as a justification for court intervention if status offenders have special needs or need to be protected by the court from youths involved in criminal activity or from adults in their lives. If there are no differences or if status offenses lead naturally to more serious delinquent activity, this too could serve as a justification for court intervention (the public needs protection from "potential criminals").

The extent of the differences between status offenders and youths involved in criminal behaviors is a matter of some debate. The majority of studies on this issue conducted before 1980 reported few differences between these two groups.[54] According to these studies, juvenile offense patterns represent a **cafeteria-style approach to delinquency** in which youths engage in a variety of delinquent (status and criminal law-violating) behaviors over time.[55] This finding supports the strategy of treating status and criminal offenders in the same way.

cafeteria-style approach to delinquency
The tendency of a youth to engage in a variety of status and criminal law-violating behaviors over time.

Several other studies, however, have reported that some status offenders may not be like other types of delinquent youths and should be treated differently. In one early effort to shed light on this debate, Charles Thomas studied the court records of juveniles who appeared for the first time in two Virginia juvenile courts between 1970 and 1974. Of these youths, 50.3% were charged with misdemeanors, 22.3% were charged with felonies, and 27.3% were charged with status offenses. In addition, he collected follow-up data on these youths to see if they returned to the court on subsequent charges.[56]

Thomas found that most of the first-time referrals were nonrecidivists during the follow-up period. However, he found that those charged with status offenses were more likely than youths charged with misdemeanors or felonies to engage in further delinquency. He also found that 12% of felony recidivists were referred a second time for a status offense, and 17% of misdemeanor recidivists were referred a second time for a status offense. Among youth initially referred for a status offense, 40% were referred a second time for a status offense and 60% were referred a second time for a criminal offense. Thus, Thomas' research suggests that some status offenders may be different in some ways, yet many others appear to be similar to other types of youths who appear before juvenile courts. Moreover, he found little evidence that court processing of status offenders led to their subsequent involvement with the court for more serious offenses.[57]

Similar findings regarding the similarities between status offenders and youths who commit criminal offenses were produced in a study by Howard Snyder of the National Center for Juvenile Justice. In this study, Snyder analyzed the court careers of more than 69,000 youths referred to juvenile courts in Utah and in Maricopa County (Phoenix), Arizona, and found that more than half of all youths referred to the juvenile court for a status offense were also referred for a criminal offense. Moreover, more than one-quarter of the youths referred to the juvenile court at some time for a delinquent offense also had a referral for a status offense. Snyder found little evidence of offense specialization. He did find that youths charged with status offenses were less likely to escalate to more serious delinquent behaviors, especially violent behaviors, than youths who did not have status offenses in their offense careers.[58]

Somewhat different findings, however, resulted from a study conducted by Steven Clarke, who examined differences between types of juvenile offenders in Philadelphia. Clarke found that, among 3,475 males who were born in Philadelphia in 1945 and had acquired a police record before their 18th birthday, 23.4% were initially involved with the police for a status offense. Moreover, when Clarke compared these youths to those who were first arrested for a criminal offense, he found the following: (1) those first arrested for a criminal offense became recidivists at twice the rate of status offenders (61% of the criminal offenders and 30% of the status offenders had at least two offenses), and (2) 21% of the criminal offenders, but only 10% of the status offenders, were chronic offenders (i.e., had five or more arrests). In addition, he found that these patterns were the same for white and nonwhite males.[59]

Similar findings have been reported by Thomas Kelly. After examining the offense patterns of more than 2,000 juveniles referred to the juvenile court in Wayne County (Detroit), Michigan, over a five-year period, Kelly found that status offenders and criminal offenders often have different offense careers. He found that status offenders were no more likely to engage in subsequent offending than those charged with a criminal offense—a finding at odds with the research done by Thomas. He also found that status offenders who returned to court often returned for a more serious offense than the one they had committed the first time, though typically for a less serious offense than other delinquent youths. These results indicate that juvenile court processing, rather than assisting status offenders, appears to encourage recidivism. Kelly concluded that status offenders are distinct from other delinquent offenders and should be removed from juvenile court jurisdiction.[60]

In another study, which examined 863 first referrals to the Clarke County (Las Vegas) juvenile court in 1980, Randall Shelden, John Horvath, and Sharon Tracy found differences in the offense careers of different types of status offenders. For example, they found little evidence that status offenders specialize in such behaviors or that the majority of youths who are initially referred to juvenile courts become serious offenders. However, they also found that, among status offenders who became recidivists, females were more likely to have a subsequent referral for a status offense than males. In addition, they found that status offenders who were referred to the juvenile court for running away or being unmanageable were less likely to be referred a subsequent time for a criminal offense than those petitioned for truancy, curfew infractions, or liquor law violations.[61]

Research by Solomon Kobrin and his associates, who studied status offenders in eight federal programs that focused on the deinstitutionalization of status offenders, also found that status offenders are not a homogeneous group. These researchers concluded that status offenders consist of three relatively distinct groups: status offenders who display little inclination to engage in more serious delinquent acts; status offenders whose records indicate a preponderance of delinquent (criminal) acts; and youths who had one citation for a status offense but had no prior or subsequent arrests for any offense (the largest group of the three).[62]

Still more recent findings also indicate that at least some status offenders may represent a distinct group of youths who are different in some ways than other types of juvenile offenders. For example, a study of more than 12,000 youths that included over 2,400 truants involved in the juvenile justice process in South Carolina found that truants represented a distinct group compared to other youths. Truants were less likely to be referred for serious crimes later in their youth. However, study results also indicated that truants tended to be younger than other offenders at the time of first referral, typically had become involved in the juvenile court at an earlier age, and had a higher number of lifetime referrals. In addition, the research indicated that truants who were male, were members of a minority group, were younger at the time of their first offense, had been placed in special education, had a history of drug use, or had come from families with criminal histories, were more likely to be recidivists.[63]

Overall, these findings indicate that many youths involved in status offense behaviors are similar (and in many respects identical) to youths who engage in delinquent or

BOX 13-1 Interview: Linda Vogler, Status Offender Caseworker—Family Court

Q: What is your educational background?

A: I have a B.A. in social work from Western Michigan University in 1979.

Q: What is your employment history, and do you have any special training with status offenders?

A: I have worked with status offenders and diversion cases since I was hired at the court in August of 1979. My training has been on the job. Prior to my court employment, I was in the clerical staff for the FBI in 1973. From 1974 to 1977, I was on active duty in the Navy doing a variety of jobs from investigator to plane captain. I left the Navy as an E4. While at Western Michigan University, I worked at the Calhoun County Juvenile Home.

Q: How big is your caseload, and what percentage of it is status offenders?

A: At any one time I have between 32 and 38 kids on my caseload. Virtually all of them are status offenders, but 50% may have delinquency charges.

Q: What special work or programming do you do with status offenders?

A: We focus on two fundamental concepts: rewarding kids for positive behavior and, thus, enhancing their self-esteem; and we hold kids accountable when they make poor choices. We contract with many of the kids, and we have a number of activities: art, sports, camps, modeling, and so on, that we get the kids involved in. These activities improve kids' self-esteem and the close supervision over them aids in holding them accountable.

Q: What parts of your job do you like best?

A: I really enjoy the personal contact with the kids, the ability to interact with them one on one and, hopefully, have a positive effect on their decision making and attitudes.

Q: What parts of your job do you like the least?

A: Sometimes, working in the court "system" can be frustrating because status offenders are at the bottom of the intervention ladder. It's hard when I see kids with needs that cannot be met due to the fact that they are not "bad" enough. In reality, we don't want them to get "bad" enough; but, without help, they might get there!

Q: What special skills or abilities do you believe you bring to the job that help you in working with these difficult and troubled youths?

A: First of all, I'm good at listening to these kids and communicating with them at their level. This really helps me identify their needs. I have a good grasp of the available community resources, which I can then match to their needs.

Q: What are your major day-to-day challenges?

A: Dealing with school systems that often take a zero tolerance attitude and can be rigid in their approach to kids. We need to be more creative by working together and looking at individual children's situations. My second biggest challenge is parents who don't discipline their children, who don't hold them accountable. If kids don't learn accountability and responsibility at home, it's difficult for society to teach it to them. Even formal court action and incarceration don't always help kids to understand that they must take individual responsibility for their actions. Parents who fail in this regard fail big time!

Q: Who do you report to?

A: The director of intake.

Q: What is your compensation?

A: About $43,000 a year.

Q: What qualities or skills would you recommend that a young person interested in the field of juvenile justice in general and in working with status offenders in particular should possess?

A: First, you must like kids as people and not be in any way judgmental. Leave that to the judges and referees. You must be creative in looking at alternatives to help kids and find creative ways of holding them accountable. You must like people in general and be systems wise. You need to understand how the court system interconnects with the welfare system, the school systems, and other agencies.

(*continues*)

Q: What is your "wish list" for effective intervention with status offenders?

A: More than anything, our community needs two group homes: one for boys and one for girls. They need a place where they can feel safe, get a timeout, and get supportive counseling to address home and school problems; but they should not be intermingled with criminal offenders.

Q: Do you believe that courts should be involved with status offenders?

A: Yes, I believe that courts should be involved with status offenders. Early intervention can prevent later serious delinquency in my experience. Kids also need to know that they will be ultimately responsible to someone, and sometimes that someone needs to be the court. When they won't listen to their parents or me, they still need to listen.

Q: Is there anything else that you want our readers to know?

A: I like my job, even after all this time. Teenage status offenders are quite a challenge, and every day brings a new one. The successes are important and lead to life-long friendships.

criminal offenses. Yet, there is also a population of youths who tend to be exclusively involved in status offense behaviors, even though chronic status offenders are rare. Consequently, status offenders are more appropriately seen as constituting a diverse category of juvenile offenders. Moreover, many of these youths require different responses than those that are typically used for youths involved in criminal offenses. Indeed, a number of studies have indicated that various populations of status offenders have a variety of social and psychological needs to be addressed.

Although at least some status offenders appear to constitute a unique group of juvenile offenders, there are some practical and philosophical problems with using the coercive power of the court to intervene in their lives. In many instances, youths before the court for status offenses are responding to disruptive and unhealthy family relationships. Given the nature of these problems, it is reasonable to question the ability of the juvenile court to resolve them.

Ultimately, a juvenile court must rely on its coercive powers as a legal institution and the expertise of other community agencies when it tackles family problems. Also, the efforts of a juvenile court to effect change in a family may feel the same to the child, whether they are made under the guise of protection or correction. From the child's perspective, removal from a disruptive family environment and placement in detention, foster care, or some other out-of-home placement may still be perceived as punishment, regardless of the court's reasons for removal. By using the power of the court to intervene in their lives and attempt to change behavior by sanctions, the state, at least philosophically, treats both delinquents and status offenders the same. Although the

COMPARATIVE FOCUS

Many European Countries Do Not Recognize Status Offenses

Many European countries, such as the Netherlands, Germany, and Sweden, do not recognize status offenses. Instead, youth behaviors, such as truancy, running away, incorrigibility, and alcohol use, are considered to be problem behaviors that are dealt with through child welfare institutions.[64]

MYTH VS REALITY

Juvenile Courts Have Limited Resources

Myth—Juvenile courts can use their enormous power to coerce families into solving their problems.

Reality—Juvenile courts possess limited methods and resources for resolving family problems. Although juvenile courts may play a key supportive role in helping families resolve their problems and can act as service brokers, they are not in a position to provide the types of interventions that families often require.

Juvenile Justice and Delinquency Prevention Act has discouraged the incarceration of status offenders, nothing prohibits the states from using a range of responses to treat status offenders, including probation, foster care placement, and institutionalization. In these circumstances, the distinction between status offenders and delinquent youths delineated in the *Walker* case becomes a distinction with no difference. Unfortunately, many status offenders are still treated like youths charged with criminal violations.

Attempts by juvenile courts to intervene in status offense cases, while clearly within their legal authority, are understandable, given the *parens patriae* focus of the courts. Indeed, many juvenile court decision makers would like to help families that are facing problems. As noted in earlier chapters, youths who come to the attention of juvenile justice agencies often face a variety of problems. They are typically poor and have been unsuccessful in school. Many come from neglectful, abusive, and dysfunctional homes. Not surprisingly, many of these youths are in conflict with their parents and display complete disregard for parental and adult authority.

Conflict between children and their parents often leads to involvement with the juvenile court under four conditions. First, parents may seek relief from the court in their effort to establish parental authority. Second, other authority figures, such as school personnel and police, often come into contact with these youths and seek assistance or punishment from the court. Third, when these youths are out on the streets with no visible means of support, they often resort to criminal activities, ranging from prostitution to robbery, which may lead to juvenile court involvement. Fourth, the juvenile court serves as the social service agency of last resort for the poor, who often lack other resources (e.g., counseling and treatment services) that might help them resolve their problems. The juvenile court becomes the default agency within the community and is forced to try to resolve a range of family and community problems that occur in poor neighborhoods. However, the court, lacking sufficient resources and expertise, will often be unable to resolve these problems, which sets the stage for more coercive responses to status offenders.

Factors Influencing Juvenile Justice Responses to Status Offenders

Under the legal philosophy of *parens patriae*, which has traditionally guided juvenile justice practice, there has been little reason to distinguish between status and criminal (delinquent) offenders. Moreover, any distinction that juvenile courts have made between these two groups has typically had no discernible effect on juvenile court practice. Given the lack of concern with due process for children, juvenile courts have often intervened without giving much thought to how the youths it deals with came to court. The result has frequently been similar treatment for status and delinquent offenders.

Despite the history of similar treatment, juvenile courts and legislatures have begun to take a much different approach to the status offender in some jurisdictions. Perhaps the most important factor motivating change in this area is the pressure on juvenile courts to deal with increasing numbers of serious juvenile offenders. In response to this pressure, courts have directed resources away from status offenders, and many courts no longer accept parental or school complaints as the sole basis for court intervention.

Another factor influencing the courts' handling of status offenders is the Juvenile Justice and Delinquency Prevention Act. Under the act, states have been encouraged to divert status offenders from the juvenile justice process. They have also been encouraged to refrain from placing them in detention or correctional facilities or risk losing JJDP Act grant funds.[65] As a result, almost every state has enacted laws designed to implement the mandates of the Juvenile Justice and Delinquency Prevention Act, such as the deinstitutionalization of status offenders and the removal of status offenders from adult jails.[66] Similarly, many juvenile courts have begun to examine a range of diversionary responses to status offenders.

Still another factor is the recognition by many juvenile justice practitioners, including judges, that juvenile courts lack the resources and expertise to successfully intervene in many status offense cases. This recognition has led courts to call for other community agencies to play a more prominent role in handling child and family problems within the community. Although courts understand that some status offenders may eventually become involved in more serious types of delinquent activity, many courts refuse to assume formal jurisdiction over them until they do. The tendency to avoid intervention in the lives of status offenders has no doubt been partly fostered by the advocacy of various groups such as the American Bar Association Juvenile Justice Standards Project, which has even proposed the elimination of general juvenile court jurisdiction over status offenders.[67]

Effective Interventions with the Status Offender

Juvenile justice agencies, including juvenile courts, have supported a wide range of responses to status offenders. Historically, these responses have ranged from institutionalization to avoidance of involvement in status offense cases. To a large extent, the range of responses reflects the ambivalence juvenile justice agencies display toward these often difficult cases. Although more juvenile courts might be inclined to intervene in these cases if they were confident of success, they must confront the reality that intervention in status offense cases is often ineffective. Indeed, in some instances, juvenile justice intervention is actually harmful.

Although juvenile justice agency involvement in status offense cases can hardly be called an unmitigated success, there are some indications that effective policies geared toward status offenders are possible. One policy that has received some support is decriminalization,[68] the removal of legal prohibitions against status offenses from juvenile codes, or *divestiture*, the termination of the juvenile court's jurisdiction over status offense behaviors. The goal of decriminalization or divestiture is to keep status offense cases from falling under juvenile court jurisdiction. This approach would have some obvious advantages for juvenile justice agencies. By removing such cases from the juvenile justice process, concerns about ineffective or harmful juvenile justice responses would be alleviated.

MYTH VS REALITY

Supporters of Decriminalization or Divestiture Do Not Believe That Children's Problems Should Be Ignored

Myth—Support for the decriminalization of status offenses or divestiture would essentially mean turning our back on child and family problems.

Reality—Many supporters of decriminalization or divestiture recognize that children and families need assistance. However, they believe that child and social welfare agencies are the agencies that are best equipped for dealing with many of the problems presented by children and families.

Those who favor continuing juvenile court involvement in status offense cases contend that the effect of discontinuing court involvement would be to leave the real problems that confront families and communities around the country unaddressed. However, advocates of continued intervention in status offense cases do not always agree on how much intervention is appropriate. Some support more limited forms of juvenile justice intervention, such as diversion.

Programs designed to divert status offenders from the formal juvenile justice process appear to have met with some success in a number of jurisdictions. For example, a program developed in Sacramento, California, that was intended to divert status offenders from detention and to avoid subsequent referrals to the juvenile court has been found to be moderately successful.[69] This program, called the Sacramento County 601 Diversion Project, began operation in 1970. It focused on status offenders who were referred to the juvenile court by police, parents, and school officials. These youths were typically white, poor, female, younger than 15 years of age, and were referred to the program because of family problems.

When a youth was referred to the court, specially trained probation officers would contact the juvenile's parents and request that they come to Juvenile Hall for an immediate counseling session. When the parents arrived at Juvenile Hall, a probation officer read the juvenile his or her Miranda warning and explained that participation was voluntary. The officer indicated, however, that if the juvenile did not wish to participate in an immediate counseling session, his or her case would be referred to the court. The officer added that if the juvenile agreed to participate in a counseling session, he or she could return for as many as five sessions. In most cases, the parents and the juvenile agreed to participate. After the juvenile waived his or her rights, a counseling session began in which the probation officer used family intervention techniques. The counseling techniques were intended to get the family to look at the problem as a family problem that should be addressed by the entire family rather than by simply focusing attention on the youth.

The Sacramento County 601 Diversion Project was evaluated using an experimental design. The results indicated that the project was able to significantly reduce the number of youths being detained and petitioned to court. In addition, a follow-up of youths 12 months after their involvement in the project found that 46% of those who had gone through the project had another court referral and that 22% of these youths were referred to the court for a criminal offense. In comparison, 54% of youths handled by regular probation officers were subsequently referred to court, and almost 30% of those were referred for a criminal offense.[70]

Given the apparent success of the Sacramento County 601 Diversion Project, other jurisdictions in California developed similar programs and began to evaluate them in the mid-1970s. These programs were similar to the Sacramento project in a number of ways. For example, they substituted short-term family and individual counseling for detention and referral to juvenile court. However, there were also some important differences. For example, more of the youths involved in these projects were male, Hispanic, and African American, and a number of these youths were diverted after committing criminal offenses. The evaluators of these projects employed quasi-experimental designs and found that those youths who were diverted were arrested slightly less often during a six-month follow-up period than similar youths who were not diverted.[71]

school within a school
This consists of a smaller grouping of students and teachers within a school. The group may have a unique curriculum and may operate on a different schedule from other students in the school.

A truancy reduction program that has been found to produce positive effects on truancy is the career academy. Career academies consist of a **school within a school** where students and teachers work in "small learning communities" for three to four years during high school. Each community is designed to develop a close personalized relationship between students and teaching staff. In addition, the career academy integrates academic and vocational work and strives to provide students with a range of career development and work-based skills through partnerships with local businesses. An evaluation of one program revealed significant improvements in attendance, increases in academic course taking, and reductions in dropping out among youths at high risk for dropping out. Few significant effects, however, were found among youths at low risk of dropping out.[73]

Another truancy reduction program that has produced positive evaluation results is the Chronic Truancy Initiative (CTI). CTI is designed for elementary-age students and consists of a series of progressive interventions targeted at chronic truants. The interventions begin with a letter to the parents. If this does not produce an improvement in attendance, additional interventions consist of a referral to a school attendance officer, referral to a social service agency, then a home visit by a uniformed police officer and the attendance officer. If these interventions are not successful, the family may then be referred to the juvenile court. An evaluation of the program indicated that improvements in truancy were found after each of the first two intervention stages. Significant improvements, however, were not noted after the final two interventions.[74]

In addition to the previous interventions, programs designed to improve the family and social adjustment of "antisocial" youths and status offenders have also shown some promise.[75] For example, programs designed to train parents and teachers to employ child management techniques at school and in the home have resulted in reductions in problem behaviors within educational and family settings.[76] One approach that emphasizes working with the child's family in an effort to reduce problem behaviors is

FYI

Truancy Is the Most Common Status Offense Handled by Juvenile Courts

In 2005, truancy accounted for 35% of the status offense cases formally handled by courts, followed by liquor law violations (19%), ungovernability (15%), running away (14%), and curfew offenses (9%). Miscellaneous offenses accounted for the remaining cases.[72]

multisystemic therapy. In this approach, counselors attempt to collaborate with the family in developing treatment goals and plans intended to help the family meet those goals. Treatment may also involve working directly with uncooperative family members, teachers, school administrators, and others in an effort to actively involve them in treatment. An important aspect of multisystemic therapy is that it focuses on identifying strengths and problem areas within various social systems, such as the family, school, workplace, and community, and encouraging the family to take action to resolve problems rather than just talking about them.[77] Although this approach has been used effectively with serious juvenile offenders, it appears to offer considerable promise for use with status offenders as well.[78]

Legal Issues

A fundamental question in juvenile justice is whether juvenile courts should exercise jurisdiction over juveniles who are only status offenders. As mentioned earlier, court involvement in the lives of status offenders is clearly justified under the doctrine of *parens patriae*, but is it appropriate or even feasible for the juvenile courts to take responsibility for all "troubled children"? Of course, some children who are chronically disobedient to parents, teachers, and other authority figures and who are continually running away from home will engage in more serious forms of delinquency over time and will become victims of crime. Moreover, because juvenile courts serve as social service agencies of last resort for many poor people, they will continue to feel pressure to respond to these cases.

There are, however, strong arguments against formal court involvement. Certainly one of the most compelling is that juvenile courts should focus their time and limited resources on the most seriously delinquent youths in order to protect the community. Devoting scant court resources, such as caseworkers' time or limited bed space, to deal with status offenders who are, at most, only dangerous to themselves squanders limited resources. Furthermore, juvenile court efforts to resolve the problems underlying status offenses are often ineffectual. In order to be truly effective, the courts need a range of resources and therapeutic techniques that are often unavailable to them. In view of these introductory arguments, what the courts have said about the legal issues involving status offenders needs to be explored. We will begin by examining the 1972 North Carolina Supreme Court case *In the Matter of Walker*.[79]

In re Walker

Facts

On August 2, 1971, Mrs. Katherine Walker, the mother of Valerie Lenise Walker, filed a petition alleging that Valerie was an "undisciplined" child. The petition alleged that Valerie was younger than 16 years of age and regularly disobedient to her parents, in that she came and went without permission, kept late hours, associated with persons objected to by her parents, and went places where her parents told her not to go. Mr. and Mrs. Walker had seven children, both parents worked, and Valerie and the children were left home during working hours. Valerie was often away when her parents got home from work. At the time of the filing of the petition, Valerie was 14 years old. A hearing was held on the petition. Valerie and her mother were present, along with a court counselor. Valerie did not have an attorney. Valerie was found to be an "undisciplined" child and placed on probation.

On September 21, 1971, Ann Jones, the court counselor (and apparently a probation officer), filed another petition concerning Valerie. This petition asked that the case be reconsidered, apparently because of problems at school. At this hearing, Valerie was represented by a public defender, who moved to vacate the original findings because Valerie did not have an attorney at the previous hearing. That motion was denied, as was a motion that status offense jurisdiction violates the Equal Protection Clause of the Fourteenth Amendment to the U.S. Constitution. In addition, after the hearing, the judge determined that Valerie was now a delinquent child and committed her to the North Carolina Board of Juvenile Correction for out-of-home placement.

Court Decision

There were two issues raised before the court: (1) Did Valerie have a constitutional right to an attorney at the initial hearing that resulted in her coming under the court's jurisdiction? (2) Does the status of being an "undisciplined" child violate the Equal Protection Clause because it subjects the child to probation and the potential risk of removal from home and incarceration when the child has committed no criminal offense (whereas adults are subject to these sanctions only after a criminal conviction)?

In addressing the first issue, the court said the following:

Appellant would have this court go further than *Gault* requires. She argues for the right to counsel at the hearing of an "undisciplined child" petition on the theory that such a hearing is a critical stage in the juvenile process since it subjects the child to the risk of probation and since a violation of probation means the child is "delinquent" and subject to commitment. In such fashion, appellant seeks to engraft upon the juvenile process the "critical stage" test used by the United States Supreme Court in determining the scope of the Sixth Amendment right to counsel in criminal prosecutions. We find no authority for such engraftment. Whatever may be the proper classification for a juvenile proceeding in which the child is alleged to be undisciplined, it is certainly not a criminal prosecution within the meaning of the Sixth Amendment which guarantees the assistance of counsel.[80]

In addressing the equal protection argument, the court said the following:

The purpose of the Juvenile Court Act is not for the punishment of offenders but for the salvation of children..... The Act treats delinquent children not as criminals, but as wards, and undertakes ... to give them the control and environment that may lead to their reformation, and enable them to become law-abiding and useful citizens.... The state must exercise its power as parens patriae to protect and provide for the comfort and well-being of such of its citizens as by reason of infancy ... are unable to take care of themselves.[81]

The court went on to say the following:

The classification here challenged is based on differences between adults and children; and there are so many valid distinctions that the basis for challenge seems shallow. These differences are reasonably related to the purposes of the Act.... [It] is our view that the desire of the state to exercise its authority as parens patriae and provide for the care of children supplies a "compelling rational" justification for the classification.[82]

Although the *Walker* case is over 20 years old, it clearly and forcefully reiterates the *parens patriae* philosophy of the juvenile court. The North Carolina Supreme Court chose to maintain the distinction between the juvenile court and adult court and, consequently, to limit juveniles' right to due process protections in status offense cases. In many respects, the Walker case reflects the continuing debate over the legal rights of children, a debate also reflected in the case of *State ex rel, Harris v. Calendine.*[83]

State ex rel, Harris v. Calendine

Facts

On April 9, 1976, Gilbert Harris, a 15-year-old, was petitioned to the juvenile court for irregular school attendance. At the time of the petition, he had missed 50 days of school. He and his mother and stepfather were summoned to appear in the Calhoun County (West Virginia) Juvenile Court, and a hearing was held on May 17, 1976. At that hearing, Gilbert appeared with his mother and an attorney. He did not contest the allegations and was adjudicated a delinquent child and committed to an industrial school for boys. The legal action that brought this case to appeal was based upon a petition for ***habeas corpus*** due to his being incarcerated for almost one year after his 16th birthday, which was in July 1976, and in view of the fact that at 16 years of age school attendance was no longer required in West Virginia.

habeas corpus Literally, that you have the body. The term refers to a writ or petition to a court requesting review of a person's incarceration to determine whether it is proper.

Court Decision

The habeas corpus action was based on a broad allegation that the West Virginia Juvenile Code was inherently unconstitutional because it combined status offenders and criminal offenders. The Supreme Court of Appeals of West Virginia did not determine the statute to be unconstitutional, but set forth parameters that limited the application of parts of the law. The court began by saying, "The statutes under consideration, in the absence of guidelines for their application, fail to meet the equal protection, substantive due process, and the cruel and unusual punishment standards because they permit the classification and treatment of status offenders in the same manner as criminal offenders."[84]

The court further said the following:

We are, however, concerned with incarceration of children for status offenses.... The Legislature has vested the juvenile court with jurisdiction over children who commit these status offenses so that the court may enforce order, safety, morality, and family discipline within the community. The intention of the law is laudable; however, the means employed to accomplish these ends are unconstitutional insofar as they result in the commitment of status offenders to secure prison-like facilities which also house children guilty of criminal conduct, or needlessly subject status offenders to the degradation and physical abuse of incarceration.[85]

The court ended by stating the following:

Finally, it should be noted that status offender legislation discriminates invidiously against females.... Our society tends to condemn female promiscuity more severely than male promiscuity, and this tendency may explain why females often are unfairly classified and treated as status offenders.... Furthermore the court finds no rational connection between the legitimate legislative purpose of enforcing family discipline, protecting children, and protecting society from uncontrolled children and the means by which the state is permitted to accomplish these purposes, namely incarceration of children in secure, prison-like facilities. It is generally recognized that the greatest colleges for crime are prisons and reform schools.[86]

In the *Calendine* ruling, the West Virginia Supreme Court of Appeals recognized that status offenders were as likely to be harmed by treatment as helped. Removal from home and family was to be a last resort, for use after all other alternatives were exhausted; and even then, placement could not be in a facility housing delinquent youths. Importantly, the court required the state to view status offenders as a separate group from delinquents and provide for them accordingly.

Although there are good arguments for limiting the types of responses available to juvenile courts or even removing status offenders from juvenile court jurisdiction, such options leave unaddressed the very real problems that children and families experience. Clearly, some families are seriously troubled and need assistance. Likewise, some status offenders are seriously troubled and need help. Moreover, leaving decisions for assistance up to the family may not be feasible, nor is it always in the best interests of the child or the community. Often, courts wait until a status offender has committed a delinquent offense before actually intervening, yet intervention at an earlier stage may prevent the child from engaging in criminal behavior and obtaining a delinquent or criminal record.

A similar issue concerns children who suffer from some form of mental disability that causes them to act out in delinquent ways or behave like a status offender. Most mental health services for children must be obtained for them by their parents and are in large part voluntary. But what if parents do not seek such help, and what if youths resist their parents' efforts to provide assistance? Should mentally challenged children be compelled to get treatment? In reality, many never get the help they need until they commit a delinquent act and are brought to the attention of the local juvenile court. In many communities, services for children who experience serious mental disturbance are severely limited or nonexistent. With seriously ill children, effective care often waits for juvenile court intervention.

Chapter Summary

As noted in Chapter 3, many families face obstacles that challenge their ability to provide a healthy environment for their children. This, of course, is not a new problem. Indeed, one of the primary reasons for the development of the juvenile court was to establish a legal mechanism for protecting children and controlling a host of predelinquent behaviors, such as incorrigibility, running away from home, and failing to attend school.[87] The result has been a court that has two, often conflicting, roles. On the one hand, the juvenile court is expected to control youths and protect the community, which often requires the court to take coercive measures. On the other hand, the needs of youths often require the court to act as a nurturing and helpful parent.[88]

Historically, juvenile court intervention into the lives of status offenders has been supported for a variety of reasons. Considerable pressure has been placed on juvenile courts to support parental authority, help families deal with vexing problems, and check juvenile behaviors that are believed to be precursors to more serious criminality. The responses of juvenile courts have ranged from inaction to detention and institutionalization.

Not all of the interventions employed by juvenile courts have been effective, and indeed some have produced questionable results. Some have even been patently harmful. Part of the explanation for this is that the juvenile courts have not always recognized that some status offenders have different needs than youths who have committed criminal offenses.

In an effort to limit some of the more harmful responses to status offenders, Congress passed the Juvenile Justice and Delinquency Prevention Act of 1974, which, along with subsequent amendments to that act, discouraged states from incarcerating status offenders. As a result of this federal effort, considerable progress in the deinstitutionalization of status offenders has been made in some states. Despite this progress, however, status offenders are still incarcerated in some jurisdictions.

To a large extent, the continuing involvement of juvenile courts in status offense cases reflects the historical conflict inherent in the *parens patriae* doctrine upon which the juvenile courts were founded. Juvenile courts often feel an obligation to control or help children and their families, but they must confront the reality that they do not always possess the resources or expertise needed to intervene effectively. In some jurisdictions, the realization that juvenile courts are sometimes ill equipped to deal with status offense cases has led to policies preventing court involvement in these cases. In other jurisdictions, juvenile courts continue to intervene, and there is evidence that effective interventions for some troubled families exist. The key for the juvenile court is to determine to what extent it can play a role in helping families and children, and at the same time avoid reliance on coercion and incarceration in those cases in which its efforts to help fail, or consider how it might support the efforts of child welfare agencies to effectively deal with the complex problems presented by many status offenders.

Key Concepts

cafeteria-style approach to delinquency: The tendency of a youth to engage in a variety of status and criminal law-violating behaviors over time.

CHINS: Children in need of supervision (also abbreviated as CINS). This term is used to refer to status offenders in some states.

cruel and unusual punishment: Punishment handed down by a court that is too harsh for the offense committed. The Eighth Amendment of the U.S. Constitution explicitly prohibits the imposition of cruel and unusual punishment.

decriminalization: The process of removing proscriptions from legal codes (in the context of juvenile justice, the process of removing status offenses from the juvenile code).

deinstitutionalization: The removal of offenders from institutional settings. In this chapter, the term is used to refer to the removal of status offenders from detention centers and other secure institutional settings.

equal protection under the law: The constitutional requirement that all laws be applied to all persons fairly and without regard to race, creed, color, national origin, sex, or age. The Fourteenth Amendment of the U.S. Constitution prohibits denial of equal protection of the laws to anyone.

***habeas corpus*:** Literally, that you have the body. The term refers to a writ or petition to a court requesting review of a person's incarceration to determine whether it is proper.

***parens patriae*:** The doctrine that the state has a legal obligation to provide for the care, protection, treatment, and rehabilitation of children if the children's parents are unable to do so.

PINS: Persons in need of supervision. This term is used to refer to status offenders in some states.

school within a school: This consists of a smaller grouping of students and teachers within a school. The group may have a unique curriculum and may operate on a different schedule from other students in the school.

status offender: A youth who comes within the jurisdiction of a juvenile court for a status offense rather than a delinquent or criminal offense. Alternative terms include *children in need of supervision* (*CINS* or *CHINS*), *undisciplined children*, *incorrigible youths*, and *persons in need of supervision* (*PINS*).

status offense: "Any offense committed by a juvenile that would not be a crime if committed by an adult, according to the statutes or ordinances of the jurisdiction in which the offense was committed, and which is specifically applicable to juveniles because of their status as a juvenile." Status offenses consist of two types: (1) acts and behaviors only possible for juveniles, such as incorrigibility, running away, and truancy, and (2) acts and behaviors that are legal for adults but are illegal for juveniles, such as smoking cigarettes, drinking alcohol, and sexual intimacy.

Review Questions

1. Should juvenile courts exercise jurisdiction and authority over status offenders?
2. Does society have a "compelling interest" in the control of status offenders?
3. Should status offenders be detained for status offenses? Should juveniles be detained for running away from home?
4. Are status offenders similar enough to juvenile delinquents that they should be handled by the same system?
5. How does early court intervention at the status offending stage affect a youth's chances of becoming a delinquent?
6. Could juvenile court intervention contribute to the eventual delinquency of status offenders?
7. Should status offenders be afforded the same legal rights at juvenile court hearings as delinquent offenders?
8. What alternative options are available to the community for addressing the issue of status offenses?
9. What are the arguments for and against the incarceration of status offenders?
10. Is it true that exposing status offenders to delinquent youths will cause the status offenders to become delinquents?
11. Where do mentally ill children fit in the spectrum between status offenders and delinquents?
12. How should we handle youths when status offending or delinquent acting out is the result of mental illness?

Additional Readings

Gendreau, P. & Ross, B. (1979). Effective correctional treatment: Bibliotherapy for cynics. *Crime and Delinquency*, *25*, 463–489.

Kelly, T. M. (1983). Status offenders can be different: A comparative study of delinquent careers. *Crime and Delinquency*, *29*, 365–380.

Kobrin, S., Hellum, F. R., & Peterson, J. W. (1980). Offense patterns of status offenders. In D. Shichor & D. H. Kelly (Eds.), *Critical issues in juvenile delinquency.* Lexington, MA: Lexington Books.

Logan, C. H. & Rausch, S. P. (1985). Why deinstitutionalizing status offenders is pointless. *Crime and Delinquency, 31,* 501–517.

Lundman, R. J. (2001). *Prevention and control of juvenile delinquency* (3rd ed.). New York: Oxford University Press.

Shelden, R. G., Horvath, J. A., & Tracy, S. (1989). Do status offenders get worse? Some clarifications on the question of escalation. *Crime and Delinquency, 35,* 202–216.

Notes

1. Marcelli, R. J., Foster, J., & White, J. L. (1977). *Juvenile facilities: Functional criteria* (p. 41). Lexington, KY: Council of State Governments.
2. Weis, J. G. (1980). Jurisdiction and the elusive status offender: A comparison of involvement in delinquent behavior and status offenses. *Assessment Center series* (p. vii). Washington, DC: Office of Juvenile Justice and Delinquency Prevention.
3. Weis, 1980.
4. Weis, 1980.
5. Rubin, H. T. (1985). *Juvenile justice: Policy, practice, and law* (2nd ed.). New York: Random House.
6. Rubin, 1985.
7. Weis, 1980.
8. Rubin, 1985.
9. Rubin, 1985.
10. Rubin, 1985.
11. Rubin, 1985.
12. Rubin, 1985.
13. Gough, A. R. (1980). Beyond-control youth in the juvenile court—The climate for change. In H. T. Rubin (Ed), *Juveniles in justice: A book of readings.* Santa Monica, CA: Goodyear Pub. Co.
14. Thomas, C. W. (1976). Are status offenders really so different? A comparative and longitudinal assessment. *Crime and Delinquency, 22,* 438–455.
15. Rubin, 1985.
16. Gough, 1980.
    Weis, 1980.
17. Gough, 1980.
18. Rubin, 1985.
19. Rubin, 1985.
20. Thomas, 1976.
    Rubin, 1985.
21. Gough, 1980.
22. Rubin, 1985.
23. Rubin, 1985.

24. Bazelon, D. L. (1970). Beyond control of the juvenile court. *Juvenile Court Journal, 21*, 44.

25. Rubin, 1985.

26. Rubin, 1985.

27. Rubin, H. T. (2003). *Juvenile justice: Policies, practices, and programs.* Kingston, NJ: Civic Research Institute.

28. Puzzanchera, C. & Sickmund, M. (2008). *Juvenile court statistics, 2005.* Pittsburgh, PA: National Center for Juvenile Justice.

29. Janson, C. (2004). Youth justice in Sweden. In M. Tonry and A. N. Doob (Eds.), *Youth crime and youth justice: Comparative and cross-national perspectives* (Vol. 31). Chicago: University of Chicago Press.

30. Platt, A. M.(1977). *The child savers: The invention of delinquency* (2nd ed.). Chicago: University of Chicago Press.

31. Rubin, 1985.

32. Handler, J. (1965). The juvenile court and the adversary system: Problems of function and form. *Wisconsin Law Review, 7.*

33. Platt, 1977, p. 138.

34. Senna, J. J. & Siegel, L. J. (1992). *Juvenile law: Cases and comments* (2nd ed.). St. Paul, MN: West Publishing Co.

35. Bernard, T. J. (1992). *The cycle of juvenile justice.* New York: Oxford University Press.

36. Weis, 1980.

37. Andrews, R. H., Jr. & Cohn, A. H. (1974). Ungovernability: The unjustifiable jurisdiction. *Yale Law Journal, 83*, 1383.

38. N. C. Gen. Stat., Chapter 7A-289.6 (1988).

D.C. Code 1981, Sec. 16-2301 (8) (A) (I)-(iii).

N.Y. Family Court Act, Sec 712 (b) (McKinney Supp 1978).

Mich. Comp. Laws 712A.2a (1), (2), (3), & (4).

39. Deuteronomy 21:18–21.

40. Platt, 1977.

41. Platt, 1977.

42. Senna & Siegel, 1992, p. 1.

43. Bernard, 1992.

44. Miller, J. G. (1991). *Last one over the wall: The Massachusetts experiment in closing reform schools.* Columbus, OH: Ohio State University Press.

45. Holden, G. A. & Kapler, R. A. (1995). Deinstitutionalizing status offenders: A record of progress. *Juvenile Justice, 2*(2), 3–10.

46. Sanborn, J. B., Jr. & Salerno, A. W. (2005). *The juvenile justice system: Law and process.* Los Angeles: Roxbury.

47. Holden & Kapler, 1995.

Logan, C. H. & Rausch, S. P. (1985). Why deinstitutionalizing status offenders is pointless. *Crime and Delinquency, 31*, 501–517.

48. *In re Darlene C.*, 278 S.C. 644, 301 S.E. 2d 136 (1983).

49. Mnookin, R. H. & Weisberg, D. K. (1995). *Child, family and state: Problems and materials on children and the law* (3rd ed.). Boston: Little, Brown & Co.

50. *W.M. v. State of Indiana*, 437 N.E. 2d 1028 (1982).

51. Holden & Kapler, 1995.

52. Sickmund, M., Sladky, T. J., Kang, W., & Puzzanchera, C. (2008). *Easy access to the census of juveniles in residential placement.* Retrieved January 27, 2009, from http://ojjdp.ncjrs.gov/ojstatbb/ezacjrp/.

53. *In the Matter of Walker*, 282 N.C. 28, 191 S.E. 2d 702 (1992).

54. Kobrin, S., Hellum, F. R., & Peterson, J. W. (1980). Offense patterns of status offenders. In D. Shichor & D. H. Kelly (Eds.), *Critical issues in juvenile delinquency.* Lexington, MA: Lexington Books.

Erickson, M. L. (1979). Some empirical questions concerning the current revolution in juvenile justice. In L. T. Empey (Ed.), *The future of childhood and juvenile justice.* Charlottesville, VA: University of Virginia Press.

Klein, M. W. (1971). *Street gangs and street workers.* Englewood Cliffs, NJ: Prentice-Hall.

Thomas, 1976, pp. 438–455.

55. Kobrin, Hellum, & Peterson, 1980.

56. Thomas, 1976.

57. Thomas, 1976.

58. Snyder, H. N. (1988). *Court careers of juvenile offenders.* Washington, DC: Office of Juvenile Justice and Delinquency Prevention.

59. Clarke, S. H. (1975). Some implications for North Carolina of recent research in juvenile delinquency. *Journal of Research in Crime and Delinquency, 12*, 51–60.

60. Kelly, T. M. (1983). Status offenders can be different: A comparative study of delinquent careers. *Crime and Delinquency, 29*, 365–380.

61. Shelden, R. G., Horvath, J. A., & Tracy, S. (1989). Do status offenders get worse? Some clarifications on the question of escalation. *Crime and Delinquency, 35*, 202–216.

62. Kobrin, Hellum, & Peterson, 1980.

63. Zhang, D., Katsiyannis, A., Barrett, D. E., & Willson, V. (2007). Truancy offenders in the juvenile justice system: Examinations of first and second referrals. *Remedial and Special Education, 28*, 244–256.

64. Albrecht, A. (2004). Youth justice in Germany. In M. Tonry and A. N. Doob (Eds.), *Youth crime and youth justice: Comparative and cross-national perspectives* (Vol. 31). Chicago: University of Chicago Press.

Kyvsgaard, B. (2004). Youth justice in Denmark. In M. Tonry and A. N. Doob (Eds.), *Youth crime and youth justice: Comparative and cross-national perspectives* (Vol. 31). Chicago: University of Chicago Press.

Janson, 2004.

Junger-Tas, J. (2004). Youth justice in the Netherlands. In M. Tonry and A. N. Doob (Eds.), *Youth crime and youth justice: Comparative and cross-national perspectives* (Vol. 31). Chicago: University of Chicago Press.

65. Holden & Kapler, 1995.

66. Brown, J. W. (1995). Beyond the mandates. *Juvenile Justice, 2*(2), 22–24.

67. American Bar Association Criminal Justice Section. (2007). ABA Youth at Risk Commission: Juvenile status offenders. *Policy Update, February 2007*, 4–5.

Shepherd, R. E., Jr. (1996). *JJ standards: Anchor in the storm*. Retrieved September 11, 2009, from http://www.abanet.org/crimjust/juvjus/cjstandards.html.Mnookin & Weisberg, 1995, p. 942.

68. Rausch, S. P. & Logan, C. H. (1983). Diversion from juvenile court: Panacea or Pandora's box? In J. R. Kluegel (Ed.), *Evaluating juvenile justice.* Beverly Hills, CA: Sage.

69. Lundman, R. J. (2001). *Prevention and control of juvenile delinquency* (3rd ed.). New York: Oxford University Press. This resource reviews the Sacramento 601 Diversion Project.

70. Baron, R. & Feeney, F. (1976). *An exemplary project: Juvenile diversion through family counseling.* Washington, DC: U.S. Department of Justice.

71. Palmer, T., Bohnstedt, M., & Lewis, R. (1978). *The evaluation of juvenile diversion projects: Final report.* Sacramento, CA: California Youth Authority.

72. Puzzanchera & Sickmund. 2008.

73. National Center for School Engagement. (2007). *Blueprints for violence prevention programs that reduce truancy and/or improve school attendance.* Denver, CO: National Center for School Engagement.

74. National Center for School Engagement, 2007.

75. Palmer, T. (1992). *The re-emergence of correctional intervention.* Newbury Park, CA: Sage.

76. Gendreau, P. & Ross, B. (1979). Effective correctional treatment: Bibliotherapy for cynics. *Crime and Delinquency, 25*, 463–489.

77. Henggeler, S. W. & Borduin, C. M. (1990). *Family therapy and beyond: A multisystemic approach to treating the behavior problems of children and adolescents.* Pacific Grove, CA: Brooks/Cole.

78. Henggeler, S. W., Melton, G. B., & Smith, L. A. (1992). Family preservation using multisystemic therapy: An effective alternative to incarcerating serious juvenile offenders. *Journal of Consulting and Clinical Psychology, 60*, 953–961. This resource provides an example of the use of multisystemic therapy with serious juvenile offenders.

79. *In the Matter of Walker*, 282 N.C. 28, 191 S.E. 2d 708 (1972).

80. *In the Matter of Walker*, 282 N.C. 28, 191 S.E. 2d 709 (1972).

81. *In the Matter of Walker*, 282 N.C. 28, 191 S.E. 2d 702 (1972).

82. *In the Matter of Walker*, 282 N.C. 28, 191 S.E. 2d 702 (1972).

83. *State ex rel, Harris v. Calendine*, 160 W. Va. 172, 233 S.E. 2d 318 (1977).

84. *State ex rel, Harris v. Calendine*, 160 W. Va. 172, 233 S.E. 2d 318 (1977).

85. *State ex rel, Harris v. Calendine*, 160 W. Va. 172, 233 S.E. 2d 324 (1977).

86. *State ex rel, Harris v. Calendine*, 160 W. Va. 172, 233 S.E. 2d 325 (1977).

87. Platt, 1977.

88. Smith, C. P., Berkman, D. J., Fraser, W. M., & Sutton, J. (1980). *A preliminary national assessment of the status offender and the juvenile justice system: Role conflicts, constraints, and information gaps.* Washington, DC: Office of Juvenile Justice and Delinquency Prevention.

89. N. C. Gen. Stat., Chapter 7A-289.6 (1988).
D.C. Code 1981, Sec. 16-2301 (8) (A) (I)-(iii).
N.Y. Family Court Act, Sec 712 (b) (McKinney Supp 1978)
Mich Comp Laws, 712A.2a (1), (2), (3) & (4).

90. Marcelli, Foster, & White, 1977.

Juvenile Justice and the Serious, Chronic, or Violent Juvenile Offender

CHAPTER OBJECTIVES

After studying this chapter, you should be able to

- Define violent juvenile offender, serious juvenile offender, and chronic juvenile offender and distinguish among these kinds of youths
- Describe the methods and outcomes of the Philadelphia cohort studies and the role they played in understanding the chronic juvenile offender
- Describe the extent of juvenile violence in the United States
- Describe the relationship between gender and violent juvenile crime
- Describe the relationship between race and violent juvenile crime
- Describe recent trends in violent juvenile offending
- Explain how the cultural, political, economic, community, peer group, family, and individual context influences juvenile violence
- Describe the characteristics of recent violence prevention efforts initiated around the United States
- Describe different types of juvenile justice responses to violent juvenile behavior

CHAPTER OUTLINE

- Introduction
- Defining the Violent Juvenile Offender
- Gender and Violent Juvenile Offending
- Race and Violent Juvenile Offending
- Trends in Violent Juvenile Offending
- Explaining Violent Juvenile Crime
- Responses to Violent Juvenile Crime
- Legal Issues
- Chapter Summary
- Key Concepts
- Review Questions
- Additional Readings
- Notes

Introduction

Reports of serious or violent crimes committed by juveniles are an all-too-common feature of contemporary urban life. In cities like Los Angeles, Chicago, Houston, Detroit, and New York City,[1] reports of gang-related violence and other serious forms of juvenile and adult crime are regular features of nightly news broadcasts and local newspapers. In many instances, the stories of violence have become routine. Occasionally a sensational event will occur, like the "Central Park jogger" incident, in which more than 30 teenagers, on a night of "wilding," allegedly attacked anyone they saw, including a 26-year-old female investment banker who was beaten with bricks and a lead pipe before being gang raped and left for dead. In December 1997, a 14-year-old high school student opened fire on classmates at a Paducah, Kentucky, high school, killing three and injuring five others. In Littleton, Colorado, two youths killed 13 people at Columbine High School. Although events such as these are uncommon, they undoubtedly symbolized what many perceive to be a stark reality—that no place is immune from violent juvenile crime. Moreover, the events that unfolded in Paducah and Columbine may have sparked even deeper fears in others—that violence is not relegated to urban minority youths, and that perpetrators and victims can be white and affluent. Indeed, reports by the media have regularly, even nightly, documented what some have referred to as an epidemic of teenage murder and violence. As former Secretary of Health and Human Services Louis Sullivan noted in a 1991 article published in U.S. News and World Report, "During every 100 hours on our streets we lose more young men than were killed in 100 hours of ground war in the Persian Gulf."[2] The sobering reality is that there is a significant amount of violent crime committed by youths in the United States and headlines like these give the impression that serious juvenile violence is a pervasive problem in the United States.

Serious and violent juvenile crime is a significant problem in the United States, and constant media reminders of juvenile violence ensure that the problem is brought to the attention of the public. Although juvenile violence is a problem, serious violence is more of a problem in certain communities than others. Moreover, overall levels of violent juvenile crime decreased substantially between 1994 and 2004, although they began to increase again in 2005.[3] Violent juvenile crime trend projections made during the early 1990s indicated that if the trend during that period persisted, juvenile arrests for violent crimes would double by the year 2010.[4] Other researchers, such as James Fox of Northeastern University, predicted an increase in violent juvenile crime throughout the 1990s based on increases in the number of poor teenagers who lacked adequate parental supervision and guidance.[5] Fortunately, those predictions did not turn out to be correct. Nevertheless, there are clear indications that juvenile violence has been increasing and is beginning once again to be a major social problem in some communities.[6] Moreover, further increases in juvenile crime, including violent juvenile crime, may be seen in the near future, depending on cultural, economic and political trends.[7]

This chapter examines the social and individual contexts of violent juvenile crime and the juvenile justice response to violent offenses. In addition, it considers several critical issues that are important for understanding violent juvenile crime and the juvenile justice response to it. For example, it defines the serious violent offender, explores the extent of juvenile violence in the United States, identifies recent trends in violent juvenile crime, and discusses recent policy proposals for reducing the levels of violent juvenile crime in the United States. The chapter ends by examining several important legal issues related to the processing of **violent juvenile offenders**.

violent juvenile offender
A juvenile who has committed a violent criminal offense against a person, such as rape, murder, aggravated assault, or robbery.

■ Defining the Violent Juvenile Offender

A variety of terms are used to refer to youths who engage in serious or violent juvenile crime. Indeed, it is not uncommon to hear terms such as *violent offender*, *chronic offender*, or *serious offender* used to denote youths who have committed offenses that result in bodily injury to others. Although these terms are used indiscriminately by the media and members of the public, they are not synonymous.

A **serious offense** can be directed at a person or at property. Juveniles do commit crimes such as murder, rape, and aggravated assault, which are clear examples of serious offenses involving violence. However, youths may also engage in serious offenses that do not cause bodily harm, but do considerable damage to or result in significant loss of property. Moreover, the victims (the property owners) may suffer significant financial and psychological harm as a result of these offenses.

serious offense An offense that results in significant personal injury, property loss, or damage.

A *chronic offender* engages in repeated acts of delinquency. These acts need not be seriously harmful or violent. Although many chronic offenders do engage in serious and violent juvenile crime, some chronic status offenders merely skip school, fail to obey their parents, or run away from home regularly. Such behaviors are not violent, nor do they represent a significant threat to others, although youths who run away may place themselves at risk for victimization. Similarly, some youths may engage in a number of less serious forms of delinquency, such as shoplifting or minor assaultive behavior. These forms of delinquency are not terribly serious as far as crime goes, although assaultive behavior is clearly violent in nature, even though no serious bodily harm results.

Research on chronic offending has been done since at least the 1930s, when Sheldon and Eleanor Glueck of Harvard University began conducting research on delinquent careers.[8] Their research led them to the conclusion that early involvement in delinquent behavior, as well as a variety of personal and family factors such as body type, personality, intelligence, the quality of discipline provided by parents, and emotional ties to parents, were good predictors of delinquent careers.[9] The term **chronic delinquent offender** was popularized later by Marvin Wolfgang, Robert Figlio, and Thorsten Sellin, as part of their landmark publication *Delinquency in a Birth Cohort* (see Chapter 2). After conducting a **cohort study** of 9,945 boys born in Philadelphia in 1945, Wolfgang and his associates discovered that about one-third (35%) of the youths in the cohort had some police contact between the time they were born and their 18th birthdays. They also found that about half (54%) of those youths had more than one police contact, and these youths were further categorized into two groups. Members of one of the groups were called *nonchronic recidivists*. These were youths who had more than one police contact but less than five. Members of the other group were called *chronic recidivists*. These were youths who had five or more arrests. Chronic recidivists, who comprised a mere 6% of the entire birth cohort, were responsible for a disproportionate share of all delinquent activity committed by the youths in the study. For example, chronic recidivists accounted for 52% of all arrests attributable to the cohort. An even more striking finding was that this 6% accounted for 71% of the homicides, 73% of the rapes, 82% of the robberies, and 69% of the aggravated assaults committed by members of the cohort.[10]

chronic delinquent offender A juvenile who, over time, accumulates a number of delinquency charges and convictions. The related term *chronic recidivist* was used by Wolfgang and colleagues to refer to juvenile offenders who had five or more police contacts.

cohort study A study that looks at a group of people born at a specific time in a particular location (a cohort).

Similar findings were produced in a subsequent study by Wolfgang, Figlio, and Paul Tracy. These researchers looked at a cohort of 27,160 males and females born in Philadelphia in 1958. Again, they found that chronic recidivists accounted for a disproportionate share of crime, including serious violent crime, attributable to the cohort. For example, of the males in the cohort, 7.5% were classified as chronic recidivists (i.e., they had five or more arrests). The chronic recidivists accounted for 61% of all arrests of cohort members, as well as 61% of the homicides, 76% of the rapes, 73% of the robberies, and 65% of the aggravated assaults.[11] Research conducted by Lyle Shannon, which looked at cohorts of youths born in Racine, Wisconsin, in 1942, 1949, and 1955, also found that a minority of youths were responsible for a disproportionate percentage of delinquent behavior.[12]

official data Data collected by juvenile justice agencies such as the police, courts, and correctional agencies.

Franklyn Dunford and Delbert Elliott analyzed data from the National Youth Survey (NYS), a large-scale study that employs self-reports of delinquent behavior, to understand criminal careers. Their research indicated that previous studies probably underestimated the actual size of the chronic offending population because they relied solely on **official data**. Dunford and Elliott defined serious career offenders as youths who had committed at least three felonies, such as aggravated or sexual assault, gang fighting, car theft, or strong-arm robbery, two or more years in a row. Their results indicated that only 24% of serious career offenders had ever been arrested.[13] Thus, it is likely that the chronic offender population is somewhat larger than the 6–8% of the juvenile population identified by Wolfgang and his colleagues, although the exact size of the population is not known.

violent offense An offense directed at the person or property of another that is intended to inflict injury on another and/or deprive them of their property. Violent offenses, however, do not always result in significant personal injury to the victim.[158]

The violent offender is still another distinct type of offender. In the broad sense, violence "is any act, event, or condition resulting from human behavior that causes death, injury, or some other damage to the physical or psychological well-being of one or more individuals."[14] However, although violence can be thought of as including actions that produce only psychological trauma, the common conception of violence is that it necessarily involves face-to-face interactions that produce physical harm or present the threat of physical harm. This concept of violence is reflected in conventional definitions of violent crime and, consequently, a range of behaviors, from homicide, robbery, and rape to simple assault, are considered **violent offenses**. Certainly, violent offenses can result in serious injury or even death. However, as the case of robbery illustrates, violent crimes do not always result in the physical injury of the victim because many robberies are carried out without anyone being hurt. Furthermore, violent offenses are not always serious offenses. For example, a simple assault can result from a schoolyard conflict, but all participants might escape injury.

Data on violent juvenile crime arrests indicate that violent juvenile crime increased substantially between 1988 and 1994 after being quite stable for the 15 years prior to 1988, but declined substantially between 1995 and 2004. Since 2004, however, it has been increasing once again (see Trends in Violent Juvenile Offending later in this chapter). Although juvenile crime has been decreasing, for many it continues to be at an unacceptably high level, particularly in some communities. Moreover, the overrepresentation of

MYTH VS REALITY

There Is a Small Population of Chronic Violent Offenders

Myth—Youths who engage in chronic delinquent behavior also engage in violent crime.

Reality—There is a small population of chronic violent offenders. However, some youths are merely chronic status offenders or engage in minor forms of delinquency and are not involved in violent or serious criminality.

minority youths, as well as an increase in the proportion of violent offenses attributed to females in recent years, raises important questions for policy makers and juvenile justice practitioners.[15]

Regardless of the trends in violent juvenile offending, it must be viewed within the total context of delinquent behavior. Even during the late 1980s and the first half of the 1990s, violent juvenile offending still accounted for only a very small proportion of juvenile arrests. For example, in 2007 only about 16% of arrests of persons younger than 18 years of age were for an Index violent offense.[16]

The relatively small size of the violent juvenile offender population has also been noted in research that has focused on the offense careers of violent delinquents. One early study was carried out by Donna Hamparian and associates at Ohio State University. This study, called the "Dangerous Offender Project," examined a cohort of 1,138 youths who were born in Franklin County (Columbus), Ohio, in 1956 and 1960 and had been arrested for at least one violent offense before reaching 18 years of age. The researchers found that only about 2% of the youths in Columbus had been arrested for a violent offense, and those who were arrested for a serious violent offense made up only about .5% of the cohort. The researchers also found that, although members of the cohort averaged about four arrests before reaching 18 years of age, very few of them were arrested for repetitive acts of violence.[17] A subsequent analysis of the same data noted that the chronic violent offender is rare, and that most violent juvenile crimes do not involve the use of weapons.[18]

Similar findings were brought to light by an analysis of the delinquent careers of more than 151,000 youths who turned 18 years of age between 1980 and 1995 in Maricopa County (Phoenix), Arizona. This research, conducted by Howard Snyder of the National Center for Juvenile Justice, found that 5 in 6 youths referred to the juvenile court for a violent offense were never referred again on a violent charge. Moreover, this same study found that the frequency with which juveniles engage in violence had not changed in recent years prior to the study.[19] Although too many youths engage in violent behavior, the average violent offender today is no more violent than in the past.

Although violent juvenile arrests account for a relatively small proportion of juvenile arrests each year, research that has employed self-report methods indicates that youth involvement in violent offending is more widespread than is suggested by arrest data. One of the most comprehensive **self-report studies** to date is the *Causes and Correlates of Delinquency Study*, which consists of three longitudinal projects supported by the Office of Juvenile Justice and Delinquency Prevention (OJJDP). These projects—which are being conducted in Rochester, New York; Denver; and Pittsburgh—are based on samples of more than 4,000 youths. The research found that, before reaching their late teen years, approximately 40% of the males and between 16% and 32% of the females reported

self-report study A study in which the participants report on their own behavior.

MYTH VS REALITY

Violent Crimes Do Not Always Result in Serious Injury to Victims

Myth—Youths who are arrested for violent crimes have engaged in actions causing serious physical injury to their victims.

Reality—Some youths who are arrested for violent offenses have not seriously harmed their victims.

BOX 14-1 Interview: Sandy Losota, Victim Advocate and Witness Coordinator, Kalamazoo County Prosecutor's Office—Family Division

Q: What is your full job title, and for whom do you work?

A: I am the victim advocate and witness coordinator for the Family Division of the Kalamazoo County prosecutor's office. In Michigan, the prosecuting attorney has the primary responsibility, under the law, for enforcing victims' rights. My direct employer is the Kalamazoo County Prosecuting Attorney.

Q: What are your job responsibilities?

A: I have a number of job duties when it comes to victims; most of them are interrelated and fall under the broad category of advocacy/communication. I send out victim impact statements; notify victims of upcoming court dates; send out information to victims regarding restitution, reimbursement, and/or compensation; monitor all court cases involving victims and follow up on the collection of restitution; attend hearings, sometimes with victims and sometimes on their behalf; and I enter all victim information and data into the office computer system.

Q: What is your educational background?

A: I have a B.S. in criminal justice from Michigan State University in 1975.

Q: What is your employment history that has led up to your present position?

A: Upon graduation from college, I went to work in a private law office as a secretary. I then moved to the Kalamazoo County Prosecutor's Office, where I have held a number of positions. I have been the word processing supervisor; intake supervisor, screening warrant requests and typing the actual warrants from the police, as well as entering them into the computer. I have spent the last year and a half in my present position.

Q: What is your usual contact with victims of juvenile crime?

A: Very few victims come into court until a formal hearing takes place. My initial contact with them, sending out the victim impact statement, generates quite a few phone calls; but unless the matter gets on the formal court docket, I would not have in person contact with them. Once having met them, my job is to shepherd them through the system.

Q: What do you see as the major concern of victims?

A: Virtually all the victims that I see want the juvenile offender punished. They even want this more than restitution in most cases. I am often told by them that their children could not get away with something similar without punishment, and they don't want the offender to get away with anything!

Q: What are the most difficult challenges you face in your job?

A: Not being able to hold juveniles accountable. What is difficult is that the court's focus seems to be primarily on rehabilitation, not punishment. Victims want punitive action. In this way, the court is out of touch with victims; but, in fairness, the court does a good job in communicating with victims and addressing their concerns when they are made known to the court.

Q: What do you find especially rewarding about your job?

A: Being able to satisfy victims; letting them know that the prosecuting attorney's office cares about them. I enjoy the special human touch that I can bring to these cases, and securing an actual restitution award is also very gratifying.

Q: What part of your job do you least care for?

A: It hurts me to have to tell victims that I cannot help them. Often the juvenile and his/her family have no money and restitution is not available. For people that may also be facing economic hardship, this can be very difficult for them, having a loss that is not their fault but for which they will receive nothing.

Q: How much do you get paid?

A: My salary is $38,000 a year.

(continues)

Q: What advice would you give victims in dealing with their victimization?

A: Don't give up! Force the communication with those persons and agencies that are legally responsible for helping you. Try to find a person in a police department, prosecutor's office, or court that will listen to you. But most of all, keep trying. Victims deserve help!

Q: What would you tell someone who wanted to become involved in your line of work?

A: Learn about the court system, how police work, and what prosecutors do. You cannot help victims if you are unable to see how they "fit" into the overall system. It is also very important to know who the key people are in the local police departments, courts, and related agencies. Effective advocacy is done between people. Be nice to everyone, establish your credibility by always going the extra mile, and accumulate favors so that you can use them for your clients.

Q: What are the major challenges you see for victim advocacy in the future?

A: The overwhelming problem is, and will continue to be, too many victims and not enough advocates. There is just so much work to do and not enough people to do it.

involvement in one or more serious violent offenses, such as aggravated assault, robbery, rape, and gang fighting. It also found, however, that serious violent behavior tends to be intermittent rather than part of a consistent pattern. For example, after more than 5 years of research in Denver, it was found that 42% of violent offenders were actively involved in violence during only one year out of five.[20] Furthermore, caution needs to be exercised in generalizing the findings to other localities. Youths in many other communities may be less involved in violence than youths in the three cities studied. In addition, some of the delinquent activity reported in these studies may be less serious than it appears. It is probably safe to say, however, that juvenile involvement in serious violent behavior is more widespread than arrest data suggest—although still not as common as some members of the public may believe.

Certainly, one type of offense that draws considerable public attention is homicide. Like many other types of violent offenses committed by juveniles, homicides by youths younger than age 18 years increased from 1988 through 1994, but declined substantially after 1994, and were at generational lows by 2003.[21] However, even during periods when there were substantial increases in homicide, the actual number of homicides was relatively small. Moreover, the problem was not widely dispersed and tended to be located in certain communities. In fact, most communities within the United States do not experience juvenile homicides.[22]

MYTH VS REALITY

Serious Juvenile Violence Tends to Be Concentrated in Certain Communities in the United States

Myth—Serious violent juvenile crime is a significant problem across the country.

Reality—Rather than being spread evenly across communities within the United States, violent juvenile crime tends to be located in certain areas. For example, in 2002, only 17% of the 3,141 counties in the United States had Violent Crime Index rates above the national average and almost one-fourth of the counties in the United States reported no juvenile violent arrests during the year.[23]

FYI

Girls' Involvement in Violence Has Been Increasing

Girls' share of arrests of persons younger than 18 years of age for Crime Index violent offenses increased in recent years, from 11% in 1983, to 15% in 1996, to more than 18% by 2003.[26] However, it declined slightly to 17.5% of juvenile arrests for violent offenses in 2007.[27]

As noted earlier, serious juvenile offenders, chronic juvenile offenders, and violent juvenile offenders can and should be distinguished from each other. It is perhaps more important, however, to distinguish between the types of offenses committed, such as serious offenses and violent offenses, because juvenile offenders, even the small number who can be considered chronic violent offenders, often engage in a range of illegal actions rather than specialize in a particular type.[24]

Gender and Violent Juvenile Offending

Arrest and self-report data indicate that males are much more likely to be involved in violent criminal activities than females. For example, in 2002, the male arrest rate for Index violent offenses was 4.2 times the female rate. However, a growing number of females have been arrested for violent offenses in recent years. Indeed, the steep decline in juvenile Index violent offense arrest rates in recent years have been primarily due to decreases in the male rate. For example, between 1994 and 2003 the male arrest rate for Index violent offenses decreased 51%, but only 32% for females. The result has been a noticeable decrease in the gap between male and female arrests for Index violent crimes in recent years.[25]

Self-report studies also indicate that male juveniles are much more likely to engage in violent offenses than females. Although more recent self-report studies indicate that levels of offending for males and females are similar for a number of minor offenses, such as truancy, driving without a license, running away from home, and minor forms of theft, differences in violent offenses are pronounced. Indeed, self-report studies indicate that a number of violent offenses, such as gang fighting, carrying a hidden weapon, strong-arming students and others, aggravated assault, hitting students, and sexual assault, can be considered male-dominated offenses.[28]

FYI

Males Engage in Higher Levels of Serious Violence than Females

There is some indication that more females are engaging in violent crimes today than at any time in recent history. For example, an increasing number of females have been arrested in recent years for assaultive behaviors.[29] However, as Meda Chesney-Lind and Randall Shelden note in their book, *Girls, Delinquency and Juvenile Justice*, although there has been a general increase in juvenile arrests for violent crime over the past two decades, the greatest increases for violent crime have "been for relatively minor assaults and boys have clearly been in the lead. The male of the species still has a considerable lead in committing violent acts."[30]

Race and Violent Juvenile Offending

As noted in Chapter 2, African American youths are disproportionately represented in UCR arrest data for Index violent offenses. For example, in 2007, African American juveniles accounted for approximately 50.7% of all juvenile arrests for Index violent crimes.[31] In addition, more recent self-report studies indicate that black youths are more likely to report involvement in more serious crimes than white youths, although the data suggest that the differences in offending may not be as great as would be expected based on arrest data. To some extent, differences in the arrest of blacks and whites for involvement in violent offenses may be a product of police bias. However, they also seem to be partly a product of behavioral differences. In a review of the influence of race on justice entitled *The Color of Justice*, Samuel Walker, Cassia Spohn, and Miriam DeLone found that differences in arrest rates between blacks and whites for at least some offenses, such as robbery and assault, are the product of differences in offending rather than selection bias or racism on the part of law enforcement agencies.[32]

Trends in Violent Juvenile Offending

Several types of data sources can be used to understand recent trends in juvenile violence. These data sources include UCR arrest data, **clearance data**, and victimization data. Arrest trends for Crime Index violent offenses for persons younger than the age of 18 years have changed since the late 1980s. Arrests of persons younger than 18 years for violent crimes increased from 1988 through 1994—a full 70%.[34] Between 1994 and 2003, however, they declined 48%.[35] However, they began to increase again in 2005, although they still remain lower than they were during most of the 1980s and 1990s. The overall trend in arrest rates of persons younger than 18 years for Index violent offenses is shown in Figure 14-1.

clearance data Data on cleared offenses. A cleared offense is an offense for which at least one person has been arrested, charged with the commission of the offense, and referred to court.

Another way to investigate trends in violent juvenile crime is to examine violent crime clearance data (i.e., data on the crimes cleared or solved by law enforcement agencies). Law enforcement agencies clear an offense "when at least one person is arrested, charged with the commission of the offense, and turned over to the court for prosecution."[37] If two persons commit a robbery and both are arrested for the offense, the UCR counts this as two arrests, but clearance data would record this as one crime

MYTH VS REALITY

Racial Differences in Violent Female Offending Are Small

Myth—Black females are much more delinquent than white females. Moreover, black girls engage in much higher levels of violent or "masculine" forms of delinquency.

Reality—As Meda Chesney-Lind and Randall Shelden note, "Race differences in girls' offending are not as marked as some might expect. Especially questionable are notions that black girls are far more delinquent than their white counterparts and that their delinquency is far more 'masculine' in content. In fact, although there are some differences between black and white girls in delinquency content, the differences tend to be less marked as the girls mature. More to the point, black girls, like their white counterparts, are still quite likely to be arrested for traditionally female offenses."[33]

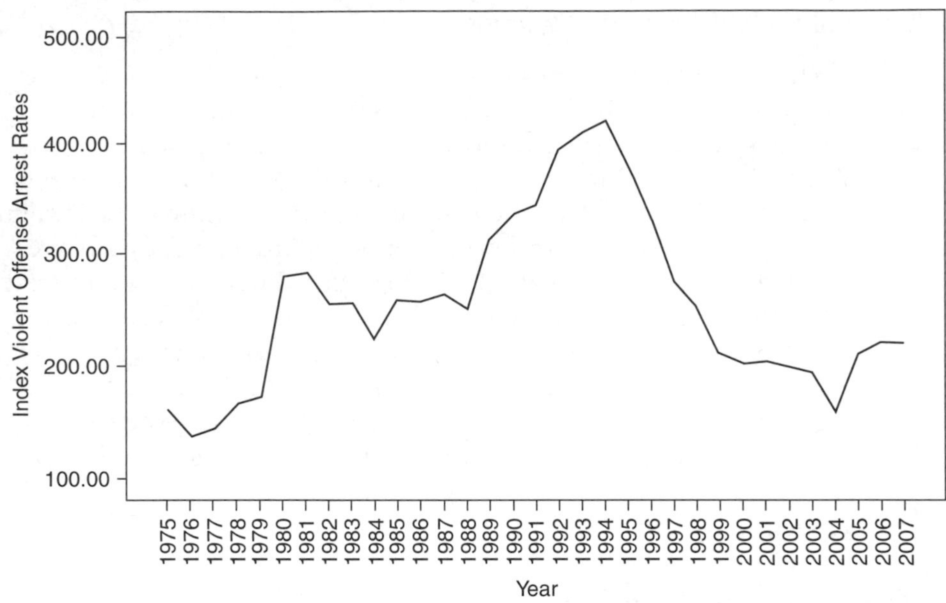

Figure 14-1 Index Violent Offense Arrest Rates (Ages 10–17 Years), 1975–2007

Source: Data from the United States Department of Justice, Federal Bureau of Investigation. (1976–2008). *Crime in the United States, 1975–2007*. Washington, DC: U.S. Department of Justice; Puzzanchera, C., Sladky, A., & Kang, W. (2008). *Easy access to juvenile populations, 1990–2007*. Retrieved from http://ojjdp.ncjrs.gov/ojstatbb/ezapop/; U.S. Census Bureau. (1982, 1992). *Current population reports*. Washington, DC: U.S. Government Printing Office.

cleared or solved by an arrest. Thus, clearance data give a better picture of the amount of crime committed by juveniles.

When clearance trends for violent crimes are examined, they look much like the trends for violent arrests. However, the proportion of violent crimes cleared by the arrest of youths younger than 18 years indicates that the proportion of violent crimes accounted for by persons younger than 18 years of age has increased over time. For example, in the 1980s, persons younger than 18 years accounted for about 9% of violent crime clearances. This number, however, increased to 14% in 1994 and declined to 12% between 1997 and 2007.[38] These data indicate that juveniles commit about 1 in 8 violent offenses known to the police.[39] The percentages of violent crimes cleared by the arrest of a juvenile since 1975 are shown in Figure 14-2.

Additional information on violent juvenile crime can be gleaned from data collected as part of the National Crime Victimization Survey (NCVS). As noted in Chapter 2, the

MYTH VS REALITY

Arrests of Youths for Violent Offenses Have Been Generally Stable Over Time

Myth—Arrests of persons younger than 18 years for Crime Index violent offenses have increased steadily over time.

Reality—Arrests of persons younger than 18 years for Crime Index violent offenses were quite stable from 1975 through 1987. However, they increased from 1988 to 1994, then they declined from 1995 to 2004.[36] Unfortunately, they began increasing again in 2005.

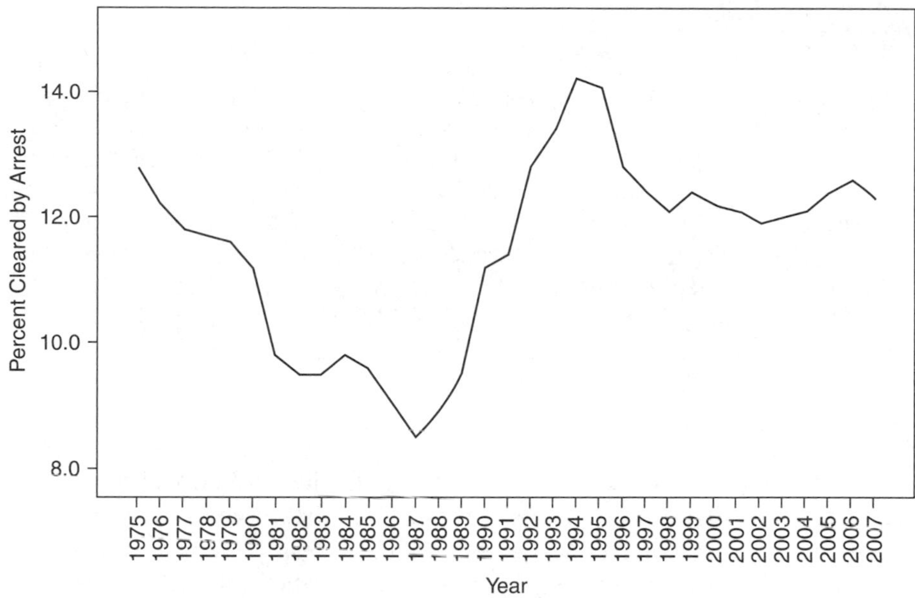

Figure 14-2 Percent of Index Violent Offenses Cleared by Arrest of Persons Younger Than 18 Years of Age, 1975–2007

Source: Data from the United States Department of Justice, Federal Bureau of Investigation. (1978–2008). *Crime in the United States, 1975–2007.* Washington, DC: U.S. Department of Justice.

NCVS, which is a **victimization study**, asks a nationally representative sample of persons 12 years or older about the characteristics of offenders. Thus, it provides some insight on the level of violent juvenile crime. Importantly, the NCVS data indicate a downward trend in violent juvenile victimization between 1993 and 2002 with some increase noted in 2003 among youths 12 to 17 years of age. Indeed, the rate of violent victimizations for teens dropped from about 130 per 1,000 youths in 1993 to 60 per 1,000 youths in 2003. Nevertheless, the rate of violent victimization of teens during this period was about two and one-half times that of adults.[40]

victimization study
A study in which the participants are asked to report their victimization experiences.

Victimization data are of limited utility for describing levels of juvenile criminality because the age of offenders is not always known. Nevertheless, NCVS data indicate that, in most instances of violence against teens, the perpetrator is an adult. Indeed, a juvenile is the perpetrator in only about 1 in 4 violent victimizations of teens. Moreover, this pattern has remained quite stable over time.[41] In addition, data on crimes known to law enforcement (remember not all victims report their experiences to the police) indicate that adults are the perpetrators in 60% of all violent victimization incidents involving teens.[42]

FYI

Family and Community Factors are Strong Predictors of Teen Victimization

Research by Janet Lauritsen that examined NCVS data found that youths who lived in single-parent families experienced a 50% greater risk of violence than youths in two-parent families. She also found that youths living in disadvantaged communities and who had recently moved into these communities were at greater risk of violent victimization.[43]

COMPARATIVE FOCUS

Serious Juvenile Crime Is Not Necessarily a Characteristic of Modern Industrialized Societies

Japan, which is a modern society, has a relatively low rate of violent juvenile crime. For example, there were 105 homicide and 1,668 robbery arrests of juveniles in Japan in 2000. By comparison, there were 641 homicide and 15,310 robbery arrests of juveniles in the United States during that same year. A comparison of the homicide and robbery arrest rates in the two countries for 2000 shows that the homicide rate in the United States was almost twice that of Japan (.89 per 1000 youths in the United States versus .46 in Japan) and the robbery rate in the United States was nearly three times greater than the rate in Japan (21.2 per 1,000 youths in the United States versus 7.3 per 1,000 youths in Japan).[47]

Arrest, clearance, and victimization data on violent juvenile crime present a similar picture. Recent arrest and victimization data indicate that violent juvenile crime rates have been decreasing.[44] However, some evidence indicates a larger number of juveniles are being arrested for repeat violent offenses in some jurisdictions.[45] Moreover, clearance data indicate that youths today continue to be responsible for a significant proportion of violent crime, and while overall levels of violent juvenile crime declined between the mid-1990s and mid-2000s, there is some indication that it may be increasing once again.

It must be borne in mind, however, that some communities experience far more violent juvenile crime than others, and that the high levels of violent crime experienced in the United States are primarily the result of adult, not juvenile, violence. Indeed, in 2007, almost 88% of Crime Index violent crimes cleared by arrest were committed by adults.[46]

■ Explaining Violent Juvenile Crime

There is not one single cause of juvenile violence. Like other types of illegal behavior, violent juvenile crime is best understood by considering the interrelationships among the cultural, political, economic, community, peer group, family, and individual context within which it occurs (see Chapter 3 for a discussion of this context). Also, like other types of illegal behavior, it has been explained using a variety of theories. Although it is not possible to discuss thoroughly every explanation of violent juvenile behavior that has been proposed, some of the more recent explanations are presented in the following sections.

The Cultural Context of Violent Juvenile Crime

Violence has always played a large role in the cultural make-up of the West and of the United States in particular. Even before the founding of the country, violence was seen as an acceptable way to subjugate and "civilize" native peoples. Later, it was viewed as a primary means of achieving the nation's "manifest destiny," of protecting corporate interests from strikers, and of protecting the country's international interests. Indeed, the history of the United States has been filled with ongoing conflict between a variety of groups, including, but hardly limited to, capitalists and workers, men and women, minorities and majorities, environmentalists and corporate executives, civil rights activists and segregationists, anti-war protestors and pro-war supporters, and the young and the old. Because people are more connected to the present than the past, many people feel that violence has reached an unprecedented level today. In reality, violence has always been a regular feature of American culture.

Perhaps the pervasiveness of violence is best understood by considering the television viewing habits of youths and the content of TV programs. Systematic viewing of television begins at 2.5 years of age and remains at a high level throughout the teen years. According to a national study conducted by the Kaiser Family Foundation, children 8 to 18 years old spend an average of 3 hours per day watching TV and about 1.75 additional hours each day listening to the radio, CDs, tapes, or MP3 players. They also spend more than an hour a day on the computer (not engaged in homework); slightly less than 50 minutes a day playing video games; and about 43 minutes per day reading books, magazines, and other publications that are not part of their homework. Thus, on an average day, youths spend about 6.5 hours engaged with various types of media.[48] Furthermore, a considerable amount of the content of this media contains violence. Consider television programming in which violence is depicted in a variety of forms, whether it is Saturday morning cartoons or prime-time dramas. Indeed, some research indicates that 80% of television programs contain some degree of violence, and it is estimated that, by the time a child's secondary education is completed, he or she will have witnessed 13,000 violent deaths on television and viewed more than 200,000 violent acts.[49]

A number of research studies have found a relationship between exposure to TV violence and aggression among children,[50] and subsequent aggression in young adulthood.[51] Indeed, the American Psychological Association and the National Institute of Mental Health have gone on record in support of the evidence showing a link between media violence and **interpersonal violence.**[52] Nonetheless, a direct causal link has not been firmly established. Perhaps the best that can be said at this point is that the literature indicates that media violence is a reinforcement for youths who have been exposed to a number of violent models. Of course, TV violence is only one type of media violence to which children are exposed. Numerous violent models are also found in movies, comic books, and video games.

interpersonal violence
Face-to-face interaction that produces physical harm or presents a threat of physical harm.

Violence is also a part of many other American institutions. A considerable amount of violence is found in sports, in schools, in communities, and in families (an issue examined later in this chapter). The effect of pervasive violence on youths is not completely understood. Nevertheless, some researchers maintain that regular exposure to violence on the part of others in the environment and in the media, particularly when such violence is rewarded, encourages violent *modeling* on the part of youths.[54] Others believe that regular exposure to violence leads to a more negative and cynical view of the world, while still others maintain that youths may actually develop *biological adaptations* that make them more susceptible to aggressive responses to the violence they experience.

FYI

Poor Youths Watch More TV than Affluent Youths

Approximately 98% of households in the United States have a television set, and many of these have multiple sets. Moreover, television viewing occupies more of a child's waking time than any other nonschool activity. However, time spent in front of the television is not evenly distributed across the youth population. For example, the heaviest viewers of television are poor youths, because they often lack alternative leisure time opportunities.[53]

The Political Economy of Violent Juvenile Crime

As noted in Chapter 3, the *political economy* of a society at a particular point in time plays a critical role in the production of delinquency, because it determines the ways in which economic and political resources are developed, managed, and distributed, which, in turn, affect the ability of other socializing institutions (e.g., families, schools, and communities) to meet people's needs.[55] For example, political and economic decisions made by corporations and government representatives influence the distribution of jobs, the pay and benefits earned by workers, and the conditions in which people work. Similarly, political and economic decisions influence the quality of schools and the education that children receive as well as the types and quality of the TV programs that they view. Obviously, the decisions made by government and businesses have a profound effect on the quality of life experienced by all Americans.

Although there are exceptions, it is clear that violent juvenile offenders typically come from those groups who are the least advantaged—minorities and the poor. Self-report studies indicate that lower-class youths report more involvement in serious and violent juvenile crimes than middle-class youths.[56] Although poverty affects large numbers of youths of all races, it is particularly prevalent among minority youths. According to the U.S. Census Bureau, some 13.3 million children, or 18% of all youths younger than 18 years, lived in poverty in 2007.[57] However, 33.7% of African American children and 28.6% of Hispanic children were living in families below the poverty level in 2007.[58] Consequently, it is not surprising that UCR, self-report, and juvenile court data indicate that African American youths are overrepresented among the perpetrators of violent juvenile crime.[59]

The Community and Peer Group Contexts of Violent Juvenile Crime

According to a number of different theoretical explanations of delinquency, community factors play an important role in delinquent behavior. Research indicates that delinquency rates appear to be influenced by the economic and social conditions found in communities. Today, some 13 million children live in poverty, and the poverty rate for children younger than 18 years has been trending upward since 2000.[60] Although poverty is a problem for youths of all racial groups, it is especially problematic for African American and Hispanic youths. Levels of poverty are important because the economic situation in many communities is believed to contribute to urban violence.[61]

A major problem in inner-city areas, which often have predominately minority populations, is that many youths no longer have access to entry-level jobs in factories and shops. These jobs either have disappeared as the economy has become more service oriented or have moved to other locations, often in foreign countries, where wages are low. The jobs that do remain are typically low paying and offer few opportunities for advancement. Because of the lack of legitimate options for earning income, many youths in urban and other poor areas turn to drug dealing, theft, and violence in order to make money and handle the strains of urban life.[62]

ecological approach to delinquency An approach to understanding delinquency that focuses on the geographic distribution of delinquent behaviors and the factors that are related to this distribution.

Importantly, **ecological approaches to delinquency** (i.e., attempts to understand delinquency that focus on the geographic distribution of delinquent behaviors and the factors related to this distribution) maintain that community characteristics are related to delinquency in important ways. For example, factors such as population density, poverty, job loss, income inequality, transience, dilapidation, the existence of deviant subcultures, the mixed use of property (i.e., interspersion of residential and commercial land), and a lack of participation by residents in community organizations have been found to be associated with increases in delinquency.[63]

A number of researchers that have examined high-delinquency areas have found that many of the factors previously mentioned have a significant impact on the social organization of the community. For example, research indicates that communities characterized by conditions such as poverty, population density, mixed use of land, transience, and dilapidation experience **social disorganization** (i.e., an inability to exert effective social control over community members). Research on children in disorganized communities indicates that they tend to be less involved with conventional social institutions. As a result, such youths are more vulnerable to participation in delinquent peer groups and are more likely to engage in interpersonal aggression.[64] Moreover, the lack of employment opportunities for youths and adults in poor communities is associated with high levels of predatory crime. In these communities, the scarcity of jobs tends to destabilize families, and children from these families are more likely to use violence and aggression as a means of dealing with their limited options.[65]

social disorganization The state of a community or society characterized by a lack of social control over community members.

Youths who live in socially disorganized communities experience a variety of stresses that have been found to be related to delinquency.[66] Studies of inner-city youths indicate that substantial numbers have witnessed violence within their environment. For example, surveys of middle and high school students in high crime areas in Chicago found that as many as 35% had witnessed a stabbing, 39% had witnessed a shooting, and 24% had seen someone killed. In the majority of these cases, the children indicated that they knew the victims, and 47% of the victims were characterized as friends, family members, classmates, or neighbors. Other studies have produced similar findings.[67] Importantly, children's exposure to violence can have an adverse effect on their mental health. One problem experienced by many youths exposed to violence is **posttraumatic stress disorder** (an affliction common among combat veterans). In addition, youths exposed to violence can have difficulty forming close personal relationships, believe that they will not reach adulthood, suffer from lowered self-esteem, perform poorly in school, and display aggression toward others.[68]

posttraumatic stress disorder A disorder in which a significant traumatic event is reexperienced by the individual and produces intense fear, anxiety, helplessness, and avoidance of stimuli associated with the trauma.

Youths in socially disorganized neighborhoods also make various social adaptations to their environment. These adaptations can include carrying weapons for protection and joining delinquent peer groups such as gangs. For example, a study of 835 male inmates in correctional facilities in California, New Jersey, Illinois, and Louisiana found that the major reason youths carried guns was for self-protection.[69] Indeed, weapon carrying is not uncommon, particularly in some communities. In a 2007 survey of a nationally representative sample of students in grades 9 through 12, 18.0% reported carrying a weapon at least once in the previous month, and 5.2% reported carrying a gun during that period. Moreover, 5.9 % also reported carrying a weapon on school grounds at least once in the 30 days prior to the survey.[70] The ready accessibility of guns is another factor that has contributed to high levels of violent crime in some communities.

FYI

The Juvenile Arrest Rate for Murder Declined Significantly Between 1993, When It Peaked, Until 2004
Since 2004, however, it has been increasing once again. Nevertheless, it was still lower in 2007 than in any year in the 1980s and almost every year in the 1990s.[71] Trends in the juvenile arrest rate for murder can be seen in Figure 14-3.

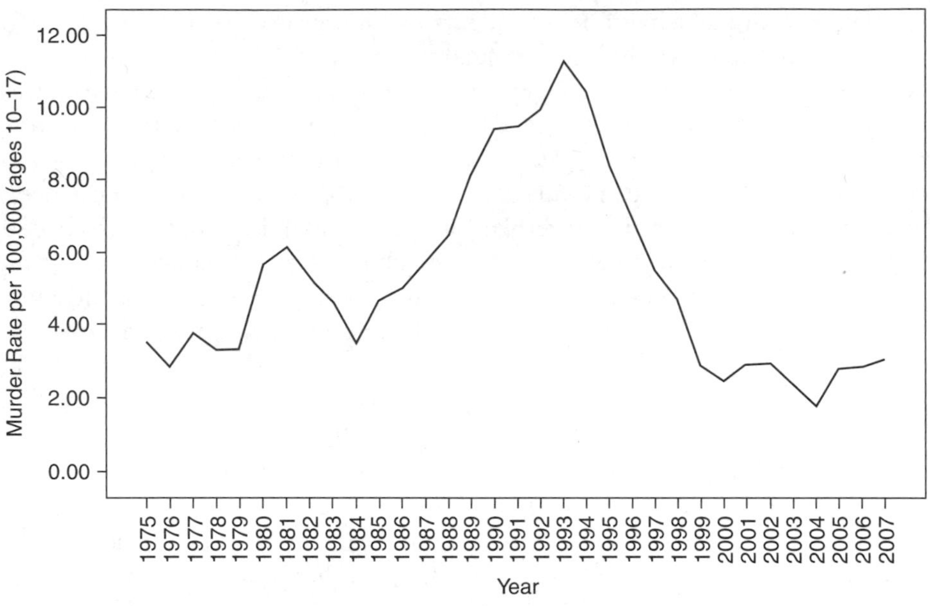

Figure 14-3 Murder Arrest Rate per 100,000 (Ages 10–17 Years), 1975–2007

Source: Data from the United States Department of Justice, Federal Bureau of Investigation. (1976–2008). *Crime in the United States, 1975–2007.* Washington, DC: U.S. Department of Justice; Puzzanchera, C., Sladky, A., & Kang, W. (2008). *Easy access to juvenile populations, 1990–2007.* Retrieved from http://ojjdp.ncjrs.gov/ojstatbb/ezapop/; U.S. Census Bureau. (1982, 1992). *Current population reports.* Washington, DC: U.S. Government Printing Office.

The social and economic conditions in many communities have also been factors leading to the proliferation of youth gangs. According to Pamela Irving Jackson, the development of gangs in many areas is a natural consequence of the shift from a manufacturing economy, with many high-paying jobs, to a low-wage service economy. As well-paying manufacturing jobs leave communities, youths face diminished economic opportunities. Under these conditions, family stability is eroded and gangs become attractive alternatives for many youths.[72]

Youths join gangs, like they develop other affiliations, for a variety of reasons. In some instances, gang involvement is part of a cultural family tradition passed from one generation to the next. For many youths, a gang serves as a substitute family, where they find a sense of belonging and status, or as a social outlet, where they can meet with friends, have fun, and "hang out."[74] For others, gang membership may be seen as a practical way to protect themselves in a hostile environment—a means of achieving at least marginal

MYTH VS REALITY

Gangs Can Be Found in Various Geographic Locations

Myth—Youth gangs are only a problem in large urban areas.

Reality—Although a substantial percentage of large cities report gang problems, gangs are also found in smaller cities, and in some suburban and rural areas. According to the *2004 National Youth Gang Survey*, which is a survey of a nationally representative sample of law enforcement agencies, approximately 80% of large city agencies, 40% of suburban agencies, 28% of small city agencies, and 12% of rural agencies reported gang problems in 2004.[73]

MYTH VS REALITY

Some Gang Members Are Not Heavily Involved in Violence

Myth—Today's gangs are much more oriented toward violence than gangs in the 1950s or 1960s.

Reality—Although belonging to a gang increases the chance that a youth will engage in violence,[78] some gang members are not heavily involved in violence. It is true that gang violence has increased in severity. There is evidence that some gangs recruit members who own guns, and many other youths obtain firearms after they join a gang.[79] As a result, many gang members possess firearms, including automatic weapons, which increases the likelihood of serious injury or death. However, it is a mistake to assume that all gang members are involved in violence or that all gangs are engaged in violent activity. In fact, some gangs engage in relatively little serious or violent delinquent behavior. In a study of gangs in Chicago, San Diego, and Los Angeles, Jeffrey Fagan found that some gangs are more oriented toward social activities than delinquency.[80]

safety within the confines of their neighborhood.[75] Or a gang may seem to offer a way of achieving economic rewards not available through more conventional means.[76] For youths with serious psychopathologies, a gang may provide an environment where their enjoyment of violence is rewarded and respected.[77]

Gang membership is important because it, like involvement in other types of groups, plays an important role in shaping the behavior of many youths. For example, in one study of juvenile gang members in Rochester, New York, researchers found that many youths who joined gangs were no more delinquent before they joined than nongang youths. After they became involved in the gangs, however, their rate of delinquency increased, particularly their involvement in crimes against persons.[81] Indeed, violent criminal activities are encouraged in some gangs because group norms call for violent responses to those who insult gang members, encroach on their territory, or present a threat to their illegal activities.[82] Furthermore, violence may be condoned as a tactic for settling internal gang disputes, as a means of intimidating neighborhood residents, or as part of a gang initiation. In addition, violence may help strengthen gang solidarity and may serve as an outlet for more aggressive members to exercise their skills and develop and/or maintain reputations.[83]

The Family Context of Juvenile Violence

The family should be understood as only one part of the mosaic of social institutions that influence children. It does, however, play a critical role in child development. Consequently, it is not surprising that research indicates that the family also plays an important

FYI

Gang Membership Encourages Delinquency

One important study that examined gang member involvement in delinquency, the Rochester Youth Development Study, found that 30% of a representative sample of seventh and eighth graders in Rochester, New York, indicated they belonged to a gang. The study also revealed that gang youths were disproportionately involved in juvenile crime. Although gang members accounted for only 30% of the sample population, over a four-year period they accounted for 65% of all delinquent acts and 69% of all violent acts committed by youths in the sample.[84]

child neglect Failure of a caregiver to provide for the physical, mental, or educational needs of a child.

role in delinquency.[85] Indeed, a number of family-related factors, such as family conflict and breakup, **child neglect** and abuse, inconsistent discipline, parental deviance, parental economic status, family structure, and a lack of emotional ties between parents and children, have been found to be related to delinquent behavior.[86] Unfortunately, many children's lives are influenced by a combination of these factors. For example, a child may live with a single unemployed parent who has a serious substance abuse problem. This same child may have siblings who are involved in criminal activities, and the child may be a victim of abuse and neglect. Moreover, as noted in Chapter 3, the family itself is part of an even larger social context that influences the quality of family life. Indeed, the wider political and economic contexts within which families are located determine the types of jobs available to adults and youths, the pay associated with those jobs, the health benefits that are available, the quality of schools that youths attend, and the types of recreational and leisure activities that can be easily accessed by family members.

One factor that has been found to be closely related to juvenile aggression is the quality of the interactions among family members. Research indicates that violent interactions within the family encourage subsequent aggression and violence, not only within the family, but also in other social contexts. After looking at the discipline practices of parents, Murray Straus found a strong relationship between physical punishment within the family and later aggression.[89] Physical punishment not only may serve as a model of aggressive behavior for children, but also may weaken the bond between the parent and child, lower the child's self-esteem, result in self-labeling as a "bad kid," and undermine the child's sense of justice.

child physical abuse Actions by a caregiver that result in a nonaccidental injury to a child.

The difference between spanking and **child physical abuse** is one of degree, not kind. Moreover, the precise point at which spanking and other types of physical punishment become abuse is difficult to determine. Consequently, there is a lack of agreement among social scientists and others as to what constitutes physical abuse. Nevertheless, child physical abuse is generally defined as actions that result in a nonaccidental injury to a child by a caregiver.[90]

Although it is not possible to provide an accurate estimate of the extent of child physical abuse and neglect in the United States, the available evidence indicates that it is a significant social problem. Indeed, 3 million reports of alleged abuse and neglect, involving some 5 million children, are received by state or county child protective services each year.[91] The actual number that occur each year, however, may even be higher because many instances of abuse and neglect go unreported. Unfortunately, child welfare agencies often lack the necessary resources to adequately investigate each alleged incident of abuse or neglect.[92]

Although even mild forms of violence against children may encourage aggression, it is clear that physical abuse is strongly related to a variety of childhood adjustment problems, including violent juvenile crime.[93] The victimization of children can have

FYI

Youths Have a High Rate of Violent Victimization

Children and teenagers suffer more victimization than adults in almost every category of violent crime, including physical abuse, sibling assaults, bullying, sexual abuse, and rape.[87] In 2005, the rate of violent victimization of youths from 12 to 19 years of age was higher than those of any other age group except for those between 20 and 24 years of age and substantially higher than for persons 25 years old and older.[88]

a profound influence on their development.[94] Obviously, physical abuse can result in physical injury and even death. It can also lead to emotional and psychological trauma, which can affect the individual throughout his or her life. Research indicates that problems such as chronic anxiety, lowered self-esteem, depression, school problems, aggressive peer relations, pregnancy, and emotional and mental health problems are associated with child abuse.[95] Furthermore, children who have been abused are three times more likely than their peers to become substance abusers.[96]

There also appears to be a strong relationship between child abuse and later delinquency and adult criminality. In one study that examined the relationship between child abuse and subsequent offending, Cathy Spatz Widom found that child abuse increases the odds of future juvenile and adult criminality by 40%. Widom followed the offending careers of 908 youths who had been reported as victims of abuse and neglect between 1967 and 1971, and she compared them to a control group of 667 youths who had not been maltreated. She found that 29% of those who had histories of abuse or neglect had adult criminal records, compared with 21% for the control group. Widom also found that age, race, and gender affected the likelihood that a youth with a history of abuse would become involved in criminal activity. For example, her analysis indicated that older black males who had been abused were more likely to become involved in adult criminality than other youths. Indeed, almost 67% of this group had adult criminal records, compared with only 4% of white females with no history of abuse. According to Widom, being abused or neglected as a child increased the likelihood of having a juvenile arrest by 53%, it increased the likelihood of being arrested as an adult by 38%, and it increased the likelihood of being arrested for a violent crime by 38%. Indeed, children who had been the victims of physical abuse had a violent crime arrest rate that was twice that of the controls. Interestingly, Widom discovered that children who had been neglected also had a higher rate of violent arrests than the controls.[97] Subsequent research by Widom indicates that neglect is as damaging to children as physical abuse.[98]

A study that examined a representative sample of 1,000 seventh and eighth graders in Rochester, New York, also found a strong link between child maltreatment and violent delinquency. The researchers interviewed the caretakers of sample subjects and collected data from police, Department of Social Service, school, and other agency records at different points in time. They found that children from disadvantaged families and those that had at least one missing biological parent were more likely to experience maltreatment. They also found that maltreated youths (i.e., youths who had been the victim of some form of abuse or neglect) were significantly more likely to become involved in delinquency, including violent delinquency, than other youths.[99]

Child abuse is only one type of violence found in American families. Significant numbers of families also experience spouse abuse, sibling–sibling abuse, and abuse of a parent by a child. Indeed, it is estimated that between 2 and 6 million women each year

FYI

Homicide Is a Leading Cause of Death Among Youths

According to the National Center for Injury Prevention and Control, in 2005, homicide was the second leading cause of death for young people between 10 and 24 years of age.[100]

multiassaultive family A family whose members exhibit ongoing patterns of physical aggression.

FYI

Multiassaultive Families

Research indicates that at least 7% of all intact families can be classified as **multiassaultive families**. These families are characterized by ongoing patterns of physical aggression among their members. Research indicates that the patterns of assault that are learned in these families are carried into other contexts as well. For example, males in these families are five times more likely to have assaulted a nonfamily member than males from nonassaultive families.[106]

family violence Violence that occurs between persons who are living together as a family unit. Often this violence, which can be emotional as well as physical, takes place in the family dwelling.

are abused by their partners, and battery is the major cause of injury to women in the United States.[101] Although little research has been done on violence between siblings, such violence is another form of **family violence**,[102] and it is one that likely goes undetected by authorities. The abuse of a parent by a child is still another type of family violence. In an analysis of family violence, Richard Gelles found that 3 out of every 100 teenagers aged 15 to 17 years had kicked, bit, punched, hit with an object, beat up, threatened, or used a gun or knife against a parent.[103] Moreover, it is estimated that between 700,000 and 1 million elderly persons may be abused each year.[104]

Clearly, being the victim of abuse or neglect is strongly associated with a variety of youth adjustment problems as well as violent behavior. However, there is evidence that witnessing violence also leads to a variety of developmental problems among children.[105]

The Individual Context of Juvenile Violence

Although it is important to understand the political, economic, family, community, and peer group contexts of juvenile violence, there is also an individual dimension that must be considered. Clearly, individuals do make decisions to act, although these decisions are not always carefully thought out. Sometimes, youths plan criminal activities and take precautions to avoid detection; however, in other instances, delinquent activity is best viewed as a type of "pick-up game." In other words, it is the product of the coming together of individuals at a certain time, in a certain place, and under certain conditions.[107] Moreover, although there is no evidence that the majority of youths who engage in delinquent activities, including violent criminal activities, are suffering from an identifiable biological or psychological pathology, there is evidence that a small proportion of violent juveniles have a biological and/or psychological abnormality that influences their behavior.

A number of researchers have maintained that there is a genetic or biological basis for delinquent behavior. However, even these researchers often argue that a combination of environmental and biological factors cause delinquency. They tend to posit that biological and genetic factors are important because they influence central nervous system (i.e., brain) functioning in ways that lead to antisocial and criminal behaviors. In their view, how the brain responds to environmental stimuli is determined by one's biology.

A number of approaches have been taken by researchers to uncover a biological or genetic basis for criminality. Some researchers have conducted twin studies that investigate similarities in delinquent behavior between identical and fraternal twins. These studies have typically found that identical twins are more similar in their behavior than fraternal twins, indicating there may be a biological basis for some criminality.[108]

However, research has also shown that identical twins are more likely to be treated alike, so these studies have been unable to disentangle the unique influences of the environment and genetics on delinquent behavior.

In an effort to overcome the methodological problems of twin studies, other researchers have studied children who have been removed from their biological parents. The goal of their research is to compare the level of criminality of the children, that of their biological parents, and that of their adoptive parents. If the level of criminality of the adopted children is like that of their biological parents and not like that of their adoptive parents, this would indicate that a genetic basis for criminality exists. The results of these studies, like the results of the twin studies, indicate that genetics possibly plays some role in criminality.[109] However, those studies that have been large enough to examine violent and property crimes separately have found that genetic factors seem to play a larger role in property crimes than in violent crimes. As Adrian Raine notes, "This seems counterintuitive since one would imagine that violent, more serious offending, being more extreme, would be more 'hard wired' and have a higher heritability than property offending, which one could imagine is more driven by social and economic factors."[110]

Another approach to investigating delinquency focuses on intelligence and its relationship to delinquent behavior. Some research has found a weak to moderate inverse relationship between IQ test scores and delinquency (i.e., as IQ test scores decrease to the low normal range, delinquency increases).[111] Other research has failed to find a strong relationship between IQ scores and delinquency.[112] The debate on the relationship of IQ to delinquency rests on assumptions about what IQ tests actually measure. Some maintain that intelligence is a heritable trait that can be measured by standardized IQ tests. These researchers believe that a strong inverse relationship between IQ scores and delinquency is evidence that innate intelligence is inversely related to delinquent behavior. However, others maintain that IQ tests are culturally biased and that they do not measure innate intelligence but rather cultural and educational background.

In addition to the previously mentioned studies, a variety of other efforts have been made to uncover a biological basis for criminal behavior. For example, studies have been done positing a relationship between delinquency and factors such as body build; temperament; neurological dysfunctions, such as attention deficit hyperactivity disorder (ADHD), other learning disabilities, and minimal brain dysfunction (MBD); and hormonal imbalances. Moreover, some neurological dysfunctions, such as MBD and learning disabilities, have been linked to impulsive and aggressive behaviors among youths.[113] So have hormonal imbalances.[114] This general approach has been used to explain what Terrie Moffitt calls **life-course-persistent (LCP) delinquents**—individuals

life-course-persistent (LCP) delinquent
A person who engages in antisocial behavior at a very early age; gradually becomes involved in increasingly more serious criminal activity, including violent criminal activity; and continues to commit criminal offenses well into his or her adult years.

FYI

There Is Considerable Debate Over the Meaning of the Relationship Between IQ and Delinquency
Some researchers suggest that the relationship between IQ and delinquency is indirect. According to their view, low IQ is related to school failure, which in turn leads to delinquent behavior. There is indeed considerable evidence of a strong link between school failure and delinquency. However, it is important to recognize that this view is still predicated on the assumption that IQ tests measure innate intelligence. If they do not, then it may be more accurate to state that a deprived cultural and educational background is related to school failure, which in turn is related to delinquency.

who begin involvement in antisocial behaviors at a very early age and engage in criminal activity of increasing seriousness, including violent criminal activity, well into their adult years. According to Moffitt, many of these youths display minor neuropsychological problems during their childhoods, including difficult temperaments, attention deficit hyperactivity disorders, and learning difficulties in school. When these youths come into contact with mental health professionals during childhood, they are often diagnosed as conduct disordered. Contacts made during their adult years often result in a diagnosis of antisocial personality disorder.[115]

Some researchers focus on psychological abnormalities, while admitting that these abnormalities could well have a biological basis. For example, some think that youths with psychopathic personalities suffer from biological defects that influence their behavior. Psychopaths typically are unreliable; fail to develop strong attachments to others; repeatedly engage in antisocial, often violent behavior; and lack remorse. Some researchers suggest that psychopaths suffer from a defect in the autonomic nervous system (ANS) that leads to low arousal. Low arousal, in turn, causes the individuals to act out in an effort to receive stimulation. However, because of low arousal, they have difficulty in learning from their behaviors.[116] Finally, some researchers focus on the personality, and these tend to think that antisocial or criminal personalities are the product of emotionally disturbed parents, a lack of parental affection, parental rejection, and inconsistent discipline.[117]

The literature on violent juvenile crime does contain some evidence indicating a link between a number of biological and psychological factors and violent offending. However, there are several weaknesses with the strategy of looking for such links. First, the relationships found between biological and/or psychological abnormalities and delinquency are often weak. This is particularly true for the studies that are more rigorous and have employed sound methodologies.[118] Second, some of the supposed links between biological and/or psychological abnormalities are counterintuitive, such as the link between biological factors and property crime.[119] Third, existing research indicates that only a small proportion of juvenile crime, including violent juvenile crime, is the product of biological or psychological abnormalities. This does not mean that biological or psychological factors are not important. It seems reasonable to believe that they play a major role in a small proportion of the violent crimes committed by juveniles each year. However, many violent juvenile crimes are committed by youths who appear to suffer from no biological or psychological abnormality. Instead, violence by these youths, in some cases at least, seems like a normal and expected response to the family, community, peer, political, economic, and cultural contexts within which they live.

Responses to Violent Juvenile Crime

Violent juvenile crime represents a significant problem in many communities around the country. Not surprisingly, a variety of responses have been implemented, from programs designed to prevent violent juvenile crime to legislation making it easier to try juveniles in adult courts. Nearly every state has enacted legislation changing the way that juvenile justice operates, and many of the reforms have focused specifically on the violent juvenile offender.

Violence Prevention

In an effort to take a proactive approach to violent juvenile crime, some states, as well as some local communities, have implemented violence prevention initiatives, in some cases with the support of the federal government through the Office of Juvenile Justice and Delinquency Prevention (OJJDP). In recent years, OJJDP has been in the process of developing what it refers to as a "Comprehensive Strategy for Serious, Violent, and Chronic Offenders." This strategy is predicated on **risk-focused delinquency prevention**, an approach that attempts to identify and then address the following factors that contribute to delinquent behavior:

risk-focused delinquency prevention An approach to delinquency prevention in which factors that contribute to delinquent behavior are identified and addressed.

1. Individual characteristics, such as alienation, rebelliousness, and lack of bonding to society
2. Family influences, such as parental conflict, child abuse, and family history of problem behavior (e.g., substance abuse, criminality, teen pregnancy, and school dropouts)
3. School experiences, such as early academic failure and lack of commitment to school
4. Peer group influences, such as friends who engage in problem behavior (e.g., minor criminality, gangs, and violence)
5. Neighborhood and community factors, such as economic deprivation, high rates of substance abuse and crime, and low neighborhood attachment[120]

By encouraging states and communities to develop effective responses to these factors, OJJDP hopes to play a role in reducing levels of violent juvenile crime.

States have also taken steps to support violence prevention initiatives. For example, the Illinois Violence Prevention Act of 1995 established the Illinois Violence Prevention Authority, which is designed to coordinate statewide violence prevention programs, raise funds for state and community organizations involved in comprehensive violence prevention efforts, and provide assistance to communities and organizations attempting to develop, implement, and evaluate violence prevention strategies. The act also established a violence prevention fund, which is a repository for appropriations and grants from federal, state, and private sources earmarked for violence prevention efforts. Missouri and Oregon have completed comprehensive juvenile justice reforms that offer tax breaks for individuals and organizations that become involved in delinquency prevention activities. Missouri's Youth Opportunities and Violence Prevention Act provides tax credits for individuals or businesses that make monetary or physical contributions to programs that provide new or expanded educational opportunities for youths. The Oregon program provides tax incentives for programs that hire at-risk youths. In Mississippi, legislation has been enacted that is intended to foster collaboration between schools, families, and local agencies involved in youth development activities and in programs that attempt to provide cost-effective responses to youth-related problems before they become more serious.[121] Also, by 2007 Massachusetts was slated to implement the Massachusetts Youth Violence Prevention Program (MYVPP) designed to address a variety of factors that contribute to youth violence.[122]

In addition to state-level support for delinquency and juvenile violence prevention efforts, a number of local communities have developed and implemented their own programs. In some instances, these programs have been state or federally supported; in other instances they have been "locally grown." For example, a number of communities in Maryland have implemented late-night basketball programs, and these are believed to be associated with a 60% decrease in drug-related crimes in that state.[123] In Allegheny County, Pennsylvania, a countywide comprehensive antiviolence effort was launched in

FYI

Model Prevention Programs

To examine a list of model prevention programs identified by the Office of Juvenile Justice and Delinquency Prevention, see the *OJJDP Model Programs Guide* at http://www.dsgonline.com/mpg2.5/mpg_index.htm.

the early 1990s, after a dramatic increase in violent juvenile crime between 1989 and 1993. This effort can be best characterized as a large-scale attempt to respond to youth violence through the collaboration of law enforcement agencies, other government agencies, juvenile justice agencies, educational institutions, the religious community, grass-roots community activists, the media, health and human services, and community representatives. Among the accomplishments to date are reductions in turf warfare between agencies, better coordination of law enforcement efforts, the development of over 20 family support centers, the establishment of youth sports leagues in 12 communities, a summer youth jobs program, and neighborhood antiviolence initiatives. According to 1995 arrest data, total juvenile arrests had declined from 1994 levels by 13%, and arrests of juveniles for violent offenses had dropped by 30%. When these decreases are compared with a 2% statewide decline in the number of total juvenile arrests and a 9% statewide decline in the number of violent crime arrests, Allegheny County's efforts appear to have paid off.[124]

Another approach taken by some communities has been to implement curfews as a way of reducing juvenile violence. Although curfews have been in existence around the country since the 1890s, they have been receiving renewed attention in recent years. Although there are differences in curfew laws, they typically require juveniles to be at home between 11 p.m. and 6 a.m., with exceptions for weekends and during the summer.[125] Despite their growing popularity, curfew laws do have opponents. First, some of these ordinances have been challenged on constitutional grounds. The American Civil Liberties Union (ACLU) has challenged curfew laws because it feels they violate the rights of children and parents to free speech and association. Others maintain that these laws give police undue powers to detain and question children without probable cause. Still others are concerned that these laws may act to widen the net of social control by subjecting a new category of children, often minority children, to police and court intervention.[126] There is also the question of whether curfews are effective in reducing delinquency. A few evaluations have found that curfews appear to play a significant role in reducing juvenile crime, including violent juvenile crime. Other studies indicate that curfews have little

MYTH VS REALITY

Curfews Have a Minimal Effect on Violent Juvenile Crime

Myth—Curfew laws reduce levels of violent juvenile crime because most of this crime occurs late in the evening or very early in the morning.

Reality—Curfews may have little effect on violent juvenile crime because, in many communities, most crime of this type occurs during the afternoon and early evening. A study of FBI crime data from eight states found that on school days the peak time for violent juvenile crime was 3 p.m. (just after school), and on weekends, it was 8 p.m.[128]

impact on delinquency or produce a **displacement effect** (i.e., juvenile crime decreases during curfew hours, but increases during other times of the day).[127]

displacement effect Moving an activity of a certain kind to another location or another time. Curfews can have the effect of causing a decrease in juvenile crime during curfew hours and an offsetting increase in juvenile crime during other times of the day.

Gang prevention and intervention efforts have also been implemented in some communities in order to reduce levels of juvenile violence. One such prevention program is the Gang Resistance Education and Training (GREAT) program. This program targets sixth and seventh graders who may be exposed to gangs. GREAT consists of a curriculum that stresses ways to resist peer pressure, improve self-esteem, and avoid gang influences and violence. The program curriculum is taught by uniformed police officers within a school setting. However, one evaluation of this program indicated that it was only moderately successful at changing youths' attitudes about gangs.[129] Also, a more recent evaluation found that it had no significant impact on youths' gang participation, although participation in GREAT was associated with less victimization, less risk-taking behavior, more positive attitudes toward the police, and more negative views of gangs.[130]

Other community approaches to dealing with gangs range from assigning youth workers to gangs (in an effort to redirect youth gangs and their members to more conventional activities) to providing intensive supervision to gang members who are on probation for minor offenses. In addition, grassroots programs have been developed in a number of communities. For example, in Philadelphia, Sister Falaka Fattah, after learning that her 16-year-old son had become involved in gang activity, invited gang members to live in her home, which she named the House of Umoja (umoja means "unity" in Swahili). Her idea was to provide a refuge and supportive network for gang members, who, in turn, were required to avoid illegal activity and participate in activities within the home. These activities included performing regular household chores, participating in an early morning conference at which they set goals for the day, and attending school. Over time, the program spread to include other homes in the community; connections were established with the juvenile courts, which began to place youths there for care; and additional assistance was provided by various sources of support.[131]

One other approach to dealing with serious and violent juvenile offenders that has begun to receive considerable attention is multisystemic therapy (MST), which is being pioneered by Scott Henggeler at the Medical University of South Carolina. One of the goals of this approach is to address the factors that lead to problem behaviors among youths. Consequently, MST focuses attention on the individual, family, school, peer group, and community context of delinquency. Indeed, at the heart of this approach is the recognition that the family is a key institution in a child's life. Direct services are provided by a therapist who has a caseload of four to six families and provides approximately 60 hours of service over approximately 4 months. In addition, the direct service provider is supported by a supervisor. Services consist of individualized interventions that address family, school, peer, or other issues related to antisocial behavior. Evaluations of MST programs indicate that this approach has considerable potential. For example, one study that compared a randomly selected group who received MST services and a control group who received regular services showed that the former had significantly fewer arrests (.87 arrests on average, versus 1.52) over a 59-week postreferral follow-up. In another study, violent offenders who received MST services had a four-year **recidivism** rate of 22%, compared with a rate of 72% for youths receiving individual therapy and a rate of 87% for youths who refused to participate in the program.[132]

recidivist A person who engages in further criminal activity following arrest, court conviction, or release from incarceration.

There are also a number of programs that have been found to have considerable potential in efforts to prevent violence. One good example of these programs is nurse visitation programs such as the ones operated in Montreal (Montreal Home Visitation Study) and in the Appalachian region of New York State (University of Rochester Nurse Home Visitation Program). An essential feature of these programs is the provision of a range of health, infant care, and child development and supportive services to working class or low income mothers. Both of these programs were associated with a variety of positive outcomes such as fewer child injuries, the development of a more positive home environment, improved parenting skills, and lower levels of child abuse.[133] Similar outcomes have also been found in programs that provide a range of educational, health, safety, and support services to pregnant women and to their preschool children.[134]

In addition to programs for preschool youths, several programs for school-aged children have received favorable evaluations. One of the more promising of these programs is the Seattle Social Development Project (SSDP). The SSDP is a comprehensive approach to violence prevention that uses a variety of interventions designed to strengthen youths' bonds to their family, school, and community groups by providing them with opportunities for active involvement, and giving them the skills to participate successfully. This is accomplished through a variety of interventions, including classroom management and interactive and cooperative instructional practices, along with parent training. Interviews with youths 6 years after program completion revealed that students who had successfully completed the program were less likely to report involvement in violent behavior, to engage in heavy drinking, to participate in sexual intercourse, and had greater school achievement than members of a control group.[135]

Juvenile Justice Responses to Violent Juvenile Crime

As a result of the decline in juvenile arrests in recent years, the number of juvenile court cases, including cases involving violent juvenile crime, has also decreased. In 2005, juvenile courts handled an estimated 429,497 offenses classified as person offenses.[136] As a response to concerns about juvenile crime, particularly violent juvenile crime, state legislatures have enacted new laws designed to change the ways that juvenile courts respond to violent juvenile offenders. In addition, juvenile courts themselves have developed new policies and programs in an effort to deal with those youths who are charged with violent offenses. In the period between 1992 and 1995, legislatures in 47 states passed laws that toughened state responses to juvenile crime.[137] What is significant about these changes is that they represented a significant change in the philosophy of the juvenile court. Indeed, the basic rationale behind the reforms was to "punish, hold accountable, and incarcerate for longer periods of time those juveniles who, by history or instant offense, passed a tolerable threshold of tolerated juvenile law violating behavior."[138] In effect, this legislation pushed the juvenile courts away from their traditional *parens patriae* philosophy toward a more punitive criminal court philosophy.

State legislative changes in juvenile law designed to take a tougher stance toward juvenile crime have been directed at five basic areas of juvenile court and juvenile justice operation: (1) jurisdictional authority, (2) sentencing authority, (3) confidentiality, (4) victims' rights, and (5) correctional programming.[139] For example, 41 states passed laws that removed juvenile court jurisdiction over certain categories of juvenile offenders charged with violent offenses and gave jurisdiction over these cases to criminal courts. Twenty-five states expanded the sentencing options available to juvenile and criminal courts with respect to juvenile offenders. Forty states passed legislation that made juvenile court records and proceedings more open. In 22 states, laws were passed that gave victims an increased role in the juvenile justice process. In response to these legislative changes, juvenile correctional administrators have felt increasing pressure to develop new programs for serious and violent offenders.[140]

One of the most significant changes in juvenile justice has been the passage of legislation that makes it easier to try some juveniles as adults. As noted in Chapter 9, each state has a mechanism—typically called waiver, transfer, bind over, certification, or remand—by which juveniles may be tried in adult courts. Indeed, since 1992, every state except Nebraska has changed their laws regarding transfer to make it easier to try juveniles as adults. These changes in law involved implementing statutory exclusions, which remove certain juveniles from juvenile court jurisdiction if they meet specific age and offense criteria, lowering the age at which a juvenile can be transferred to adult courts, and expanding the list of offenses for which a juvenile could be waived. In addition, other states passed **presumptive waiver** statutes requiring certain youths to be waived unless they can prove they are suitable for juvenile justice services, others passed legislation that allows some youths to be transferred if they have prior records, and several sates lowered the upper age limit for original juvenile court jurisdiction, thereby bringing larger numbers of youths under adult court jurisdiction.[141] The result of these changes is to remove increasing numbers of youths from juvenile court jurisdiction and place them under the control of the criminal courts. However, research in Florida comparing the recidivism rate for youths transferred to criminal courts and the rate for a matched group of youths handled by juvenile courts found that waiver to adult court tended to increase recidivism.[142] Moreover, a more recent review of research studies that have examined the effectiveness of transfer conducted by the federal Centers for Disease Control and Prevention found that transfer typically increases the likelihood of recidivism for a violent offenses.[143] Thus, there is strong evidence that transfer is harmful to youths and threatens public safety.

presumptive waiver
A statute-mandated waiver of a youth to criminal court that can be countermanded if the youth is able to prove he or she is suitable for juvenile justice services.

A number of states, as well as individual juvenile courts, have also made substantial changes in the types of dispositions and sentencing options employed in juvenile cases. Since 1992, a number of states have passed statutes requiring mandatory sentences for certain types of violent or serious offenses, while others have expanded the dispositional age limit (usually to 21 years of age), which allows courts in those states to maintain jurisdiction over juvenile offenders for a longer period of time. In addition, a number of states passed laws that make **blended sentences** possible. Blended sentences allow courts to impose juvenile and/or adult correctional sanctions on some categories of offenders, such as youths designated as "aggravated juvenile offenders" or "youthful offenders." For example, in New Mexico, certain youths can be sentenced to an adult or a juvenile correctional program by a juvenile court. In Colorado, Massachusetts, Rhode Island, South Carolina, and Texas, the juvenile court may make a disposition that remains in force beyond the age of its extended jurisdiction, at which time the youth may

blended sentence
A sentence handed down by a juvenile court that encompasses juvenile and adult correctional sanctions. Blended sentences are allowed in some jurisdictions for certain categories of offenders, such as those designated as "aggravated juvenile offenders" or "youthful offenders."

FYI

There Are Various Mechanisms by Which Youths Can Be Transferred to Adult Courts

Overall, the number of cases being judicially waived to adult court decreased between1994 and 2001, when the number of cases transferred to adult courts began to increase once again.[144] However, because there are a number of mechanisms whereby juveniles can be transferred to adult court, the actual number of juveniles subjected to adult court jurisdiction is difficult to determine.

be transferred to the adult correctional system. In California, Colorado, Florida, Idaho, Michigan, and Virginia, youths tried in criminal courts may be sentenced to an adult or a juvenile correctional program.[145] The result of allowing blended sentences has been to expand the range of dispositional options available in juvenile cases, give juvenile courts control over youths for longer periods, and increase criminal court involvement in a wider range of juvenile cases.

Although a "get tough" approach to juvenile crime is seen as good politics by many legislators and prosecutors, it often produces the opposite of what its supporters claim. This is best illustrated in Cook County (Chicago), Illinois, when Richard M. Daly was the state's attorney. In the late 1980s, his "political" response to the rise of teenage violence was to prosecute 97% of all cases forwarded to him by the police. This effort, although good "copy" for the media, had several detrimental effects:

- Almost 70% of the cases were dropped as not prosecutable due to witness problems and lack of necessary information. As a result, many youths lost respect for a system that made threats that it couldn't enforce.
- The system got "clogged" with kids who did not pose a serious threat to the community. As a result, precious and limited resources were expended on nonserious cases, while many serious cases failed to receive the necessary attention.
- Because many courts had limited resources, probation became overused, caseloads became too high, and effective supervision over probationers became impossible, contributing further to the disrespect felt by juveniles toward the juvenile justice process.[146]

Although court practices in recent years have shifted more toward accountability, punishment, and a concern for community safety and away from rehabilitation,[147] a variety of promising correctional programs have begun to be developed for violent and serious juvenile offenders. Many of these interventions are community based. As noted earlier, MST appears to be a very promising community-based intervention that can be employed with many youths charged with violent offenses. There is some indication that intensive supervision programs can be as effective as placement in state training schools for some violent offenders. In an evaluation of three home-based intensive supervision programs in Wayne County, Michigan, William Barton and Jeffrey Butts found that intensive community-based programs were as effective as state placement for some serious and violent juvenile offenders, and they were more cost-effective.[148]

Day treatment programs represent another community-based option that can be used effectively with some violent offenders. The Bethesda Day Treatment Center Program in West Milton, Pennsylvania, provides a range of services to youths, including intensive supervision, counseling, study skills development, cultural enrichment, life and job skills training, and coordination of other services intended to meet youths' needs. Preliminary results suggest that many youths could benefit from this and similar programs.[149]

Institutional placement is another viable option for dealing with violent youths. Offering the most promise are small, secure facilities that have good leadership and well-trained staff. One such program is the Florida Environmental Institute (FEI), also known as "The Last Chance Ranch." This small program, which is located in a remote area of the Everglades, accepts some of Florida's most violent and serious offenders. It is operated by staff members who focus on education, hard work, developing caring relationships with youths, and providing quality aftercare to youths after they return to

COMPARATIVE FOCUS

While the French Approach to Juvenile Justice Shares Many Similarities With the Approach Taken in the United States, There Are Important Differences.

In France, the age of criminal responsibility is 13 years and youths are considered adults subject to adult prosecution when they turn 18 years of age. However, juveniles are tried in either a juvenile court (for youths between 13 and 15 years of age) or a court of Assizes (for youths who are 16 or 17 years of age). The primary difference between these two courts is that the juvenile court employs more of a social welfare approach to cases than the court of Assizes that handles older juveniles. However, French law is decidedly more rehabilitation oriented than many state laws in the United States. Also, when custodial sentences are made in France, the law requires that the sentence must be less than one-half what an adult would receive for the same offense.[153]

the community. Despite the large percentage of violent and serious offenders among the participants, evaluations of the "ranch" indicate that only about 30% of youths who went through this program became recidivists, compared with 70% to 75% of youths who went through other institutional programs.[150]

For youths who are placed in out-of-home settings, intensive aftercare is essential to helping them make a successful transition to the community and preventing recidivism. Unfortunately, aftercare has been given little attention within juvenile justice, despite the fact that it is widely recognized as an important component of rehabilitation. To some extent, the lack of attention given to aftercare programs is being remedied in some jurisdictions. Moreover, there is a developing body of research that is spelling out the key ingredients of successful aftercare programs. This research indicates that effective programs begin planning for release while youths are incarcerated; provide a structured transition to the community, involving institutional and aftercare staff; and provide long-term reintegrative services.[151] There is some evidence that aftercare programs that are well implemented and that provide intensive supervision reduce the probability of rearrest. Importantly, good aftercare programs can shorten the time youths spend in institutions, thus saving precious resources without endangering public safety.[152]

Clearly, a number of promising programs are being developed for violent offenders. These programs tend to offer a number of potential benefits: (1) they address problems related to youths' involvement in delinquency, (2) they are associated with reductions in delinquency, (3) they protect community safety, and (4) they are cost effective. Despite the development of these programs, however, the juvenile corrections field still faces a number of challenging problems. Too many juvenile corrections programs still lack any of the previously listed benefits, and consequently fail to achieve their basic mission, which is to assist youths and protect community safety.

■ Legal Issues

The problem of violent juvenile crime has led to a reexamination of the philosophical underpinnings of juvenile justice. Are violent offenses less serious because they are committed by juveniles? Because of their age, do juveniles really lack the capacity of adults to control their behavior? If, in fact, they lack the same capacity, should they not be treated less severely? These questions need to be systematically raised and discussed. In trying to reach meaningful answers to these questions, careful consideration needs to be given to the fact that not all juveniles are alike. They vary not only in terms of age, but also in terms of mental capacity, level of maturity, emotional development, and social circumstances.

Even if the issue of juveniles' responsibility for their actions can be resolved, the appropriate response to juvenile offenses remains an open question. Historically, the official focus of the juvenile court has been on treatment rather than punishment. Nevertheless, punishment and treatment often go hand in hand. Being placed in an institution for purposes of treatment may seem a lot like punishment. Moreover, placement in an institutional setting for an indeterminate time period may result in the juvenile doing more time than a comparable adult defendant. The possibility of disproportionate punishment is an issue that needs to be addressed by juvenile courts and state legislatures.

The final legal issue concerns due process protection for juveniles. Regardless of how competent an attorney is, do 13- and 14-year-olds really understand what they are facing when they are transferred to criminal court? Is it absolutely certain that they are competent to stand trial as adults? Given that juveniles are not likely to be reformed in adult prisons, does society's interest in retribution outweigh its *parens patriae* obligations? How much due process protection will be enough when juveniles are tried as adults? Is it necessary to ensure that they understand what is happening, or will additional due process procedures only make matters more confusing? In reality, all of the due process rights guaranteed to juveniles by *In re Gault* and subsequent cases are meaningless if the juvenile is incapable of exercising them in a competent way.

The issue of competency has been thoroughly researched over the last 20 years as it applies to adults but little attention has been given to juveniles. In one sample of 112 pretrial juvenile defendants who underwent competency evaluations, 14% were judged incompetent to stand trial and 61% were found to have one or more deficits that could lead to a finding of incompetence by the court. Thus, 75% of the surveyed juveniles were or could potentially be found to be incompetent.[154]

Competency is increasingly being raised in adult courts with the more frequent use of waiver or transfer, as well as in juvenile courts, but very few appellate courts have spoken to this issue. One court that has addressed the issue is the Georgia Court of Appeals in the case of *In the Interest of S.H., A Child*, 469 S.E. 2d 810 (Ga. Ct. App., 1996). The court noted in its opinion, that all of the "rights" that were recognized as applicable to juveniles, such as notice, the right to an attorney, and the privilege against self-incrimination, to name just a few, were meaningless if the juvenile did not understand their importance and was not capable of exercising them. What good does an attorney do, the court asked, if the juvenile cannot assist them in their defense? The Georgia Court of Appeals concluded that youths in delinquency proceedings could be tried only if found competent to stand trial.

This determination, however, leads to an additional series of considerations when the entire concept of competency is considered. For adults, competency has generally consisted of three components:

1. The ability to comprehend what they are charged with and the possible consequences of those charges
2. The ability to help defense counsel in defending their case
3. The ability to make decisions to exercise or waive important rights[155]

The adult judicial system presumes that adult criminal defendants possess the preceding abilities, absent a showing of mental deficit or mental illness. The adult criminal justice system in every state has a set procedure for determining adult competency, but very few states have a legal framework to determine a juvenile's competency to stand trial.

In reality, the considerations for competency differ greatly between adults and juveniles.[156] Most adults, for example, understand that a "right," like the right to counsel, is an entitlement. Studies have shown, however, that among children under the age of 13 years, only about 25% understood that a "right" was something to which they were entitled.[157] Also, children's level of development influences their ability not only to comprehend the charges against them and their legal rights, but it also influences their ability to effectively communicate with their attorney. Indeed, communication between adult attorneys and children can be difficult for a variety of reasons, including the following:

- Fear of adults in general
- Lack of comprehension about who the attorney is and what he or she does
- Mental retardation or mental deficits
- Low IQ
- Learning disabilities
- ADHD

Finally, decision making is crucial to having a meaningful defense. In any criminal or juvenile proceeding the defendant or juvenile must be able to (1) foresee the possible consequences of a decision, (2) imagine hypothetical situations, (3) envision situations that could result from a decision, and (4) evaluate potential outcomes. These abilities require highly developed reasoning skills that many adults do not possess let alone young adolescents. Certainly, children who are younger are less likely to have these skills, but because some state laws make possible the waiver of youths who are 12 years of age or even younger, great pressure is placed on children to make life-altering decisions that they are not competent to make.

In a spirit of fundamental fairness, there are two solutions to this problem of juvenile competency. First, raise the age of adult waiver or eliminate it altogether. Second, modify laws so that there is the presumption that all juveniles younger than a certain age are not competent to stand trial, even as delinquents, and require the state to prove that they are competent. States need to enact legislation to provide a procedure for determining competency for juvenile offenders and alternatives for youths found not competent.

Chapter Summary

Violent juvenile crime has received considerable attention in recent years. As a result, a number of significant changes in state laws and juvenile court practices have been implemented. These changes signal an important shift away from the traditional juvenile court focus on treatment toward a new focus on punishment and accountability. This shift is evident in legislation that is intended to remove many youths from juvenile court jurisdiction by lowering the age limits of juvenile court jurisdiction, by making it easier to transfer youths to criminal courts, by increasing the time that juvenile courts may maintain jurisdiction over youths, and by increasing the severity of sanctions imposed by juvenile courts. The result of such legislation is that increasing numbers of youths are falling under the jurisdiction of juvenile and adult courts for longer periods of time. Although "getting tough" on juvenile crime continues to be a politically popular stance, the reduction in crime promised by "get tough" advocates remains unrealized.

Several problems presently confront the juvenile justice process in its efforts to effectively deal with violent juvenile crime. First, in order to reduce violent juvenile crime, juvenile justice agencies and local community leaders must be willing to commit the resources to address the social, political, economic, community, family, and individual contexts of youth violence. Second, strong prevention and early intervention programs need to be developed and implemented. Third, each of the components of the juvenile justice process (police, courts, and corrections) must be given the resources and support it needs to accomplish its mission. Fourth, state agencies, local communities, and juvenile courts need to be actively involved in developing and implementing a wider range of effective options for violent offenders.

Despite the lack of effective interventions for violent juvenile offenders in many jurisdictions, a growing body of research indicates that viable programs for these youths can be developed, including violence prevention programs and community-based and institutional programs that treat existing violent offenders. Indeed, the development of cost-effective programs that address the factors that contribute to violent behavior holds some promise as a way of dealing with juvenile violence while safeguarding the ideal of the juvenile court—to serve the best interests of children and simultaneously protect public safety.

Key Concepts

blended sentence: A sentence handed down by a juvenile court that encompasses juvenile and adult correctional sanctions. Blended sentences are allowed in some jurisdictions for certain categories of offenders, such as those designated as "aggravated juvenile offenders" or "youthful offenders."

child neglect: Failure of a caregiver to provide for the physical, mental, or educational needs of a child.

child physical abuse: Actions by a caregiver that result in a nonaccidental injury to a child.

chronic delinquent offender: A juvenile who, over time, accumulates a number of delinquency charges and convictions. The related term *chronic recidivist* was used by Wolfgang and colleagues to refer to juvenile offenders who had five or more police contacts.

clearance data: Data on cleared offenses. A cleared offense is an offense for which at least one person has been arrested, charged with the commission of the offense, and referred to court.

cohort study: A study that looks at a group of people born at a specific time in a particular location (a cohort).

displacement effect: Moving an activity of a certain kind to another location or another time. Curfews can have the effect of causing a decrease in juvenile crime during curfew hours and an offsetting increase in juvenile crime during other times of the day.

ecological approach to delinquency: An approach to understanding delinquency that focuses on the geographic distribution of delinquent behaviors and the factors that are related to this distribution.

family violence: Violence that occurs between persons who are living together as a family unit. Often this violence, which can be emotional as well as physical, takes place in the family dwelling.

interpersonal violence: Face-to-face interaction that produces physical harm or presents a threat of physical harm.

life-course-persistent (LCP) delinquent: A person who engages in antisocial behavior at a very early age; gradually becomes involved in increasingly more serious criminal activity, including violent criminal activity; and continues to commit criminal offenses well into his or her adult years.

multiassaultive family: A family whose members exhibit ongoing patterns of physical aggression.

official data: Data collected by juvenile justice agencies such as the police, courts, and correctional agencies.

posttraumatic stress disorder: A disorder in which a significant traumatic event is reexperienced by the individual and produces intense fear, anxiety, helplessness, and avoidance of stimuli associated with the trauma.

presumptive waiver: A statute-mandated waiver of a youth to criminal court that can be countermanded if the youth is able to prove he or she is suitable for juvenile justice services.

recidivist: A person who engages in further criminal activity following arrest, court conviction, or release from incarceration.

risk-focused delinquency prevention: An approach to delinquency prevention in which factors that contribute to delinquent behavior are identified and addressed.

self-report study: A study in which the participants report on their own behavior.

serious offense: An offense that results in significant personal injury, property loss, or damage.

social disorganization: The state of a community or society characterized by a lack of social control over community members.

victimization study: A study in which the participants are asked to report their victimization experiences.

violent juvenile offender: A juvenile who has committed a violent criminal offense against a person, such as rape, murder, aggravated assault, or robbery.

violent offense: An offense directed at the person or property of another that is intended to inflict injury on another and/or deprive them of their property. Violent offenses, however, do not always result in significant personal injury to the victim.[158]

Review Questions

1. What are the differences between chronic, serious, and violent juvenile offenders?
2. Many people feel that violent crime is much worse now than at any time in our country's history. Is this impression true?
3. How much violence is committed by chronic juvenile offenders?
4. In what ways has violent juvenile crime changed in recent years?
5. In what ways should society respond to the serious offender and the chronic offender?
6. What is the relationship between gender and violent juvenile crime?
7. What is the relationship between race and violent juvenile crime?
8. In what ways do cultural, political, economic, community, peer, family, and individual factors influence juvenile violence?
9. What factors need to be considered in the development of violence prevention initiatives for youths?

10. What types of youth violence prevention programs have been developed in recent years?
11. What are recent trends in the juvenile justice response to chronic and serious offenders?
12. What types of programs appear to hold the most promise in the battle to reduce violent juvenile crime?
13. Should more juveniles be transferred to adult courts? Why or why not?

Additional Readings

Hamparian, D. M., Schuster, R., Davis, J., & White, J. (1985). *The young criminal years of the violent few.* Washington, DC: Office of Juvenile Justice and Delinquency Prevention.

Henggeler, S. W. (1997). Treating serious anti-social behavior in youth: The MST approach. *Juvenile Justice Bulletin.* Washington DC: Office of Juvenile Justice and Delinquency Prevention.

Howell, J. C. (2009). *Preventing and reducing juvenile delinquency: A comprehensive framework* (2nd ed.). Los Angeles: Sage.

Klein, M. W., Maxson, C. L., & Miller, J. (Eds.). (1995). *The modern gang reader.* Los Angeles: Roxbury.

Loeber, R. & Farrington, D. P. (Eds.). (1998). *Serious and violent juvenile offenders: Risk factors and successful interventions.* Thousand Oaks, CA: Sage Publications.

National Criminal Justice Association. (1997). *Juvenile justice reform initiatives in the states: 1994–1996.* Washington, DC: Office of Juvenile Justice and Delinquency Prevention.

Office of Juvenile Justice and Delinquency Prevention. (1993). *Comprehensive strategy for serious, violent, and chronic juvenile offenders: Program summary.* Washington, DC: Office of Juvenile Justice and Delinquency Prevention.

National Institute of Justice. (1996). The cycle of violence revisited. *Research Preview.* Washington, DC: National Institute of Justice.

Notes

1. Sickmund, M., Snyder, H. N., & Poe-Yamagata, E. (1997). *Juvenile offenders and victims: 1*997 update on violence. Washington, DC: Office of Juvenile Justice and Delinquency Prevention.
2. Witkin, G., et al. (1991, April 18). Kids who kill. U. S. News and World Report, pp. 26–32.
3. Snyder, H. N. (2008). Juvenile arrests 2006. Juvenile Justice Bulletin. Washington, DC: Office of Juvenile Justice and Delinquency Prevention, 2008.

   Snyder, H. N. & Sickmund, M. (1999). Juvenile offenders and victims: 1999 national report. Washington, DC: Office of Juvenile Justice and Delinquency Prevention.

   United States Department of Justice, Federal Bureau of Investigation. (2004). *Crime in the United States, 2003.* Washington, DC: U.S. Department of Justice.
4. Snyder, H. N. & Sickmund, M. (1995). *Juvenile offenders and victims: A focus on violence.* Washington, DC: Office of Juvenile Justice and Delinquency Prevention.

5. Fox, J. A. (1996). *Trends in juvenile violence: A report to the United States attorney general on current and future rates of juvenile offending.* Washington, DC: Bureau of Justice Statistics.

6. Page, C. (n.d.). Commentary: Juvenile violence on rise in America. *The Daily Advertiser*. Retrieved February 2, 2009, from http://www.theadvertiser.com/article/20090113/OPINION/901130305.

Fenton, J. (n.d.). Summit seeks answers to juvenile violence. *The Baltimore Sun.* Retrieved February 2, 2009, from http://www.baltimoresun.com/news/local/baltimore_city/bal-md.ci.summit22jan22,0,4229140.story.

Snyder, 2008.

7. Benson, M. L. & Fox, G. L. (2004). When violence hits home: How economics and neighborhood play a role. *NIJ Research in Brief.* Washington, DC: National Institute of Justice.

8. Laub, J. H. & Sampson, R. J. (1991). The Sutherland-Glueck debate: On the sociology of criminological knowledge. *American Journal of Sociology, 96*, 1402–1440.

9. Glueck, S. & Glueck, E. (1967). *Predicting delinquency and crime.* Cambridge, MA: Harvard University Press.

10. Wolfgang, M. E., Figlio, R. M., & Sellin, T. (1972). *Delinquency in a birth cohort.* Chicago: University of Chicago Press.

11. Tracy, P. E., Wolfgang, M. E., & Figlio, R. M. (1985). *Delinquency in two birth cohorts: Executive summary.* Washington, DC: U.S. Department of Justice.

12. Shannon, L. W. (1982). *Assessing the relationship of adult criminal careers to juvenile careers: A summary.* Washington, DC: U.S. Department of Justice.

13. Dunford, F. W. & Elliott, D. S. (1984). Identifying career offenders using self-report data. *Journal of Research in Crime and Delinquency, 21*, 57–86.

14. Michalowski, R. J. (1985). *Order, law, and crime: An introduction to criminology* (pp. 277–278). New York: Random House.

15. Snyder, H. N. & Sickmund, M. (2006). *Juvenile offenders and victims: 2006 national report.* Washington, DC: Office of Juvenile Justice and Delinquency Prevention.

16. United States Department of Justice, Federal Bureau of Investigation. (2008). Table 38. *Crime in the United States 2007.* Retrieved February 5, 2009, from http://www.fbi.gov/ucr/cius2007/data/table_38.html.

17. Hamparian, D. M., Dinitz, S. & Schuster, R. (1978). *The violent few: A study of dangerous juvenile offenders.* Lexington, MA: Lexington Books.

18. Hamparian, D. M., Schuster, R., Davis, J., & White, J. (1985). *The young criminal years of the violent few.* Washington, DC: Office of Juvenile Justice and Delinquency Prevention.

19. Sickmund, Snyder, & Poe-Yamagata, 1997, p. 24. This resource cites H. N. Snyder.

20. Kelley, B. T., Huizinga, D., Thornberry, T. P., & Loeber, R. (1997). Epidemiology of serious violence. *Juvenile Justice Bulletin.* Washington, DC: Office of Juvenile Justice and Delinquency Prevention.

21. Snyder & Sickmund, 2006.

22. Snyder & Sickmund, 1999.

23. Snyder & Sickmund, 2006.

24. Loeber, R., Farrington, D. P., & Waschbusch, D. A. (1998). Serious and violent juvenile offenders. In R. Loeber & D. Farrington (Eds.), *Serious and violent juvenile offenders: Risk factors and successful interventions.* Thousand Oaks, CA: Sage Publications.

Hamparian, Schuster, Davis, & White, 1985.

Snyder, & Sickmund, 1995.

25. Snyder & Sickmund, 2006.

26. United States Department of Justice, Federal Bureau of Investigation. (1997). *Crime in the United States 1996.* Retrieved September 12, 2009, from http://www.fbi.gov/ucr/Cius_97/96CRIME/96crime4.pdf.

Poe-Yamagata, E. & Butts, J. A. (1996). *Female offenders in the juvenile justice system: Statistics summary.* Washington, DC: Office of Juvenile Justice and Delinquency Prevention.

United States Department of Justice, Federal Bureau of Investigation, 2004.

27. United States Department of Justice, Federal Bureau of Investigation. (2008). Table 35. *Crime in the United States 2007.* Retrieved February 6, 2009, from http://www.fbi.gov/ucr/cius2007/data/table_35.html.

28. Chesney-Lind, M. & Shelden, R. G. (2004) *Girls, delinquency and juvenile justice* (3rd ed.). Belmont, CA: Thompson/Wadsworth.

29. Chesney Lind & Shelden, 2004.

30. Chesney-Lind & Shelden, 2004, p. 15.

31. United States Department of Justice, Federal Bureau of Investigation. (2008). Table 43. *Crime in the United States 2007.* Retrieved February 6, 2009, from http://www.fbi.gov/ucr/cius2007/data/table_43.html.

32. Walker, S., Spohn, C., & DeLone, M. (1996). *The color of justice: Race, ethnicity, and crime in America.* Belmont, CA: Wadsworth Pub. Co.

33. Chesney-Lind & Shelden, 2004, p. 28.

34. Sickmund, Snyder, & Poe-Yamagata, 1997.

35. Calculations based on data from the United States Department of Justice, Federal Bureau of Investigation, *Crime in the United States reports for the years 1994 through 2003.*

36. Cook, P. J. & Laub, J. H. (1986). The (surprising) stability of youth crime rates. *Journal of Quantitative Criminology*, *2*, 265–277.

Osgood, D., O'Malley, P. M., Bachman, J. G., Johnston, L. D. (1989). Time trends and age trends in arrests and self-reported illegal behavior. *Criminology*, *27*, 389–417.

Snyder, H. N. & Sickmund, M. (1995). *Juvenile offenders and victims: A national report.* Washington, DC: Office of Juvenile Justice and Delinquency Prevention.

37. United States Department of Justice, Federal Bureau of Investigation. (2008) *Crime in the United States 2007.* Retrieved September 12, 2009, from http://www.fbi.gov/ucr/cius2007/offenses/clearances/index.html.

Snyder & Sickmund, 2006.

38. United States Department of Justice, Federal Bureau of Investigation. (2008). Table 28. *Crime in the United States 2007*. Retrieved February 6, 2009, from http://www.fbi.gov/ucr/cius2007/data/table_28.html.

Snyder & Sickmund, 2006.

39. Snyder & Sickmund, 2006.

40. Baum, K. (2005). Juvenile victimization and offending: 1993–2003. *Bureau of Justice Statistics Special Report*. Washington, DC: U.S. Department of Justice.

41. Baum, 2005.

42. Snyder & Sickmund, 2006.

43. Lauritsen, J. L. (2003). How families and communities influence youth victimization. *Juvenile Justice Bulletin*. Washington, DC: Office of Juvenile Justice and Delinquency Prevention.

44. Baum, 2005.

United States Department of Justice, Federal Bureau of Investigation. (2008). *Crime in the United States 1980–2007*. Retrieved February 11, 2009, from http://www.fbi.gov/research.htm.

45. Sickmund, Snyder, & Poe-Yamagata, 1997, p. 24. This resource cites Howard N. Snyder.

46. United States Department of Justice, Federal Bureau of Investigation. (2008). Table 28. *Crime in the United States 2007*. Retrieved February 11, 2009, from http://www.fbi.gov/ucr/cius2007/data/table_28.html.

47. Elrod, P. & Yokoyama, M. (2006). Juvenile justice in Japan. In P. C. Friday & X. Ren (Eds.), *Delinquency and juvenile justice systems in the non-Western World*. Monsey, NY: Criminal Justice Press.

48. Roberts, D. F., Foehr, U. G., & Rideout, V. (2005). *Generation M: Media in the lives of 8–18 year–olds*. Menlo Park, CA: The Henry J. Kaiser Family Foundation.

49. Gerbner, G., Gross, L., Signorielli, N., Morgan, M., & Jackson-Beeck, M. (1979). The demonstration of power: Violence profile no. 10. *Journal of Communication, 29*, 177–196.

Huston, H. C., Donnerstein, E., Fairchild, H., Feshbach, N. D., Katz, P. A., & Murray, J. P., et al. (1992). *Big world, small screen: The role of television in American society*. Lincoln, NE: University of Nebraska Press.

50. Bushman, B. J. & Anderson, C. A. (2001). Media violence and the American public: Scientific facts versus media misinformation. *American Psychologist, 56*, 477–489.

Friedrich-Cofer, L. & Huston, A. H. (1986). Television violence and aggression: The debate continues. *Psychological Bulletin, 100*, 364–371.

Johnson, J. G., Cohen, P., Smailes, E. M., Kasen, S., & Brook, J. S. (2002). Television viewing and aggressive behavior during adolescence and adulthood. *Science, 295*, 2468–2471.

Murray, J. P. (2008). Media violence: The effects are both real and strong. *American Behavioral Scientist, 51*, 1212–1230.

51. Huesmann, L. R., Moise-Titus, J., Podolski, C., & Eron, L. D. (2003). Longitudinal relations between children's exposure to TV violence and their aggressive and violent behavior in young adulthood: 1977–1992. *Developmental Psychology, 39*, 201–221.

52. American Psychological Association. (1985). *Violence on TV: Social issue release from the Board of Social and Ethical Responsibility for Psychology.* Washington, DC: American Psychological Association.

National Institute of Mental Health. (1982). *Television and behavior: Ten years of scientific progress and implications for the eighties.* Washington, DC: U.S. Government Printing Office.

53. Donnerstein, E., Slaby, R. G., & Eron, L. D. (1994). The mass media and youth aggression. In L. D. Eron, J. H. Gentry, & P Schlegel (Eds.), *Reason to hope: A psychosocial perspective on violence and youth.* Washington, DC: American Psychological Association.

Kubey, R. & Csikszentmihalyi, M. (1990). *Television and the quality of life: How viewing shapes everyday experience.* Hillsdale, NJ: Lawrence Erlbaum.

54. Bandura, A. & Walters, R. (1963). *Social learning and personality development.* New York: Holt, Rinehart & Winston.

Carlson, B. E. (1986). Children's beliefs about punishment. *American Journal of Orthopsychiatry, 56*, 308–312.

Perry, D. G., Perry, L. C., & Rasmussen, P. (1986). Cognitive social learning mediators of aggression. *Child Development, 57*, 700–711.

55. Schwendinger, H. & Schwendinger, J. S. (1985). *Adolescent subcultures and delinquency.* New York: Praeger Publishers. This resource contains an excellent examination of the relationship between the political economic context and delinquency.

56. Elliott, D. S. & Ageton, S. S. (1980). Reconciling race and class differences in self-reported and official estimates of delinquency. *American Sociological Review, 45*, 95–110.

57. DeNavis-Walt, C., Proctor, B. D., & Smith, J. C. (2008). *Income, poverty, and health insurance coverage in the United States: 2007.* (U.S. Census Bureau, Current Population Reports, pp. 60–235). Washington, DC: U.S. Government Printing Office.

58. U. S. Census Bureau. (2008). *Current population survey (CPS).* Retrieved February 14, 2009, from http://pubdb3.census.gov/macro/032008/pov/new01_100.htm.

59. United States Department of Justice, Federal Bureau of Investigation, 2008.

Huizinga, D. & Elliott, D. S. (1987). Juvenile offenders: Prevalence, offender incidence, and arrest rates by race. *Crime and Delinquency, 33*, 206–223.

Snyder & Sickmund, 2006.

Walker, Spohn, & DeLone, 1996.

60. DeNavis-Walt, Proctor, & Smith, 2008.

61. Brooks-Gunn, J. & Duncan, G. J. The effects of poverty on children. *The Future of Children, 7*, 55–71.

Carroll, L. & Jackson, P. I. (1983). Inequality, opportunity, and crime rates in central cities. *Criminology, 21*, 178–194.

National Research Council. (1989). *A common destiny: Blacks and American society.* Washington, DC: National Academy Press.

Wallace, R. (1991). Expanding coupled shock fronts of urban decay and criminal behavior: How U. S. cities are becoming "hollowed out." *Journal of Quantitative Criminology*, *7*, 333–356.

62. Wilson, W. J. (1991). Studying inner-city social dislocations: The challenge of public agenda research. *American Sociological Review*, *56*, 1–14.

63. Blau, J. R. & Blau, P. M. (1982). The cost of inequality: Metropolitan structure and violent crime. *American Sociological Review*, *47*, 114–129.

Block, R. (1979). Community, environment, and violent crime. *Criminology*, *17*, 46–57.

McGahey, R. (1986). Economic conditions, neighborhood organization, and urban crime. In A. J. Reiss & M. Tonry (Eds.), Communities and crime: Vol. 8. *Crime and justice*. Chicago: University of Chicago Press.

Parker, K. F., & Maggard, S. R. (2005). Structural theories and race-specific drug arrests: What structural factors account for the rise in race-specific drug arrests over time? *Crime and Delinquency*, *51*, 521–547.

Rosenfeld, R. (1986). Urban crime rates: Effects of inequality, welfare dependency, region, and race. In J. Byrne & R. Sampson (Eds.), *The social ecology of crime*. New York: Springer-Verlag.

Sampson, R. J. (1985). Structural sources of variation in race–age-specific rates of offending across major U.S. cities. *Criminology*, *23*, 647–673.

Schuerman, L. & Kobrin, S. (1986). Community careers in crime. In A. J. Reiss & M. Tonry (Eds.), Communities and crime: Vol. 8. *Crime and justice*. Chicago: University of Chicago Press.

Shaw, C. R. & McKay, H. D. (1972). *Juvenile delinquency and urban areas* (rev. ed.). Chicago: University of Chicago Press.

Simcha-Fagan, O. & Schwartz, J. E. (1986). Neighborhood and delinquency: An assessment of contextual effects. *Criminology*, *24*, 667–703.

Stark, R. (1987). Deviant places: A theory of the ecology of crime. *Criminology*, *25*, 893–910.

64. Gottfredson, D. C., McNeil, R. J., & Gottfredson, G. D. (1991). Social area influences on delinquency: A multilevel analysis. *Journal of Research in Crime and Delinquency*, *28*, 197–226.

65. Benson & Fox, 2004.

McGahey, 1986.

66. Agnew, R. (1985). A revised strain theory of delinquency. *Social Forces*, *64*, 151–167. This resource discusses the relationship between strain and delinquent behavior.

Agnew, R. (1992). Foundation for a general strain theory of crime and delinquency. *Criminology*, *30*, 47–87.

67. Bell, C. C. & Jenkins, E. J. (1991). Traumatic stress and children. *Journal of Health Care for the Poor and Underserved*, *2*, 175–188.

68. Bell & Jenkins, 1991.

69. Sheley, J. F. & Wright, J. D. (1993). Motivations for gun possession and carrying among serious juvenile offenders. *Behavioral Sciences and the Law*, *11*, 375–388.

70. Eaton, D. K., Kann, L., Kinchen, S., Shanklin, S., Ross, J., Hawkins, J., et al. (June 6, 2008). Youth risk behavior surveillance—United States, 2007. *MMWR Surveillance Summaries, 57*. Retrieved February 14, 2009, from http://www.cdc.gov/mmwr/preview/mmwrhtml/ss5704a1.htm.

71. U. S. Census Bureau. (1982). *Current population estimates and projections* (p. 25, No. 917). Washington, DC: U. S. Government Printing Office.

U. S. Census Bureau. (1992). *Current population reports: Population projections for the United States by age, sex, race, and Hispanic origin, 1992–2050.* Washington, DC: U.S. Government Printing Office.

United States Department of Justice, Federal Bureau of Investigation. (n.d.). *Uniform Crime Reports 1975–1994.* Washington, DC: Federal Bureau of Investigation.

United States Department of Justice, Federal Bureau of Investigation, *Crime in the United States, 1995–2007.* Retrieved February 20, 2009, from http://www.fbi.gov/ucr/ucr.htm.

Puzzanchera, C., Sladky, A., & Kang, K. (2008). Easy access to juvenile populations: 1990–2007. Retrieved February 20, 2009, from http://www.ojjdp.ncjrs.gov/ojstatbb/ezapop/.

72. Jackson, P. I. (1991). Crime, youth gangs, and urban transition: The social dislocations of postindustrial economic development. *Justice Quarterly*, *8*, 379–397.

73. Egley, A. Jr., & Ritz, C. E. (2006). Highlights of the 2004 National Youth Gang Survey. *OJJDP Fact Sheet.* Washington, DC: Office of Juvenile Justice and Delinquency Prevention.

74. Miller, W. B. (1958). Lower class culture as a generating milieu of gang delinquency. *Journal of Social Issues*, *14*, 5–19.

Moore, J. W. (1991). *Going down to the barrio: Homeboys and homegirls in change.* Philadelphia: Temple University Press.

Spergel, I. A. (1995). *The youth gang problem: A community approach.* New York: Oxford University Press.

Thrasher, F. M. (1927). *The gang.* Chicago: University of Chicago Press.

Vigil, J. D. (1988). *Barrio gangs.* Austin, TX: Texas University Press.

75. Spergel, 1995.

76. Jackson, 1991.

Sullivan, M. L. (1990). *Getting paid: Youth crime and work in the inner city.* Ithaca, NY: Cornell University Press.

77. Klein, M. W. (1995). *The American street gang: Its nature, prevalence, and control.* New York: Oxford University Press.

Yablonsky, L. (1966). *The violent gang.* Baltimore: Penguin.

78. Browning, K., Thornberry, T. P., & Porter, P. K. (1999). Highlights of findings from the Rochester Youth Development Study. *OJJDP Fact Sheet.* Washington, DC: Office of Juvenile Justice and Delinquency Prevention.

79. Bjerregaard, B. & Lizotte, A. J. (1995). Gun ownership and gang membership. *Journal of Criminal Law and Criminology*, *86*, 37–53.

80. Fagan, J. (1989). The social organization of drug use and drug dealing among urban gangs. *Criminology, 27,* 633–669.

81. Thornberry, T. P., Krohn, M. D., Lizotte, A. J., & Chard-Wierschem, D. (1993). The role of juvenile gangs in facilitating delinquent behavior. *Journal of Research in Crime and Delinquency, 30,* 55–87.

82. Sanders, W. B. (1994). *Gangbangs and drive-bys: Grounded culture and juvenile gang violence.* NewYork: Aldine de Gruyter.

83. Campbell, A. (1990). Female participation in gangs. In R. C. Huff (Ed.), *Gangs in America.* Beverly Hills, CA: Sage Publications.

Sanders, 1994.

Short, J. F. Jr. & Strodtbeck, F. L. (1963). The response of gang leaders to status threats: An observation on group process and delinquent behavior. *American Journal of Sociology, 68,* 571–578.

84. Thornberry, T. P. & Burch, J. H. II, (1997). Gang members and delinquent behavior. *Juvenile Justice Bulletin.* Washington, DC: Office of Juvenile Justice and Delinquency Prevention.

85. Gove, W. R. & Crutchfield, R. D. (1982). The family and juvenile delinquency. *Sociological Quarterly, 23,* 301–319.

Simons, R. L., Simons, L., & Wallace, L. (2004). *Families, delinquency, and crime: Linking society's most basic institution to antisocial behavior.* Los Angeles, CA: Roxbury Publishing Company.

86. Dornbusch, S. M., Carlsmith, J. M., Bushwall, S. J., Ritter, P. L., Leiderman, H., Hastorf, A. H., et al. (1985). Single parents, extended households, and the control of adolescents. *Child Development, 56,* 326–341.

Farrington, D. P. (1989). Early predictors of adolescent aggression and adult violence. *Violence and Victims, 4,* 79–100.

Henggeler, S. W. (1989). *Delinquency in adolescence.* Newbury Park, CA: Sage Publications.

Patterson, G. R. (1982). *Coercive family process.* Eugene, OR: Castalia Publishing Co.

Tolan, P. H. & Lorion, R. P. (1988). Multivariate approaches to the identification of delinquency proneness in adolescent males. *American Journal of Community Psychology, 16,* 547–561.

Wells, L. E. & Rankin, J. (1991). Families and delinquency: A meta-analysis of the impact of broken homes. *Social Problems, 38,* 71–93.

87. Eron, L. D. & Slaby, R. G. (1994). Introduction. In L. D. Eron, J. H. Gentry, & P Schlegel (Eds.), *Reason to hope: A psychosocial perspective on violence and youth.* Washington, DC: American Psychological Association.

88. Catalano, S. M. (2006) Criminal victimization, 2005. *Bureau of Justice Statistics Bulletin.* Washington, DC: U.S. Department of Justice.

89. Gershoff, E. T. (2002). Corporal punishment by parents and associated child behaviors and experiences: A meta-analytic and theoretical review. *Psychological Bulletin 128,* 539–579.

Straus, M. A. (1991). Discipline and deviance: Physical punishment of children and violence and other crime in adulthood. *Social Problems, 38,* 133–154.

90. Clement, M. (1997). *The juvenile justice system: Law and process.* Boston: Butterworth-Heinemann.

91. Gibbons, A., Moore, A., Jaffe, K., & Chalk, R. (2004). *Media handbook: Child abuse and neglect.* Washington, DC: Child Trends.

92. Chalk, R., Gibbons, A., & Scarupa, H. J. (2002). The multiple dimensions of child abuse and neglect: New insights into an old problem. *Child Trends Research Brief.* Washington, DC: Child Trends.

93. Simons, Simons, & Wallace, 2004.

Straus, 1991.

94. Boney-McCoy, S. & Finkelhor, D. (1995). Psychosocial sequelae of violent victimization in a national youth sample. *Journal of Consulting and Clinical Psychology,* 63, 726–736.

Chalk, Gibbons, & Scarupa, 2002.

95. Boney-McCoy & Finkelhor, 1995.

Briere, J. N. (1992). *Child abuse trauma: Theory and treatment of the lasting effects.* Newbury Park, CA: Sage Publications.

Chalk, Gibbons, & Scarupa, 2002.

Kelley, B. T., Thornberry, T. P., & Smith, C. A. (1997). In the wake of childhood maltreatment. *Juvenile Justice Bulletin.* Washington, DC: Office of Juvenile Justice and Delinquency Prevention.

96. Chalk, Gibbons, & Scarupa, 2002.

Finkelhor, D. & Dziuba-Leatherman, J. (1994). Victimization of children. *American Psychologist, 49,* 173–183.

Scott, K. D. (1992). Childhood sexual abuse: Impact on a community's mental health status. *Child Abuse and Neglect, 16,* 285–295.

97. Widom, C. S. (1992). *The cycle of violence.* Washington, DC: National Institute of Justice.

98. National Institute of Justice. (1996). The cycle of violence revisited. *Research Preview.* Washington, DC: National Institute of Justice.

99. Kelley, Thornberry, & Smith, 1997.

100. Centers for Disease Control and Prevention. (2009). Understanding youth violence. *Fact Sheet.* Retrieved February 21, 2009, from http://www.cdc.gov/ViolencePrevention/pdf/YV-FactSheet-a.pdf.

101. Browne, A. (1987). *When battered women kill.* New York: The Free Press.

102. Ohlin, L. & Tonry, M. (1989). Family violence in perspective. In L. Ohlin & M. Tonry (Eds.), Family Violence: Vol. 11. *Crime and Justice.* Chicago: University of Chicago Press.

103. Gelles, R. J. (1982). Domestic criminal violence. In M. E. Wolfgang & N. A. Weiner (Eds.), *Criminal violence.* Beverly Hills, CA: Sage Publications.

104. Pillemer, K. & Finkelhor, D. (1988). The prevalence of elder abuse: A random sample survey. *The Gerontologist, 28,* 51–57.

105. Eron, L. D., Gentry, J. H. & Schlegel, P. (Eds.). *Reason to hope: A psychosocial perspective on violence and youth.* Washington, DC: American Psychological Association. This resource cites E. L. Feindler and J. V. Becker.

106. Hotaling, G. T., Straus, M. A., & Lincoln, A. J. (1989). Intra-family violence, and crime and violence outside the family. In L. Ohlin & M. Tonry (Eds.), Family Violence: Vol. 11. *Crime and Justice*. Chicago: University of Chicago Press.

107. Gold, M. (1970). *Delinquent behavior in an American city*. Belmont, CA: Brooks/Cole.

108. Raine, A. (1993). *The psychopathology of crime: Criminal behavior as a clinical disorder.* San Diego, CA: Academic Press.

109. Hutchings, B. & Mednick, S. A. (1977). Criminality in adoptees and their adoptive and biological parents: A pilot study. In S. A. Mednick & K. O. Christiansen (Eds.), *Biosocial bases of criminal behavior*. New York: Gardner Press.

110. Raine, 1993, p. 66.

111. Gordon, R. A. (1987). SES versus IQ in the race-IQ delinquency model. *International Journal of Sociology and Social Policy*, *7*, 29–96.

Hirschi, T. & Hindelang, M. J. (1977). Intelligence and delinquency: A revisionist review. *American Sociological Review*, *42*, 571–586.

Wilson, J. Q. & Herrnstein, R. J. (1985). *Crime and human nature.* New York: Simon & Schuster.

112. Menard, S. & Morse, B. J. (1984). A structuralist critique of the IQ-delinquency hypothesis: Theory and evidence. *American Journal of Sociology*, *89*, 1347–1378.

113. Moffitt, T. E. (1993). The neuropsychology of conduct disorder. *Development and Psychology*, *5*, Special issue.

114. Fishbein, D. H. (1990). Biological perspectives in criminology. *Criminology*, *28*, 27–72.

Buchanan, C. M., Eccles, J. S., & Becker, J. B. (1992). Are adolescents the victims of raging hormones? Evidence for activational effects of hormones on moods and behavior at adolescence. *Psychological Bulletin*, *111*, 62–107.

115. Moffitt, T. E. (1993). Adolescence-limited and life-course-persistent antisocial behavior: A developmental taxonomy. *Psychological Review*, *100*, 674–701.

116. Mednick, S. A. & Christiansen, K. O. (Eds.). (1977). *Biosocial bases of criminal behavior.* New York: John Wiley.

117. Rathus, S. (1996). *Psychology,* New York: Holt, Rinehart & Winston.

118. Walters, G. D. (1992). A meta-analysis of the gene-crime relationship. *Criminology*, *30*, 595–613.

119. Raine, 1993.

120. Office of Juvenile Justice and Delinquency Prevention. (1993). *Comprehensive strategy for serious, violent, and chronic juvenile offenders: Program summary* (p. 13). Washington, DC: Office of Juvenile Justice and Delinquency Prevention.

121. National Criminal Justice Association. (1997). *Juvenile justice reform initiatives in the states: 1994–1996.* Washington, DC: Office of Juvenile Justice and Delinquency Prevention.

122. Bureau of Family and Community Health. *Violence prevention and intervention services.* Retrieved February 21, 2009, from http://www.mass.gov.

123. National Criminal Justice Association, 1997.

124. Hsia, H. M. (1997). Allegheny County, PA: Mobilizing to reduce juvenile crime. *Juvenile Justice Bulletin.* Washington, DC: Office of Juvenile Justice and Delinquency Prevention.

125. National Criminal Justice Association, 1997.

126. Ruefle, W. & Reynolds, K. M. (1995). Curfews and delinquency in major American cities. *Crime and Delinquency, 41*, 347–363.

127. Hirschel, J. D., Dean, C. W., & Dumond, D. (2001). Juvenile curfews and race: A cautionary note. *Criminal Justice Policy Review, 12*, 197–214.

Males, M. A. (2000). Vernon, Connecticut's juvenile curfew: The circumstances of youths cited and effects on crime. *Criminal Justice Policy Review, 11*, 254–267.

Ruefle & Reynolds, 1995.

128. Snyder & Sickmund, 2006.

129. Palumbo, D. J. & Ferguson, J. L. (1995). Evaluating gang resistance education and training (GREAT): Is the impact the same as that of drug abuse resistance education (DARE)? *Evaluation Review, 19*, 597–619.

130. Esbensen, F., Osgood, D. W., Taylor, T. J., Peterson, D., & Freng, A. (2001). How great is G.R.E.A.T? Results from a longitudinal quasi-experimental design. *Criminology and Public Policy, 1*, 87–118.

131. Binder, A., Geis, G., & Dickson, D. B. (1997). *Juvenile delinquency: Historical, cultural, and legal perspectives* (2nd ed.). Cincinnati, OH: Anderson.

132. Henggeler, S. W. (1997). Treating serious anti-social behavior in youth: The MST approach. *Juvenile Justice Bulletin.* Washington DC: Office of Juvenile Justice and Delinquency Prevention.

133. Larson, C. P. (1980). Efficacy of prenatal and postpartum home visits on child health and development. *Pediatrics, 66*, 191–197.

Olds, D. L., Henderson, C. R. Jr., Tatelbaum, R., & Chamberlin, R. (1988). Improving the life-course development of socially disadvantaged mothers: A randomized trial of nurse home visitation. *American Journal of Public Health, 78*, 1436–1445.

134. Berrueta-Clement, J. R., Schweinhart, L. J., Barnett, W. S., Epstein, A. S., & Weikart, D. P. (1984). *Changed lives: The effects of the Perry Preschool Program on youths through age 19.* Ypsilanti, MI: The High/Scope Press.

Johnson, D. L. & Walker, T. (1987). Primary prevention of behavior problems in Mexican-American children. *American Journal of Community Psychology, 15*, 375–385.

Lally, J.R., Mangione, P.L., & Honig, A.S. (1988). The Syracuse University Family Development Research Project: Long-range impact of an early intervention with low-income children and their families. In D. R. Powell (Ed.), *Annual advances in applied developmental psychology*. Norwood, NJ: Ablex Publishing Corporation.

135. Wasserman, G. A. & Miller, L. S. (1998). The prevention of serious and violent juvenile offending. In R. Loeber & D. Farrington (Eds.), *Serious and violent juvenile offenders: Risk factors and successful interventions.* Thousand Oaks, CA: Sage Publications.

136. Sickmund, M., Sladky, A., & Kang, W. (2005). *Easy access to juvenile court statistics: 1985-2005*. Retrieved February 21, 2009, from http://ojjdp.ncjrs.gov/ojstatbb/ezajcs/.

137. Sickmund, Snyder, & Poe-Yamagata, 1997.

138. Sickmund, Snyder, & Poe-Yamagata, 1997, p. 28. This resource cites Torbet, P., Gable, R., Hurst, H., IV, Montgomery, I., Szymanski, L., & Thomas, D. (1996). *State responses to serious and violent juvenile crime: Research report.* Washington, DC: Office of Juvenile Justice and Delinquency Prevention.

139. Sickmund, Snyder, & Poe-Yamagata, 1997, p. 28.

140. Sickmund, Snyder, & Poe-Yamagata, 1997, p. 28.

141. Snyder & Sickmund, 2006.

142. Winner, L., Lanza-Kaduce, L., Bishop, D. M., & Frazier, C. E. (1977). The transfer of juveniles to criminal court: Reexamining recidivism over the long term. *Crime and Delinquency*, *43*, 548–563.

143. McGowan, A., Hahn, R., Liberman, A., Crosby, A., Fullilove, M., Johnson, R., et al. (2007). Effects on violence of laws and policies facilitating the transfer of juveniles from the juvenile justice system to the adult justice system: A systematic review. *American Journal of Preventive Medicine*, *32*, 7–21.

144. Sickmund, Sladky, & Kang, 2005.

145. Sickmund, Snyder, & Poe-Yamagata, 1997

Snyder & Sickmund, 2006.

146. Butterfield, F. (1997, July 21). With juvenile courts in chaos, critics propose their demise. *New York Times*, pp. A–13.

147. Sickmund, Snyder, & Poe-Yamagata, 1997.

148. Barton, W. H. & Butts, J. A. (1990). Viable options: Intensive supervision programs for juvenile delinquents. *Crime and Delinquency*, *36*, 238–256.

149. Coordinating Council on Juvenile Justice and Delinquency Prevention. (1996). *Combating violence and delinquency: The National Juvenile Justice Action Plan.* Washington, DC: Office of Juvenile Justice and Delinquency Prevention.

150. Howell, J. C. (Ed.). (1995). *Guide for implementing the comprehensive strategy for serious, violent, and chronic juvenile offenders.* Washington, DC: Office of Juvenile Justice and Delinquency Prevention.

151. Altschuler, D. M. & Armstrong, T. L. (1996). Aftercare not afterthought: Testing the IAP model. *Juvenile Justice, 3*, 15–22.

152. Fagan, J. A. (1990). Treatment and reintegration of violent juvenile offenders: Experimental results. *Justice Quarterly*, *7*, 233–263.

153. Zalkind, P. & Simon, R. J. (2004). *Global perspectives on social issues, Juvenile justice systems.* Lanham, MD: Lexington Books.

154. McKee, G. R. & Shea, J. J. (1999). Competency to stand trial in family court: Characteristics of competent and incompetent juveniles. *Journal of the American Academy of Psychiatry and Law*, *27*, 65–73.

155. Courtesy of Thomas Grisso, PhD., Director of Psychology and Director of the Law-Psychiatry Program at the University of Massachusettes Medical School.

Dusky v. United States, 362 U.S. 402 (1960).

Drope v. Missouri, 420 U.S. 162 (1975).

156. Virginia Juvenile Competency Statute, Virginia Code, Sec. 16.1 to 16.35 et. Seq.

157. Grisso, T. (1980). Juveniles' capacities to waive *Miranda* rights: An empirical analysis. *California Law Review*, *68*, 1134–1166.

158. Beers, M. H. & Berkow, R. (1999). *The Merck manual of diagnosis and therapy* (17th ed.). New York: John Wiley & Sons.

Present Conditions and Future Directions in Juvenile Justice

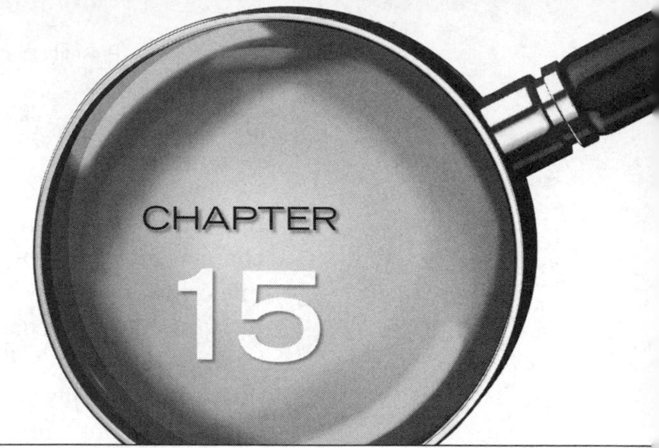

CHAPTER OBJECTIVES

After studying this chapter, you should be able to

- Describe historical changes in juvenile justice practice, including the significance of the legal reform years, that began in the mid-1960s and to the "get tough" movement that began in the mid-1970s
- Explain why rehabilitation has played an important role in juvenile court history
- Describe the problems that confront contemporary juvenile courts
- Describe the problems that presently confront juvenile corrections
- Explain why some critics of the juvenile court feel that it should be abolished
- Explain why some critics of the juvenile court feel that it should be reinvented or fundamentally reformed
- Explain why some people would like to maintain the status quo in juvenile justice
- Describe what is meant by the balanced approach in juvenile justice
- Describe the problems that are likely to confront juvenile justice in the near future
- Describe why the authors feel there are good reasons to be optimistic about the future of juvenile justice

CHAPTER OUTLINE

- Introduction
- Changes in the Quality of Juvenile Justice Practice
- The Resiliency of Treatment and Rehabilitation in Juvenile Justice
- Contemporary Problems and Issues in Juvenile Justice
- The Future of Juvenile Justice
- Problems That Will Continue to Influence America's Youth, Families, Communities, and Juvenile Justice Agencies
- Legal Issues
- Chapter Summary
- Key Concepts
- Review Questions
- Additional Readings
- Notes

Introduction

Since the development of the first juvenile court in Cook County, Illinois, more than 100 years ago, juvenile justice has undergone a number of changes. Some of these changes represent fundamental differences in how juvenile justice agencies view juvenile crime, how they are organized, and how they respond to youth problems. In other respects, however, changes in juvenile justice are more cosmetic than real. Indeed, it can be argued that juvenile justice has changed very little over time. Today, many youths involved in the juvenile justice process and the people who work with them face many of the same problems that were confronted by juvenile justice clients and practitioners more than 100 years ago. Moreover, given the present state of juvenile justice practice, it appears that many of the problems that have historically confronted juvenile justice will continue to affect the quality of services received by juvenile justice clients as well as the working conditions within juvenile justice agencies. Yet, despite the many problems that continue to challenge juvenile justice professionals and the children they work with, there are hopeful signs for the future. This chapter examines some of the major changes that have occurred in juvenile justice since the first juvenile court was established. In addition, it explores problems that presently affect the quality of juvenile justice practice, and it makes several predictions about future developments in juvenile justice practice in the early part of the 21st century.

Changes in the Quality of Juvenile Justice Practice

Juvenile justice has changed in a number of ways in the last 100 years. Indeed, changes in juvenile law, the structure and operation of juvenile courts, and the quality and quantity of services offered to youths and their families have occurred in this period. Many of the changes have undoubtedly improved the ability of juvenile justice agencies to meet the needs of youths, families, and the communities they serve. Other changes, however, have had little or no positive effect on children, families, or community safety. Unfortunately, today, as in the past, children and their families are sometimes harmed as a result of their involvement in juvenile justice. Indeed, some have argued that one distinguishing characteristic of juvenile justice since its inception has been its resistance to fundamental change.[1]

Early Reforms: The Development of the Juvenile Court

The early juvenile courts were developed in response to a particular set of problems confronting urban citizens during the late 1800s. One of these problems was the large and ever-growing number of neglected, dependent, and wayward youths, many of whom were also engaged in illegal behavior. From the perspective of many city dwellers, these children—more than a few were the offspring of immigrants who did not speak English—were felt to constitute a "dangerous class" that threatened the established order.[2] Julia Lathrop, a noted 19th-century **child saver** stated the following:

> There are at the present moment in the state of Illinois, especially in the city of Chicago, thousands of children in need of active intervention for their preservation from physical, mental, and moral destruction. Such intervention is demanded, not only by sympathetic consideration for their well-being, but also in the name of the commonwealth, for the preservation of the State. If the child is the material out of which men and women are made, the neglected child is the material out of which paupers and criminals are made.[3]

child savers Affluent 19th-century reformers who were concerned about controlling and training youths so that the youths could become productive, law-abiding citizens. The Child Savers played a major role in the establishment of the juvenile courts.

Although wealthy citizens in developing urban areas desired to control the increasing number of problem children, they confronted several major obstacles. First, criminal courts had no jurisdiction over youths unless they engaged in criminal activity. Second, the existing criminal justice system was ineffective at dealing with these youths, because it punished youth offenders either too severely or not at all.[4] For example, when young children were brought before criminal courts, the outcomes were often less than satisfactory. In many instances, young children were placed in adult jails and prisons, despite the fact that it was generally recognized that placing children with hardened criminals was hardly an effective way to protect children or community safety. The judges who felt this way tended to balk at jailing children for minor criminal behaviors. As a result, many citizens began to see adult courts as an ineffective mechanism for controlling problem youths.

In addition to criminal courts, a number of other mechanisms were established throughout the 1800s to deal with problem youths, including houses of refuge, the placing out system, probation, training schools, and orphanages. None of these mechanisms, however, proved to be particularly effective. Furthermore, the Illinois Supreme Court, in *People v. Turner* (1870), ruled that a child could not be institutionalized without having committed an offense. The Court noted that the institutionalization and punishment of youths who had not committed a criminal offense were violations of basic **due process** protections.[5] Although this case did not affect efforts to control youths outside of Illinois, it prevented the institutionalization of youths who had not committed crimes in that state and required that, prior to commitment, youths be given due process protection. In effect, this case stopped the long-standing practice of institutionalizing problem youths who had not been charged with a crime—a practice that many people favored because criminal courts could not legally control these children.

The inability of criminal courts to effectively deal with a large population of wayward youths, along with the Illinois Supreme Court's decision in *Turner*, presented a major obstacle to child savers in Chicago, who were interested in controlling and reforming problem children. To circumvent the *Turner* decision and to reestablish control over wayward children, the child savers invented a new legal institution designed to control and reform troublesome youths. This new legal institution was the juvenile court.[6]

After the first juvenile court was established in Chicago, the idea of a court designed especially for children became so appealing that by 1920 all but three states had developed juvenile courts.[7] These courts were designed to be different from adult courts. They were essentially **chancery courts** (civil courts) founded on the legal concept of ***parens patriae*** (i.e., the idea that the state must act as a protector of children).[8] The goals of the new courts were, at least in theory, to serve the best interests of children and to protect the community. What distinguished the juvenile courts from adult criminal courts was their concern for the interests of the children who came before the bench. As Judge Julian Mack of the Cook County (Chicago) Juvenile Court wrote in 1909, "The problem for the determination by the judge is not, Has this boy or girl committed a specific wrong, but What is he, how has he become what he is, and what had best be done in his interest and in the interest of the state to save him from a downward career."[9] As this statement makes clear, the focus of the juvenile courts was on training and reform as opposed to punishment.

In order to facilitate their efforts to reform children, the new courts were given broad discretionary powers over a range of youths—not only youths involved in criminal

due process
An established course of proceedings designed to protect individual rights and liberties and to ensure that people are treated with fairness before the law. Although juveniles have most of the due process protections afforded adults in the criminal courts, they do not have all of these rights. Moreover, because of the informality found in many juvenile courts, legal protections that are theoretically available may in fact be unavailable.

chancery court English court concerned primarily with property rights. Over time, these courts became more involved in the welfare of families and children, and served as a model for the development of the juvenile court.

parens patriae A legal doctrine according to which the state has an obligation to protect and serve the interests of children.

behaviors, but also neglected, dependent, and other types of wayward or unfortunate youths. They were also given expanded powers to intervene in the relationships between parents and their children.[10] **Informal juvenile court** procedures, another characteristic that distinguished juvenile courts from criminal courts, were believed to give them the flexibility they needed to handle a range of child and family problems. Importantly, due process protections were felt to hinder the ability of the juvenile court to serve the best interests of children. As a result, juvenile court hearings were typically held in an office, representation by counsel was rare, few records of hearings were made, specific charges were avoided in favor of general charges of delinquency, and witnesses and complainants were rarely present.

informal juvenile court
A court that places little emphasis on due process protections for youths. Because juvenile courts exist, in theory, to protect the best interests of children, legal protections for children are felt to be unnecessary within an informal court.

The particular approach taken by juvenile courts in exercising their powers over children and families was influenced by reformers' beliefs about the causes of delinquency. The reformers who were instrumental in the establishment of the juvenile courts thought that several factors contributed to delinquent behavior, including biologic and mental defects as well as poverty and inappropriate parenting.[11] However, poverty and inappropriate parenting were seen as types of individual pathologies rather than the products of the political and economic conditions within which people lived. Thus, the responses to youth crime and waywardness developed by juvenile courts focused on controlling, treating, and reforming the individual rather than on changing harmful social conditions. Barry Krisberg and James Austin note the following in their book *Reinventing Juvenile Justice*:

> The thrust of Progressive Era reforms was to found a more perfect control system to restore social stability while guaranteeing the continued hegemony of those with wealth and privilege. Reforms such as the juvenile court are ideologically significant because they preserved the notion that social problems (in this case delinquency, dependency, and neglect) could be dealt with on a case-by-case basis, rather than through broad-based efforts to redistribute wealth and power throughout society. The chief dilemma for advocates of the juvenile court was to develop an apparently apolitical or neutral system, while preserving differential treatment for various groups of children. The juvenile court at first lacked a core of functionaries who could supply the rationale for individualized care for wayward youth, but soon these needs were answered by the emergence of psychiatry, psychology, and criminology, as well as by the expanding profession of social work.[12]

The belief that youth problems were products of individual pathology led juvenile justice agencies to focus on control, treatment, and rehabilitation rather than other types of correctional responses, at least until the late 1970s, when more people began to believe that crime was the product of individual choice.[13]

MYTH VS REALITY

Responses to Juvenile Crime Are Based on Theoretical Ideas

Myth—Theories of delinquent behavior have no practical significance.

Reality—Every response to juvenile crime is based on some theory as to the causes of delinquency. For example, reformers who developed the juvenile court created responses to juvenile crime that were consonant with their understanding of delinquent and problem behaviors. Similarly, contemporary responses to juvenile crime are also based on theories of delinquent behavior.

The development of the juvenile court represented a significant development in efforts to control and reform children. It became the legal mechanism by which power over children and their parents could be exercised. However, the court itself, even in instances in which it adhered to a reform and treatment orientation, had little influence on other institutions that had been established to deal with troublesome youths. The development of the juvenile court did nothing to change, for example, the often appalling conditions found in the facilities—the houses of refuge, reform schools, jails, and prisons—that typically housed delinquent youths. Moreover, the informal procedures of the juvenile court lent themselves to abuses by judges, who had few checks on their power.

The Legal Reform Years: The Supreme Court's Focus on the Juvenile Court, 1966–1971

Until the mid-1960s, juvenile court operations continued to be characterized by the same informality found in the first juvenile courts. Despite this informality (or perhaps because of it), juvenile courts did provide assistance to many children and their families. By the mid-1960s, however, juvenile court practices began to be more carefully scrutinized by the U.S. Supreme Court. Beginning with the case of *Kent v. United States* (1966), the Supreme Court made it clear that it was going to take a closer look at the due process protections afforded youths in the juvenile court.[14] In subsequent decisions, such as those in the landmark cases *In re Gault* (1967)[15] and *In re Winship* (1970),[16] the Supreme Court expanded the due process rights of children appearing before a juvenile court. As a result of these cases, much of the traditional informality of the courts was replaced with procedural requirements designed to give children enhanced legal protections. By 1971, youths had the right to counsel, the right to adequate and timely written notice, the right to confront and cross-examine witnesses, and the privilege against self-incrimination in instances when they were subject to incarceration.

Closer scrutiny of juvenile court practice by the Supreme Court led to more formality in juvenile courts and increased concern with the due process rights of juveniles. These Supreme Court cases can be seen as important milestones in the development of the juvenile court. However, in another important case, *McKeiver v. Pennsylvania* (1971), the Supreme Court ruled that juveniles do *not* have a constitutional right to a jury trial.[18]

FYI

The Supreme Court Case *In re Gault* Addressed the Informal Nature of Juvenile Court

The following excerpts from the *Gault* decision clearly demonstrate the U.S. Supreme Court's concern with the informality that characterized juvenile court until that time. In the opinion, the majority justices noted the following:

> The early reformers were appalled by adult procedures and penalties, and by the fact that children could be given long prison sentences and mixed in jails with hardened criminals. They were profoundly convinced that society's duty to children could not be confined by the concept of justice alone.... The rules of criminal procedure were therefore altogether inapplicable.... [However] Juvenile Court history has again demonstrated that unbridled discretion, however benevolently motivated, is frequently a poor substitute for principle and procedure.... The absence of procedural rules based upon constitutional principle has not always produced fair, efficient, and effective procedures. Departures from established principles of due process have frequently resulted not in enlightened procedure, but in arbitrariness. ...Under our Constitution, the condition of being a boy does not justify a kangaroo court.[17]

This case indicated that the Supreme Court was not willing to grant juveniles all of the due process protections given to adults in criminal courts. In the majority opinion, the justices argued that jury trials would make the juvenile courts indistinguishable from criminal courts and would create a more adversarial type of proceeding, to the disadvantage of children.[19]

The *McKeiver* case gave notice that the Supreme Court was not willing to completely remove the informality that has been the hallmark of juvenile courts. Overall, the cases heard by the Supreme Court between 1966 and 1971 that dealt with juvenile court practices indicated that juvenile courts should be more concerned about due process rights, and limited the blatant informality common in juvenile courts until that time. Yet, the Supreme Court also indicated that it continued to support some degree of informality, which it believed was necessary if the juvenile courts were to accomplish their mission of serving the best interests of children and protecting the community.

The "Get Tough" Movement in Juvenile Justice

The period between 1966 and 1971, during which the Supreme Court focused attention on the need for due process protections in the juvenile courts, was immediately followed by another important development in juvenile justice history—the appearance of the **"get tough" movement**. The "get tough" movement is part of a conservative reform movement that began to dominate the national debate over juvenile crime by the late 1970s[20] and that continues to play a significant role in juvenile justice policy development. According to advocates of a "get tough" approach to juvenile crime, the primary obligation of juvenile justice agencies is to protect citizens from youthful offenders. Conservatives have called for a number of changes in the juvenile justice process, particularly in how juvenile justice agencies deal with chronic, serious, and violent juvenile offenders. These changes include developing a more adversarial model of juvenile justice, strengthening the position and improving the effectiveness of the prosecuting attorney in the juvenile justice process, developing mandatory sentencing guidelines, opening access to juvenile

"get tough" movement
A legal reform movement that began during the late 1970s. The goals of the movement included developing a more punitive juvenile court orientation or abolishing the juvenile courts so that more severe sanctions could be leveled against juvenile offenders.

COMPARATIVE FOCUS

The English and American Approach to Juvenile Justice Are Similar in Some Respects, But There Are a Number of Important Differences

One thing that distinguishes the American juvenile justice process from the process in most countries is that there are actually multiple juvenile justice "systems" in the United States. In the United States, each state, the District of Columbia, the federal government, and each U.S. territory has its own juvenile justice process. In most countries, including England, there is only one juvenile justice process for the entire country, although there are aspects of the English system that make possible a more "local flavor" to juvenile justice operations. Also, both the English and American approaches to juvenile justice have shifted to more of an accountability model over time that places more responsibility on youths and their parents. In England, as in the United States, more youths have been referred to formal court over time and there has been an increase in the number of youths placed in custody. However, the English approach still maintains more emphasis on the prevention of juvenile crime and the treatment and rehabilitation of youths than juvenile justice in many jurisdictions in the United States. In England, parents and their children are required to play an active role in the rehabilitation of children involved in the juvenile justice process and there is considerable emphasis on identifying and addressing the causes of a youth's illegal behavior, not simply punishing children for their misdeeds.[26]

court records and procedures, and developing mechanisms for identifying potentially violent or chronic offenders.[21] From a correctional standpoint, "get tough" reformers are concerned with accountability, deterrence, and punishment as opposed to treatment and rehabilitation.[22]

The "get tough" movement has pushed legislators to make a number of substantive changes in the operation of juvenile justice. Initially, a number of states implemented reforms intended to fine-tune juvenile justice operations. More recently, there has been more of a focus on implementing comprehensive juvenile justice reforms.[23] As a result, a number of states have given prosecutors a more prominent role in the juvenile justice process, penalties for juvenile offenders have been stiffened, mandatory sentencing guidelines have been established, new mechanisms for transferring juveniles to criminal courts have been created and existing procedures simplified, and increasing numbers of youths have been incarcerated in juvenile and adult correctional facilities.[24] In short, many juvenile courts have become more concerned with protecting the community and with punishing offenders than with their traditional mission—serving the best interests of children.[25]

The Resiliency of Treatment and Rehabilitation in Juvenile Justice

Despite the reforms just mentioned, treatment and **rehabilitation** continue to play a significant role in juvenile justice in many jurisdictions. Today, a wider range of treatment and rehabilitation programs exist for juvenile offenders than at any other time in our nation's history. To begin with, there are a wide variety of community-based programs, institutions, and interventions, such as diversion programs, probation, intensive supervision programs, Multisystemic Therapy, house arrest, group homes, shelter facilities, day treatment programs, treatment foster care, restitution, mediation, and community service programs, outdoor experiential programs, intensive **aftercare** programs, and short-term substance abuse treatment programs. In addition, a number of residential treatment programs for youths exist around the country, some of which provide highly specialized treatment for chronic, serious, and violent juvenile offenders. Moreover, other programs are presently being developed.

rehabilitation The process of restoring an individual's ability to play a constructive role in society through education, training, or therapy.

aftercare Services delivered to youths after they are released from a residential setting and while they are still under the jurisdiction of the state or juvenile court.

There continues to be widespread support for rehabilitation and treatment among the general public and those who work in juvenile justice. Public opinion surveys have consistently found high levels of support for the treatment and rehabilitation of juvenile offenders. For example, a national survey conducted in the 1980s found that 73% of the respondents felt that treatment and rehabilitation, rather than punishment, should be the primary mission of the juvenile courts. Similar findings are evident in polls conducted

MYTH VS REALITY

Most Citizens Favor the Treatment of Juvenile Offenders

Myth—Most Americans believe that the goal of juvenile courts should be the punishment of youths who violate the law.

Reality—Research on public attitudes toward crime and justice has consistently found that most Americans believe that treatment and rehabilitation are the appropriate goals of juvenile justice and they are willing to support effective rehabilitation programs.[27]

in California, Illinois, and Michigan in the 1980s and early 1990s.[28] In a 1995 national poll conducted as part of the survey research program at Sam Houston State University, approximately 48% of respondents indicated that the most important purposes to be achieved in sentencing juveniles were the training, education, and counseling of offenders. In comparison, only 30% indicated that the purpose of juvenile court sentencing should be to give offenders the punishment that they deserved.[29] Also, in a 2005 survey in Pennsylvania that examined respondents' willingness to pay for various approaches to juvenile crime, respondents indicated that they were as willing to pay for rehabilitative approaches as for more punitive approaches and tended to favor more rehabilitative efforts to reduce juvenile crime.[30] Similar support for corrective treatment of juvenile offenders has also been found in state-level surveys and among judges and correctional administrators.[31] In a study that examined a sample of Florida judges' support for the treatment ethic, Gordon Bazemore and Lynette Feder found that the majority of juvenile division judges supported rehabilitation, although these judges felt that crime control was also important.[32] Similarly, a survey of chief administrators of state-operated secure correctional facilities in 1992 found that 84% of the respondents indicated that treatment and rehabilitation characterized the correctional philosophy employed at their facility.[33]

Juvenile justice practice has changed in a number of ways over the last 100 years. The informality of earlier juvenile courts has been mitigated by the inclusion of formal procedures intended to ensure due process. Moreover, the *parens patriae* philosophy, with its focus on serving the best interests of the child, has been supplanted in many instances by a punishment and deterrence orientation. Yet, despite these changes, many juvenile courts continue to retain a great deal of informality. Although there has been a clear movement toward punishment, rehabilitation continues to be a key ingredient in many court and correctional programs in many communities. The following section examines how the opposing orientations—informality and rehabilitation versus formality and punishment—influence the operation of contemporary juvenile justice practice.

Contemporary Problems and Issues in Juvenile Justice

Today, juvenile justice faces a number of problems and issues, some of which call into question the continued existence of separate courts for juveniles. Can the juvenile courts protect the due process rights of children? Are juvenile corrections programs able to rehabilitate youths and protect the community? Should the juvenile courts be abolished? These are some of the important questions that presently confront those who work in the field of juvenile justice.

Problems Confronting Juvenile Courts

The problems currently confronting the juvenile courts are not new. In fact, they have existed since the development of a separate justice process for children. Nevertheless, the ability of the juvenile courts to successfully deal with these problems may well determine their very future. These problems include a failure to protect the due process rights of children, a lack of appropriate responses to status offenders, the processing of many youths who could be diverted from juvenile justice agencies, the lack of coordinated resources for responding to troubled youths, and the failure to recognize the limits of juvenile justice in solving youth problems.

Juvenile courts have often failed to protect the due process rights of children. An ongoing conflict continues between the state's caretaking responsibilities and its obligation to ensure that families and children are treated fairly. The state is increasingly called upon to intervene in the lives of serious delinquents and exercise "control" over them. If juvenile courts are going to continue to incarcerate larger numbers of youths and transfer more youths to adult courts, they must take children's due process protections more seriously to ensure that justice is done.

Although Supreme Court rulings have changed juvenile court practices, it is debatable how fundamental these changes have been.[34] Overall, juvenile courts have become more cognizant of the need to protect juveniles' due process rights. Nevertheless, juvenile courts are still characterized by considerable informality. In reality, juveniles often lack adequate representation (or any representation at all) in juvenile proceedings.[35] As noted in Chapter 10, courts must do more than pay passing attention to due process if it is to become an integral part of juvenile court operations. In order to receive the benefits of due process, youths and their parents or guardians must know what their rights are and feel comfortable in exercising those rights. However, there is evidence that many of those who come before juvenile courts neither understand their rights nor feel comfortable in exercising them.[36]

Since the inception of the juvenile courts, their procedural informality has been seen as one of their strengths. However, this alleged strength has also been a major weakness. When juveniles and their parents do not understand their rights, do not feel comfortable exercising those rights, or do not have adequate representation, they run the risk of being harmed rather than helped by juvenile courts.

The inability or unwillingness of juvenile courts to adequately protect the due process rights of juveniles and parents has led some to call for the abolishment of traditional juvenile courts. One proponent of abolishment is Professor Barry Feld of the University of Minnesota Law School. Feld maintains that juveniles have not received adequate protection of their rights within the juvenile court, nor is he optimistic that such protection will be afforded juveniles in the future. He argues that juveniles would be better served in criminal courts, where there is more emphasis on due process protections. Because of their age, Feld maintains that juveniles deserve special due process protections. For example, he argues that juveniles should automatically receive the assistance of counsel when they are taken into custody and should be allowed to consult with counsel before the right to counsel can be waived or before any interrogation takes place.[38]

Feld also believes, however, that before adult courts should be allowed to handle younger defendants, they need to be changed. One important change would be to give

FYI

Abuses of Power by Juvenile Justice Decision Makers Continues Today in Some Communities

For example, in 2006, two juvenile court judges in Luzerne County, PA, pleaded guilty to a kickback scheme in which they received $2.6 million from privately operated juvenile correctional facilities for placing youths in their "care," sometimes over the objections of their probation officers, for offenses that, in many instances, would have been diverted in many communities. As a result of their plea, the two judges were sentenced to more than seven years in a federal correctional facility.[37] This case not only raises questions about the behavior of the judges in question, but about the entire juvenile justice process in this community that failed to protect the rights of children and families.

youths a "discount" in recognition of their diminished culpability. According to Feld, "Criminal courts can provide shorter sentences for reduced culpability with fractional reductions of adult sentences in the form of an explicit 'youth discount.' For example, a 14-year-old might receive 33% of the adult penalty, a 16-year-old 66%, and an 18-year-old the adult penalty, as is presently the case."[39]

youth discount
A stratagem for taking into account the supposed diminished culpability of youths (based on their immaturity) in the sentencing process. The use of youth discounts would generally result in shorter sentences for juveniles than for adults who committed identical offenses.

From Feld's point of view, a **youth discount** would remove many of the inconsistencies and injustices presently found in juvenile court practices. For example, currently a youth who is one day shy of his or her 18th birthday is tried as a juvenile, while a youth who has just turned 18 years old is tried as an adult. In addition, by trying juveniles in adult courts, costly and time-consuming waiver hearings would be eliminated.[40]

Unlike Feld, who is worried about issues of justice, some juvenile court critics are essentially concerned with holding youths accountable when they break the law. These critics maintain that the criminal courts are better than the juvenile courts for punishing juvenile offenders because juvenile courts have traditionally paid too much attention to the needs of the youths and ignored the needs of victims and the community.

Another problem facing juvenile courts is the lack of appropriate responses to status offenders. The juvenile court was established, in part, as a legal mechanism for controlling and reforming status offenders. However, the ability of juvenile courts to control these youths or to "fix" the problems that contribute to their behavior is often limited. Historically, they have never had the resources needed to deal effectively with the large numbers of youths and families who need assistance of various kinds. Moreover, they may not be particularly well equipped to respond to problems such as truancy, running away from home, and incorrigibility. Too often courts have resorted to coercion and incarceration in dealing with these problems, even though the wisdom of such treatment is suspect.

Recognizing the potentially harmful effects of the incarceration of status offenders, Congress passed the Juvenile Justice and Delinquency Prevention Act of 1974, which encouraged states to remove status offenders from secure detention and correctional facilities and to separate juvenile offenders from adult offenders. In addition, the federal government encouraged the development of diversionary programs as a way of responding to status offenders. However, neither of these approaches has been completely successful. Although the number of status offenders detained in secure facilities has decreased since the passage of the Juvenile Justice and Delinquency Prevention Act, status offenders continue to be detained in secure facilities in some jurisdictions. Moreover, diversionary programs, which are used for many status offenders, also face a number of the following problems.

There is a tendency to process many youths who could be diverted from juvenile justice. Status offenders are obvious candidates for diversion, but diversion could reasonably be used for many youths who commit nonserious criminal offenses. Because of

COMPARATIVE FOCUS

In France, a "Youth Discount" Is Used

Under French law, when a custodial sentence is imposed on a youth, a juvenile can only receive a sentence that is less than half of the sentence that could be imposed on an adult.[41]

the pervasive nature of delinquency, juvenile justice agencies cannot possibly respond to each act of delinquency that comes to their attention. Moreover, even if they could respond to each act, the appropriateness of doing so would be questionable for at least three reasons. First, juvenile justice agencies lack the technologies and resources to deal effectively with each case of delinquent behavior. As noted in Chapter 13, although juvenile courts can play a role in helping youths and families respond to many problems that are brought to their attention, the courts themselves, as well as other juvenile justice agencies, are limited in their ability to respond to many of the problems that underlie delinquent behavior. Second, there is evidence that juvenile justice intervention can encourage additional delinquency rather than prevent it.[42] Third, most youths who commit delinquent acts eventually desist from delinquent behavior even though they are not identified and processed by juvenile justice agencies. Consequently, it is important that juvenile justice agencies divert cases from the juvenile justice process when it is possible and appropriate. Unfortunately, it is not always easy to agree about which cases are appropriate for diversion.

Diversion, of course, has its own problems, and the development of diversion programs needs to be approached with caution. As noted in Chapter 7, diversion programs have the potential to increase the costs associated with case processing, they can be stigmatizing for those involved, they can be coercive and deny youths and parents basic due process rights, and they can widen the net of social control by increasing the number of youths under correctional supervision.[43] All of these outcomes are at odds with the typical goals of diversion programs.

Many communities also lack a coordinated set of responses that allow courts to effectively deal with youth problems. Juvenile courts have historically been responsible for dealing with the needs of a diverse population of youths. Indeed, the youths who come to the attention of juvenile courts vary in a number of ways, such as in their level of psychological, emotional, and physical maturity, and their economic, family, and social background. Consequently, it is important that juvenile courts have access to a **continuum of quality services** that they can draw on. Courts in many jurisdictions, however, have relatively few resources. In some communities, juvenile court judges and other juvenile justice professionals have taken an active role in developing and coordinating services for youths, but in others, juvenile court judges and juvenile justice professionals have instead been content to maintain the status quo. Juvenile courts and other juvenile justice agencies often fail to recognize that they possess limited technologies and resources for responding to youth problems. They play an important role in dealing with delinquency, of course, but without the support of other institutions, they are unable to substantially change the behavior of many youth offenders. Although the juvenile courts possess considerable power, the ability to influence behavior ultimately

continuum of quality services A range of effective services consisting of prevention and correctional responses for delinquent youths.

MYTH VS REALITY

The Court's Ability to Solve Many Family Problems Is Limited

Myth—Courts are the most effective institutions for controlling problem youths.

Reality—Courts can play an important role in helping youths and families deal with their problems. However, courts often have limited resources for assisting families. As a result, they often rely on their coercive powers to solve problems, which limit their effectiveness in many instances.

rests on coercion, which has limited utility in reforming behavior. The courts, therefore, need to develop strong linkages with other community agencies and to determine when it is appropriate to use coercive responses, when to use noncoercive responses, and when to avoid responses altogether.

Problems Confronting Juvenile Corrections

Since the development of the first specialized correctional institutions for youths in the 1800s, treatment and rehabilitation have been primary goals of juvenile corrections—in word if not in deed. By the 1960s and into the 1970s, rehabilitation dominated progressive correctional thinking. However, in the late 1960s and mid-1970s, several literature reviews appeared that raised questions about the effectiveness of many correctional interventions.[44] In what is undoubtedly the most often cited of these reviews, Robert Martinson stated that "with few and isolated exceptions, the rehabilitative efforts that have been reported so far have had no appreciable effect on recidivism."[45]

Research that questioned the effectiveness of juvenile corrections programs proved to be powerful ammunition for conservatives who wished to "get tough" with juvenile offenders. Moreover, it struck a chord with liberals who recognized that correctional programs were, in many cases, designed to punish rather than rehabilitate youths. Doubts about the effectiveness of corrections programs were supported by additional research done during the 1980s. For example, in an extensive review of evaluations of juvenile corrections programs published in professional journals from 1975 to 1984, John Whitehead and Steven Lab concluded that these studies indicated that existing correctional treatments had "little effect on recidivism."[46]

Certainly, a number of questions have been raised about the utility of many corrections programs for juvenile offenders. Yet there continues to be strong evidence that some programs are effective in assisting youths and protecting the community. Indeed, during the 1970s and 1980s, a period in which a number of articles appeared questioning the effectiveness of corrections programs, other studies and reviews indicated that some programs did work.[47] For example, a **meta-analysis** of studies of corrections programs completed between 1960 and 1983 by Carol Garrett found that there were few negative effects of treatment. Instead, she reported that treatment of delinquents in institutional and community settings generally was associated with positive effects, even though, in many instances, these effects were not large. Potentially promising types of interventions uncovered by Garrett included some that employed a cognitive-behavioral approach or incorporated a family therapy component.[48] Furthermore, more recent research by Mark Lipsey and his colleagues indicated that many juvenile justice interventions with serious offenders are also effective.[49] Moreover, articles that questioned the effectiveness of corrections programs in general acknowledged that some programs did reduce recidivism. Even Robert Martinson's research indicated that some programs may work.[50]

meta-analysis
A statistical technique that permits statements about the effects of interventions across a number of different studies.

Interestingly, although there was a definite shift in many juvenile courts toward deterrence, punishment, and protecting community safety during the 1990s, there has also been increasing evidence of the viability of treatment and rehabilitation in juvenile corrections.[51] There is good empirical evidence that a number of community-based and institutional juvenile corrections programs serve the best interests of youths and are effective at reducing recidivism, even in the case of serious and violent juvenile offenders.[52] Moreover, our understanding of the characteristics of effective corrections programs and our knowledge of which types of interventions are most effective with which types of youths have improved considerably in recent years.[53]

Although good programs do exist, many other programs fail to provide quality services to youths, are poorly managed, have poorly trained and poorly paid staff, and do little to protect community safety. Some jurisdictions have what might be called "file drawer" probation in which overworked probation officers are responsible for large caseloads that preclude effective supervision or assistance to clients. Parole or aftercare programs are often similarly overburdened. Moreover, the quality of care and services provided in many group homes, foster homes, shelter-care facilities, detention centers, and institutional placements are far less than youths need or deserve.

Many correctional personnel lack the education, training, pay, and motivation needed to perform their jobs well. Part of the explanation for this is the dearth of basic and follow-up training programs for correctional staff. Although some state systems require basic training and certification of juvenile justice personnel and provide good training programs for this purpose, ongoing training of staff is rare. It is also possible to find staff who are in their positions because of who they know, not what they know.[54] One result is the extensive use, in many correctional programs, of "boob tube" therapy—the relegation of youths to watching television most of the time.[55]

People's penchant for fads and panaceas in responding to juvenile crime needs to be recognized. One problem that confronts juvenile corrections is that those in policy making positions often promote faddish and simplistic responses to juvenile crime that have little or no correctional utility. In fact, such approaches may actually be counter-productive. Nevertheless, faddish approaches are often easily "sold" as effective because they have intuitive appeal. Unfortunately, intuitive appeal is not always associated with true effectiveness. According to criminologist James Finckenauer, many people are looking for simple and inexpensive solutions to the problems of crime and delinquency.[56] Thus, responses like the Scared Straight! program, in which hardened adult offenders attempt to frighten youths away from a life of crime, or the passage of tougher laws come to be seen, at least for a short period, as panaceas. Many people want to believe that such approaches to delinquency work. Moreover, programs like Scared Straight! are much less costly than attempting to respond to the problems faced by families and the lack of good education, employment, and recreation opportunities for youths. However, as research by Finckenauer has shown, youths who went through the Scared Straight! program actually engaged in more new offenses than juveniles in a control group.[57] Furthermore, the passage of tougher laws appears to have had no appreciable effect on delinquency. In fact, violent juvenile crime actually increased after states began to "get tough" on juvenile crime.

Perhaps the most favored recent panacea is the juvenile boot camp.[58] Despite the growing popularity of boot camps, however, research on their effectiveness at reducing

MYTH VS REALITY

Commonsense Approaches to Delinquency Are Not Always Effective

Myth—If we just use "common sense," we can solve the delinquency problem.

Reality—A problem with commonsense solutions to crime is that seemingly good ideas do not always translate into effective interventions. If they did, the "delinquency problem" would have been solved long ago. In reality, delinquency is a complex phenomenon that is not always amenable to simple solutions. As a result, programs need to be carefully evaluated to assess their effectiveness.

recidivism has not produced encouraging results. For example, an evaluation of three federally funded boot camp programs in Cleveland, Denver, and Mobile indicated that youths who participated in these programs were no less likely to relapse than a comparable group of youths who did not attend boot camps. Indeed, in Cleveland, the percentage of recidivists among the boot camp participants was much higher (72%) than for the control group (50%). In Denver and Mobile, the percentages of boot camp participants and controls who relapsed were roughly comparable.[59] Thus, there is no evidence that these programs have a significant impact on recidivism. Boot camp participants in Cleveland and Mobile did improve their academic skills,[60] but these types of improvements are found in a variety of correctional programs.

Although boot camps may have considerable appeal, especially to those with prior military experience, the theoretical basis for using military-style training to rehabilitate youth offenders appears to be suspect. As Merry Morash and Lila Rucker noted in a critique of boot camps, "The very idea of using physically and verbally aggressive tactics in an effort to 'train' people to act in a prosocial manner is fraught with contradiction. The idea rests on the assumption that forceful control is to be valued. The other unstated assumption is that alternative methods of promoting prosocial behavior, such as development of empathy or a stake in conformity (e.g., through employment), are not equally valued."[61]

Another serious problem is the lack of attention given to the needs of female juvenile offenders. This problem is not new, but it is becoming more severe because of the increased number of girls who are entering the juvenile justice process.[62] Women face a number of unique problems within our society, and these problems are even more difficult for young girls. Physical abuse, sexual abuse, neglect, and exploitation are often found in the backgrounds of female juvenile offenders,[63] but few resources have been available for girls who face these problems. As a result, girls have often been forced to flee from abusive situations, but in leaving these situations, they sometimes find themselves victimized again by juvenile justice agencies, which either force them to return home or incarcerate them. Those placed in juvenile or adult correctional facilities might then suffer further abuse, neglect, and exploitation—this time at the hands of those who are charged with caring for their needs.[64]

If the needs of female offenders are to be met, policy makers and those responsible for implementing programs must begin to take seriously the reality that gender does influence one's life experiences. As Meda Chesney-Lind and Randall Shelden note, "Girls, in short, experience a childhood and adolescence heavily colored by their gender. It is simply not possible to discuss their problems, their delinquency, and what they encounter in the juvenile justice system without considering gender in all its dimensions."[66] Recognition of the importance of gender is required for the development of the types of interventions

FYI

The Number of Minority Females Formally Handled by Juvenile Courts Has Been Increasing

For example, in 2005, approximately 223,200 females were formally processed by juvenile courts. This number represented a 4.5% increase since 2000. However, between 2000 and 2005, the number of black females being formally processed by juvenile courts increased 14% and the number of Asian and American Indian females increased by 9% compared to a 1.2% decline in the number of white females.[65]

FYI

Many Minority Groups Are Overrepresented at Various Stages of the Juvenile Justice Process. Although black, Latino, Native American, Asian, and Pacific Island youths do not appear to be overrepresented at every stage of the juvenile justice process when national-level data are examined,[69] this is not the case when data for individual states and jurisdictions are examined.[70]

that can help girls deal with issues such as sexual and physical abuse, neglect, exploitation, pregnancy, parenting, substance abuse, and family dysfunction. It is also required for the development of the types of aftercare and support services that females need. For example, some girls do not have homes or families that want them or can care for their needs. Consequently, it is important that support networks be developed for girls, as well as housing and employment options that will help them live successfully on their own.[67]

Careful attention needs to be given to disproportionate minority contact (DMC) at each stage of the juvenile justice process. The reality is that minority youths are more likely to be arrested, detained, and referred to juvenile court than white youths, and they are more likely to be waived to adult courts. Also, once referred to juvenile court, they are more likely to have a formal petition filed. Moreover, although minority youths are less likely than white youths to be adjudicated, when they are adjudicated, they are more likely to be placed out of their home and less likely to be placed on probation than white youths.[68]

Of course, the disproportionate overrepresentation of minority youths in juvenile justice could be the result of their disproportionate involvement in crime. Research that has examined the extent to which incarceration rates reflect crime rates has found little support for this hypothesis. In a large-scale study of the relationship between offense behavior and incarceration, David Huizinga and Delbert Elliott drew the following conclusion:

> A summary of the findings would suggest that differences in incarceration rates among racial groups cannot be explained by differences in offense behavior among these groups.... There is some indication of differential arrest rates for serious crimes among the racial groups, but further investigation of the relationship of race to arrest and juvenile justice system processing is required if reasons underlying differences in incarceration are to be more fully understood.[71]

A similar conclusion was reached in a study of juvenile incarceration in California conducted by the National Council on Crime and Delinquency. As the researchers noted, "In broad terms, this analysis unveils a picture of persistent differential treatment for some minority groups after having accounted for prereferral factors such as offense and prior record. This leads one to the observation that some ethnic disparities in detention and sentencing outcomes are limited to African-Americans and cannot be fully explained by the juvenile justice attributes of that group."[72] These findings raise serious concerns because they indicate that, at least in some jurisdictions, institutional racism or racial bias influences decisions about which youths should be incarcerated and results in the disproportionate incarceration of minority youths.

There will, of course, be instances when youths will require out-of-home placement. Consequently, it is imperative that a continuum of services be available for youths who, for at least a short time, or for periods of the day, require more structured placements than their own homes. This continuum of services should consist of primarily community-based programs such as day treatment, foster care, and group home placement, as well as small secure treatment programs for serious juvenile offenders. Unfortunately, too many youths are placed in large locked facilities that provide little effective treatment and almost no aftercare upon their release. Moreover, many community-based programs lack the staff and resources to provide quality services to youths and families. Fortunately, however, there are good community-based and secure residential programs that do more than warehouse youths, and these programs can serve as models for effective correctional intervention.

Strong aftercare programs must also be developed in order for correctional programs to reduce recidivism. Certainly, one of the most significant problems facing juvenile corrections today is the need to develop sound aftercare programs for youths placed in residential settings. Unfortunately, aftercare has too often been an afterthought in juvenile corrections. In far too many instances, youths have been placed in institutional settings, then released without careful preparation and planning. Once in the community again, they have often received very little in the way of services, because those responsible for aftercare have large caseloads that preclude meaningful supervision or assistance.

The transition of a youth from a residential setting to the community often requires considerable adjustment on the part of the youth, the parents or guardian, and others in the community, such as teachers, neighbors, and victims. Moreover, in most cases, the environment to which the youth is returning has played a significant role in his or her delinquent behavior. Indeed, chaotic, neglectful, and/or abusive family situations; poor schools; and a scarcity of conventional economic and leisure activities for children are critical problems that need to be addressed. In addition, youths often face a number of system barriers that prevent them from becoming immediately involved in positive structured activities within the community.[73] Consequently, it is important that steps are taken to prepare youths to deal effectively with their environment and to support their efforts to avoid delinquent behavior.

In an effort to support the development of effective aftercare services for youths, the Office of Juvenile Justice and Delinquency Prevention began the Intensive Community-Based Aftercare Programs (IAP) initiative in 1988, under the direction of David Altschuler of Johns Hopkins University and Troy Armstrong of California State University, Sacramento. The IAP model, as developed thus far, focuses attention on three areas: (1) planning and preparation for release during placement, (2) implementing a structured transition that requires the participation of institution and aftercare staff prior to and following community reentry, and (3) engaging in long-term reintegrative activities that provide adequate levels of services to clients and ensure social control.[74] The IAP model includes the following objectives:

- preparing juveniles for progressively increased responsibility and freedom in the community
- facilitating interaction and involvement between juveniles and the community
- working with offenders and targeted community support systems (families, peers, schools, and employers) on the qualities needed for constructive interactions that advance the juvenile's reintegration into the community
- developing new resources and support as needed
- monitoring and testing juveniles and the community on their capacity to deal with each other productively[75]

BOX 15-1 Interview: George Page, Southern Regional Administrator, Kentucky Baptist Homes for Children

Age: 57
Education: MSW, University of Tennessee
Salary: $59,800

Q: You have been involved in youth-serving organizations for a number of years. Could you say something about how long you have worked with juvenile offenders and in what capacity?

A: I have been working with delinquent youths for 29 years. I began my career with the Kentucky Department of Child Welfare in 1969 as assistant superintendent at the Frenchburg Boys Center. I also served as the superintendent of Woodsbend Boys Camp and as superintendent of the Lake Cumberland Boys Camp. From 1991 through 1994, I was the assistant director of treatment for the Kentucky Department of Social Services, Residential Services. I was responsible for 49 different state programs, including residential, day treatment, group home, and clinical programs. I presently serve as the director, Southern Region, Kentucky Baptist Homes for Children.

Q: What are your present job responsibilities with KBHC?

A: My primary responsibilities include developing and overseeing a variety of youth-serving programs in 27 Kentucky counties, which include an 18-bed residential/shelter program, a 24-bed wilderness treatment program, 31 foster care homes, a family preservation and reunification program, and an evening reporting center. I am also responsible for developing and monitoring an annual budget of approximately $3.2 million, raising funds, coordinating the building of new facilities, and overseeing staff development and training.

Q: What are the biggest challenges you face in your work?

A: Probably the biggest challenge we face is finding appropriate placements for youths when they leave our residential programs. In an effort to deal with this, we have been developing more treatment foster homes for youths to help them make the transition from placement back to the community. Another challenge I face is developing an annual budget that allocates funds in such a way that the best possible services are delivered to youths in our care, and then staying within that budget.

Q: Over the course of your career, what changes have you seen in how we deal with delinquent youths?

A: Kids today seem to have more emotional and chemical dependency issues. In Kentucky, basically, we are still doing the same thing today that we were doing in 1965 with small residential programs. There are, however, a few more options today than in the past, such as day treatment and family preservation and reunification. Also, more private child care agencies are willing to take delinquents than in the past and treatment planning is more sophisticated today.

Q: Are there any things that you don't particularly like about your job?

A: I don't like dealing with allegations of abuse by staff. Fortunately, this does not occur often.

Q: What do you like about this type of work?

A: I enjoy the challenges of this job. It requires a great deal of creativity to meet the needs of the children and families that we deal with, but I feel that what we do really does make a difference. It's fun.

Q: What qualities does a person need to successfully work with youths in an institutional setting?

A: To be good in this work you have to have two ears that work, lots of patience, a great deal of empathy, a creative mind, lots of energy, and a strong faith. You must also sincerely like kids and have an unquestionable belief that kids can and want to change. Finally, you can't give up on kids.

The IAP model represents a comprehensive approach to aftercare. Unfortunately, in far too many instances, youths do not receive the types of aftercare services they need and deserve.

Juveniles' Right to Treatment in Correctional Institutions

The most frequently used justification for placing juvenile offenders in correctional institutions is that they can receive the help they need to prevent subsequent offending. Therefore, many people maintain that juvenile correctional institutions have an obligation to provide meaningful treatment. In fact, some attorneys argue that juveniles have a legal right to treatment. As noted earlier, meaningful treatment is often lacking in residential placements.

The legal right of children to receive humane and appropriate treatment in juvenile corrections institutions has been the focus of a number of lawsuits brought on behalf of incarcerated youths since the early 1970s. The first of these cases was *Inmates of Boys' Training School v. Affleck* (1972).[76] This case brought to light the inhumane conditions that existed in some juvenile institutions, such as the use of solitary confinement, the use of strip cells, and a lack of educational opportunities. The decision of the lower court in this case was that conditions in the institutions violated the prohibition against cruel and unusual punishment spelled out in the U.S. Constitution.[77]

Inhumane conditions have also been pointed out in several subsequent cases. For example, in the 1974 case of *Nelson v. Heyne*, the Seventh Circuit Court of Appeals in Indiana ruled that the use of unnecessary and excessive corporal punishment, the use of tranquilizing drugs to control behavior, and the failure to provide rehabilitative treatment to youths violated the youths' Eighth and Fourteenth Amendment rights.[78] This case was the first appellate court decision to affirm juveniles' right to rehabilitative treatment. In another important 1974 case, *Morales v. Turman* the court found that the Texas Youth Council violated the Eighth Amendment.[79] Problems addressed by the court included the use of tear gas, the segregation and solitary confinement of residents, enforced silence, and the use of excessive and unnecessary corporal punishments, among others. The court found that youths were denied their right to treatment and indicated that the state has an obligation to rehabilitate delinquent youths.[80]

Although these cases remedied some abuses in juvenile correctional facilities, such problems continue to emerge. In a 1987 case, *Gary H. v. Hegstrom*, a federal judge found that isolation punishments employed at the MacLaren School for Boys in Oregon were inappropriate and that juveniles were being denied their right to treatment.[81] More recently, in a 1995 South Carolina case, *Alexander v. Boyd*, the plaintiffs argued that four institutional facilities violated youths' constitutional rights in a number of ways, including: (1) they relied on lock-up units and CS gas (a potent form of tear gas) for handling discipline problems; (2) they placed youths in padlocked cells in violation of fire safety regulations; (3) they served youths food containing cockroaches and other foreign substances; (4) they failed to provide good medical care; (5) they failed to provide adequate living space; and (6) they failed to provide minimally adequate programming. In its ruling, the district court noted that the purpose of the facilities was to rehabilitate juvenile offenders, and existing practices, in failing to do this, violated constitutional and statutory requirements. As a result, the court ordered that an acceptable plan be developed in order to remedy existing deficiencies.[82]

These cases document the continued denial of quality care and treatment in some juvenile correctional facilities. Importantly, they have helped rectify many problematic

FYI

Problems in Juvenile Correctional Facilities Continue

A report released by *The Dallas Morning News* in 2007, revealed systematic corruption in juvenile correctional facilities in Texas that involved the sexual exploitation and beating of residents, sometimes by facility correctional administrators, and efforts to cover up reports of the abuse, including the intimidation of whistleblowers when they attempted to bring the corruption to light.[83]

situations in institutions around the country. However, the federal courts have, to date, dealt with these problems on a case-by-case basis and have refused to make sweeping rulings that would affect a wide range of institutions around the country. Moreover, they have refused to make mandatory current professional association standards for the care of youths.

Besides lawsuits brought on behalf of individual youths in institutions or even groups of youths, a mechanism that can be used to protect the rights of youths in institutional settings is the Civil Rights of Institutionalized Persons Act (CRIPA). This act was authorized in 1980, and it gives the Civil Rights Division within the U.S. Department of Justice the power to take legal action against states or local governments for "violating the civil rights of persons institutionalized in publicly operated facilities."[84] Moreover, the act has been used on a number of occasions to initiate investigations regarding the treatment of children in various juvenile correctional facilities.

The privatization of juvenile corrections needs to be carefully scrutinized. Although privately operated facilities have always played an important role in juvenile justice, they have come to play an increasingly important role in recent years. For example, from 1991 through 1999, the number of residents committed to public facilities increased 41% while the number committed to privately operated facilities increased 79%.[85] Since 2001, however, the number of youths committed to public and private correctional facilities has declined. This decline is primarily attributable to reductions in admission to publicly operated facilities. There has been very little decline in admissions to privately operated facilities. As a result, commitments to private facilities made up approximately 39% of the commitments to juvenile correctional facilities in 2006, up from about 34% in 1997.[86]

There are many privately operated juvenile correctional facilities around the country. Some of these facilities contain the most innovative and effective programs for juvenile offenders in the country. Consequently, these institutions represent an important element in the continuum of care for youths committed to residential settings. Nevertheless, it is a mistake to assume that, just because a program is privately operated, it provides quality care. Indeed, there is considerable evidence that some privately operated programs fail to provide quality care to their clients.[87] For example, in one study that examined private facilities in Southern California, David Shichor and Clemens Bartollas found that many agencies that place children in private facilities do not have the staff to adequately monitor the quality of services provided and that many staff members in private facilities lacked appropriate qualifications. In addition, they found that the desire of some institutions "to make money from kids" resulted in lowering the number and quality of staff in some programs.[88] A potential problem with private facilities is that the desire to make a profit or keep the program operating acts as an incentive to cut corners and acts as a disincentive to help the residents complete treatment in a timely manner.

FYI

Jails Are Not Appropriate Placements for Juveniles

The suicide rate for juveniles in adult jails is estimated to be 4.6 times higher than the rate for juveniles in the general population and 7.7 times higher than the rate for youths in juvenile detention facilities. Moreover, there is some evidence that it may even be higher.[89]

Placing juveniles in adult jails is harmful. As noted in Chapter 12, the incarceration of juveniles in adult jails creates a number of problems for both juveniles and jail administrators. Many jails have difficulty meeting the needs of adult inmates, much less the needs of incarcerated juveniles. Sadly, there is a long history of abuse of juveniles in these facilities. Moreover, there is clear evidence that placing juveniles in adult jails places them at risk for self-destructive behaviors.

In recognition of the problems associated with the jailing of juveniles, the federal government has encouraged states to remove juveniles from jails through passage of the Juvenile Justice and Delinquency Prevention Act. Despite the encouragement, jail removal has not been completely successful. Although progress has been made in many jurisdictions, juveniles are still being incarcerated in adult jails. In fact, there is some pressure from those who argue for a "get-tough" approach to juvenile crime to place more juveniles in adult facilities.[90] As a result of such pressure, a failure by the states to develop juvenile detention facilities, and an unwillingness to develop appropriate alternatives to the jailing of juveniles, 6,837 persons younger than 18 years of age were housed in adult jails at midyear 2007.[91]

The lack of adequate program development and evaluation hinders the development of effective juvenile corrections programs. Few juvenile corrections programs engage in a systematic and ongoing process of evaluating the extent to which program interventions are being implemented as intended and achieving their stated goals and objectives. As a result, many program managers have little information to help them modify and improve the services they provide to youths, and they have little or no information about their programs' effects on recidivism. Many program managers in juvenile corrections would like to have good information, but they face several obstacles in trying to collect it. For one thing, they often lack the necessary expertise to collect and analyze information about program quality and effectiveness. For another, they rarely have adequate budgets for contracting with outside organizations that have the expertise to assist them. Then, there are those program managers who seem to have little interest in collecting data and scrutinizing program operations. The end result is an appalling lack of good information about the quality and effectiveness of services provided to youths placed in juvenile correctional programs.

Juveniles' Right to be Safe in Correctional Facilities

Youths' concern about their safety is an ever present reality for juveniles who are incarcerated. Will their physical safety be protected and assured while they are in placement ? Recently, there have been an increasing number of injuries and deaths of juveniles in correctional settings that could have been prevented.

In 2003, two residents of the Pinellas County, Florida Juvenile Detention Center died, one from a burst appendix that was not medically treated, even though the youth had complained of stomach pain for three days prior to his death. The other juvenile was killed by another resident in a fight.[92] In Texas, there is an ongoing investigation, starting in 2005, into serious injuries of residents, including suspicious long-bone fractures due to the handling of juveniles by guards in certain facilities. This investigation continued as recently as April 2006 when medical authorities in Texas identified 60 suspicious breaks of arms and other bones that merited further investigation and follow-up. There was no question in the minds of the medical authorities doing the investigation that these injuries were not "sports related" injuries.[93]

In South Dakota "arbitrary and inhuman" disciplinary practices have been identified as far back as 1999. A fourteen-year-old girl collapsed during a forced run at the state's boot camp for girls. She eventually became unresponsive after lying where she fell for three and a half hours before she was finally transported to a hospital. When she arrived at the hospital, her temperature was 108 degrees Fahrenheit. She never revived.[94]

In Fulton County, Georgia as recently as 1998, the juvenile detention facility was characterized as "unconstitutional, dangerous, abusive and overcrowded." Staff routinely stripped residents of their clothing, removed sleeping mattresses from their rooms, and had residents spend days in solitary confinement.[95] In Michigan, in 2005, fourteen detention centers were cited for repeated violations. These violations included failing to protect residents from attacks by other residents; criminal sexual conduct on residents by staff; and extreme violence toward residents by staff, including choking or using improper restraints.[96] In a Rockville, Maryland facility, a guard had organized residents into a gang known as the "family" whose initiation was to strip boys naked and hit them in the groin. The guard used these boys to keep order in the facility and to do his bidding.[97]

There are numerous anecdotal examples of situations in which youths' safety is endangered by their placement in juvenile correctional facilities. In the past seven years, juvenile facilities in 11 states have been the focus of federal government reviews because of possible civil rights violations.[98] What is important, however, is to look deeper into the causes of these abusive and dangerous situations and develop viable solutions.

One reason why juveniles are at risk in some juvenile facilities is that they constitute inappropriate placements because they house youths who do not need placement and/or the facility does not have the staff or other resources to effectively meet residents' needs. A study done in 2004 found that there were 15,000 mentally ill children residing in detention facilities only because they were awaiting mental health services.[99] Mixing mentally ill children with seriously delinquent youths endangers both. The mentally ill acting-out youth is very likely to encounter a delinquent acting-out youth and the results can be dangerous to both youths as well as other residents and staff. Also, the staff in many juvenile correctional facilities are not sufficiently trained, if they are trained at all, to deal with youths who have serious mental health issues. Mentally ill youths take an increasing amount of staff time and attention, which makes staff less available for dealing with the challenges presented by delinquent youths, some of whom are dangerous. Without proper training, staff cannot recognize the health needs of residents or know when situations are becoming dangerous. They also are not prepared to intervene in ways that de-escalate conflicts. The job of a corrections officer in a juvenile facility is a demanding one that requires a well-trained staff conversant in passive restraint techniques, child psychology, due process, and crisis de-escalation. The reality is that many times these positions pay much less than they should, given the complex demands of the job.[100]

Another problem that threatens some youths' safety is lack of access to adequate health care. If staff do not or cannot recognize health issues, children are placed at risk. Many youths entering facilities have high rates of basic health problems, including asthma, poor dental hygiene, vision problems, and sexually transmitted diseases, just to name a few. Many have been exposed to domestic violence and were themselves physically or sexually abused. A high percentage of youths in the juvenile justice system, possibly as high as 80%, are under the influence of drugs or alcohol when committing their crimes. Nearly two-thirds of incarcerated children with substance abuse disorders have at least one other mental health disorder. These co-occurring disorders require careful diagnosis and special treatment, which all too often is lacking.[101]

Too many facilities have inadequate policies and procedures that do not address these serious problems, thus leaving staff without guidance or direction in how to respond. In the Pinellas County, Florida situation in which the 17-year-old youth died after a burst appendix, there was no procedure in place for having detention workers call 911 in an emergency![102]

Inadequate programming also places juveniles at risk in juvenile correctional facilities. These facilities should have education, vocation, and recreation programming to keep residents occupied and to support their positive social development. Obviously, providing adequate health care and counseling, and ensuring the safety of youths and staff are critical features of correctional programming. Unfortunately, in too many instances, correctional programming often falls short in one or all of these areas.

What can be done to reduce the risk and improve the conditions for juveniles in correctional facilities? First, advocates for children need to be familiar with the Civil Rights of Institutionalized Persons Act (CRIPA). Congress enacted CRIPA in 1980 to provide the Department of Justice with statutory authority to bring cases to federal courts to protect institutionalized persons. This use of the federal legal system to effectuate changes at the state level can be very effective, but it is underutilized. In the first 17 years that CRIPA has been available, there have been only 73 investigations into juvenile correctional institutions. As of June 2007, there were 22 ongoing juvenile detention and treatment center investigations. The Justice Department is also monitoring consent decrees in Kentucky, New Jersey, and Puerto Rico. The consent decree in Kentucky includes 13 juvenile facilities in that state.[103]

Second, states and even local governments can establish a "child welfare" ombudsman to investigate and correct abuses of children in state and local institutions. This person could monitor key issues such as

- staff training;
- facility policies and procedures;
- health care availability;
- worker qualifications and compensation.

This individual could provide a voice for children who are detained in public facilities, as well as investigate complaints, report abuses, advocate for change, and monitor changes that are implemented.[104]

Third, the position of correctional officer, child guidance, or child care worker needs to be enhanced and the pay for these staff members needs to be significantly increased in many jurisdictions. This will require increasing the education requirements for persons in these positions and improving the quality of their training. This will also require that states and local governments match rhetoric with resources and refrain from "low balling" salaries for these positions.

Finally, a shift in thinking needs to take place on the part of local and state government leaders. For too long, it was thought that detained youths did not deserve anything more than punishment. If juvenile justice is to be truly about rehabilitation, then adequate resources need to be devoted to all parts of the system, including detention and correctional facilities.

■ The Future of Juvenile Justice

Predicting the future is fraught with problems. Nevertheless, if present circumstances are any indication of future trends, then it is possible to make some reasonably educated guesses about what juvenile justice will look like in the near future and the challenges that it will face.

Perhaps the most important issue confronting juvenile justice is whether the juvenile courts should continue to exist. Some have called for the elimination of the juvenile courts on the basis that they are either unwilling or unable to protect the due process rights of juveniles. Critics of the juvenile courts, such as law professor Barry Feld, maintain that the *parens patriae* focus of the juvenile courts has led to court practices that deny juveniles the same due process protections afforded adults, even though the juvenile courts today operate much like adult courts, and juveniles may face the same sanctions as adults. According to Feld, this means that youths receive neither treatment nor justice in the juvenile courts.[105]

Other critics of the juvenile courts claim that they represent an ineffective response to delinquent behavior, especially serious or violent delinquent behavior. According to these critics, the criminal courts are better equipped to deal with chronic, serious, and violent juvenile offenders.

Some juvenile justice experts agree with these critics to an extent—that is, they believe that adult and serious juvenile offenders are not vastly different—but they argue that the juvenile court model should be extended to the criminal courts. Travis Hirschi and Michael Gottfredson, for example, maintain that juvenile and adult offenders have many similarities and that many adult crimes are neither more nor less serious than many juvenile offenses. In addition, they argue that juvenile and adult criminality is the product of low self-control and that adults are as amenable to reform as juveniles. Consequently, they call for one criminal justice system, but one that is constructed along the lines of the juvenile justice process. From their point of view, such a system not only would be more cost efficient, but also would provide more consistent responses to offenders and possibly focus more resources on the prevention of criminal behavior.[106]

The debate over the future of juvenile justice will undoubtedly continue for some time. However, in the near future, it is unlikely that the juvenile courts will be abolished, and it is equally unlikely that the criminal courts will be reconstructed according to the juvenile court model. What is probable is that, in some quarters, considerable energy will be directed toward protecting the status quo, and, in others, the push will continue to reform juvenile justice practice.[107] Because juvenile courts are political institutions, judges, being elected officials in many cases, are often hesitant to take risks or act in ways that can be perceived as soft on crime. Hence, juvenile courts, like other courts, tend to be rather conservative institutions that approach change cautiously.

Critics on the reform end of the scale include Barry Krisberg and James Austin, who are in favor of "reinventing" juvenile justice. The kind of reinvention they have in mind involves four steps. First, a public health approach to serious juvenile crime would be taken. This approach would place greater emphasis on the environmental factors that contribute to delinquency and on community organizing and self-help strategies in dealing with youth crime. Second, knowledge about child development would be incorporated into juvenile justice policies and procedures. Third, the juvenile justice process would be altered to fully protect youths' legal rights. Fourth, "the whole child in his or her family and community context" would be treated by developing coordinated and comprehensive approaches to youth problems.[108]

Other advocates of reform, such as Preston Elrod and Daryl Kelley, have also called for a critical approach to juvenile justice. Elrod and Kelley argue that neither liberal nor conservative reforms are capable of effectively meeting the needs of youths or effectively responding to juvenile crime. Consequently, they call for a different approach, which includes the development of cooperative relationships between social researchers and communities, including the juvenile justice community, in an effort to devise responses to youth problems that are more humane, more effective, and, at the same time, sensitive to the rights of children.[109] Elrod contends that traditional, liberal, and conservative approaches to juvenile crime share similar flaws that invariably will produce ineffective juvenile justice policies. For example, one major flaw is the failure of traditional approaches to address the cultural, economic, political, and social factors that contribute to delinquent behavior. Elrod further argues that traditional approaches have employed a very narrow range of methodologies in their efforts to understand delinquency and develop correctional responses. Consequently, he calls for juvenile justice responses that (1) focus attention on the needs of youths and families, (2) work to integrate offenders into the community, (3) limit juvenile justice involvement when possible, (4) protect the rights of children, and (5) are regularly evaluated to determine their effectiveness.[110]

Some change will undoubtedly occur in the future because of the perceived ineffectiveness of existing juvenile justice policies. Moreover, it is possible that some communities will implement some fundamentally different responses to juvenile crime. One response likely to receive attention is the so-called **balanced approach to juvenile justice**. The balanced approach is based on the recognition that three important community needs exist: the need to sanction crime, the need to rehabilitate offenders, and the need to ensure public safety. As a result, the balanced approach emphasizes the following three goals for juvenile justice:

balanced approach to juvenile justice
An approach based on the recognition that a community has three needs: to sanction crime, to rehabilitate offenders, and to ensure public safety. The balanced approach emphasizes three corresponding goals for juvenile justice: accountability, competency, and public safety.

1. *Accountability*. The goal is for offenders to take responsibility for their crimes and the harm caused to victims by making amends and restoring losses. Moreover, accountability requires that victims and offenders be involved in the sanctioning process, when feasible.
2. *Competency*. The goal is to require that youths measurably improve skills that will allow them to function as productive adults (successful rehabilitation). These skills include various education, social, work, and civic skills.
3. *Public safety*. The goal is for communities to invest more heavily in crime prevention, which includes ensuring that youth offenders placed in institutions spend large amounts of time preparing for their return to the community and that the youths develop connections with community resources that will help make their return successful.[111]

Because of its focus on the needs of offenders, victims, and the community at large, the balanced approach to juvenile justice will, perhaps, serve as an important bridge to more humane and productive ways of responding to juvenile crime while simultaneously protecting community safety. One thing is for sure, however: Juvenile courts will continue to handle a variety of problems presented by a diverse and growing population of youths over the next decade.

Problems That Will Continue to Influence America's Youths, Families, Communities, and Juvenile Justice Agencies

One of the more obvious, but insidious, problems that many young people face is the breakup of their families. Moreover, the single parents (usually the mothers) who are left with the primary responsibility for raising these children will confront a number of difficulties. Many of these parents will struggle financially to provide the necessities for their children, who themselves will often suffer hardship. In addition, these single parents will have difficulty supervising their children because working will take them out of the home. Without a significant adult present, these children will be exposed to a variety of questionable influences.

In other families, youths will be exposed to violence and in many instances will be the victims of domestic conflict. Many children will learn that it is "okay" for men and women to hit one another, and that using violence is an acceptable way to resolve interpersonal disputes. Sadly, significant numbers of children will live in homes where their educational, psychological, and emotional needs will not be met. The scars associated with their treatment will not always be visible, but the resulting psychological and social impairment will play an important role in how these youths adapt to their environment. Because children learn how to be parents from their own experiences, harmful parenting practices will be repeated from generation to generation unless the cycle is broken by effective intervention. In the future, youths will continue to face important decisions about sexual relationships, drug use, education, and careers. Many youths will choose drugs as a form of self-medication to escape from difficult family situations or as a way of dealing with other problems. Along with this drug use will come addiction, serious health problems, and even death. The emotional retardation that can result from serious drug use will take years to overcome. When youths use drugs to keep from feeling painful emotions—emotions that people need to learn about as part of maturing—they fail to benefit from necessary life experiences. Remaining in this drug-induced emotional immaturity can have effects that will carry over into adulthood and impair their ability to parent their own children. Drug use will also bring many young people into contact with individuals involved in a range of illegal activities and will place them in situations in which there is an increased probability of violence and arrest.

Many youths will have to deal with the consequences of unprotected sex and unwanted pregnancies. Sexual encounters carry with them a number of serious risks. Sexually transmitted diseases and the emotional entanglements that intimacy can bring will take their toll on many youths. Romantic relationships must be based upon more than physical attraction; when they are not, they are likely to go bad within a short time. In addition, the burden of unwanted or unplanned pregnancies resulting from failed relationships will continue to fall primarily on young girls and their parents or guardians. Having a child is the most important decision that two people can make together, regardless of age or marital status. Many children, however, will continue to have children of their own, but without the skills and maturity necessary to provide them the love and nurturing they need.

Many young people will attend schools that possess few resources to meet their varied educational needs. Still others will be exposed to messages, often subtle but sometimes explicit, that they are not capable learners. Far too many youths will find learning boring and school irrelevant. They will likely experience school failure and face diminished prospects for well-paying and meaningful employment. Regarding employment prospects, the developing global economy promises tremendous opportunities for some, but there is a real possibility that many others will find themselves competing with low-paid workers in foreign lands who are employed by corporations that have more loyalty to profit than to their employees and communities. Still others will find their lives changed as a result of the trend toward corporate downsizing. Companies are getting "leaner and meaner," eliminating blue-collar and white-collar jobs across the board and requiring the remaining employees to do more and varied tasks.

Economic and political arrangements will ensure that poverty will be a major problem confronting many Americans, both young and old. Consequently, large numbers of individuals will lack access to legal, medical, or educational resources, not to mention the kinds of social services that could help them deal with family or child problems. In these circumstances, many juvenile courts will continue to serve as the only available social service agencies for large numbers of citizens, yet it is doubtful that they will have the resources or expertise to deal with all of their clients' problems.

Gangs will continue to be a serious problem in some communities, especially in urban areas. They will continue to serve as substitute families for many youths. For its members, a gang provides a sense of belonging, a kind of status, and an identity, valuable commodities in situations characterized by uncertainty and instability.

Clearly, juvenile justice will face a number of serious challenges throughout the 21st century; however, these problems should not be viewed as insurmountable. In fact, there is considerable room for optimism. For example, the extent of support by public and juvenile justice agencies for rehabilitation is an encouraging sign. Although there has been some movement toward making the juvenile court more punitive, it should be remembered that there has always been a punitive side to juvenile justice. Nevertheless, rehabilitation continues to have many adherents in juvenile justice, and we believe that support for it will become even more pronounced in the future.

We believe several factors will increase the amount of attention paid to the treatment of juvenile offenders in the future. First, juvenile justice will become increasingly more professional in its approach to youths, partly because juvenile justice personnel will be better educated and better trained. Second, delinquency will increasingly be seen as a community problem that requires a coordinated and comprehensive community response. This coordinated response will lead to more efforts directed toward delinquency prevention and to the recognition that effective correctional responses are based on the existence of sound preventive mechanisms. To a large extent, the development of more coordinated and comprehensive responses to youths will be driven by the failure of punitive responses to solve juvenile crime and by the realization that they are not cost-effective. Third, the continued professionalization of juvenile justice will lead to more emphasis on program development and evaluation, which in turn will help juvenile justice agencies deal more effectively with their clients. Fourth, continued program development and evaluation will expand our knowledge about what works and for whom, which will increase the ability of juvenile justice to provide effective treatment to youths.

■ Legal Issues

Three significant legal issues will have an ever-increasing impact on juvenile delinquents, their families, and the juvenile justice system. The first is the development of the family court and the resulting changes in process, procedure, and attitude toward dealing with youths that is associated with the family court model. The second issue is the recognition that therapeutic jurisprudence and restorative justice are concepts that need to be integrated into the juvenile justice process. Third, but not at all least, is the lack of access to justice faced by many families and children. The reality is that the costs of private or court-appointed legal representation continue to increase, and this jeopardizes the availability and quality of attorney representation made available to youths and their families.

The idea of a "family court" where all legal matters involving families can be taken care of (i.e., divorce, custody, child support, visitation, delinquency, and neglect) is not a new idea, but one that is gaining increasing acceptance. Combining courts with similar jurisdictions to handle both domestic and delinquency matters can often save money in the long run because of staff reductions and increased efficiency. This is possible because delinquency, child protection, and other domestic relations cases revolve around common issues and require similar types of services.

One of several key characteristics that need to be found in a family court for it to be effective is that the court needs to be outcome-focused without ignoring due process. Services need to be available to treat the multitude of issues that children and families bring to the court. Without the ability to effectively intervene and encourage change, the delinquent and his or her family will continue in their old, dysfunctional patterns of behavior. Second, family problems need to be identified as early as legally possible, but no later than disposition or sentencing. To accomplish this, the court must have a knowledgeable and experienced intake department so that assessments can be made early in the process. Access to psychological and substance abuse services is also important in developing an accurate understanding of the family. A mechanism has to be in place that encourages the extended family, parents, siblings, and influencing relatives to cooperate with assessment and issue identification. Focusing dispositional orders on the juvenile alone is often a waste of time. Children are much more likely, on a day-to-day basis, to be influenced by family and family associations than by a judge, referee, or probation officer. Third, the creation of this type of problem identification–centered court requires dedicated leadership from the judiciary. Judges elected to or assigned to the family court must want to be there, and hopefully are not looking to use it as a "stepping stone" to a higher or more publicly visible court. These judges must also be dedicated to staying in family court so that over time they can develop the necessary expertise to be fully effective.

Understanding the subtle connections between delinquency and child neglect, abuse and delinquency, and domestic violence requires experience that can be gained only by dealing with individual cases. These same judges need to be active in their communities in advocating for services and must be creative in utilizing both the public and the private sector. One of the catch phrases of the family court is "one judge; one family." This means that by handling all matters involving a single family, from divorce to delinquency, the judge can develop an understanding of each family's individual dynamics and does not need to be educated anew for each case. The only drawback to this approach is that it can be emotionally draining and frustrating to see the same family repeatedly, not to mention the second and third generations of that family.

A family court judge must be able to maintain empathy for families that are repeatedly before the court, while at the same time holding all of them accountable. This intersection of accountability and compassion is an important stop on the road map of therapeutic jurisprudence and restorative justice. **Therapeutic jurisprudence** is defined as "the use of social science to study the extent to which a legal rule or practice promotes the psychological and physical well-being of the people it affects."[112] In simple terms, it means that when the court interacts with those persons who come before it, it needs to be aware of both the psychological and legal impact it can have on those people and make certain that the impact is a positive one. It involves the use of court orders directing people to make positive life changes after ensuring that there is a supportive system to help those changes occur. It also involves a willingness to use the power of the court (e.g., contempt and detention) to confront persons with the immediate need to change or face court-imposed sanctions.[113]

therapeutic jurisprudence The use of the legal process to promote the psychological and physical well-being of the people it serves.

A common example of courts engaging in therapeutic jurisprudence is juvenile drug courts. These courts focus on identifying risk factors associated with the individual's illegal activity (i.e., drug use) and then dealing with these risk factors. Those involved in drug courts are encouraged by consequences, detention, or placement in residential treatment facilities, to make behavioral changes, and are treated in a supportive and affirming manner by the judge and court personnel rather than in a harsh, punitive fashion.

One of the keys to effective intervention is accountability. Accountability has two parts—the individual's accepting responsibility for his or her acts and understanding the impact on the community and the victim, if any, and the court's willingness to hold the person accountable. The court can obviously play a key role in holding the person accountable by available sanctions. But just as important, if not more important, is the role the court can play in helping the juvenile accept responsibility for his or her behavior. If the juvenile enters a plea, the court can refuse to allow a no contest plea, forcing the juvenile to admit, in open court in front of his or her parents and possibly the victim, to what he or she did. Without an up-front acknowledgment of wrongdoing or misbehavior, it is virtually impossible to convince a person that there is some behavior that needs to be changed. Constant reminders of prior inappropriate behavior may also be necessary to continue the momentum necessary for change.

Including victims in the dispositional process is essential in many instances. Youths involved in illegal behaviors need to be able to understand the impact of their actions on others. They need to know and understand that more than just a law was broken. The peace of their community was violated and real people were hurt. This can be accomplished in a number of ways:

- victim impact statements available to the court
- requirements for restitution by the juvenile and the parent, guardian, or custodian responsible for supervision when the criminal act occurred
- community service restitution
- group or offender educational programming
- face-to-face victim offender mediation
- focus on restoring a balance in the community

One of the key components of restorative justice is the idea that the offender must also be "restored." The offender must be helped back into balance with his or her community, the victim, and the family. By problem solving to determine the causes of the crime and by addressing those causes, the juvenile offender can "leave" the system in a better position than when he or she entered it. Indeed, the victim, the community, and the offender are all integral players in the idea of restorative justice. "Paying back" the victim and the community restores the offender as a productive citizen, and hopefully increases the offender's knowledge about how his or her behavior influences the community.

The third legal issue that needs to be discussed is access to justice. This issue has two parts, one involving the problems surrounding the actual litigant–court interaction and the other involving access to legal representation. In order for courts to be accessible, they need to be in a physical location that can be conveniently accessed by all. They need to be on a bus line, and have adequate parking and handicap access. As much as possible, they should provide as many services in one place as possible so that people who have limited means and difficulties with transportation do not have to unnecessarily spend precious time and money to get to court. Court staff members need to be knowledgeable and helpful when citizens have difficulties or need assistance with legal issues. An intake system needs to be available so that clients seeking or needing assistance are directed to the correct person or place. Court staff members need to understand that courts are created for the citizenry, not for the employees! Making people feel that the courts are available to help them resolve their legal issues is too often an overlooked part of the "access to justice" issue.

The second part of the "access to justice" issue is the availability of attorneys for litigants, including juveniles who cannot afford to retain counsel. The U.S. Supreme Court case of *Gideon v. Wainwright* (1963) held that: "in our adversary system of criminal justice, any person hauled into court, who is too poor to hire a lawyer, cannot be assured a fair trial unless counsel is provided for him."[114] At first, the *Gideon* precedent was limited to felony offenses, but now it applies to all criminal offenses (see *Alabama v. Shelton*, 535 U.S. 654, 2002).

The legacy of *In re Gault*, *Gideon*, and *Shelton* would seem to be undeniable—that juveniles charged with any delinquent offense are entitled to counsel. Counsel for minors would seem to be even more of a necessity than for adults in similar circumstances because of the lack of life experience and recognized immaturity. However, the promise of the case law is not always kept in practice. Because the finding of appointed counsel for indigent adult defendants and juvenile offenders is often left to local funding units, the compensation rates for attorneys can vary widely. Equal protection would seem to require equal compensation, but that is not always the case. The great majority of persons arrested in the United States cannot afford an attorney and must rely on the government to provide one. For youths, the rate is arguably higher because the presumption is that *all* juveniles are indigent because, by law, they cannot own property. However, funding for indigent defense varies greatly across jurisdictions and is directly tied to the economic health of communities. As a result, many communities, particularly in economic downturns, have difficulty providing adequate resources for indigent defense. Although most people recognize that criminal defendants and juvenile offenders deserve representation, it is hard to convince many politicians that they should spend more than the bare minimum on "criminals" who they believe are probably guilty anyway.

Who provides legal representation can also vary. It may be members of the private bar who are paid an hourly rate. That rate may be limited to "in court time," so that legal research and case preparation are not compensated. Some jurisdictions use public defenders who are salaried rather than paid by the hour or by the case. Often, public defenders are overwhelmed by case loads and cannot give each defendant or juvenile offender the time needed. If the attorney is paid by the case (e.g., $500 for a simple or class 1 felony), then there is more incentive to plea bargain than go to trial. All of the compensation methods have flaws that affect the quality of legal representation and, thus, the legitimate access to justice for the accused.

Problem-solving family courts, where families can go to have their problems addressed, are clearly a goal for the future. These courts must be litigant friendly, easy to access, and all in one location. They should be staffed by judges and court personnel who treat everyone with civility, dignity, and respect; who are focused on protecting rights and identifying problems; and who believe in accountability and a person's ability to change. Through these "new" courts, the historical purpose of the juvenile court may be realized. In such courts, children would be given a more realistic opportunity to overcome youthful mistakes so that they can become contributing adult citizens.

Chapter Summary

The purpose of this chapter was to (1) examine historical changes in the ways that juvenile courts have dealt with problem youths, (2) review problems that currently confront contemporary juvenile justice agencies, and (3) consider issues that are likely to confront juvenile justice in the near future. Clearly, juvenile justice practice has changed over time. Although juvenile courts were developed as informal chancery courts and had little concern for due process protections, U.S. Supreme Court decisions during the mid-1960s and early 1970s required juvenile courts to take due process more seriously. Nevertheless, the Supreme Court balked at giving children all of the due process protections afforded adults. As a result, juvenile courts continue to be more informal than criminal courts, causing some critics to call for their reform or abolishment.

One type of reform effort that began in the late 1970s was the "get tough" movement. This movement was led by reformers who urged the abandonment of the juvenile court's traditional *parens patriae* philosophy, with its focus on serving the best interests of children. Instead of the traditional *parens patriae* approach to juvenile justice, "get tough" reformers sought to substitute a more punitive philosophy, and some even called for the abolishment of the juvenile court. The "get tough" movement continues to influence the development of juvenile justice policy today, although rehabilitation and treatment still maintain a strong foothold in juvenile justice, at least in rhetoric if not always in actual practice, in many jurisdictions around the country.

Today, the juvenile courts, despite numerous operational changes, continue to be plagued by many long-standing problems: the failure to protect the due process rights of children, a lack of appropriate responses to status offenders, the processing of many youths who could be diverted from juvenile justice, the lack of coordinated resources for responding to problem youths, and the failure to recognize the limits of juvenile justice in solving youth problems. Juvenile corrections programs also continue to face significant problems: the existence of ineffective programs; inadequately trained, paid, and motivated staff; a penchant for relying on simplistic panaceas; a paucity of interventions that address the special needs of girls; the disproportionate processing and incarceration of minority youths; a shortage of good aftercare programs for youths; a failure to give sufficient attention to youths' right to treatment; a reliance on the privatization of juvenile corrections; the continued jailing of juveniles; and the lack of sound **program development and evaluation**.

program development and evaluation A systematic process of collecting and using data to determine program effectiveness, with the goal of making changes that will improve the level of effectiveness.

Perhaps the most important issue currently facing juvenile justice is whether the juvenile courts should continue to exist. As noted earlier, some have called for the elimination of the juvenile courts, arguing that they are either unwilling or unable to protect the due process rights of youths. Still others hold that they do not constitute an effective response to juvenile crime, particularly serious juvenile crime. There are those who argue that the juvenile court needs fundamental change, whereas others argue that adult courts should be patterned after the juvenile court model.

Although the future will continue to see considerable debate and conflict over the juvenile courts, it seems likely that they will continue to exist with some changes. One possible change is a move toward a more balanced approach to juvenile justice, one that stresses accountability for crimes, helping offenders develop competencies to enable them to lead productive lives, and protecting public safety. Another possible change is toward more formality in juvenile court operations.

No matter what changes in their operation, juvenile courts will continue to serve as primary social agencies responsible for dealing with a variety of problems that confront children, families, schools, and communities. These problems include family dissolution; child abuse and neglect; drug abuse; poor education, recreation, and employment opportunities; and poverty. The good news is that juvenile justice will become more professional, it will employ better educated and trained personnel, more emphasis will be placed on preventing delinquency, and the linkage between prevention and effective corrections will become clearer. Moreover, more attention will be given to program development and evaluation, which will expand our knowledge of effective juvenile justice interventions.

There are those who doubt the continued existence of juvenile justice. Our belief is that, on the contrary, juvenile justice is likely to become even more vital in the near future. In any case, one thing is certain: The future of juvenile justice ultimately depends on its ability to deliver on its promise of serving the best interests of children and protecting the community.

Key Concepts

aftercare: Services delivered to youths after they are released from a residential setting and while they are still under the jurisdiction of the state or juvenile court.

balanced approach to juvenile justice: An approach based on the recognition that a community has three needs: to sanction crime, to rehabilitate offenders, and to ensure public safety. The balanced approach emphasizes three corresponding goals for juvenile justice: accountability, competency, and public safety.

chancery court: English court concerned primarily with property rights. Over time, these courts became more involved in the welfare of families and children, and served as a model for the development of the juvenile court.

child savers: Affluent 19th-century reformers who were concerned about controlling and training youths so that the youths could become productive, law-abiding citizens. The Child Savers played a major role in the establishment of the juvenile courts.

continuum of quality services: A range of effective services consisting of prevention and correctional responses for delinquent youths.

due process: An established course of proceedings designed to protect individual rights and liberties and to ensure that people are treated with fairness before the law. Although juveniles have most of the due process protections afforded adults in the criminal courts, they do not have all of these rights. Moreover, because of the informality found in many juvenile courts, legal protections that are theoretically available may in fact be unavailable.

"get tough" movement: A legal reform movement that began during the late 1970s. The goals of the movement included developing a more punitive juvenile court orientation or abolishing the juvenile courts so that more severe sanctions could be leveled against juvenile offenders.

informal juvenile court: A court that places little emphasis on due process protections for youths. Because juvenile courts exist, in theory, to protect the best interests of children, legal protections for children are felt to be unnecessary within an informal court.

meta-analysis: A statistical technique that permits statements about the effects of interventions across a number of different studies.

***parens patriae*:** A legal doctrine according to which the state has an obligation to protect and serve the interests of children.

program development and evaluation: A systematic process of collecting and using data to determine program effectiveness, with the goal of making changes that will improve the level of effectiveness.

rehabilitation: The process of restoring an individual's ability to play a constructive role in society through education, training, or therapy.

therapeutic jurisprudence: The use of the legal process to promote the psychological and physical well-being of the people it serves.

youth discount: A stratagem for taking into account the supposed diminished culpability of youths (based on their immaturity) in the sentencing process. The use of youth discounts would generally result in shorter sentences for juveniles than for adults who committed identical offenses.

Review Questions

1. In what ways have juvenile justice practices changed since the development of the first juvenile courts?
2. What were the concerns of early child savers, and how did these concerns lead to the development of the juvenile courts?
3. What was the position taken by the U.S. Supreme Court during the mid-1960s and early 1970s toward the informal character of the juvenile courts?
4. What is the "get tough" movement, and how has it affected juvenile justice practice?
5. How has rehabilitation fared in the face of the "get tough" movement?
6. What are some of the most important problems that presently face the juvenile courts?
7. What are some of the most important problems that presently face juvenile corrections?
8. What are the essential components of the Intensive Aftercare Program (IAP) model?

9. Do juveniles have a right to treatment when they are placed in juvenile corrections programs?
10. Do juveniles have a right to be safe when placed in juvenile corrections programs?
11. What mechanisms exist or should be developed to protect youths' safety in correctional programs?
12. What arguments support the abolishment of the juvenile courts?
13. What arguments support the continuation of the juvenile courts?
14. What are the primary assumptions of the balanced approach to juvenile justice?
15. Do you agree with the authors' prediction that the balanced approach to juvenile justice will receive substantial attention in the future?
16. What problems will continue to confront the juvenile courts? How will these problems affect the juvenile courts?

Additional Readings

Altschuler, D. M. & Armstrong, T. L. (1996). Aftercare not afterthought: Testing the IAP model. *Juvenile Justice, 3*, 15–22.

Antonowicz, D. H. & Ross, R. R. (1994). Essential components of successful rehabilitation programs for offenders. *International Journal of Offender Therapy and Comparative Criminology, 38*, 97–104.

Bazemore, G. & Day, S. E. (1996). Restoring the balance: Juvenile and community justice. *Juvenile Justice, 3*, 3–14.

Bernard, T. J. (1992). *The cycle of juvenile justice*. New York: Oxford University Press.

Chesney-Lind, M. & Shelden, R. G. (2004) *Girls, delinquency and juvenile justice* (3rd ed.). Belmont, CA: Thompson/Wadsworth.

Elrod, P. (2009). The potential for fundamental change in juvenile justice: Implementing an alternative approach to problem youth. In J. I. Ross (Ed.), *Cutting the edge: Current perspectives in radical/critical criminology and criminal justice* (2nd ed.). New Brunswick, NJ: Transaction Publishers.

Elrod, P. & Kelley, D. (1995). The ideological context of changing juvenile justice. *Journal of Sociology and Social Welfare, 22*, 57–75.

Feld, B. C. (1993). Juvenile (in)justice and the criminal court alternative. *Crime and Delinquency, 39*, 403–424.

Gendreau, P. & Goggin, C. (2000). Correctional treatment: Accomplishments and realities. In P. Van Voorhis, D. Lester, & M. Braswell (Eds.), *Correctional Counseling and Rehabilitation* (4th ed.). Cincinnati, OH: Anderson.

Krisberg, B. (2005). *Juvenile justice: Redeeming our children*. Thousand Oaks, CA: Sage.

Platt, A. M. (1977). *The child savers: The invention of delinquency* (2nd ed.). Chicago: University of Chicago Press.

Puritz, P., Burrell, S., Schwartz, R., Soler, M., & Warboys, L. (1995). *A call for justice: An assessment of access to counsel and quality of representation in delinquency proceedings*. Washington, DC: American Bar Association.

Ryerson, E. (1978). *The best laid plans: America's juvenile court experiment*. New York: Hill and Wang.

Notes

1. Elrod, P. (2009). The potential for fundamental change in juvenile justice: Implementing an alternative approach to problem youth. In J. I. Ross (Ed.), *Cutting the edge: Current perspectives in radical/critical criminology and criminal justice* (2nd ed.). New Brunswick, NJ: Transaction Publishers.
2. Platt, A. M. (1977). *The child savers: The invention of delinquency* (2nd ed.). Chicago: University of Chicago Press.
3. Mennel, R. M. (1973). *Thorns and thistles: Juvenile delinquents in the United States 1825–1940* (p. 129). Hanover, NH: University Press of New England. This resource contains a statement made by Julia Lathrop.
4. Bernard, T. J. (1992). *The cycle of juvenile justice*. New York: Oxford University Press.
5. *People v. Turner*, 55 Ill. 280 (1870).
6. Bernard, 1992.
7. Ryerson, E. (1978). *The best laid plans: America's juvenile court experiment*. New York: Hill and Wang.
8. Bernard, 1992.
9. Mack, J. W. (1909). The juvenile court. *Harvard Law Review*, *23*, 104–122.
10. Ryerson, 1978.
11. Bernard, 1992.

    Krisberg, B. & Austin, J. F. (1993). *Reinventing juvenile justice*. Newbury Park, CA: Sage.
12. Krisberg & Austin, 1993, p. 31–32.
13. Elrod, 2009. This resource contains a brief historical perspective of juvenile justice policy.
14. *Kent v. United States*, 383 U.S. 541, 86 S. Ct. 1045, 16 L. Ed. 2d 84 (1966).
15. *In re Gault*, 387 U.S. 1, 87 S. Ct. 1428, 18 L. Ed. 2d 527 (1967).
16. *In re Winship*, 397 U.S. 358, 90 S. Ct. 1068, 25 L. Ed. 2d 368 (1970).
17. *In re Gault*, 387 U.S. at 15, 18, 19, and 28 (1967).
18. *McKeiver v. Pennsylvania*, 403 U.S. 528, 91 S. Ct. 1976, 29 L. Ed. 2d 647 (1971).
19. Bernard, 1992.
20. Krisberg & Austin, 1993.
21. Elrod, 2009.

    Elrod, P. & Kelley, D. (1995). The ideological context of changing juvenile justice. *Journal of Sociology and Social Welfare*, *22*, 57–75.
22. Elrod, 2009.

    Krisberg & Austin, 1993.
23. Krisberg, B., Schwartz, I., Litsky, P., & Austin, J. (1986). The watershed of juvenile justice reform. *Crime and Delinquency*, *32*, 5–38.

    National Criminal Justice Association. (1997). *Juvenile justice reform initiatives in the states: 1994–1996*. Washington, DC: Office of Juvenile Justice and Delinquency Prevention.

Schwartz, I. M. (1989). *(In)justice for juveniles: Rethinking the best interests of the child.* Lexington, MA: Lexington Books.

24. Krisberg & Austin, 1993.

Rubin, H. T. (1980). The emerging prosecutor dominance of the juvenile court intake process. *Crime and Delinquency, 26,* 299–318.

Sickmund, M., Snyder, H. N., & Poe-Yamagata, E. (1997). *Juvenile offenders and victims: 1997 update on violence.* Washington, DC: Office of Juvenile Justice and Delinquency Prevention.

25. Feld, B. C. (1993). Juvenile (in)justice and the criminal court alternative. *Crime and Delinquency, 39,* 403–424.

26. Hirschel, D., Wakefield, W., & Sasse, S. (2008). *Criminal justice in England and the United States* (2nd ed.). Sudbury, MA: Jones and Bartlett Publishers.

27. Cullen, F .T., Golden, K. M., & Cullen, J. B. (1983). Is child saving dead? Attitudes toward juvenile rehabilitation in Illinois. *Journal of Criminal Justice, 11,* 1–13.

Krisberg & Austin, 1993.

Maguire, K. & Pastore, A. L. (Eds.). (1996). *Sourcebook of criminal justice statistics – 1995.* Washington, DC: U.S. Government Printing Office.

Nagin, D. S., Piquero, A. R., Scott, E. S., & Steinberg, L. (2006). Public preferences for rehabilitation versus incarceration of juvenile offenders: Evidence from a contingent valuation survey. *Criminology and Public Policy, 5,* 627–651.

28. Cullen, Golden, Cullen, 1983.

Krisberg & Austin, 1993.

29. Maguire & Pastore, 1996.

30. Nagin, Piquero, Scott, & Steinberg, 2006.

31. Bazemore, G. & Feder, L. (1997). Rehabilitation in the new juvenile court: Do judges support the treatment ethic? *American Journal of Criminal Justice, 21,* 181–212.

Cullen, Golden, Cullen, 1983.

Moon, M. M., Sundt, J. L., Cullen, F. T., & Wright, J. P. (2000). Is child saving dead? Public support for juvenile rehabilitation. *Crime and Delinquency, 46,* 38–60.

32. Bazemore & Feder, 1997.

33. Elrod, P., et al. (1992). *Toward an understanding of state operated juvenile institutions: A preliminary descriptive analysis of a survey of chief correctional administrators.* Paper presented at the American Society of Criminology annual meeting, New Orleans.

34. Bernard, 1992.

35. Feld, B. C. (1988). In re Gault revisited: A cross-state comparison of the right to counsel in juvenile court. *Crime and Delinquency, 34,* 393–424.

Puritz, P., Burrell, S., Schwartz, R., Soler, M., & Warboys, L. (1995). *A call for justice: An assessment of access to counsel and quality of representation in delinquency proceedings.* Washington, DC: American Bar Association.

Sanborn, J. B. Jr. (1994). Remnants of parens patriae in the adjudicatory hearing: Is a fair trial possible in juvenile court? *Crime and Delinquency, 40,* 599–615.

36. Bortner, M. A. (1982). *Inside a juvenile court: The tarnished ideal of individualized justice.* New York: New York University Press.

Puritz, Burrell, Schwartz, Soler, & Warboys, 1995.

37. Chen, S. (2009, February 24). Pennsylvania rocked by "jailing kids for cash" scandal. *CNN.* Retrieved February 27, 2009 from http://www.cnn.com/2009/CRIME/02/23/pennsylvania.corrupt.judges/.

Democracy Now! (2009, February 17). Penn. judges get kickbacks for placing youths in privately owned jails. *Democracy Now! The War and Peace Report.* Retrieved February 27, 2009 from http://www.democracynow.org/2009/2/17/penn_judges_plead_guilty_to_taking.

38. Feld, 1993.

Feld, B. C. (1993). *Justice for children: The right to counsel and the juvenile courts.* Boston: Northeastern University Press.

39. Feld, 1993, p. 418.

40. Feld, 1993.

41. Zalkind, P. & Simon, R. J. (2004). *Global perspectives on social issues: Juvenile justice systems.* Lanham, MD: Lexington Books.

42. Schur, E. M. (1973). *Radical nonintervention: Rethinking the delinquency problem.* Englewood Cliffs, NJ: Prentice Hall.

43. Bullington, B., Sprowls, J., Katkin, D., & Phillips, M. (1978). A critique of diversionary juvenile justice. *Crime and Delinquency, 24*(1), 59–71.

Dunford, F. W., Osgood, D. W., & Weichselbaum, H. F. (1981). *National evaluation of diversion projects: Final report.* Washington, DC: Office of Juvenile Justice and Delinquency Prevention.

Palmer, T. & Lewis, R. V. (1980). *Evaluation of juvenile diversion.* Cambridge, MA: Oelgeschlager, Gunn and Hain.

44. Bailey, W. C. (1966). Correctional outcome: An evaluation of 100 reports. *The Journal of Criminal Law, Criminology, and Police Science, 57,* 153–160.

Martinson, R. (1974). What works? Questions and answers about prison reform. *The Public Interest, 35,* 22–54.

Robison, J. & Smith, G. (1971). The effectiveness of correctional programs. *Crime and Delinquency, 17,* 67–80.

Wright, W. E. & Dixon, M. C. (1977). Community prevention and treatment of juvenile delinquency. *Journal of Research in Crime and Delinquency, 14,* 35–67.

45. Martinson, 1974, p. 23.

46. Whitehead, J. T. & Lab, S. P. (1989). A meta-analysis of juvenile correctional treatment. *Journal of Research in Crime and Delinquency, 26,* 276–295.

47. Gendreau, P. & Ross, B. (1979). Effective correctional treatment: Bibliotherapy for cynics. *Crime and Delinquency, 25,* 463–489.

Gendreau, P. & Ross, R. R. (1984). Correctional treatment: Some recommendations for effective intervention. *Juvenile and Family Court Journal, 34,* 31–39.

Genevie, L., Margolies, E., & Muhlin, G. L. (1986). How effective is correctional intervention? *Social Policy, 16,* 52–57.

48. Garrett, C. J. (1985). Effects of residential treatment on adjudicated delinquents: A meta-analysis. *Journal of Research in Crime and Delinquency*, *22*, 287–308.

49. Lipsey, M. W. (1995). What do we learn from 400 research studies on the effectiveness of treatment with juvenile delinquents? In J. McGuire (Ed.), *What works: Reducing reoffending: Guidelines from research and practice*. New York: Wiley.

Lipsey, M. W. & Wilson, D. B. (1998). Effective interventions for serious juvenile offenders: A synthesis of research. In R. Loeber & D. B. Farrington (Eds.), *Serious and violent juvenile offenders: Risk factors and successful interventions*. Thousand Oaks, CA: Sage.

50. Palmer, T. (1975). Martinson revisited. *Journal of Research in Crime and Delinquency*, *12*, 133–152.

51. Lipsey, 1995.

Lipsey & Wilson, 1998.

Palmer, T. (1992). *The re-emergence of correctional intervention*. Newbury Park, CA: Sage.

52. Grissom, G. R. & Dubnov, W. L. (1989). *Without locks and bars: Reforming our reform schools*. New York: Praeger Publishers.

Henggeler, S. W. (1997). Treating serious anti-social behavior in youth: The MST approach. *Juvenile Justice Bulletin*. Washington DC: Office of Juvenile Justice and Delinquency Prevention.

Howell, J. C. (Ed.). (1995). *Guide for implementing the comprehensive strategy for serious, violent, and chronic juvenile offenders*. Washington, DC: Office of Juvenile Justice and Delinquency Prevention.

Howell, J. C. (2009). *Preventing and reducing juvenile delinquency: A comprehensive framework* (2nd ed.). Los Angeles: Sage.

Schneider, A. L. (1986). Restitution and recidivism rates of juvenile offenders: Results from four experimental studies. *Criminology*, *24*, 533–552.

53. Antonowicz, D. H. & Ross, R. R. (1994). Essential components of successful rehabilitation programs for offenders. *International Journal of Offender Therapy and Comparative Criminology*, *38*, 97–104.

Gendreau, P. & Goggin, C. (2000). Correctional treatment: Accomplishments and realities. In P. Van Voorhis, D. Lester, & M. Braswell (Eds.), *Correctional Counseling and Rehabilitation* (4th ed.). Cincinnati, OH: Anderson.

54. Miller, J. G. (1998). *Last one over the wall: The Massachusetts experiment in closing reform schools* (2nd ed.). Columbus, OH: Ohio State University Press.

55. Silberman, C. E. (1978). *Criminal violence, criminal justice*. New York: Random House.

56. Finckenauer, J. O. (1982). *Scared Straight! and the panacea phenomenon*. Englewood Cliffs, NJ: Prentice Hall.

57. Finckenauer, 1982.

58. Peters, M., Thomas, D., & Zamberlan, C. (1997). *Boot camps for juvenile offenders: Program summary*. Washington, DC: Office of Juvenile Justice and Delinquency Prevention.

59. Peters, Thomas, & Zamberlan, 1997.

60. Peters, Thomas, & Zamberlan, 1997.

61. Morash, M. & Rucker, L. (1990). A critical look at the idea of boot camp as a correctional reform. *Crime and Delinquency, 36*, 204–222.

62. Snyder, H. N. & Sickmund, M. (2006). *Juvenile offenders and victims: 2006 national report*. Washington, DC: Office of Juvenile Justice and Delinquency Prevention.

63. Chesney-Lind, M. & Shelden, R. G. (2004) *Girls, delinquency and juvenile justice* (3rd ed.). Belmont, CA: Thompson/Wadsworth.

64. Chesney-Lind, M. (1988). Girls in jail. *Crime and Delinquency, 34*, 150–168. Schwartz, 1989.

65. Sickmund, M., Sladky, A., & Kang, W. (2005). *Easy access to juvenile court statistics: 1985–2005*. Retrieved March 1, 2009, from http://ojjdp.ncjrs.gov/ojstatbb/ezajcs/.

66. Chesney-Lind, M. & Shelden, R. G. (1998). *Girls, delinquency, and juvenile justice* (2nd ed.). (pp. 241–242). Belmont, CA: Wadsworth Publishing Company.

67. Chesney-Lind & Shelden, 2004.

68. Puzzanchera, C. & Adams, B. (2008). *National disproportionate minority contact databook*. Retrieved March 1, 2009, from http://ojjdp.ncjrs.gov/ojstatbb/dmcdb/.

69. Puzzanchera & Adams, 2008.

70. Krisberg, B. (2005). *Juvenile justice: Redeeming our children*. Thousand Oaks, CA: Sage.

71. Huizinga, D. & Elliott, D. S. (1987). Juvenile offenders: Prevalence, offender incidence, and arrest rates by race. *Crime and Delinquency, 33*, 206–223.

72. Krisberg & Austin, 1993, p. 127.

73. Lehman, C. (1997). Oregon initiative for reintegrating adjudicated youth. *Reaching Today's Youth: The Community Circle of Caring Journal, 2*, 65–68.

74. Altschuler, D. M. & Armstrong, T. L. (1996). Aftercare not afterthought: Testing the IAP model. *Juvenile Justice, 3*, 15–22.

75. Altschuler & Armstrong, 1996, p. 16.

76. *Inmates of Boys' Training School v. Affleck*, 356 F. Supp. 1354 (D.R.I. 1972).

77. Del Carmen, R. V., Parker, M., & Reddington, F. P. (1998). *Briefs of leading cases in juvenile justice.* Cincinnati, OH: Anderson Publishing Company.

78. *Nelson v. Heyne*, 491 F. 2d 353 (7th Cir. 1974).

79. *Morales v. Turman*, 383 F. Supp. 53 (E.D. Tex. 1974).

80. Hemmens, C., Steiner, B., & Mueller, D. (2004). *Significant cases in juvenile justice*. Los Angeles, CA: Roxbury Publishing Company.

81. *Gary H. v. Hegstrom*, 831 F. 2d 1430 (9th Cir. 1987).

82. *Alexander S. v. Boyd*, 876 F. Supp. 773 (D.S.C. 1995).

83. The Dallas Morning News, *Investigative Reports*. Retrieved March 1, 2009, from http://www.dallasnews.com/investigativereports/tyc/.

84. Puritz, P. & Scali, M. A. (1998). *Beyond the walls: Improving conditions of confinement for youth in custody* (p. 1). Washington, DC: Office of Juvenile Justice and Delinquency Prevention.

85. Sickmund, M. (2004). Juveniles in corrections. *Juvenile Offenders and Victims: National Report Series, Bulletin.* Washington, DC: Office of Juvenile Justice and Delinquency Prevention.

86. Sickmund, M., Sladky, T. J., Kang, W., & Puzzanchera, C. (2008). *Easy access to the census of juveniles in residential placement.* Retrieved Sept 18, 2009, from http://ojjdp.ncjrs.gov/ojstatbb/ezacjrp/.

87. Schwartz, 1989.

88. Shichor, D. & Bartollas, C. (1990). Private and public juvenile placements: Is there a difference? *Crime and Delinquency, 36,* 286–299.

89. Community Research Associates. (1985). Juvenile suicides in adult jails: Findings from a national survey of juveniles in secure detention facilities. In R. A. Weisheit & R. G. Culbertson (Eds.), *Juvenile delinquency: A justice perspective* (3rd ed.). Prospect Heights, IL: Waveland.

Memory, J. M. (1989). Juvenile suicides in secure detention facilities: Correction of published rates. *Death Studies, 13,* 455–463.

90. Schwartz, 1989, pp. 81–82.

91. Sabol, W. J. & Minton, T. D. (2008). Jail inmates at midyear 2007. *Bureau of Justice Statistics Bulletin.* Washington, DC: U.S. Department of Justice.

92. Krueger, C. (2003, September 29). Juvenile detention deaths spark public hearing. *St. Petersburg Times.* Retrieved June 7, 2007, from http://www.sptimes.com/2003/09/29northpinellas/Juvenile_detention_de.shtml.

93. Johnson, K. (2007, May 15). Injuries of teen inmates probed. *USA Today.* Retrieved June 6, 2007, from http://www.usatoday.com/news/nation/2007-05-14-texas-inmate-abuse_N.htm.

94. Human Rights Watch. (2000, March 5.) South Dakota: Stop abuses of detained kids: Governor must end inhumane practices. *Human Rights Watch.* Retrieved June 6, 2007, from http://www.hrw.org/en/news/2000/03/05/south-dakota-stop-abuses-detained-kids.

95. Martz, R. (1998, February 2). Feds condemn juvenile justice abuses: Georgia detention centers rated grim to "harmful": U. S. blisters state's juvenile justice: Young offenders abused in detention, probe finds. *The Atlanta Journal and Constitution.* Retrieved June 6, 2007 from http://nospank.net/n-b10.htm.

96. Heath, B. (2005, April 18). State finds abuses at youth homes. *The Detroit News.* Retrieved June 7, 2007, from http://www.detnews.com/2005/metro/0504/18/A01-153841.htm.

97. The Washington Post. (2005, April 10). Trouble for troubled youth. *The Washington Post.* Retrieved June 7, 2007, from http://www.washingtonpost.com/wp-dyn/content/article/2005/04/08/Ar200504-801834_pf.

98. Johnson, 2007.

99. Heath, 2005.

Burita, J. (2004). *Senator Collins chairs hearing on warehousing mentally ill children in juvenile detention centers: Releases new report with Rep. Waxman focusing on nationwide problem.* Senate Committee on Homeland Security and Governmental Affairs. Retrieved Sept. 18, 2009 from http://hsgac.senate.gov/public/index.cfm?FuseAction=Press.MinorityNews&ContentRecord_id=b1e8fe3e-177f-46e3-a1be-03dbea28557e&Region_ id=&Issue_id=d77a5735-caca-4e15-a3f6-4092c999444d.

100. The Washington Post, 2005.

101. Physicians for Human Rights. (n.d.). Health and human rights in juvenile justice. *Health and Justice for Youth Campaign Fact Sheet*. Cambridge, MA: Physicians for Human Rights.

102. Krueger, 2003.

103. Puritz, P. & Scali, M. A. (1998). Civil Rights of Institutionalized Persons Act in juvenile correctional facilities. *Beyond the walls: Improving conditions of confinement for youth in custody*. Retrieved March 7, 2009, from http://ojjdp.ncjrs.org/pubs/walls/sect-01.html.

104. Puritz, P. & Scali, M. A. (1998). Use of ombudsman programs in juvenile corrections. *Beyond the walls: Improving conditions of confinement for youth in custody*. Retrieved March 7, 2009, from http://ojjdp.ncjrs.org/pubs/walls/sect-02.html.

105. Feld, 1993.

Feld, B. C. (1990). The punitive juvenile court and the quality of procedural justice: Disjunctions between rhetoric and reality. *Crime and Delinquency*, *36*, 443–466.

106. Hirschi, T. & Gottfredson, M. (1993). Rethinking the juvenile justice system. *Crime and Delinquency*, *39*, 262–271.

107. Krisberg & Austin, 1993.

108. Krisberg & Austin, 1993, p. 186.

109. Elrod & Kelley, 1995.

110. Elrod, P. (1998). Similarities in conservative and liberal juvenile justice policies: Is there a critical alternative? In J. I. Ross (Ed.), *Cutting the edge: Current perspectives in radical/critical criminology and criminal justice*. Westport, CT: Praeger Publishers.

Elrod, 2009.

111. Bazemore, G. & Day, S. E. (1996). Restoring the balance: Juvenile and community justice. *Juvenile Justice*, *3*, 3–14.

112. Slobogin, C. (1995). Therapeutic jurisprudence: Five dilemmas to ponder. *Psychology, Public Policy, and Law*, 193–219.

113. Hora, P. F., Schma, W. G., & Rosenthal, J. T. A. (1999). Therapeutic jurisprudence and the drug treatment court movement: Revolutionizing the criminal justice system's response to drug abuse and crime in America. *Notre Dame Law Review*, *74*, 439–537.

114. *Gideon v. Wainwright*, 372 U.S. 335 (1963).

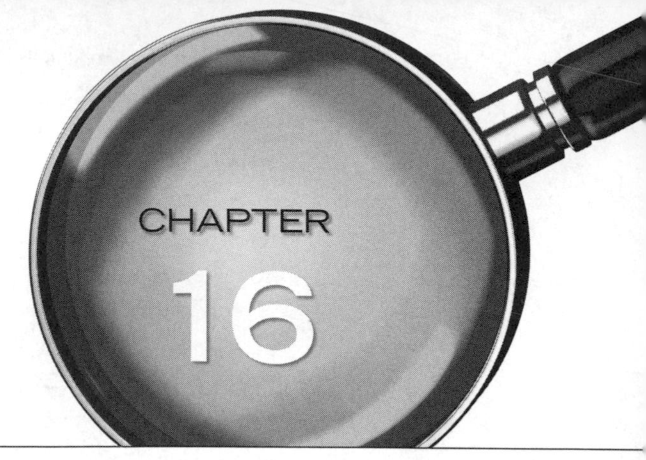

Working in Juvenile Justice

▸ ▸ CHAPTER OBJECTIVES

After studying this chapter, you should be able to

- Describe those factors that contribute to conflict between juvenile justice practitioners
- Describe how factors such as race, ethnicity, gender, social class, education, government association, and age can influence relationships between juvenile justice practitioners and their clients
- Describe how the offenses that youths commit can influence the relationship between juvenile justice practitioners and their clients
- Explain why coercive power makes juvenile justice agencies different from other social welfare agencies
- Describe the ways in which clients may be harmed through the exercise of coercive and persuasive force
- Explain why due process is important in juvenile justice, and indicate the four elements that guide practitioners in their efforts to ensure that juvenile justice clients are treated fairly and justly
- Explain what is meant by the presumption of innocence, and describe why it is important to ensure the fair treatment of youths and families
- Describe why conflicts of interest or the appearance of these conflicts should be avoided
- Explain the importance of maintaining confidentiality, and describe steps that practitioners can take to ensure confidentiality
- Describe how different types of bias can adversely influence juvenile justice practice
- Explain how economics can lead to ethical problems in juvenile justice
- Describe how ethics affect the quality of juvenile justice practice
- Describe the basic job functions of intake officers, psychologists, substance abuse counselors, court administrators, judges, and casework and child care staff
- Describe the required skills for effective juvenile justice practice

▸ ▸ CHAPTER OUTLINE

- Introduction
- Dealing with Conflict in Juvenile Justice
- Ethics in Juvenile Justice
- Economics and Ethical Dilemmas in Juvenile Justice
- Critical Positions in Juvenile Justice
- Required Skills for Effective Juvenile Justice Practice

Introduction

The purpose of this chapter is to review important jobs that are performed in juvenile justice and to examine the skills needed for effective juvenile justice practice. Although descriptions of the essential functions of many of the important positions found in juvenile justice are provided throughout this text, the objective of this chapter is to examine practical and professional requisites for successfully negotiating the everyday work of juvenile justice and to provide additional information on some of the important positions found in juvenile justice agencies. The first order of business, however, is to explore two topics that receive far too little attention in the classroom, the courtroom, and the community: the importance of dealing with conflicts of interest in juvenile justice and the role of **ethics** in juvenile justice practice. Our belief is that how people go about their jobs in juvenile justice is as important as what is achieved—that the "end" result does not always justify the "means," and that the process is an important part of the result.

ethics The principles of right conduct. These principles determine which actions are right and which are wrong.

Dealing with Conflict in Juvenile Justice

Very few people seek work in juvenile justice without a sincere desire to help youths and their families. The process of helping others, however, is often complicated by a number of factors. Persons who work in juvenile justice must, at times, deal with conflict between their own personal goals and those of others with whom they have contact, including coworkers. In addition, cultural and social factors, as well as a juvenile justice practitioner's gender, race, age, and association with government power (e.g., the power of the police, courts, and correctional institutions), also complicate the practitioner's dealings with others. How the practitioner handles these issues, which is colored by his or her individual and professional values, largely determines how effective the practitioner is in his or her job and how much satisfaction he or she receives from doing it.

Conflicts Between Juvenile Justice Practitioners

As we noted in Chapter 1, conflict is an inherent feature of juvenile justice practice. In fact, the juvenile justice process can be best conceptualized as a mechanism for resolving conflicts. Most of the conflicts are not terribly serious, and these can often (though not always) be resolved with relatively little expenditure of time and other resources. Some conflicts, on the other hand, involve serious criminal behaviors and result in serious property damage, personal injury, and even death. These conflicts usually require considerable expenditure of time and resources.

Because juvenile justice is essentially a conflict resolution process, disagreement over the best way to handle particular cases is a normal part of everyday juvenile justice

> ### MYTH VS REALITY
>
> **Status Offenses and Minor Delinquent Offenses May Require the Expenditure of Large Amounts of Resources**
>
> **Myth**—Status offenses and other minor types of delinquent behaviors require very little expenditure of time and energy on the part of juvenile justice practitioners.
>
> **Reality**—Many status offense cases and those involving minor delinquent behaviors can be quickly and effectively resolved. Many others, however, require considerable expenditure of time and other resources. For example, status offense cases that result from a long-standing family dysfunction can be very difficult and time-consuming to mediate.

practice. Such disagreement is found in interactions that take place between treatment providers and probation officers, between law enforcement officers and intake personnel, between judges and probation staff, and between defense and prosecuting attorneys, among others. For example, a probation officer may recommend that a youth be placed on probation in his or her home, believing that this is the best disposition for the youth and the community. However, the judge hearing the case may decide, in agreement with the prosecuting attorney, that a residential placement is in the best interests of the youth and the community. In another instance, a police officer called to the scene of a minor altercation between two youths may think that a formal complaint is unnecessary, but will file one anyway because the complainant demands that an arrest be made.

Conflicts such as those just described are common in juvenile justice because those involved—whether victims, offenders, witnesses, or juvenile justice professionals—do not always look at things from the same perspective. The police officer in the previous example may view the altercation as a typical disagreement between two youths that could be effectively dealt with by parental intervention. From the perspective of one of the participants, however, it may look like an assault that should be handled by the legal authorities. The other participant may see it as a matter of self-defense. The important thing to note about cases that come to the attention of juvenile justice practitioners is that rarely will all parties involved in a case look at it in the same way.

Conflicts also arise within juvenile justice practice because different practitioners have different roles and responsibilities. Indeed, effective operation of the juvenile justice process often depends on conflicting perspectives. The most obvious example of this is the conflict that occurs as a result of the different roles and responsibilities of defense and prosecuting attorneys. However, there are some role differences between any two juvenile justice positions. As a result, probation officers and treatment providers, judges and probation officers, or police and intake officers will disagree over how some cases should be handled. A probation officer and a judge may have a rather different picture of a client because of the amount of time the probation officer has spent with the youth, the family, and school officials trying to develop a good understanding of the case. Similarly, a probation officer, who sees the youth in the community, may have a very different view than a detention worker, who will have interacted with the youth only in a secure setting. This is not to say that one perspective is best. It is simply to acknowledge that differences in perspectives can be expected.

Of course, there is also some overlap in the responsibilities of those who work in juvenile justice. Everyone, regardless of position, is obligated to act with **integrity** in

> **integrity** The quality of acting in a moral, sincere, and honest way.

carrying out his or her responsibilities. Thus, the goal of the prosecuting attorney is not to win cases at any cost. The prosecuting attorney should not ignore evidence of a defendant's innocence in order to claim another successful prosecution. Nor should a probation officer omit information regarding a client's violations of probation in order to increase the chances that the judge will agree with the probation officer's **dispositional recommendation**. Regardless of the fact that these individuals have different roles and responsibilities, they both have an obligation to act with integrity and to uphold the basic legal and ethical principles upon which juvenile justice is founded.

dispositional recommendation
A recommendation made to the court about how a case should be handled.

Even when individuals act with integrity, however, different outcomes are possible. In the example in which the judge and the probation officer disagree about the proper placement of a youth, each may have considered the options carefully and become fully convinced of the correctness of his or her opinion. The decision by the judge to place a youth in a residential setting despite a different recommendation by the probation officer may be because the judge feels that a residential placement is the best placement for the youth and the community, which is exactly what drove the probation officer to recommend placement at home. Because of the judge's position of authority, the judge's wishes will prevail, but this is not to say that the judge's opinion is the only one that a reasonable person could have arrived at. Clearly, two different people acting with integrity may make different recommendations about a case. Juvenile justice decision making, it should be clear, is fraught with conflicts. Moreover, these conflicts can arise between people who are acting with integrity and who are following what they believe to be the most appropriate course of action. The question then arises as to how these conflicts should be dealt with. There is no simple answer, other than to say that juvenile justice practitioners have an obligation to act with professional integrity.

Of course, respecting another's differing opinion is usually more palatable when one is confident that the other person is well meaning, competent, and honest. If the person is felt to lack these qualities, respecting his or her opinion can be difficult. Unfortunately, juvenile justice, like other social domains, contains some people who are neither competent nor honest. Indeed, the history of juvenile justice, like the history of criminal justice, is replete with individuals who harmed youths in their care and engaged in unprofessional and even illegal behavior.

Malevolence does exist in juvenile justice, although it is probably a less widespread problem than incompetence. Fortunately, decisions based on sound logic and strong evidence will usually prevail over those predicated on faulty arguments. Moreover, incompetence is often easily revealed, although not always easily dealt with. In some situations, individuals are doing the best that they can, given their level of education and training.

MYTH VS REALITY

Differences of Opinion Are Common in Juvenile Justice

Myth—It is necessary for everyone to agree about the best way to handle a case.

Reality—It is important that juvenile justice practitioners recognize that honest, decent individuals acting with integrity do not always reach the same conclusions. Because individuals are called upon to provide honest and candid appraisals of issues that arise in juvenile justice practice, disagreements are to be expected. When disagreements between persons acting with integrity occur, they must be respected.

In others, individuals' performance is unacceptable despite their education and training. Dealing with these situations is never easy. One solution is to terminate the individuals in question. Another solution, and one that is often used, is to move the individual to another position where they are less likely to cause problems. Still another response is to provide the individual with additional training and require them to improve their performance or risk termination. Ideally, those in supervisory positions will take the appropriate actions. However, in some instances, those in supervisory positions are the problem.

Conflicts Between Juvenile Justice Practitioners and Others

The interactions between youths who have engaged in illegal behaviors (or behaviors that are alleged to be unlawful) and social control agents, such as police officers, court personnel, probation officers, and other corrections workers, are often conflictive. Even relationships between defendants and their defense attorneys can be problematic, especially if there is a lack of trust and poor communication. Similarly, relationships between probation officers and their clients, and between treatment program staff and their clients, can be difficult to establish and maintain, even though probation and treatment personnel believe their role is to help youths.

One obstacle to working with others is the difficulty people have in communicating with those who are different—those who come from different backgrounds and have different life experiences. This obstacle is prevalent within juvenile justice, because juvenile justice practitioners usually differ in a number of ways from the youths with whom they work. Juvenile justice practitioners are often of a different race, ethnic group, gender, or social class than their clients, and they are generally more educated and older. And even when these differences do not apply, there is one important distinguishing feature: juvenile justice practitioners are representatives of the juvenile justice process. As a result, they likely have a different perspective of the world and its possibilities than their clients, they frequently speak a language filled with professional jargon, and they typically live in different communities. Not surprisingly, communication between practitioners and their clients is often problematic.

Another obstacle to working with youths and their significant others is the difficulty of developing trust. Many Americans have the perception that government should not be trusted—a perception enhanced by the media emphasis on the lack of personal accountability and ethical commitment of persons in government. Moreover, at least some of those who come into contact with social welfare agency employees have not been helped but, instead, have encountered "red tape," condescension, excuses, and sometimes insults and abuse. Unfortunately, the phrase "I'm from the government, and I'm here to help" is often interpreted as "I'm from the government and I'm here to help myself!"

FYI

Public Confidence in a Number of Institutions Has Declined in Recent Years

According to Gallup polls taken in recent years that ask citizens how much confidence they have in selected social institutions, the percentage of respondents indicating that they have a "great deal of confidence" has been declining. For example, the percentages of citizens who reported that they had a great deal of confidence in the Congress, the presidency, the criminal justice system, and the police showed substantial declines between 2004 and 2007.[1]

Of course, there are good reasons for youths not to trust juvenile justice practitioners. After all, those who work in juvenile justice, from police to judges to parole officers, are representatives of "the authorities." For many youths and their parents, such individuals are associated with hassles, arrests, violence, and loss of freedom. And in some cases, of course, youths are involved in activities they wish to hide from juvenile justice practitioners.

Juvenile justice practitioners face another problem in dealing with youth offenders. Some of these youths have engaged in actions that caused serious physical or psychological trauma to their victims, or even resulted in the death of another human being. Confronting the anguish of the victims can be difficult, but it is imperative that juvenile justice practitioners are able to effectively communicate with their clients and provide quality services congruent with their job responsibilities. Even when a youth has engaged in behaviors that a practitioner finds abhorrent, the practitioner must conscientiously perform his or her duties. This does not mean that the practitioner has to condone the offender's behavior. It simply means that the practitioner needs to take his or her responsibilities seriously in all cases, not just in those cases in which it is easy.

As the previous discussion indicates, juvenile justice practitioners confront a number of conflicts and difficulties in performing their jobs. Many of these conflicts and problems can be considered "normal" aspects of juvenile justice practice, and practitioners must successfully handle these conflicts in ways that do not impede their ability to carry out their professional duties (and feel a sense of accomplishment at the same time). Therefore, the main question is not how to avoid such conflicts, but how to successfully negotiate these conflicts. Part of the answer is that successful, effective, and professional practice is based on integrity and honesty. Another part of the answer is that such practice also depends on adherence to some important ethical principles. These principles are spelled out in the next section.

Ethics in Juvenile Justice

power The ability to influence how other people think or act.

Juvenile justice practice always involves—sometimes subtly and at other times overtly—the exercise of **power**. Police officers, probation officers, detention personnel, attorneys, and others who work in juvenile justice are in a position to influence youths and their significant others in innumerable ways. Indeed, decisions made by juvenile justice personnel can result in orders to attend counseling, intrusion into individual and family privacy, loss of freedom, and termination of parental rights. Although the juvenile court is, in many respects, a social service agency that attempts to provide a range of helpful services to its clients, it differs from other types of social service agencies in one important respect: the court possesses coercive power that can be used to compel its clients to follow its directives.[2]

coercive force Physical force threatened or applied with the intention of influencing an individual's behavior.

Coercive force is a type of coercive power and involves the use of physical means or violence in order to influence an individual's behavior. Coercive force is often associated with law enforcement officers, but it is also employed by other juvenile justice personnel who have policing functions, such as probation officers, parole officers, and correctional and treatment staff. The use of coercive force must be taken seriously for at least five reasons:

1. The use of coercive force on a juvenile can affect the safety of the juvenile justice agent, the juvenile, and others.
2. The use of coercive force threatens the liberty interests of the individual, which are constitutionally protected.

3. The use of coercion and violence against another human being threatens the sanctity of the individual.
4. The use of coercive force can lead to negative perceptions of juvenile justice on the part of the juvenile and others.
5. The use of coercive force denies the juvenile the ability to make free decisions.

Although the use of coercive force is the most blatant example of the power wielded by juvenile justice practitioners, **coercive persuasion** is another type of coercive power, and constitutes a more subtle and more pervasive use of power. Coercive persuasion involves the use of one's position to influence the decisions and behaviors of others. For example, juvenile justice clients may be well aware that probation officers can revoke their probation unless they comply with the officers' requests. Similarly, a youth and his or her parents may agree to participate in a diversion program when given the not so subtle hint that failure to do so will result in a court hearing and the possibility of more severe sanctions. Also, many of the same problems associated with the use of coercive force can occur in instances of coercive persuasion. These include threats to the liberty interests of the individual, the development of negative perceptions of juvenile justice, and the denial of human agency.

coercive persuasion
The use of one's position to influence the decisions and behaviors of others.

Because of the power associated with their positions, those who embark on careers in juvenile justice must clearly understand the potentially harmful consequences associated with the exercise of that power. They must also act in ways that are consonant with the underlying ethical principles upon which juvenile justice practice is based. What are these principles? Perhaps one way to answer this question is to think of courts as institutions for dispensing justice. Persons who work in juvenile justice must ensure that people are treated in ways that are just, fair, and right. Because their actions are constrained by certain procedural requirements, it is helpful, as a way of approaching the topic of ethical principles, to consider these requirements. Among the most important are the following principles:

- ensuring due process
- maintaining the presumption of innocence
- avoiding conflicts of interest
- maintaining confidentiality
- avoiding bias

FYI

Coercive Force and Coercive Persuasion Must Be Used Wisely

A number of potentially harmful consequences are associated with the use of coercive force and coercive persuasion. Nevertheless, coercive force and coercive persuasion are sometimes necessary. For example, both coercive force and coercive persuasion may be used to protect the safety of clients, practitioners, or others; to assist clients in meeting legitimate treatment goals; or to accomplish other legitimate court goals. The problem here, of course, is that people will frequently disagree as to what constitutes client or practitioner safety, legitimate treatment goals, or court goals.

Ensuring Due Process

due process
An established course of proceedings designed to protect individual rights and liberties.

Due process is the foundation of our legal system. A fundamental tenet of due process is that the end does not justify the means. Consequently, those involved in the practice of juvenile justice must realize that being "fair" requires a personal commitment to making the decision-making process understandable, predictable, and reliable. Only people do justice, but a fair system or process can help ensure that the person who determines the outcomes will be fair. Crucial to due process and fair decision making are the following elements:

- reasonable notice of the issues to be addressed
- a realistic opportunity to have one's position considered by the decision maker
- a decision maker who is objective and unbiased
- a search for the truth

Every person working in juvenile justice must be aware of and understand these four essential elements of due process. Reasonable notice means that a person brought before a court must be forewarned of the charges. The notice must accurately describe the charges so that the individual has an opportunity to defend him- or herself effectively. Without reasonable notice, the individual would face a serious obstacle in trying to protect him- or herself against government power.

Reasonable notice would be irrelevant if the individual lacked an opportunity to present his or her position to the decision maker and have it considered. First, the accused must be physically present so that he or she (or his or her counsel) can personally respond to the charges. Second, the accused must be able to present information on his or her behalf and challenge information that is contrary to his or her interests. Third, disputed issues must be identified and ruled on by a person who has the expertise to do so. Fourth, the decision maker must be able to hear and understand the position of the accused.

Having one's case fully considered depends on having a decision maker who knows how the decision process works and how evidence or other information is used in making decisions, but at the same time is unbiased regarding the facts presented. When prejudiced or biased views are allowed to influence juvenile justice decision making, such decision making is no longer fair. Consequently, caseworkers and other persons working in the courts must guard against preadjudication contact with judges and referees so as not to influence decision making. Similarly, actions that may result in biased or prejudicial decisions made by other decision makers, such as intake officers and probation staff, should be avoided.

Reasonable notice, the opportunity to present one's case, and objectivity on the part of the decision maker increase the likelihood that the truth will emerge. Uncovering the

FYI

Being Heard Is Necessary for a Fair Hearing

Although it is important that a decision maker be able to hear and understand the arguments made by the accused, the decision maker does not have to agree with those arguments. Nevertheless, without the ability to consider different points of view, a fair hearing is impossible.

COMPARATIVE FOCUS

Some Countries' Approach to Juvenile Justice is Decidedly Non-Adversarial and Focused on Rehabilitation

The juvenile court in the United States is modeled on a quasi-adversarial approach to justice, which is clearly apparent in criminal courts. However, as we have noted earlier, youths are not always provided strong advocacy by those who represent them. Moreover, while juvenile justice in the United States is more treatment oriented than the adult criminal justice process, in too many instances, juvenile justice employs a punitive orientation while providing few due process protections. In contrast, some countries employ a more inquisitorial approach in which each of the participants in the process is obligated to "uncover the truth," and if the youth is found to have committed an offense, to proceed in a way that rehabilitates the offender. According to Paola Zalkind and Rita Simon in their book, *Global Perspectives on Social Issues: Juvenile Justice Systems,* one country that uses a non-adversarial approach and takes great pains to protect children and rehabilitate them is Germany. Children in Germany cannot be found criminally liable if they are younger than 14 years of age and reduced penalties for law violations are available for persons up to 21 years of age. In Germany, considerable effort is made to understand why a youth has engaged in illegal behavior and to involve parents and others in seeking a solution. Further, throughout the legal process, all individuals who work with children are specially trained in juvenile matters, and the entire system is oriented toward employing the least restrictive measures that will rehabilitate and reintegrate the offender in the community.[3]

truth is, in fact, one of the main goals of the juvenile justice process, and the search for the truth (i.e., what really happened as opposed to what the state would like to believe happened) is a fundamental principle in our legal system. Juvenile justice practitioners should be fully aware that the search for the truth is fundamental to achieving fairness and ensuring due process, and they should constantly strive to see that the truth is presented.

Maintaining the Presumption of Innocence

If all the legal principles that make up the foundation of American jurisprudence, the **presumption of innocence** is perhaps the most important. According to this principle, rather than an accused being responsible for proving his or her innocence, the state is responsible for proving the guilt of the accused. It is the belief in this principle that allows attorneys to represent the guilty with a clear conscience: They know that guilt must be proved by the state beyond a reasonable doubt, and that any person has the right to test the state's ability to prove the case, regardless of his or her culpability.

presumption of innocence The presumption that an individual accused of a crime is innocent of that crime until proven guilty. In the U.S. legal system, the government has the burden of proving an accused person is guilty rather than the accused person having the burden of proving he or she is innocent.

The importance of the presumption of innocence is typically associated with the trial or adjudication stage of the juvenile justice process. Nevertheless, this principle is equally important during the preadjudicatory decision-making stages. For example, presumption of guilt at the intake stage can result in intrusive, coercive, and inappropriate treatment being forced on a youth and his or her family without the benefit of a fair hearing. Similarly, presumption that a youth is guilty of misbehavior within an institutional setting can also serve as the basis for coercive and inappropriate actions on the part of correctional staff. Therefore, it is critical that presumption of guilt is avoided at every stage of the juvenile justice process pending a careful assessment of the facts.

Avoiding Conflicts of Interest

conflict of interest
A situation in which an individual has two interests that cannot jointly be satisfied. The term is frequently used to refer to cases in which an individual has a responsibility to act in a certain way but has other reasons, perhaps self-interested ones, for acting otherwise.

Conflict of interest is a problem because a person cannot faithfully serve two different masters when those masters have conflicting interests. Juvenile justice personnel need to be sensitive to this issue because many people today must have more than one job to make ends meet. Outside associations also can lead to conflicts of interest. Consider, for example, a juvenile court judge who teaches at a local community college part-time in the area of juvenile justice and delinquency. She teaches local law enforcement officers and probation officers. She receives remuneration from the college. One day, a juvenile who is accused of breaking into the college and doing severe damage is brought in front of the judge. Would her association with the college as an employee affect her ability to judge that juvenile without bias? Even if she was certain she would and could be fair, there is "an appearance" of a conflict, and that should be enough to cause the judge to recuse herself and have the matter heard by another judge.

Obviously, fairness is critical to the operation of juvenile justice. However, just as important as actual fairness is the appearance of fairness. Consequently, juvenile justice practitioners must be sensitive to how their friendships, relationships, and voluntary associations may lead to conflicts of interest. The best way and only way to ethically handle such a situation is to disclose the conflict and step away from the case.

Maintaining Confidentiality

Historically, juvenile court proceedings and juvenile court records have been closed to the public. The purpose of closed hearings and records was to protect youths from the stigmatization that might occur if information about them was to become known to the community. Although the secretive nature of juvenile justice proceedings may have helped youths in many instances, it prevented careful scrutiny of juvenile court operations and sometimes resulted in unfair and adverse responses to youths.

Over the past 15 years, a number of reforms have opened court operations to public view. Today, many juvenile courts are open to the public (i.e., the public can sit in the courtroom and observe the proceedings). In addition, information collected by juvenile justice agencies is becoming more readily available to outsiders for employment and licensing purposes.[4] This does not mean that juvenile justice employees are free to disclose intimate facts of juvenile cases to everyone. Indeed, many people still recognize that the stigmatization that may result from juvenile court involvement serves neither the interests of youths nor those of the community. As a result, individual jurisdictions have rules about what information is confidential and what can be released to interested members of the public. In Michigan, for example, the legal file and those documents that have been admitted into evidence are available for public view, including perusal by

MYTH VS REALITY

Can Criminal Courts Access Juvenile Court Records?

Myth—Criminal courts do not have access to juvenile court records when making sentencing decisions.

Reality—Every state allows prosecutor and/or court access to juvenile court records at some stage of the judicial process. Also, almost every state allows criminal courts to consider a defendant's juvenile record for sentencing purposes.[5]

the media, whereas other documents that contain intimate social information about the youth and his or her family and associates are not available. Some courts, unlike those in Michigan, allow some members of the public to access both legal and social files for specific purposes. For instance, many courts have policies that allow those involved in scholarly research to access files normally denied members of the public as long as the research does not identify specific youths.

Another issue that has received considerable attention in recent years is the need to share information among different agencies that work with youths who are involved in the juvenile justice process or youths who are at risk of juvenile justice involvement. For example, probation agencies, schools, social welfare programs, and mental health agencies may desire to share information on youths in order to improve the effectiveness of their services and keep youths from juvenile justice involvement. However, state and federal laws may restrict the sharing of some information among these agencies. State laws may require parents or guardians to provide written consent before agency records can be released or before agency personnel can discuss a particular case. The disclosure of students' education records is limited by the **Family Educational Rights and Privacy Act (FERPA)**, and some states have enacted similar laws.[6]

Family Educational Rights and Privacy Act (FERPA) A federal law that gives a student's parents the right to inspect and review their child's educational records, request amendments to those records, and exert considerable control over the release of those records.

FERPA is a complicated law that gives a student's parents the right to inspect and review their child's education records and to request amendments to those records, and it gives the parents considerable control over the release of information in those records. However, there are some exceptions. Educators can share information with other agencies or individuals based on their personal knowledge or observations, as long as that information does not come from educational records. Educators can also release information without prior consent if this information is given to other educators who have a legitimate educational interest in the information, if the release is for the purpose of complying with judicial orders or a lawfully issued subpoena, or if a state statute allows the disclosure of information to juvenile justice authorities in order to improve the ability of juvenile justice agencies to better meet the needs of specific youths involved in the juvenile justice process. In addition, FERPA allows "law enforcement units" within schools to collect and disseminate information to others outside the school, even the media, without parental consent.[7]

Sharing information between agencies is often necessary in order to provide effective and cost-efficient services to youths and their families. Yet, it is important that juvenile justice personnel recognize the potential deleterious effects that can result from sharing information when it is not necessary and legally mandated. For example, knowledge that a youth is involved in the juvenile justice process may be used against him or her. School and other agency personnel may refuse to give the youth the same opportunities given to others. The youth may be denied opportunities to participate in a variety of school-sponsored activities. He or she may be singled out for minor behaviors and suspended when similar behaviors by other youths would result in different responses. As a result, juvenile justice practitioners should be extremely careful about sharing information when it is not necessary and legally mandated. When it is necessary to discuss information about clients, the following rules are helpful:

- Make certain that releases are appropriately signed before obtaining reports from counselors, therapists, psychologists, and others from whom information is necessary.
- Make certain that when you request an order for any examination or for counseling, the person subject to that order knows that a report will be forwarded to the court.

- Remember that nothing constructive is ever gained by embarrassing someone.
- If there are sensitive issues in reports presented to juvenile justice decision makers, be sure to alert the decision maker, the youth, and the family prior to the time that the decision is made so that they will not be surprised if these sensitive issues are made public.
- Let the client know that as a juvenile justice employee, everything said to you must be made available to the court and that you will not be able to counsel the client as would a private counselor or therapist.
- Understand court policies and relevant laws about **confidentiality** in your jurisdiction and become familiar with the rules of agencies that you regularly deal with.
- Guard and respect other people's confidences as if they were your own.

confidentiality Maintaining information in strict privacy. Legal confidentiality requirements prohibit certain information from being made available to the general public.

New technology that is available to courts today can have a definite impact on the quality of justice received by individual juveniles. For example, many courts have the technology to arraign juveniles, advise them of their rights, and even conduct hearings by video rather than in person. Many states are experimenting with this procedure to save money and time. Juveniles could remain in the detention center or even at the police station or another secure location and not have to be transported to the court. Hearing officers would not have to leave their offices or courtrooms, and probation officers could remain at the court.

These experimental procedures do raise several legal and practical concerns, however:

- Does the impersonal viewing of the hearing officer by the juvenile lessen the potential impact of the proceeding?
- What about the rights of the parents to be present or participate in hearings? Can they meaningfully interact with their child and advise him or her through video technology?
- How does the juvenile's attorney counsel with the juvenile if not in his or her presence?

All of these questions and many others need to be answered by courts before implementing plans to use available technology.

Another by-product of increased technology is the potential impact it can have on issues of confidentiality. With the increased use of computers to hold information about youths, issues regarding limitations on the access to that information arise. Court files stored at a courthouse cannot be "hacked" into, but files stored on a computer can be compromised by enterprising individuals with computer skills. It is not beyond imagination that a delinquent youth with computer skills could gain entry into a court system and delete, destroy, or modify files.

On the other hand, the increased use of data transfer and the ability to share information can increase communication between courts and law enforcement agencies, giving the police more vital information much faster than before. Serious and careful data collection can aid in planning for the efficient use of available dollars and programming.

Avoiding Bias

bias A tendency, preference, or prejudice, especially one that inhibits objectivity.

Bias is defined as an inclination, a preconceived opinion, or a predisposition to decide a course or an issue in a certain way that renders an otherwise objective person unable to exercise impartiality.[8] Bias can be influenced by a number of factors—race or ethnic background, gender, geography, social class, or religion, to name a few. Furthermore, often two or more factors are combined. Bias is learned and is often strongly ingrained. Moreover, just because a person works for a juvenile justice agency does not mean that he or she is free of bias or prejudice. Consequently, it is imperative that court employees recognize any bias that they have and work to overcome it so that all persons can receive just and fair treatment at each stage of the juvenile justice process.

Racial and ethnic bias has a long history in this country, and the courts have unfortunately played a role in maintaining this bias. It is not, however, within the scope of this chapter to detail the struggle of racial and ethnic minorities for equality. The main point to be made here is that if justice is truly to be blind, a person's color or nationality should have nothing to do with his or her right to receive justice. Those who work in juvenile justice must face the unpleasant fact that minority youths are disproportionately represented in the juvenile justice process, particularly in juvenile courts and in secure juvenile correctional institutions.[9] They must also face the fact that the disproportionate involvement of minority youths in juvenile justice is, at least in some jurisdictions, the product of institutionalized bias among juvenile justice decision makers.[10] Despite stereotypical portrayals of inner-city minority youths as criminal offenders by the media and despite institutionalized responses to juvenile crime that focus attention on lower-class minority communities, juvenile justice decision makers must avoid making assumptions about youths based on the color of their skin.

The gender of youth offenders is also associated with differential treatment of youths within juvenile justice. As noted earlier, delinquent or incorrigible females do present a significant challenge to juvenile justice decision makers, and more young women are coming to the attention of the police and juvenile courts. Unfortunately, juvenile justice personnel often lack the resources or expertise to deal with their special needs and problems. As a result, girls have experienced considerable neglect, which has led to a lack of sound programming options and support systems that might help them cope with difficult home lives and avoid future juvenile justice involvement.

Not only have girls experienced gender bias within the juvenile justice process, but so have young women, including single mothers.[11] There is some evidence that youths from single-parent families may be at greater risk of juvenile justice processing than similar youths from two-parent homes.[12] Because the system deals with so many single parents, the vast majority of whom are women, juvenile justice decision makers must avoid biased attitudes toward them. Instead, each case must be examined on its own merits.

Even female juvenile justice staff members sometime suffer from the gender bias of their colleagues. Today, women make up a larger part of the professional personnel employed in juvenile justice than was true in the past. Female attorneys are far more numerous today than in the 1970s, 1980s, and 1990s. Similarly, many women serve as juvenile court judges, probation officers, and police officers and in various other positions within juvenile justice. Most important, these women have made significant contributions to the improvement of juvenile justice operations, which must be recognized and supported by everyone within juvenile justice.

Another type of bias that can influence juvenile justice operations is religious bias. The First Amendment to the U.S. Constitution guarantees that the government, which includes the courts, cannot support a particular religious philosophy or place any restriction on the free exercise of religious practice. From this constitutional pronouncement has come the commonly accepted idea that church and state should be kept separate. In fact, churches, synagogues, and mosques are not subject to the taxation power of the state because that power could be used to restrict or modify their practices and beliefs.

Courts encounter religion in any number of ways. Indeed, many juvenile justice employees hold religious beliefs that influence their ideas about how people should conduct their lives. However, such religious beliefs must never interfere with an individual's objective judgment in a case. In addition, religious beliefs partly determine the child-rearing practices of juvenile justice clients. Indeed, in some instances people use religion to justify child-rearing practices that run contrary to state laws. Moreover, religious beliefs may lead to a lack of cooperation with those who are charged with providing different types of services to youths. In situations like these, juvenile justice decision makers must attempt to determine what constitutes acceptable community or legal standards without violating the individual's constitutional right to practice his or her religion without interference from government authorities.

Since September 11, 2001, there is no question that many Americans have modified their views toward and relationships with individuals of Muslim faith and ethnic background. Whether juveniles who have a Middle Eastern background have been treated unfairly as a result of prejudice in the juvenile justice system is open to debate, but it is clear that such potential exists and must be recognized as bias and combated at every opportunity.

The last forms of bias that need to be discussed are geographic and social class bias. Some people are inclined to think that youths or families who reside on "the wrong side of town" or in certain communities are potentially more dangerous or less amenable to intervention. Like racial or gender bias, such thinking prevents juvenile justice practitioners from making objective decisions about cases. Geography or social class is neither a cure for nor a cause of delinquency. Where a person lives can influence peer and adult associations and the values a youth is exposed to, but location is not an indicator of criminality or receptivity to treatment. Even in the most crime-ridden communities, there are many people who avoid criminality and lead conventional and law-abiding lives.

Economics and Ethical Dilemmas in Juvenile Justice

Juvenile justice agencies, like other government agencies, are influenced to a considerable degree by what is often called the "golden rule" in government—those who have the gold, rule! By appropriating money for certain programs or to support particular policies and by withdrawing funding from other programs and policies, political decision makers have tremendous influence over the range of programs and resources that are available to juvenile justice practitioners. If officials at the local level believe that "getting tough" with juvenile offenders is an effective way to reduce juvenile crime, detention facility budgets may be increased to hire more staff or expand the number of facility beds. Moreover, a judge's views about which juvenile justice strategies are effective may influence the level of program funding before the budget gets to the spending source.

Too often, dispositional alternatives that might help children do not exist because they lack funding support. As a result, juvenile justice practitioners may face an ethical dilemma when figuring out what to recommend for their clients. Do they recommend what is truly in their clients' "best interest" regardless of cost, or do they recommend only that which they believe is financially available? Faced with this dilemma, juvenile justice practitioners must attempt to provide for clients' needs while simultaneously being a good steward of public funds. They also need to be knowledgeable about and creative with other revenue sources, such as federal dollars, grant money, and private insurance funds, in order to best meet client needs.[13]

How juvenile justice practitioners deal with the ethical issues discussed in this section will determine the extent to which juvenile justice is able to effectively meet its two main goals: serving the interests of children and protecting the community. Our belief is that each person who works in juvenile justice should give considerable thought to these issues and should conduct him- or herself in a manner that meets the highest standards of professional conduct. To do otherwise would compromise the ability of juvenile justice to deal fairly and justly with children and their families.

Critical Positions in Juvenile Justice

Throughout this text, the phrase "juvenile justice process" has been used to refer to the totality of decision-making stages at which juvenile justice practitioners make determinations about the appropriate ways of handling juvenile offenders. This term was chosen because, unlike "juvenile justice system," it does not mislead by hiding the considerable conflict that exists between and within juvenile justice agencies—conflict that often precludes practitioners from working toward common goals and objectives. As noted, some of this conflict is unavoidable, but it is also important to keep in mind that the effectiveness of juvenile justice agencies in helping children and families depends on cooperation between the individuals who work in these agencies.

This section examines some of the important positions that make up juvenile justice. These positions require different skills, but they each play a critical role in effective juvenile justice practice. Moreover, they each offer a tremendous opportunity for individuals who are looking for a challenging career that provides a chance to assist others, can contribute to the quality of community life, and offers many intrinsic rewards as well.

Intake Officers

Juvenile justice agencies, including juvenile courts, are under increasing pressure from many sides to deal effectively with juvenile crime. In response, some legislatures are limiting the power, discretion, and jurisdiction of the juvenile courts by lowering the age for waiver and expanding automatic waiver provisions. At the same time, local governments are demanding more accountability for each tax dollar spent. This often means that juvenile justice agencies are being called upon to do more with static or diminishing resources. Consequently, intake—the process of screening cases and determining which juveniles should enter the juvenile court system—is more critical today than ever before.

Although the intake process varies from jurisdiction to jurisdiction, there are some common intake practices across communities. For example, regardless of who does the screening—probation officers, referees, court-designated workers, or prosecutors—the screeners must be knowledgeable about the following elements:

- viable diversion programs, including non-court alternative interventions such as counseling programs, drug treatment programs, mediation programs, and alternative education programs
- the seriousness of the various offenses and whether they are appropriate for a court hearing

- nonjudicially ordered placement options, such as runaway shelters and temporary foster care placements
- mental health services for juveniles

Intake officers also need to communicate well with youths and their parents, law enforcement officers, judges and other hearing officers, social service agency staff, school personnel, and others. Moreover, many intake officers provide short-term crisis intervention services to children and their families in an effort to avoid formal court processing. Most of these areas of knowledge are self-explanatory, but one of them needs special emphasis. Because all courts have only limited docket or hearing time, it needs to be put to the most productive use possible. Spending court time trying cases that are poorly investigated and improperly charged is a waste of time and resources. Consequently, intake screeners must have some legal knowledge to help them evaluate the legal merits of juvenile delinquency charges. They also need the authority to require the responsible police agency to conduct further investigation when needed.

Psychologists and Substance Abuse Counselors

Many courts employ or have access to the services of a psychologist who has expertise in diagnosing children and identifying family problems. Because many youths who come to the court exhibit "acting out" behaviors that are rooted in their own personal and family problems, it is important that steps be taken to identify and respond to these problems as early as possible.

Similarly, many courts use the services of substance abuse counselors. Youths who come to the attention of juvenile justice agencies are often affected in some way by substance abuse. In many instances, it is the youth who has a substance abuse problem. In many other instances, one or both parents, or another significant other, has the problem. Consequently, it is important that courts have access to individuals who have the expertise to conduct good substance abuse assessments and provide effective interventions when appropriate.

BOX 16-1 Interview: Jim Kendrick, Assistant Professor, Western Michigan University and Director of Field Education for the Specialty Program in Alcohol and Drug Abuse (SPADA)

Education: B.A. (education); M.A. (sports administration); M.A. (clinical mental health); and a specialist's certificate in alcohol and drug abuse, all from WMU

Salary: $60,000, WMU plus between $20,000 and $25,000 for contract and consultant work each year

Q: What is your present employment position, and what do you do?

A: I am presently an assistant professor in the College of Health and Human Services. I work in the School of Community Health Services as the director of Field Education for the Specialty Program in Alcohol and Drug Abuse (SPADA). I teach classes in the SPADA program, training students to become substance abuse therapists and counselors. I am also the clinic coordinator for the WMU substance abuse clinic, which is a community clinic treating persons with substance abuse problems. In the clinic, I do direct counseling with patients. I also coordinate the administration of the substance abuse treatment programs in correctional programs for southwestern Michigan. The broad category of correctional programs includes those in many juvenile facilities.

(continues)

Q: What is your employment background that built up to your present position?

A: My first "field-related" job was from 1979 to 1981 as a delinquency caseworker for the Michigan Department of Social Services working with serious delinquent youths who had been committed to the state by the juvenile courts. From 1981 to 1984, I worked as a probation officer with the Kalamazoo County Juvenile Court and started the substance abuse assessment program for delinquent youths, which is mandatory for all delinquents now. From 1984 to 1985, I was director of a half-way house for young adult felony offenders, called the Twin County Community Probation Center. From 1985 to 1991, I worked in private practice as a clinical substance abuse and mental health therapist. Since 1991, I have held my present position at WMU.

Q: How did you begin your career that relates to juvenile justice?

A: Sports was my gateway to connecting with young people. I played all the major sports in high school—football, basketball, baseball, and ran track. I then played basketball for two years at Kellogg Community College in Battle Creek, Michigan, before going on to finish my degree at WMU. My interest in athletics, especially basketball, was a natural connector with many of the delinquent youths that I encountered working for DSS and the court.

Q: What aspects of your job do you enjoy most? What parts of your job don't you enjoy?

A: I enjoy three parts of my job a great deal. I enjoy direct therapy and the personal contact with children and families. I also enjoy creating treatment delivery systems and programs that can help deliver treatment to a larger community. I also enjoy training people in these larger systems. The thing that frustrates me continually is the institutional or bureaucratic resistance in agencies and the government. The constant concern with one's turf frustrates progress and treatment effectiveness too often. With the limited dollars available, however, some of this survival mentality has to be expected.

Q: What are the major challenges you see in the future for the juvenile justice system as it relates to your job?

A: I believe that the next 10 years will be crucial in determining whether a juvenile justice system independent of the adult criminal justice system will continue to exist. I genuinely fear that political reactions to some violent acts by juveniles will end the independent juvenile court system, and that will be a shame. I believe that children need to be treated differently from adults in most cases.

Q: What individual personal qualities do you believe a person must possess to successfully do your job?

A: Most important are honesty and personal integrity. Without these qualities, a person will not have any credibility professionally or in the community. From this personal integrity must come a solid foundation of values and beliefs. Unless a person is deeply rooted, the aberrant values they will encounter doing therapy can blow them away!

Q: What would you want to tell a person who is thinking about entering your field?

A: I would tell them that they need to be committed to a career that impacts the lives of children and families. This commitment must exceed their needs and desires for material goals. You can make a decent living in my field, but the career has some monetary limits, and they need to know that going in.

Court Administrator

The court administrator (the individual mainly responsible for the daily operation of the court) is playing an increasingly important and public role in juvenile courts. The court administrator is often responsible for the fiscal management of court programs and for presenting the court's position to other agencies and the public, and he or she may have a great deal of influence in determining the types of programs that are operated by the court.

Moreover, as competition for tax dollars increases, the court administrator must have a great deal of political savvy; in particular, he or she must know where the local government funding power is and be able to sell the court, its programs, and its mission.

The court administrator must take a lead role in formulating and defining the court's mission. To do this, the court administrator needs a comprehensive understanding of the juvenile justice process, from intake to treatment. The court administrator needs to know how the court can effectively network with other social agencies in both the public and private sectors to effectively deliver therapy, treatment, and other programming to juveniles and their families. The court administrator also must know the applicable laws and court rules governing the individual court and must be familiar with employment law and good administrative practices. The court administrator is responsible for interviewing, hiring, reviewing, and terminating employees, so knowledge of these techniques, effective communication skills, and interpersonal skills are a must. Because of the diverse legal expertise required in the administrator position, many courts are requiring formal legal training and/or a law degree for this position.

Finally, the court administrator must be computer literate so that the court can join the information age in handling its files, process, and docket. The importance of this position cannot be overstated.

BOX 16-2 Interview: Doug Slade, Administrator for the Family Division of the Circuit Court for Kalamazoo County

Q: How did you begin your career in juvenile justice, and when?

A: In 1966, while I was a police administration major at Michigan State, a probation officer's job came open in the Ingham County, Michigan, Juvenile Court. Although my career goal at that time was to work in industrial security, I was nearing graduation and my wife thought I should apply for the position. I did and was hired at the lofty salary of $6,300 per year. Two years later, I was appointed to a referee's position holding preliminary hearings, removal hearings, and doing other intake functions. In 1970, I helped persuade our judge that the court needed an intake unit, and he appointed me to run it. In 1975, I was appointed probate registrar so I went over to the Estate Division and worked there until 1977 when I was appointed court administrator as well. I held the combined positions from 1977 to 1987 when I resigned from the court.

I was the director of Central Michigan Legal Aid from 1987 to 1993, serving eight counties in central Michigan. In 1993, I was hired to be the court administrator for the Kalamazoo County Juvenile Court. As a result of the legislatively mandated combination of the juvenile court and the "family jurisdiction" of the circuit court, I am now the administrator in charge of the Family Division of the Circuit Court.

Q: What is your educational background?

A: I did graduate with a B.A. in police administration from MSU in 1966. I immediately started graduate school in that field but did not finish. My area of study was juvenile programs, but my increasing court responsibilities interfered with my continuing. In 1971, I attended the summer session of the National Judicial College at Reno, Nevada, and started law school at Cooley School of Law in Lansing, Michigan, in 1984. I finished on an accelerated program, earning my *juris doctoris* in December 1986 while working full time as court administrator. As soon as I graduated from Cooley, I began teaching there as an adjunct professor teaching probate practice. I've also attended and taught at too many workshops and training seminars to count!

(continues)

Q: What is your present job, how much do you make, and what are your major responsibilities?

A: As the court administrator for the Family Division of the Circuit Court, my annual salary is $88,000 per year. My major responsibilities are as follows:

- **a.** Leadership to court staff—not just supervising or managing behavior, my job requires vision and being proactive in solving the myriad of challenges the court faces.
- **b.** Planning—both day to day and long term. The creation of the local family court is a planning enterprise I was intimately involved in.
- **c.** Public relations for the court—along with, and in addition to, the judges, I have major responsibilities in dealing with the media, citizenry, agencies, the county government, and others.
- **d.** Community collaboration—one important duty is to identify areas in which the court and the community can work together and support each other.
- **e.** Budget and finance—taking care of the taxpayers' dollars and meeting the needs of the children and families that appear in front of the court are very important duties.
- **f.** Personnel administration—dealing with the many court employees and unions.
- **g.** Mentoring other administrators and managers—I have long believed that persons who have experience owe newcomers all the help they can give them.

Q: What do you like and/or dislike about your job?

A: I enjoy the impact that I can have, in my position, on community issues of social justice and social change. The responsibilities of leadership mean a lot to me, and I take them very seriously. At my professional level, the monetary benefits are substantial, and I do enjoy the financial rewards. I don't like dealing with negative people. There is always somebody who will see how something can't work or can't be accomplished. I don't like people who don't work hard, who don't do their jobs. They make it harder for everyone else who have to pick up what's left undone.

Q: What major challenges do you see in the future for the juvenile justice system?

A: The major challenge that I see is overcoming the media praise of the idea that kids and families are so bad and out of control that they should just be treated as adults. I believe that the very existence of the juvenile justice system is threatened by the media misrepresentation of the facts about juvenile crime and delinquency and the juvenile court's ability to intervene effectively. I'm afraid that the real independence of a separate court for juveniles is threatened.

Q: What personal qualities do you believe a person must possess to successfully do your job?

A: Let me see if I can count them off for you: patience, being proactive, a good listener, ability to see other perspectives, internally motivated, strong moral principles, quick learner, good communication skills, and humble. By humble, I mean that I need to always remind myself that I don't have all the answers or always know the best way to solve problems.

Q: What do you think a young person who is thinking about entering the juvenile justice field should know?

A: Foremost, they need to understand why kids come into the system—not because they are somehow inherently bad kids, but because their families and their community have failed them. They need to have empathy for the kids.

Q: Who are you responsible to?

A: First, I am responsible to the taxpayers whose money pays my salary and supports the court. Second, I am responsible to the presiding judge of the court.

Q: How many people do you supervise?

A: I am responsible for 164 employees: 10 managers, 67 professionals, and 87 clerical staff.

Judges

No position in the juvenile court is more important than that of a juvenile court judge. Juvenile court judges must wear many hats. First, they need to be well versed in the juvenile and family law of their particular jurisdiction. In addition, because of the increasing concern over due process in juvenile courts and because of the increasing visibility of juvenile court proceedings, judges should be well versed in procedural requirements. Increasingly, juvenile court judges will need trial experience in order to effectively perform their jobs and avoid making errors that would open the way for costly and time-consuming appeals.

In addition to having the legal acumen to make correct decisions regarding matters of law, juvenile court judges must be able to communicate well with youths, their parents, and a range of professional staff. Cases involving children and families are often the most emotionally draining cases that can be heard in courts. Consequently, juvenile court judges must have a sincere dedication to helping families and children. They must also have the ability to advocate for resources and services needed to enable the juvenile court to do its job. They may spend considerable time meeting with community groups, community leaders, and politicians in addition to the time they spend on court cases. Therefore, juvenile court judges must have vision and an understanding of how the court works and the ability to effectively articulate this to others. Lastly, they need to fully understand how the court interfaces with other public and private agencies in order to meet the needs of children and families and to protect public safety.

Juvenile court and family court judges must increasingly be able to handle heavy dockets and deal with increasing workloads. Given these difficult economic times for many state and local governments, judges have to do more work with the same or reduced staff. For example, in Michigan, since 1990 the number of abused and neglected children in out-of-home care has increased by 17%.[14] Michigan law requires 90-day review hearings for all children in care and termination of parental rights of parents if the children remain out of home for 364 consecutive days. The number of hearings involving these children has increased exponentially, decreasing the time available for dealing with delinquent juveniles. In addition to the docket pressure, the delinquency rate in Michigan has increased since 1990, requiring more court interventions and court hearings that place considerable strain on the docket. (To read an interview with a juvenile court judge, see Chapter 10.)

Appointed Judicial Officers

Regardless of whether they are called referees, magistrates, administrative law judges, masters, or commissioners, appointed judicial officers play an important role in juvenile courts. Their role may grow in importance, in fact, because they cost far less than judges. Routinely, their salaries are less and they require fewer "trappings" of office than judges. Many judges have personal secretaries, court reporters, and bailiffs. It is rare for referees to have "personal" staff. Appointed judicial officers have another advantage: unlike many judges, they are not elected, and therefore they can focus more of their attention on hearing cases. Moreover, many of them are attorneys with expertise in juvenile and family law, which makes them an important legal asset to the court. Too often, juvenile court judges see the juvenile court bench as a stepping-stone to a higher court. In contrast, judicial officers who apply for juvenile court positions often do so because they have an interest in helping families and children and are more likely to make their service a career, thus giving the court needed stability. (To read an interview with a juvenile court referee, see Chapter 10.)

Casework and Child Care Staff

Because of the increasing emphasis on the court's role in protecting the community, those juvenile justice practitioners who have the responsibility for supervising youths, providing services to youths and families, and making placement recommendations play a critical role in juvenile justice. These practitioners include several positions:

- probation and aftercare officers
- detention home workers
- detention facility administrators

Probation officers perform multiple roles. They must monitor clients' adherence to court orders (e.g., the rules of probation), they must be able to provide services to clients, they must have knowledge of effective services and make referrals to appropriate service agencies when necessary, and they must complete a variety of administrative tasks in a timely manner. Consequently, probation officers need to know how to access the court process in order to set up probation violation hearings and effectuate necessary changes in placement. Probation officers need to be part police officer, part court officer, part therapist, and part friend. (See Chapter 10 for a detailed description of the duties of probation officers. To read interviews with various casework staff, see Chapter 11.)

Detention facilities serve as short-term secure settings for a range of juvenile offenders, from youths who have committed serious crimes against persons and property to youths who have run away from another jurisdiction. They are also used for temporarily holding youths awaiting long-term placement and, increasingly, for youths serving determinate sentences handed down by juvenile courts. Their diverse population makes

BOX 16-3 Interview: John P. Dantis, MSW, Director of the Bernalillo County, New Mexico, Juvenile Detention Center

Q: What is your educational and employment background?

A: I have a bachelor's degree from the College of Santa Fe and an MSW from New Mexico Highlands University. I also spent four years in the Marine Corps with a tour of Vietnam in 1968. My professional work experience includes eight years as a juvenile probation officer, two years as the coordinator of probation services for the state of New Mexico, and 11 years as the director of Child Haven, a youth facility for delinquents.

Q: How long have you been the detention center director for Bernalillo County? What is your pay?

A: I am the director of the center, and have also been involved in the management of the county jail. I make over $80,000.

Q: What aspects of your job do you like and dislike?

A: The most important part of the job to me is the personal contact with the juveniles and their families. The real ability to impact kids and their families in positive directions gives the job real meaning for me. The part of the job that is most difficult is working in a political arena where I have to compete for limited resources with other county departments.

Q: What do you think that a college student should know about your job?

A: To understand the reality of any juvenile detention center today, the student must understand that success—the safe and productive management of these juveniles—requires them to understand cultural issues, educational issues, community resources, and mental health issues for the juvenile. They must also understand that detention workers are not "jailers" but must be counselors, teachers, social workers, and even coaches.

(continues)

Q: What are the major challenges that you face today and in the immediate future?

A: Managing the continual growth in population, and continuing to educate politicians and other community leaders that the system needs balances. The need for secure detention is real, but it is not the only answer. Community-based corrections are vital to our doing the job in managing delinquent youths. The community must understand the spectrum of correctional choices that needs to be available.

Q: What are the demographics of your detention population? How many youths can your facility hold, and what is the average daily census?

A: Our population's median age is 15–16 years old, although we have had youths as young as 8 years of age. Our ethnic breakdown at any given time is 60–65% Hispanic, 25% Caucasian, 10% Native American, and 5% African American. Our license is for 78 youths at any one time, but our population average is about 100 plus 30 to 35 youths on community release and monitoring.

Q: Who operates the detention center? Who funds it?

A: It is a county facility supported primarily by tax dollars from Bernalillo County. I report directly to the county manager.

Q: How many employees work at the detention center?

A: We employ 76 child care workers in two classifications who are responsible for the daily hands-on management of the youths. We have four program managers who are responsible for managing "clusters" of youths. Their responsibility is not 8:00 a.m. to 5:00 p.m., but they manage their clusters 24 hours a day. We have 16 to 18 teachers in the school, which has components for special education and emotionally impaired students, as well as a **GED [general education diploma]** program.

general education diploma (GED) A high school equivalency diploma awarded to adults who have not graduated from high school but have passed an official test covering a number of academic subjects. Commonly referred to by its acronym GED.

Q: What different types of programs for youths do you offer?

A: First of all, you need to understand that we are a preadjudicative facility. Our residents are presumed innocent, so there are legal limits on what we can force youths to do. Having said that, we offer programs in the following areas:

- education
- leisure activities and the management of free time
- substance abuse
- gang intervention
- communication skills (with a "Toastmasters" program)
- computer education
- family planning and AIDS education

Q: What types of offenses are you seeing juveniles detained for?

A: Primarily, youths in the facility are here for offenses against persons, not property offenses. We also receive a lot of habitual or repeat offenders. Many of our population have gang connections and suffer from substance abuse problems.

Q: What other serious issues are you seeing with these juveniles?

A: Too many of our residents are parents as well as delinquents. They need serious and relevant parenting education, including instruction in child development issues. We have, as noted before, a chronic overcrowding problem, with the political and financial implications that carries with it. Finally, because we do a good job managing the youths sent to us, there is always going to be pressure to become a sentencing alternative or residential facility. I don't see any of these pressures or issues going away soon.

Q: My last question: What are you most proud of in your accomplishments as the detention center director?

A: We are the only accredited juvenile detention facility in the state of New Mexico. This is a tribute to the hard work of our staff and the vision of our political leaders. Each staff person takes individual responsibility for the children under the nine principles of direct supervision. To help maintain the high level of care, each of our staff gets 96 hours of training a year with one-third of the staff in training each week.

the job of detention facility administrators even more difficult. Detention facility administrators must ensure the physical safety of residents and staff and maintain order. They must also provide basic education programs for youths and see that their physical and mental health needs are addressed. In addition, they are responsible for hiring, evaluating, training, and terminating staff; taking a lead role in staff and program development; and communicating with other agencies and the public regarding detention facility operations.

Detention employees, like their supervisors, must understand how to deal with a diverse resident population. This population consists of youths who have committed a wide range of offenses and who come from varied backgrounds. A major responsibility of detention staff is to ensure the safety of each resident under their care and to make certain that no one escapes. They are responsible for ensuring the safety of other staff as well. Consequently, they must be capable of forceful physical interventions. At the same time, they must understand that they are dealing with children. Although responsible for the safety and security of residents, detention staff members are also in a position to offer the residents guidance and basic counseling when appropriate. They must, therefore, have the skills to work effectively with children. In addition, they must be able to work effectively with other caseworkers, treatment staff, and detention supervisors in order to meet the needs of the youths under their care.

Required Skills for Effective Juvenile Justice Practice

The previous discussion of the critical jobs in juvenile justice should make evident some of the skills and knowledge needed to be an effective juvenile justice practitioner. Additional knowledge and skills are also needed. Everyone working for a juvenile justice agency must understand the interrelations among the agencies that make up the juvenile justice process. Each worker also must understand how his or her position interrelates with other positions within the same agency.

Given the importance of communication, computer skills are clearly essential. Courts are now using computers to store and track files, docket cases, perform routine paperwork, and do any number of other jobs. Computers enable the printing of court orders at the close of court hearings for service on those present. Judges, judicial officers, prosecutors, and attorneys who practice in the court are doing legal research by computer. As a result, law libraries will become less important in the future. In addition, more and more probation, detention, and treatment staff are using computerized systems to track cases, complete reports, communicate with other agencies, and monitor program effectiveness.

Interpersonal skills will continue to be the keys to effective job performance. Juvenile justice practitioners need the ability to communicate well, both orally and in writing. Moreover, given the diverse population served by juvenile justice personnel, they must be able to relate to and get along with many different kinds of people. Being fluent in other languages besides English, particularly Spanish, is very important, as are good listening skills and the ability to work effectively with people from varied racial, cultural, and ethnic backgrounds.

Two final prerequisites for effective juvenile justice practice are a desire to work with troubled youths and their families and a strong commitment to professional ethics. These qualities serve as the foundation upon which effective practice rests. They can be the keys to a career that is both professionally and personally rewarding.

Legal Issues

In 1974, the federal government passed the Child Abuse Prevention and Treatment Act, 42 U.S.C. 67. The original act required the individual states to enact similar statutes in order to remain eligible for certain types of federal funding involving children. Although child abuse and neglect were hardly new in 1974, this statute represented a major step forward in raising public awareness about the problem. (A thorough discussion of the connection between delinquency and future criminal behavior and prior abuse and neglect was discussed in Chapter 3.)

The federal statute contained certain mandatory requirements for the states to follow in crafting statutes focusing on abuse and neglect. Specifically, the federal government required the state statutes to contain

1. a state agency designated responsible for the investigation of all child abuse and neglect cases;
2. a "central registry" system created to maintain records of abuse and neglect perpetrators;
3. a mandatory reporting requirement that requires certain persons to report suspected abuse and neglect without the legal impediment of privileged communications;
4. a provision that abrogates all "privileged communications" but for that of attorney–client so that no privilege would interfere with the reporting of suspected child abuse and neglect.[15]

Privileged communications are communications between two people that are encouraged to take place without the scrutiny of others. For example, if a person is having personal problems and decides to go talk with a therapist, for the most part, what is talked about cannot be made public by the therapist, thus encouraging the person to fully disclose the nature of their problem. If people believed that whatever they say in confidence could be made known to other persons, they would be less likely to engage in beneficial counseling. These privileges are often recognized in statutory law or by common law principles.

The law, however, has now determined that the reporting of abuse and neglect of children is a higher priority than the unfettered communications with counselors, therapists, and other human service providers. Society's interest in protecting children from abuse is greater than protecting the communications of abusers. The federal statute addresses this issue in two of the mandatory state statute requirements. First, the state statutes must contain a list of mandatory reporters, persons who might otherwise be under the restrictions of "privileged communications," but who are required by statute to report suspected child abuse or neglect. Second, the statute must have a specific provision eliminating all privileged communications except those between an attorney and his or her client. The reason for the continuation of the attorney–client privilege is that abuse and neglect can result in serious legal consequences to the perpetrators, as it should. Abusers can be subject to civil prosecution concerning their parental rights as well as criminal prosecution. Fundamental fairness dictates that accused perpetrators have the right to legal representation in these matters and the integrity of that legal representation would be compromised if the attorney was mandated to report.

The Michigan statute is very typical of statutes that list mandatory reporters. Michigan Compiled Laws 722.623 states:

> (1) An individual is required to report under this act as follows;
>
> (a) A physician, dentist, physician assistant, registered dental hygienist, medical examiner, nurse, person licensed to provide emergency medical care, audiologist, psychologist, marriage and family therapist, licensed professional counselor, social worker, licensed bachelor's social worker, registered social worker, registered social service technician, social service technician, school administrator, school counselor or teacher, law enforcement officer, member of the clergy, or registered child care provider…

Thus, virtually anyone working in the field of juvenile justice or with a juvenile court is covered by this list of mandatory reporters. The statute goes further in requiring all employees of the Department of Human Services, the state designated agency under the statute, to be mandatory reporters also.

What this means for persons working in juvenile justice is that if you are in a situation in which you suspect that a child you are working with or who is on your caseload has been neglected or abused, you *must* make a report of your suspicions and concerns to the relevant state agency. There will be a time requirement for you to either make an oral or a written report. Failure to make a timely report can have serious legal consequences. First, many reporting statutes, if not all, make it a criminal offense to fail to report suspected abuse or neglect. In Michigan, MCL 722.633 makes the failure to report abuse or neglect a 93-day misdemeanor punishable by incarceration for up to 93 days, a $500 fine, or both. Just as significantly however, the Michigan statute at MCL 722.633(1) states: " A person who is required by this act to report an instance of suspected child abuse or neglect and who fails to do so is civilly liable for the damages caused by the failure." What this statutory provision has done is create a civil "tort" action for damages for the failure to report. If a child continues to be abused after that abuse should reasonably have been expected to stop, then the person failing to make the report can be sued for damages due to any ongoing injuries caused by the continuing abuse. Because serious physical, mental, and emotional abuse can have lifetime repercussions, the monetary damages could be substantial.

These same statutes, however, have legal protection for persons reporting. The best interests of society are to encourage reporting so those persons taking on that responsibility are protected. MCL 722.625 protects the identity of any reporter from disclosure except with that person's consent or by court order. This statute also protects any reporter's civil or criminal liability so long as the report was made "in good faith." Persons working in juvenile justice have a legal and moral obligation to report suspected child abuse and neglect. After all, they should know better than anyone the long–term disastrous consequences that abuse and neglect has on children.

Chapter Summary

This chapter began with an examination of the role of conflict within juvenile justice. As noted, conflict is an inherent characteristic of juvenile justice practice and occurs between juvenile justice professionals and between these professionals and their clients.

Indeed, conflicts between juvenile justice practitioners are to be expected, because those who work in juvenile justice have different ideas about how juvenile crime should be dealt with, because they have different responsibilities, and because they often have considerable discretion in how they handle cases. Such conflicts will invariably lead to tension and disagreements between juvenile justice practitioners, who, therefore, must learn how to negotiate disagreements effectively.

Although conflicts between juvenile justice practitioners are an inherent feature of juvenile justice practice, so too are conflicts between juvenile justice professionals and their clients. Differences in social class, education, sexual orientation, gender, race, and orientation toward authority can lead to difficulties in working together, not to mention the behaviors and attitudes exhibited by some juvenile justice clients. Nevertheless, it is important that juvenile justice practitioners recognize the potential problems presented by working with those who are different from themselves, that they have skills that allow them to work with a diverse client population, and that they act with honesty and integrity at all times.

Indeed, upholding the ethical principles upon which juvenile justice is based is fundamental to effective juvenile justice practice. Because of the coercive powers of juvenile justice agencies, juvenile justice practitioners must act to ensure due process, maintain the presumption of innocence, avoid conflicts of interest, guard confidentiality when appropriate, and avoid bias. To the extent that juvenile justice practitioners adhere to these ethical principles, they will help ensure that juvenile justice agencies live up to the mission of dealing with children and their families fairly, looking after the best interests of children, and at the same time protecting the community.

As is true of any profession, juvenile justice is made up of a diverse group of individuals who perform a variety of interrelated functions. If these individuals understand how their positions and responsibilities are related to those of others, both in their own agency and elsewhere, they will be better able to do their jobs effectively and efficiently. Their performance, whether they are intake officers, substance abuse counselors, judges, probation officers, or detention workers, determines the quality of justice given to youths and their families. The work is challenging and demanding, but it provides tremendous benefits to the community as well as to those who decide on a career in this exciting field.

Key Concepts

bias: A tendency, preference, or prejudice, especially one that inhibits objectivity.

coercive force: Physical force threatened or applied with the intention of influencing an individual's behavior.

coercive persuasion: The use of one's position to influence the decisions and behaviors of others.

confidentiality: Maintaining information in strict privacy. Legal confidentiality requirements prohibit certain information from being made available to the general public.

conflict of interest: A situation in which an individual has two interests that cannot jointly be satisfied. The term is frequently used to refer to cases in which an individual has a responsibility to act in a certain way but has other reasons, perhaps self-interested ones, for acting otherwise.

dispositional recommendation: A recommendation made to the court about how a case should be handled.
due process: An established course of proceedings designed to protect individual rights and liberties.
ethics: The principles of right conduct. These principles determine which actions are right and which are wrong.
Family Educational Rights and Privacy Act (FERPA): A federal law that gives a student's parents the right to inspect and review their child's educational records, request amendments to those records, and exert considerable control over the release of those records.
general education diploma (GED): A high school equivalency diploma awarded to adults who have not graduated from high school but have passed an official test covering a number of academic subjects. Commonly referred to by its acronym, GED.
integrity: The quality of acting in a moral, sincere, and honest way.
power: The ability to influence how other people think or act.
presumption of innocence: The presumption that an individual accused of a crime is innocent of that crime until proven guilty. In the U.S. legal system, the government has the burden of proving an accused person is guilty rather than the accused person having the burden of proving he or she is innocent.

Review Questions

1. What factors lead to conflict between juvenile justice professionals?
2. What factors lead to conflict between juvenile justice practitioners and their clients?
3. What are the potential negative outcomes associated with the use of coercive force and coercive persuasion in juvenile justice?
4. In what ways can racial, gender, religious, geographic, social class, or other types of bias affect fairness in the juvenile justice process?
5. What is meant by a conflict of interest? Give an example of how a conflict of interest could affect fairness in the juvenile justice process.
6. What is meant by the presumption of innocence? How is it fundamental to our system of justice?
7. What steps should juvenile justice practitioners take to ensure their clients' right to due process?
8. What steps should juvenile justice professionals take to ensure confidentiality in the juvenile justice process?
9. What is the "golden rule" in government and politics, and how does it affect client services?
10. If you entered the field of juvenile justice, what would you like to do? Why is the type of job you chose appealing to you?
11. Why is it important for juvenile justice practitioners to be able to work effectively with a diverse population of youths?
12. What types of skills are required for successful juvenile justice practice?
13. What are the legal obligations of juvenile justice practitioners who become aware of suspected child abuse or neglect?

Additional Readings

Albanese, J. (2008). *Professional ethics in criminal justice: Being ethical when no one is looking* (2nd ed.). Upper Saddle River, NJ: Allyn & Bacon/Prentice Hall.

Braswell, M., McCarthy, B. J., & McCarthy, B. R. (2008). *Justice, crime and ethics* (6th ed.). Cincinnati, OH: Anderson Publishing.

Carpenter, M. & Fulton, R. (2008). *A practical career guide for criminal justice professionals: Take charge of your career!* Flushing, NY: Looseleaf Law Publications.

Goodman, D. J. & Grimming, R. (2007). Work in criminal justice: An a-z guide to careers in criminal justice. Upper Saddle River, NJ: Prentice Hall.

Gordon, G. R., Hage, H. H., & McBride, R. B. (2008). *Criminal justice internships: Theory into practice* (6th ed.). Cincinnati, OH: Anderson Publishing.

Notes

1. Pastore, A. L. & Maguire, K. (Eds.). (2007). Table 2.10.2007: Reported confidence in selected institutions. In A. L. Pastore & K. Maguire (Eds.), *Sourcebook of criminal justice statistics online*. Retrieved March 9, 2009, from http://www.albany.edu/sourcebook/pdf/t2102007.pdf.
2. Bernard, T. J. (1992). *The cycle of juvenile justice*. New York: Oxford University Press.
3. Zalkind, P. & Simon, R. J. (2004). *Global perspectives on social issues: Juvenile justice systems*. Lanham, MD: Lexington Books.
4. Belair, R. R. (1997). *Privacy and juvenile justice records: A mid-decade status report.* Washington, DC: Bureau of Justice Statistics.
5. Miller, N. (1995). State laws on prosecutors' and judges' use of juvenile records. *National Institute of Justice Research in Brief.* Washington, DC: National Institute of Justice.
6. Madeiras, M. L., Campbell, E., & James, B. (1997). *Sharing information: A guide to the Family Educational Rights and Privacy Act and participation in juvenile justice programs.* Washington, DC: Office of Juvenile Justice and Delinquency Prevention.
7. Madeiras, Campbell, & James, 1997.
8. Black, H. C. (1968). *Black's law dictionary* (4th ed.). St. Paul, MN: West Publishing Co.
9. Snyder, H. N. & Sickmund, M. (1999). *Juvenile offenders and victims: 1999 national report.* Washington, DC: Office of Juvenile Justice and Delinquency Prevention. Krisberg, B. & Austin, J. F. (1993). *Reinventing juvenile justice*. Newbury Park, CA: Sage.
10. Krisberg & Austin, 1993.
11. Johnson, R. E. (1986). Family structure and delinquency: General patterns and gender differences. *Criminology*, *24*, 65–84.
12. Hirschi, T. (1969). *Causes of delinquency.* Berkeley, CA: University of California Press. Johnson, 1986.

13. Jacobs, M. D. (1990). *Screwing the system and making it work: Juvenile justice in the no-fault society*. Chicago: University of Chicago Press. This resource contains a good example of the struggles that juvenile justice practitioners encounter in order to meet clients' needs.

14. Child Welfare League of America. (2003). Michigan's juvenile justice online technology: A system of care for child welfare and juvenile justice. *The Link*, *2*, 1.

15. See Massachusetts General Laws Chapter 119, Section 51A; Indiana Compiled Laws IC 31-33; Florida Statutes Chapter 39.201–39.206; California Penal Code Section 11164–11174.3; and Michigan Compiled Laws 722.621–722.638.

Index

Boxes, figures, and tables are indicated by b, f, and t following the page number.

A

E

F

G

K

L

N

O

P

T

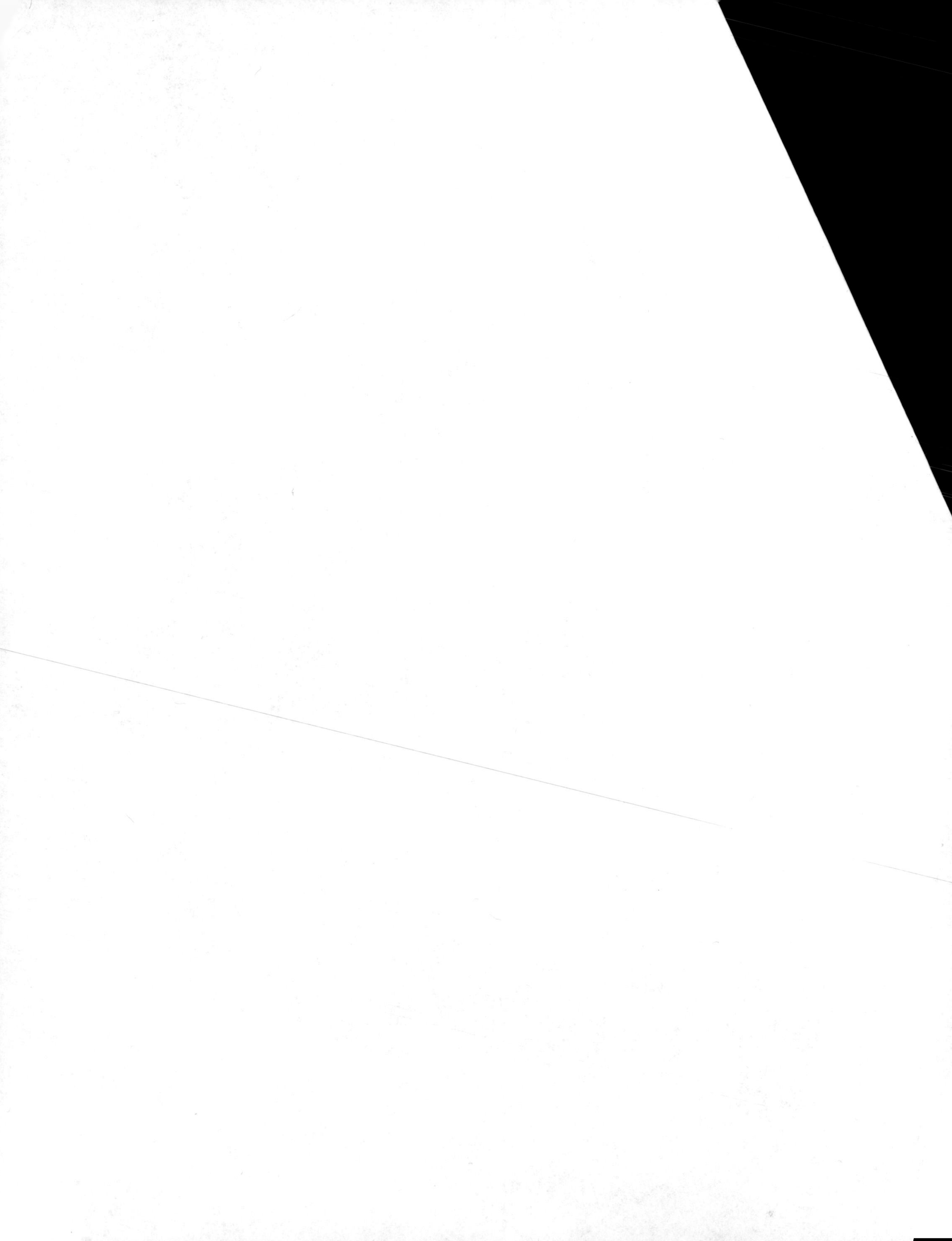

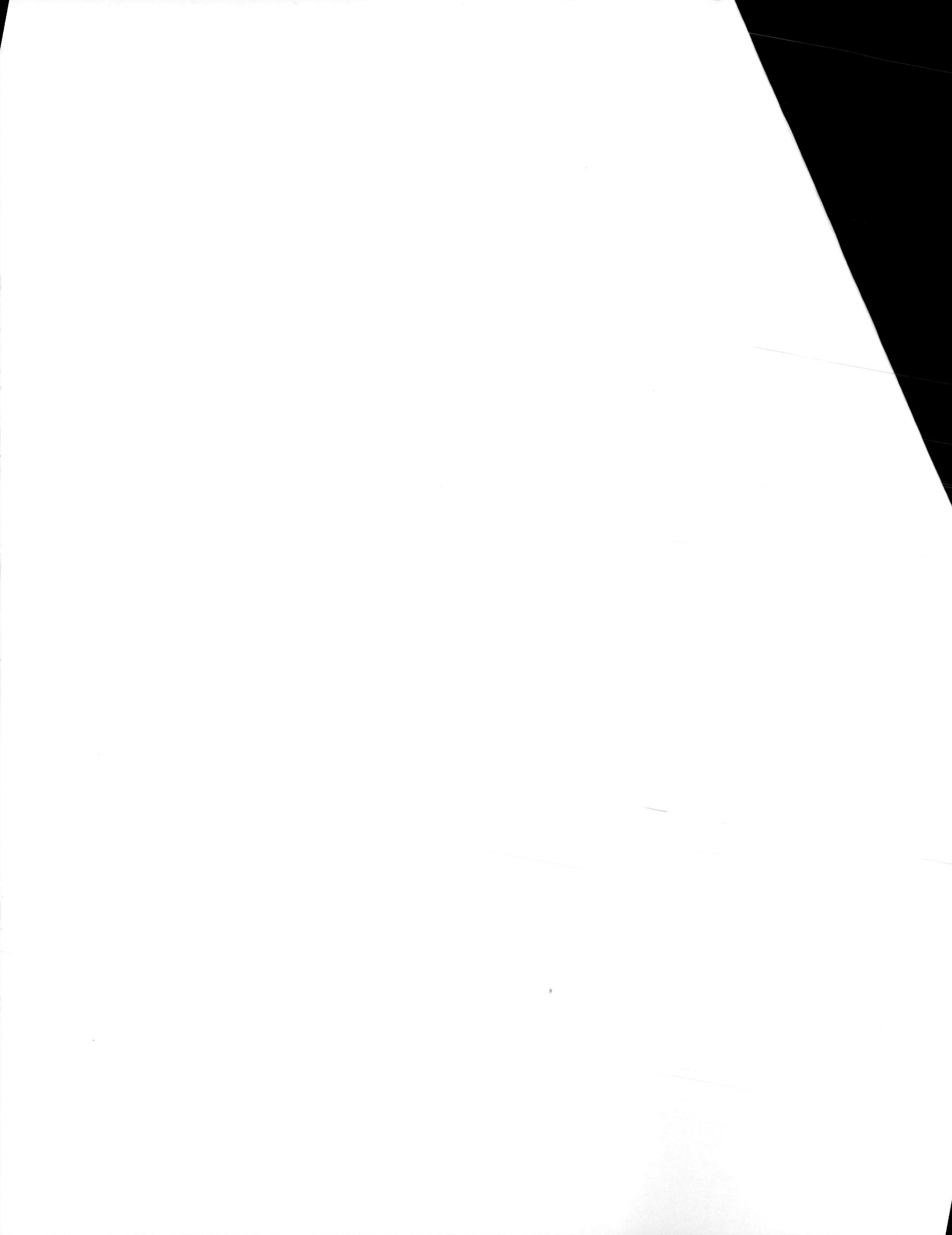